ADVANCES IN CERAMICS • VOLUME 3

SCIENCE AND TECHNOLOGY OF ZIRCONIA

Edited by A. H. Heuer
and L. W. Hobbs
Case Institute of Technology
Case Western Reserve University
Cleveland, Ohio

The American Ceramic Society, Inc.
Columbus, Ohio

Proceedings of the First International Conference on the Science and Technology of Zirconia held at the Case Institute of Technology, Case Western Reserve University, Cleveland, Ohio, June 16-18, 1980. Co-sponsored by the Ceramics Group of the Department of Metallurgy and Materials Science, Case Institute of Technology of Case Western Reserve University, and the American Ceramic Society and held in conjunction with the Case Institute of Technology Centennial Celebration

Associate Editor
Geraldine Smith

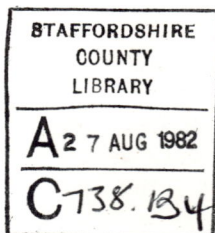

International Standard Book Number: 0-916094-42-1
Library of Congress Catalog Card Number: 80-69648
Coden: ADCEDE

Printed in the United States of America

ADVANCES IN CERAMICS • VOLUME 3

SCIENCE AND TECHNOLOGY OF ZIRCONIA

Foreword

The First International Conference on the Science and Technology of Zirconia (ZrO_2) was held on 16–18 June 1980 in Cleveland, Ohio. The present volume, number three of the series *Advances in Ceramics,* published by the American Ceramic Society, comprises the Conference Proceedings.

The meeting, sponsored and organized by the Ceramics Group of the Department of Metallurgy and Materials Science, Case Institute of Technology, Case Western Reserve University, was held in conjunction with the Centennial Celebration of Case Institute of Technology. The Organizing Committee, in addition to A. H. Heuer and L. W. Hobbs of Case, included R. J. Bratton of Westinghouse, R. J. Brook of the University of Leeds (United Kingdom), R. C. Garvie of CSIRO (Australia), and E. C. Subbarao of IIT, Kanpur (India).

As reflected in these Proceedings, the Conference provided a state-of-the-art picture of the rapidly-growing field of ceramic science and technology based on ZrO_2-containing ceramics. Interest in the field is evident from the attendance at the Conference, 175 in total, including representatives from all parts of the United States and 8 other countries. A second international conference has been tentatively scheduled for 1983 in Stuttgart.

The meeting organizers are grateful to the following orgainzations for their financial support: The National Science Foundation; Army Research Office; Norton Company, Worcester, Massachusetts; Corning Glass Foundation, Corning, New York; Magnesium Elektron, Ltd., Twickenham, United Kingdom, and Magnesium Elektron, Inc., Flemington, New Jersey; Singh Industries, Inc., Randolph, New Jersey; NL Industries, Inc., Niagara Falls, New York, and especially to the Department of Metallurgy and Materials Science, Case Institute of Technology for financial underpinning. Ms. Phyllis Witriol is thanked for her untiring efforts as Conference Secretary, and Ms. Gerry Smith of the American Ceramic Society for her yeoman editorial efforts which permitted publication of these Proceedings in a timely fashion.

A. H. Heuer
L. W. Hobbs

Contents

Zirconia—an overview

E. C. SUBBARAO

Department of Metallurgical Engineering
Indian Institute of Technology
Kanpur 208016, India

Intense investigations of the science and technology of zirconia over the last half century have propelled it into an outstanding, versatile material. A historical perspective is provided through bench marks in the development of zirconia. The crystal structures of the three polymorphs—cubic, tetragonal, and monoclinic—are reviewed. The martensitic nature of the monoclinic-tetragonal transformation is substantiated. Enhancement of strength and toughness through the phase transformation in partially stabilized zirconia, which led to important applications as extrusion dies and tool bits, is discussed. The process of stabilization, which enabled refractory applications of zirconia, is covered, followed by a discussion of the defect structure and consequent ionic conductivity of doped zirconia. The use of stabilized zirconia in oxygen sensors, fuel cells, etc., is based on its high oxygen ion conduction. Cubic zirconia, because of its high refractive index, is widely used as jewelry. The state of knowledge in each area is reviewed, and gaps in present understanding are highlighted for future studies.

Zirconia (ZrO_2) is a remarkable material, which has attracted a great deal of attention from scientists, technologists, and users. The progress in our understanding of this material and in exploiting it has been substantial. Pedagogically, ZrO_2 is a case study in materials science, since structure-property correlations have been very extensively examined. Atomic structure and microstructure, defects, phase transformations, and processing on the one hand and properties (thermal, mechanical, electrical, and optical) on the other, are intimately connected in zirconia. In this overview, a historical perspective is presented, followed by a brief mention of the sources of zirconia. The major aspects (crystal structure, phase transitions in zirconia, mechanical behavior of partially stabilized zirconia, and stabilization, ionic conductivity, and optical properties of fully stabilized zirconia) are reviewed, including the important applications of zirconia as well as the significant outstanding problems.

Historical Perspective

Ten major bench marks in the field of zirconia are listed in Table I. The discovery of baddeleyite in 1892 in Brazil, by Hussak, was followed soon thereafter by attempts at its exploitation for refractory applications. Ruff and Ebert,[1] in a classic study, used X-ray diffraction (XRD) to establish the monoclinic symmetry of zirconia at room temperature. They also studied the monoclinic-tetragonal transformation at $>1000\,°C$ by high temperature

XRD. Ruff and coworkers[2] demonstrated, in the same year, the disruptive phase change in zirconia as well as the stabilization of the material by the addition of metal oxides. Related to this is the determination of binary phase equilibria of ZrO_2 with RO and R_2O_3 by Duwez and coworkers.[3,4] Both these aspects were later studied more extensively.

While refractory applications represent the tonnage usage of stabilized zirconia, Nernst,[5] as early as 1900, used zirconia as a glower for incandescent lighting; Wagner[6] elucidated the defect structure of stabilized zirconia, and Kiukkola and Wagner[7] determined its electrolytic nature and used it as an oxygen concentration cell in thermodynamic measurements. Kingery et al.[8] confirmed the oxygen ion conduction through diffusion and conductivity measurements.

By high temperature XRD ($>2370\,°C$), Smith and Cline[9] found the expected cubic phase of zirconia, though doubts[10] have been raised about the possible contamination or reduction of zirconia at such high temperatures. Careful studies of the tetragonal→monoclinic transition in ZrO_2 prompted Wolten[11] to liken this transformation to that of martensite. Many elaborate studies which have since been undertaken support the martensitic nature of the transition.[12-14] These studies have led to the development of transformation strengthening in partially stabilized zirconia by appropriate heat treatment.[15-17]

The ionic conductivity and transformation toughening of zirconia have opened up such nonrefractory uses of zirconia as oxygen sensors, heaters, and extrusion dies. Much progress regarding transformations and processing is due to the detailed crystal structure analyses of the various phases of zirconia, using powder as well as single crystal diffraction methods.[18-20] The conference, totally devoted to zirconia, organized in 1962 at Wright-Patterson Air Force Base must be recognized as a significant landmark.

The chapters on zirconia by Ryshkewitch[21] in 1960 and by Garvie[22] in 1970 are excellent reviews on the subject. The present proceedings represents the most up-to-date and exhaustive state of our understanding of zirconia.

Occurrence

The two main sources of zirconia are: (1) baddeleyite in Brazil, which contains 80–90% ZrO_2, with TiO_2, SiO_2, Fe_2O_3, etc. as major impurities and (2) zircon, $ZrSiO_4$, which occurs as secondary deposits in Kerala (India), New South Wales (Australia), and Florida (U.S.A.). With its presence of 0.02 to 0.03% in the earth's crust, zirconia is more abundant than many metallic ores.

Crystal Structure

Cubic

The cubic phase, which is stable from $2370\,°C$ to the melting point ($2680 \pm 15\,°C$), was detected by Smith and Cline[9] by high temperature XRD. It has a fluorite-type crystal structure, in which each Zr is coordinated by eight equidistant oxygens and each oxygen is tetrahedrally coordinated by four zirconiums (Fig. 1(a)).

Tetragonal

This phase is stable between about $1170\,°C$ and $2370\,°C$. Teufer[20] has shown that Zr is surrounded by eight oxygens, four at a distance of 0.2455

2

nm and the other four at a distance of 0.2065 nm (Fig. 1(*b*)).

Monoclinic

This phase is stable at all temperatures below 1170 °C. The crystal structure of monoclinic ZrO_2 was determined by XRD by McCullough and Trueblood,[18] Smith and Newkirk,[19] and others. The main features of the structure are: (1) sevenfold coordination of Zr with a range of bond lengths and bond angles (Fig. 1(*c*)), (2) layers of triangularly coordinated O_I-Zr_3 and (distorted) tetrahedrally coordinated O_{II}-Zr_4 (Fig. 1(*d*)), (3) Zr atoms are located in layers parallel to the (100) planes, separated by O_I and O_{II} atoms on either side (Fig. 1(*d*)) and (4) the layer thickness is wider when the Zr atoms are separated by O_I atoms than when they are separated by O_{II} atoms (Fig. 1(*d*)). These interlayer distances become equal in the tetragonal phase.

Some outstanding tasks: The following list is only suggestive, but not exhaustive: (1) AX_2 compounds with a radius ratio of cation to anion between 0.414 and 0.732 have the rutile structure. However, ZrO_2 with an $r_{Zr^{4+}}/r_{O^{2-}}$ of 0.57 has the distorted fluorite structure. Why? (2) The sevenfold coordination of Zr (Fig. 1(*c*)) suggests a certain degree of covalency. Is this the reason for ZrO_2 to assume the fluorite structure? The covalency may be explored with X-ray absorption spectra and Moessbauer spectra (preferably with an Sn probe). (3) The electronic structure of ZrO_2 may be studied by photoelectron spectroscopy.

Phase Transformations

Zirconia exhibits the following transformations:

$$\text{monoclinic} \underset{}{\overset{1170\,°C}{\rightleftharpoons}} \text{tetragonal} \underset{}{\overset{2370\,°C}{\rightleftharpoons}} \text{cubic} \underset{}{\overset{2680\,°C}{\rightleftharpoons}} \text{liquid}$$

Monoclinic ⇌ Tetragonal

This transformation was first detected by Ruff and Ebert[1] in 1929 using high temperature XRD. Since then, and particularly in the last 35 years, this transformation has been intensely studied using a variety of techniques (DTA, X-ray and electron diffraction, optical and electron microscopy, electrical resistivity, spectroscopy, etc.) due to the theoretical interest in understanding this important transition and due to the practical importance of the large, disruptive volume change associated with this phase change. The extensive literature in this field was reviewed by Subbarao *et al.*[12] in 1974 and by Subbarao more recently.[14] Only the salient features are mentioned here.

Wolten[11] was the first to suggest that this transformation is martensitic. The main experimental results are:

(1) The high temperature tetragonal phase cannot be quenched to room temperature.

(2) The thermal expansion of monoclinic ZrO_2 is strongly anisotropic, with the *b* axis exhibiting negligible expansion while the expansion is substantial for the *a* and *c* axes (Fig. 2).[23] There is an abrupt change in the lattice parameters at the phase transition (Fig. 2). Zirconia ceramics undergo substantial contraction on heating and a corresponding expansion on cooling through the monoclinic ⇌ tetragonal transformation[24-26] (Fig. 3), leading to a crumbling of the ceramic.

(3) The transformation is athermal, as established by XRD, metallography, and DTA. Thus, the transformation does not take place at a

fixed temperature but over a range, i.e. the amount of transformed phase changes by varying the temperature but not as a function of time at a fixed temperature. For example, no changes in XRD intensities are observed with time under isothermal conditions.[11] High temperature metallography shows the appearance of new plates with change of temperature but not growth of existing plates.[27] In successive, interrupted DTA runs, the transformation starts each time only at the temperature to which the sample was heated in the previous cycle (Fig. 4).[27] However, fine-grained (≈ 100 nm) ZrO_2 appears to exhibit an isothermal component in the transformation kinetics.[28,29] This is attributed to the contribution of the surface energy.

(4) The transition takes place with a velocity approaching that of sound in solids, as evidenced by the growth of platelets during high temperature metallographic observations.[27] The transformation exhibits a burst-like behavior in DTA during cooling (Fig. 5).[29] Annealing out the imperfections, including strains induced at the monoclinic $\rightarrow$ tetragonal transition by heating at high temperatures for short times or at lower temperatures for longer times, enhances the burst phenomena.[29] The bursts were observed only in the reverse direction, never in the forward direction. This may be due to the different strain energy contribution and the deformation mechanisms in the two directions.

(5) The transformation exhibits a large thermal hysteresis (Fig. 6).[11] The forward transition occurs at 1170 °C and the reverse one between 850° and 1000 °C, depending on the surface and strain energies associated with the forward transformation. The hysteresis is decreased by annealing at high temperatures (1550 °C) for short times or at lower temperatures (e.g. 1260 °C) for longer times.[29]

(6) Optical and electron microscopic observations of the transformed materials reveal surface distortions due to the sudden appearance of the acicular tetragonal phase through shear-like atomic movements (Fig. 7).[13] Transformation twins are also observed.[13,30]

(7) Single crystal XRD showed a diffusionless, shear transformation in which the atoms retain their neighbors in either phase.[31] Atomic movements of less than one interatomic distance, mainly of the oxygen atoms with only minor shifts of Zr atoms, are inferred.[32,33]

(8) An orientation relationship is established between the parent (monoclinic) and product (face centered tetragonal) phases.[33,13]

$$(100)_m \parallel (110)_{fct}$$
$$[010]_m \parallel [001]_{fct}$$

At temperatures below 1000 °C, Bansal and Heuer,[13] Bailey,[34] and Smith and Newkirk[19] have also found

$$(100)_m \parallel \approx (100)_{fct}$$
$$[001]_m \parallel \approx [001]_{fct}$$

(9) The following habit planes and lattice invariant deformation have been suggested[13]:

	Habit plane normals	Lattice invariant deformation
Lenticular-type product	$(671)_m$, $(761)_m$	$(1\bar{1}0)$ [001]
Plate-shaped product	$(100)_m$	$(1\bar{1}0)$ [110]

4

While the above experimental results generally support the martensitic nature of the monoclinic-tetragonal transformation, there are many aspects which are unclear and deserve further study. These include:

(a) Detailed crystallography of the transformation, including atomic movements, orientation relationships, habit, and twinning planes. The experimental studies are made difficult by the high temperature of the transition.

(b) Thermodynamics of the transformation, including free energy changes, possibly incorporating strain and surface energy terms and effects of alloying elements.

(c) The role of imperfections, including point defects, impurities, dislocations, stacking faults, etc., on the nature of the transformation.

Tetragonal ⇌ Cubic

The transformation temperature (2370 °C) has been established, but little else is known about this phase change. While the high temperatures make experiments very difficult, a study of this transformation may throw light on the phenomenon of stabilization of the cubic phase in ZrO_2-based alloys.

Partially Stabilized Zirconia (PSZ)

Early studies on zirconia alloyed with CaO, MgO, Y_2O_3, and rare earth oxides showed that the two transition temperatures (monoclinic-tetragonal and tetragonal-cubic) are lowered. The phase diagrams of the binary oxide systems, reviewed in detail by Stubican,[35] show that a PSZ [i.e. a mixture of cubic and monoclinic (or tetragonal) zirconia] occurs (a) when the dopant is present in a concentration less than that needed for complete stabilization (i.e. for the formation of cubic, fluorite-type zirconia phase) or (b) when the fully stabilized, cubic zirconia with a suitable solute content is heat-treated under appropriate conditions of temperature and time.

The incorporation of alloying elements not only decreases the transition temperatures, but decreases the linear thermal expansion coefficient of the two-phase material as also the volume change associated with the monoclinic-tetragonal phase change (Fig. 3). The lower linear thermal expansion coefficient of PSZ than that of pure as well as fully stabilized zirconia[24-26] contributes to the better thermal shock resistance of the PSZ[36] than that of cubic (stabilized) and monoclinic (pure) zirconia.

King and Yavorsky[37] were the first to draw attention to the interesting mechanical properties of PSZ. Based on microstructural and microhardness studies, they argued that the material consists of single grains containing monoclinic and cubic domains. As the original tetragonal domain converts to the stable monoclinic form on cooling, the stresses resulting from the accompanying volume increase (over and above the stresses due to the difference in the thermal expansion coefficients of the monoclinic particles and cubic matrix), instead of rupturing the ceramic, are believed to be relieved by the plastic deformation of the cubic grains around pores and inclusions in the monoclinic-cubic hybrid phase by a slip mechanism. Fully stabilized zirconia did not show slip bands or plastic deformation and exhibited grain boundary strain.

Since then, there has been an intense interest in the mechanical behavior of PSZ. Thermal processing of PSZ to achieve maximum strength and

5

toughness was studied by Garvie and Nicholson[38] and by Green et al.[39] for Ca-PSZ and by Bansal and Heuer[16] and Porter and Heuer[40] for Mg-PSZ. For example, Porter and Heuer subjected commercial 8.1 mol% Mg-PSZ (which seems to have been sintered at < 1800 °C) to a solution anneal at 1850 °C for 4 h. While optical microscopy of the as-received material shows a single phase material (Fig. 8(A)), transmission electron microscopy reveals inter- and intragranular monoclinic particles (Fig. 8(B) and 8(C)), both of which are formed with tetragonal symmetry and transformed to the monoclinic form during cooling below the M_S (start of martensite) temperature. The intergranular particles (Fig. 8(B)) are believed to have been present during sintering, while the intragranular particles are probably formed by precipitation during the postsintering cooling (Figs. 8(C) and (D)). Solution-annealed samples essentially consist of cubic ZrO_2 solid solution, though a small amount ($\approx 3\%$) of a second phase occurs as small plate-like precipitates (≈ 5 nm) (Fig. 8(E)). The authors aged the solution-annealed samples at 1400–1500°, 1200–1300°, 1000°, and < 1000 °C.

Enhanced mechanical properties (strength and fracture toughness) were obtained by aging at 1400 °C for ≈ 4 h (Fig. 9). Under these conditions, tetragonal ZrO_2 solid solution, precipitated in a matrix of cubic ZrO_2 solid solution, is retained metastably during cooling. If the tetragonal particles were converted to the stable monoclinic phase during cooling, the mechanical properties of the material were found to be poorer. The phase assemblages resulting from the various heat treatments can be understood in terms of the phase diagram for the ZrO_2-MgO system (Fig. 10.)

Microstructural evolution of PSZ was investigated in Mg-PSZ by Bansal and Heuer[16] and by Porter and Heuer[40] and in Ca-PSZ by King and Yavorsky,[37] Garvie and Nicholson,[38] and Green et al.[39] The microstructure of PSZ generally consists of cubic ZrO_2 solid solution as the major phase with monoclinic or tetragonal ZrO_2 solid solution as the minor precipitate phase. The second phase may exist at grain boundaries either from the sintering process or by precipitation during postsintering heat treatment or during cooling, or within the cubic matrix grains. The intragranular precipitates are ellipsoidal, are oriented along {100} of the cubic phase, and have an optimum size of ≈ 0.2 μm. Particles of this size range retain the tetragonal symmetry on cooling, whereas larger particles convert spontaneously to the monoclinic form which involves a volume increase leading to microcracking. The monoclinic particles are often heavily twinned. The largest coherent tetragonal ZrO_2 solid solution precipitate particles appear to be smaller (≈ 0.1 μm) in Ca-PSZ. At optimum aging, the precipitate accounts for 25 to 30% of the volume.

Various mechanisms have been suggested to account for the strengthening and toughening of PSZ ceramics. As already mentioned, King and Yavorsky[37] proposed that the plastic deformation of the cubic matrix is responsible for the stress relief. Garvie and Nicholson[38] refer to the subgrain structure of PSZ ceramics in which 100 nm domains of pure ZrO_2 are distributed in a cubic stabilized ZrO_2. These domains undergo the tetragonal→monoclinic transformation on cooling. The volume increase associated with this phase change (which may take place at subtransformation temperatures also) produces extensive microcracking. Because of the large number of cracks, they propagate only quasi-statically and the sample retains most of its strength.

6

Green *et al.*[41] studied the fracture toughness of Ca-PSZ, in conjunction with the microstructure,[39] and drew attention to the formation of a microcrack zone at the tip of the propagating crack. The stresses developed during the transformation of the pure ZrO_2 grain present at the grain boundaries lead to separation and weakening of the grain boundaries and consequent strength degradation. The initial stable crack propagation is attributed to the increase in microcrack zone size. This zone absorbs energy by processes such as secondary crack formation, surface roughening, and crack branching. Bansal and Heuer[16] argued that the fine monoclinic precipitates dispersed coherently in the cubic grains impede crack propagation. This has been questioned by Claussen[42] and Rice,[43] who prefer to consider that the microcracks produced within or between the cubic grains by the transformation of large monoclinic particles cause subcritical propagation of the cracks, similar to the ideas of Green *et al.*[39,41] and based on results of ZrO_2-containing Al_2O_3. Porter *et al.*[44,45] refuted the above questions involving microcracking as the strengthening mechanism.

The importance of metastable tetragonal phase precipitated in a cubic stabilized zirconia for toughening PSZ was first noted by Garvie *et al.*[15] and elucidated by Porter *et al.*[40,46,47] Porter *et al.* have shown that all the precipitate particles within several micrometers of a crack had monoclinic symmetry (Fig. 8(*F*)), whereas all other particles were tetragonal (Fig. 8(*G*)). This suggests that the stresses near the crack tip had caused the particles to transform to the monoclinic symmetry by making the particles lose their coherency; additional stress is then required for crack extension. Thus, the stress-induced martensitic transformation of the metastable tetragonal particles to the stable monoclinic phase is the mechanism which absorbs energy and inhibits crack propagation, thereby strengthening and toughening PSZ. These authors have estimated the large strain energy difference between the tetragonal and monoclinic particles within a cubic matrix. This large energy difference prevents the martensitic transformation on cooling below M_S, at least for small particles. They also estimated the difference in chemical free energy, ΔG_0, as 228 MPa, in close agreement with the 226 MPa determined by DTA. Porter *et al.*[47] also mapped the crack tip transformation zones. Strength decrease due to overaging results from precipitate coarsening, since large particles spontaneously transform to the monoclinic form and hence are not available for energy absorption processes.

On the basis of the above, these materials are now more appropriately termed transformation-toughened zirconia ceramics (rather than precipitation-strengthened ceramics) and likened to TRIP steels.[15] Table II shows that the coexistence of tetragonal and cubic zirconia results in strength and fracture toughness values approximately three times those of a mixture of monoclinic and cubic phases or only cubic zirconia.

The importance of the metastable tetragonal phase in toughening PSZ was forcefully illustrated by Gupta and coworkers.[17] They showed that a nearly fully tetragonal material, prepared of ZrO_2-Y_2O_3 solid solution by fine particle technology,[48,49] exhibits high strength at room temperatures,[17,50] whereas a ZrO_2-CeO_2 solid solution, in which the tetragonal phase is stable at room temperature, is relatively weak, though the same composition exhibits enhanced strength and toughness at liquid nitrogen temperature, where the tetragonal phase is only metastable.[51]

7

Unlike most ceramic materials, the PSZ has enhanced strength when the surface is abraded, due to the transformation strengthening.[52,53]

The strong, tough PSZ components exhibit outstanding performance as extrusion dies for nonferrous metals, as tool bits, as thread guides, etc.

Transformation-toughened PSZ ceramics is likely to be an important area of investigation in the next decade, just as the ionic conductivity of zirconia electrolytes was in the last two decades. Processing technology, detailed mechanisms, T-T-T diagrams, and alloying additions are some of the areas which need attention.

Stabilized Zirconia

Stabilization

Zirconia is stabilized in the fluorite-type cubic phase when it is alloyed with an appropriate amount of di- or trivalent oxides of cubic symmetry. According to earlier binary oxide phase diagrams,[3,4] the cubic phase exists over a wide range of composition and temperature, though more recent equilibrium studies over extended times indicate that the cubic phase is not truly stable at room temperature in many cases[35] (see Fig. 10). The slow rate of the destabilization process as well as the formation of intermediate phases (e.g. $CaZr_4O_9$[22,54-57] and $Y_4Zr_3O_{12}$[56,58]) is due to the very low cation diffusion coefficients[59] which are a millionfold smaller than the diffusion coefficient of oxygen.[8,60] As a consequence, the cubic phase remains metastable for long periods at temperatures sufficiently below the decomposition temperatures. The topotactic relationship of the decomposition products to the parent phase has also been investigated.[16]

The high melting point and chemical inertness of stabilized zirconia are the basis for its extensive use as a refractory material.

In spite of its importance, the phenomenon of stabilization is not well understood. Some of the factors which may influence stabilization deserve attention: the role of anion vacancies, minimum number of vacancies, size, charge and concentration of dopant ions, crystal structure of the dopant oxide (e.g. cubic), and the role of electronic energy levels.

Defect Structure and Crystal Structure

In 1943, Wagner[6] established that stabilized zirconia contains oxygen ion vacancies. By comparing measured and calculated densities of CaO-stabilized ZrO_2, Hund[61] established that the Ca^{2+} and Zr^{4+} ions are statistically distributed over the cation sites and the electrical neutrality is achieved by the creation of oxygen ion vacancies, equal in concentration to the Ca^{2+} ions. Similar studies have since been done on ZrO_2 stabilized by Y_2O_3 and rare earth oxides. Direct evidence for this defect structure has been obtained, using XRD methods, by Tien and Subbarao[62] among others. The defect structure and defect concentration of stabilized zirconia are essentially fixed by dopant content and are independent of temperature and surrounding atmosphere.[63-67]

The crystal structure of stabilized zirconia is of the cubic fluorite type. The cubic phase extends over a wide range of dopant concentration with a correspondingly large anion vacancy concentration. Electrostatic attraction between the two oppositely charged species (Ca''_{Zr} and $V_{\ddot{O}}$ or Y'_{Zr} and $V_{\ddot{O}}$), together with the large concentration of anion vacancies, interferes with a

8

random distribution of dopant cations over cation sites and vacancies over anion sites. Defect ordering in stabilized zirconia has been examined by X-ray, electron, and neutron diffraction methods by several workers.[54-58,68-72]

The diffuse scattering and forbidden reflections observed in neutron diffraction of CaO-stabilized zirconia prompted Carter and Roth[68] to propose that the high temperature disordered state of CaO-stabilized zirconia has a cubic fluorite structure with oxygen ions displaced from the ideal fluorite lattice sites and the low temperature (<1300 K) ordered state involved cooperative ordering of the oxygen ions on the oxygen sublattice (Fig. 11). Electron diffraction studies of CaO-stabilized zirconia by Allpress and Rossell[55] were interpreted in terms of domains of ordered arrangement, ≈ 3 nm in diameter, embedded coherently in a cubic matrix. Similar work by Hudson and Moseley[69] demonstrated that the intensity of diffuse scattering increases with CaO content, while the intensity of extra (forbidden) reflections decreases with CaO content. The anion vacancies become aligned along <111> directions. They report the presence of coherent microdomains of $CaZr_4O_9$ and intragranular precipitates of monoclinic ZrO_2 in a cubic matrix, as a result of aging.

Based on a neutron diffraction study of yttria-stabilized zirconia, Steele and Fender[70] indicated that the six oxygens surrounding an oxygen vacancy relax by 0.036 nm and probably the four neighboring cations (which include Y^{3+} ions) by a smaller relaxation. Faber et al.[71] refined the neutron diffraction studies on CaO- and Y_2O_3-stabilized zirconia. They found that the oxygen ions are shifted from the ideal fluorite positions (by ≈ 0.023 nm) by an internal shear deformation of the oxygen sublattice (Fig. 12). This may (as in CaO-doped ZrO_2) or may not (as in Y_2O_3-doped ZrO_2) be accompanied by external strain. The internal rearrangement appears to be dominated by the cation-anion (and indirectly anion-anion) interactions.

Ionic Conductivity

As early as the turn of the century, Nernst recognized the electrolytic properties of yttria-doped zirconia and used it as a Nernst glower.[5] As discussed above, Wagner[6] established the defect character of doped zirconia as being composed of oxygen ion vacancies. The galvanic cell measurements of Kiukkola and Wagner[7] in 1957 and the oxygen diffusion and ionic conductivity study of Kingery et al.[8] in 1959 confirmed the defect structure of doped cubic zirconia. In the following two decades, more detailed studies of defect equilibria and ionic conductivity of these materials have been conducted as a function of composition, concentration of dopant, temperature, oxygen partial pressure, and frequency of measuring electric field. Several recent reviews of this topic are available.[63-67] A few salient features of the ionic conductivity of doped zirconia are:

(1) It follows an Arrhenius behavior over a wide temperature range with an activation energy of ≈ 1 eV (Fig. 13).

(2) Over a wide temperature range, the electrical conductivity is independent of oxygen partial pressure (P_{O_2}) over several orders of magnitude (Fig. 14). Under these conditions, the conduction is truly electrolytic, with the transport number for oxygen ions nearly unity and the transport number for electrons $\ll 1\%$.

(3) The variation of the isothermal electrical conductivity (with dopant content) of a binary oxide system based on zirconia exhibits a maximum at or near the lowest dopant concentration required to stabilize the cubic phase and decreases with increasing dopant concentration. This trend is accompanied by an increase in activation energy for conduction (Fig. 15). The decrease in conductivity with increasing dopant concentration is contrary to the dilute solution model and has not been fully accounted for in a quantitative way, though defect ordering,[68] clustering,[69,78] electrostatic interactions,[79,80] precipitation of a second phase,[69] etc. have been invoked.

The ionic conductivity of stabilized zirconia is utilized in a variety of devices as discussed in some recent reviews.[81-83] The major applications include (a) an oxygen sensor for control of automotive emissions, deoxidation of steel, combustion control of furnaces (for glassmelting, heat treatment, thermal power plants, etc.), (b) electrochemical oxygen pump, (c) high temperature fuel cell, (d) susceptor for induction heating, (e) resistance heating element, (f) electrodes for power generation by magnetohydrodynamics, etc.

The oxygen sensor and fuel cell are based on the relation between the electromotive force (E) generated by an oxygen concentration cell and the difference in the oxygen partial pressure (P_{O_2}) on the two sides of the cell:

$$E = \frac{RT}{4F} \ln (P_{O_2}/P'_{O_2}) \tag{1}$$

The measured emf indicates the unknown P_{O_2} if the reference P'_{O_2} is known for a sensor operating at a temperature T (K). The same equation, when it is used to describe the functioning of the oxygen pump, prescribes the potential to be applied for achieving a required P_{O_2} in a flowing stream relative to a reference P'_{O_2} (e.g. air with $P'_{O_2}=0.21$).

Some of the unresolved problems regarding the ionic conductivity of doped zirconia are: (a) a quantitative explanation of the variation of conductivity with dopant concentration; (b) is it coincidental that the composition corresponding to the maximum conductivity occurs at the minimum dopant needed to stabilize the cubic form? (c) clear evidence for clustering, ordering, or other structural features; (d) relationship between the ionic conductivity and ionic size of the dopant, for a given defect concentration, though the conductivity appears to be higher when the cation size of the dopant is smaller.

Optical Properties

ZrO_2, in pure as well as in the stabilized forms, has high refractive index,[84] as shown in Table III. One of the largest users of zirconia is in ceramic glazes as an opacifier; this is based on the high refractive index of zirconia. In recent years it has been possible to grow large, clear, transparent crystals of stabilized zirconia by skull melting,[84] using zirconia itself as the susceptor in a radio frequency induction furnace (Fig. 16).

Since stabilized zirconia has adequate electrical conductivity to couple to the rf field only when it is at a high temperature (>1000 °C), a few chips of zirconium metal or metal of the stabilizing oxide are introduced into the feed material to melt first. The mixed oxide powder dissolves in the melt and then the melting operation of the feed continues. The melt is contained in an envelope of the same composition. Crystals several tens of cubic centimeters in size can be harvested. The nearness of the refractive index of zirconia to that of diamond ($n = 2.42$) laid the foundation for a flourishing production of synthetic jewel grade cubic zirconia. The ability to grow large, optical quality crystals opens up also the posssibility of producing tunable laser rods composed of rare-earth-doped (e.g. Nd) yttria-stabilized zirconia.

Acknowledgment

The author's work in this field is supported, in part, by the U.S. National Bureau of Standards and the U.S. Aerospace Laboratory.

References

[1]O. Ruff and F. Ebert, "Refractory Ceramics: I, The Forms of Zirconium Dioxide," Z. Anorg. Allg. Chem., 180 [1] 19–41 (1929).

[2]O. Ruff, F. Ebert, and E. Stephen, "Contributions to the Ceramics of Highly Refractory Materials: II, System Zirconia-Lime," ibid., 215–24.

[3]P. Duwez, F. H. Brown, Jr., and F. Odell, "Zirconia-Yttria System," J. Electrochem. Soc., 98 [9] 356–62 (1951).

[4]P. Duwez, F. Odell, and F. H. Brown, Jr., "Stabilization of Zirconia with Calcia and Magnesia," J. Am. Ceram. Soc., 35 [5] 107–13 (1952).

[5]W. Nernst, "Electrolytic Conduction in Solid Substances at High Temperature," Z. Electrochem., 6, 41 (1900).

[6]C. Wagner, "Mechanism of Electric Conduction in Nernst Glower," Naturwissenschaften, 31, 265–68 (1943).

[7]K. Kiukkola and C. Wagner, "Measurements on Galvanic Cells Involving Solid Electrolytes," J. Electrochem. Soc., 104 [6] 379–87 (1957).

[8]W. D. Kingery, Jr., J. Pappis, M. E. Doty, and D. C. Hill, "Oxygen Ion Mobility in Cubic $Zr_{0.85}Ca_{0.15}O_{1.85}$," J. Am. Ceram. Soc., 42 [8] 393–98 (1959).

[9]D. K. Smith and C. F. Cline, "Verification of Existence of Cubic Zirconia at High Temperature," ibid., 45 [5] 249–50 (1962).

[10]B. C. Weber, "Inconsistencies in Zirconia Literature," ibid., [12] 614–15.

[11]G. M. Wolten, "Diffusionless Phase Transformations in Zirconia and Hafnia," ibid., 46 [9] 418–22 (1963).

[12]E. C. Subbarao, H. S. Maiti, and K. K. Srivastava, "Martensitic Transformation in Zirconia," Phys. Status Solidi A, 21, 9–40 (1974).

[13]G. K. Bansal and A. H. Heuer, "Martensitic Phase Transformation in Zirconia (ZrO_2): I, Metallographic Evidence," Acta Metall., 20 [11] 1281–89 (1972); "II, Crystallographic Aspects," ibid., 22 [4] 409–17 (1974).

[14]E. C. Subbarao, "Phase Transformations in Pure and Impure Zirconia," in Proceedings of Metal Sciences—The Emerging Frontiers. Edited by T. R. Anantharaman, S. L. Malhotra, S. Ranganathan, and P. Ramarao. Indian Institute of Metals, Calcutta (in press).

[15]R. C. Garvie, R. H. Hannink, and R. T. Pascoe, "Ceramic Steel?," Nature (London), 258, 703–704 (1975).

[16]G. K. Bansal and A. H. Heuer, "Precipitation in Partially Stabilized Zirconia," J. Am. Ceram. Soc., 58 [5-6] 235–38 (1975).

[17]T. K. Gupta; pp. 877–89 in Fracture Mechanics of Ceramics, Vol. 4. Edited by R. C. Bradt, D. P. H. Hasselman, and F. F. Lange. Plenum Press, New York, 1978.

[18]J. D. McCullough and K. N. Trueblood, "Crystal Structure of Baddeleyite (Monoclinic ZrO_2)," Acta Crystallogr., 12 [7] 507–11 (1959).

[19]D. K. Smith and H. W. Newkirk, "Crystal Structure of Baddeleyite (Monoclinic ZrO_2) and its Relation to the Polymorphism of ZrO_2," ibid., 18 [6] 983–91 (1965).

[20]G. Teufer, "Crystal Structure of Tetragonal ZrO_2," ibid., 15 [11] 1187 (1962).

[21]E. Ryshkewitch, Oxide Ceramics. Academic Press, New York, 1960; pp. 350–96.

[22]R. C. Garvie, pp.117–66 in High Temperature Oxides, Part II. Edited by A. M. Alper. Academic Press, New York, 1970.

[23]R. N. Patil and E. C. Subbarao, "Axial Thermal Expansion of ZrO_2 and HfO_2 in the Range Room Temperature to 1400 °C," J. Appl. Crystallogr., 2, 281–88 (1969).

[24]R. F. Geller and P. J. Yavorsky, "Effect of Some Oxide Additions on Thermal-Length Changes of Zirconia," J. Res. Natl. Bur. Stand., 35 [1] 87–110 (1945).

[25]B. C. Weber, "Zirconia—An Annotated Bibliography," Aerospace Res. Lab. Rept., ARL 64-205 (1964).

[26]O. J. Whittemore, Jr., and N. N. Ault, "Thermal Expansion of Various Ceramic

Materials to 1500 °C," *J. Am. Ceram. Soc.*, **39** [12] 443–44 (1956).

[27]L. L. Fehrenbacher and L. A. Jacobson, "Metallographic Observation of the Monoclinic-Tetragonal Phase Tranformation in ZrO_2," *ibid.*, **48** [3] 157–61 (1965).

[28]B. Ya. Sukharevskii and I. I. Vishnevskii, "Polymorphic Transition Kinetics of ZrO_2," *Dokl. Akad. Nauk SSSR*, **147** [4] 882–85 (1962).

[29]H. S. Maiti, K. V. G. K. Gokhale, and E. C. Subbarao, "Kinetics and Burst Phenomenon in ZrO_2 Transformation," *J. Am. Ceram. Soc.*, **55** [6] 317–22 (1972).

[30]S. T. Buljan, H. A. McKinstry, and V. S. Stubican, "Optical and X-ray Single Crystal Studies of the Monoclinic $\rightleftharpoons$ Tetragonal Transition in ZrO_2," *ibid.*, **59** [7-8] 351–54 (1976).

[31]G. M. Wolten, "Direct High-Temperature Single-Crystal Observation of Orientation in Zirconia Phase Transformation," *Acta Crystallogr.*, **17** [6] 763–65 (1964).

[32]C. F. Grain and R. C. Garvie, "Mechanism of the Monoclinic to Tetragonal Transformation in Zirconium Dioxide," *U. S. Bur. Mines Rep. Invest.*, **1965**, No. 6619, 19 pp.

[33]R. N. Patil and E. C. Subbarao, "Monoclinic-Tetragonal Phase Transition in Zirconia: Mechanism, Pretransformation, and Coexistence," *Acta Crystallogr.*, **A26** [5] 535–42 (1970).

[34]J. F. Bailey, "Monoclinic-Tetragonal Transformation and Associated Twinning in Thin Films of Zirconia," *Proc. R. Soc. London, Ser. A*, **279** [1378] 395–412 (1964).

[35]V. S. Stubican and Hellman; this volume, pp. 25–36.

[36]R. E. Jaeger and R. E. Nickell; pp. 163–84 in Ceramics in Severe Environments. Materials Science Research, Vol. 5. Plenum Press, New York, 1971.

[37]A. G. King and P. J. Yavorsky, "Stress Relief Mechanisms in Magnesia and Yttria-Stabilized Zirconia," *J. Am. Ceram. Soc.*, **51** [1] 38–42 (1968).

[38]R. C. Garvie and P. S. Nicholson, "Structure and Thermomechanical Properties of Partially Stabilized Zirconia in the $CaO-ZrO_2$ System," *ibid.*, **55** [3] 152–57 (1972).

[39]D. J. Green, D. R. Maki, and P. S. Nicholson, "Microstructural Development in Partially Stabilized ZrO_2 in the System $CaO-ZrO_2$," *ibid.*, **57** [3] 136–39 (1974).

[40]D. L. Porter and A. H. Heuer, "Microstructural Development in MgO-Partially Stabilized Zirconia (Mg-PSZ)," *ibid.*, **62** [5-6] 298–305 (1979).

[41]D. J. Green, P. S. Nicholson, and J. D. Embury, "Fracture Toughness of a Partially Stabilized ZrO_2 in the System $CaO-ZrO_2$," *ibid.*, **56** [12] 619–23 (1973).

[42]N. Claussen, "Comments on 'Precipitation in Partially Stabilized Zirconia,'" *ibid.*, **59** [3-4] 179 (1976).

[43]R. W. Rice, "Further Discussion of 'Precipitation in Partially Stabilized Zirconia,'" *ibid.*, **60** [5-6] 280 (1977).

[44]D. L. Porter, G. K. Bansal, and A. H. Heuer, "Reply to Comments on 'Precipitation in PSZ,'" *ibid.*, **59** [3-4] 179–82 (1976).

[45]D. L. Porter and A. H. Heuer, "Reply to Further Discussion of 'Precipitation in Partially Stabilized Zirconia,'" *ibid.*, **60** [5-6] 280–81 (1977).

[46]D. L. Porter and A. H. Heuer, "Mechanism of Toughening Partially Stabilized Zirconia (PSZ)," *ibid.*, **60** [3-4] 183–84 (1977).

[47]D. L. Porter, A. G. Evans, and A. H. Heuer, "Transformation Toughening in Partially Stabilized Zirconia (PSZ)," *Acta Metall.*, **27** [10] 1649–54 (1979).

[48]T. K. Gupta, J. H. Bechtold, R. C. Kuznicki, L. H. Cadoff, and B. R. Rossing, "Stabilization of Tetragonal Phase in Polycrystalline Zirconia," *J. Mater. Sci.*, **12**, 2421–26 (1977).

[49]T. K. Gupta, "Sintering of Tetragonal Zirconia and its Characteristics," *Sci. Sintering*, **10** [3] 205–16 (1978).

[50]T. K. Gupta, F. F. Lange, and J. H. Bechtold, "Effect of Stress-Induced Phase Transformation on the Properties of Polycrystalline Zirconia Containing Metastable Tetragonal Phase," *J. Mater. Sci.*, **13**, 1464–70 (1978).

[51]C. A. Anderson and T. K. Gupta, "Martensitic Transformation Toughening of ZrO_2"; for abstract see *Am. Ceram. Soc. Bull.*, **57** [3] 312 (1978).

[52]R. T. Pascoe and R. C. Garvie; pp. 714–84 in Ceramic Microstructures '76. Edited by R. M. Fulrath and J. A. Pask. West View Press, Boulder, Colo., 1977.

[53]T. K. Gupta, "Strengthening by Surface Damage in Metastable Tetragonal Zirconia," *J. Am. Ceram. Soc.*, **63** [1-2] 117 (1980).

[54]D. Michel, "Ordered States in the Fluorite-Type Solid Solutions in the System Zirconia-Calcium with the Composition $4ZrO_2 \cdot CaO$," *Mater. Res. Bull.*, **8** [8] 943–49 (1973).

[55]J. G. Allpress and H. J. Rossell, "A Microdomain Description of Defective Fluorite-Type Phases $Ca_xM_{1-x}O_{2-x}$ (M = Zr, Hf; x = 0.1-0.2)," *J. Solid State Chem.*, **15** [1] 68–78 (1975).

[56]S. P. Ray and V. S. Stubican, "Fluorite Related Ordered Compounds in the ZrO_2-CaO and ZrO_2-Y_2O_3 Systems," *Mater. Res. Bull.*, **12**, 549–56 (1977).

[57]V. S. Stubican and S. P. Ray, "Phase Equilibria and Ordering in the System ZrO_2-CaO," *J. Am. Ceram. Soc.*, **60** [11-12] 534–37 (1977).

[58]V. S. Stubican, R. C. Hink, and S. P. Ray, "Phase Equilibria and Ordering in the System ZrO_2-Y_2O_3," *ibid.*, **61** [1-2] 17–21 (1978).

[59]W. H. Rhodes and R. E. Carter, "Cationic Self-Diffusion in Calcia-Stabilized Zirconia," ibid., **49** [5] 244–49 (1966).
[60]L. A. Simpson and R. E. Carter, "Oxygen Exchange and Diffusion in Calcia-Stabilized Zirconia," ibid., [3] 139–44.
[61]F. Hund, "Fluorite Phases in the System ZrO_2-CaO; Its Defect Structure and Electrical Conductivity," Z. Phys. Chem., **199** [1-3] 142–51 (1952).
[62]T. Y. Tien and E. C. Subbarao, "X-ray and Electrical Conductivity Study of the Fluorite Phase in the System ZrO_2-CaO," J. Chem. Phys., **39** [4] 1041–47 (1963).
[63]T. H. Etsell and S. N. Flengas, "The Electrical Properties of Solid Oxide Electrolytes," Chem. Rev., **70** [3] 339–76 (1970).
[64]T. Takahashi; pp. 989–1051 in Physics of Electrolytes, Vol. 2. Edited by J. Hladik. Academic Press, New York, 1972.
[65]W. E. Worrell; pp. 143–68 in Solid Electrolytes. Edited by S. Geller. Springer-Verlag, Berlin, 1977.
[66]R. M. Dell and A. Hooper; pp. 291–312 in Solid Electrolytes. Edited by P. Hagenmuller and W. van Gool. Academic Press, New York, 1978.
[67]C. B. Choudhary, H. S. Maiti, and E. C. Subbarao; pp. 1–80 in Solid Electrolytes and Their Applications. Edited by E. C. Subbarao. Plenum Press, New York, 1980.
[68]R. E. Carter and W. L. Roth; pp. 125–44 in Electromotive Force Measurements in High Temperature Systems. Edited by C. B. Alcock. Institution of Mining and Metallurgy, London, 1968.
[69]B. Hudson and P. T. Moseley, "On the Extent of Ordering in Stabilized Zirconia," J. Solid State Chem., **19**, 383–89 (1976).
[70]D. Steele and B. E. F. Fender, "The Structure of Cubic ZrO_2:$YO_{1.5}$ Solid Solutions by Neutron Scattering," J. Phys. C, **7**, 1–11 (1974).
[71]J. Faber, Jr., M. H. Mueller, and B. R. Cooper, "Neutron-Diffraction Study of $Zr(Ca,Y)O_{2-x}$: Evidence of Differing Mechanisms for Internal and External Distortions," Phys. Rev., **B17** [12] 4884–88 (1978).
[72]J. B. Cohen, J. Faber, Jr., and M. Morinaga; this volume, pp. 37–46.
[73]J. W. Patterson, E. C. Borgen, and R. A. Rapp, "Mixed Conduction in $Zr_{0.85}Ca_{0.15}O_{1.85}$ and $Th_{0.85}Y_{0.15}O_{1.85}$ Solid Electrolytes," J. Electrochem. Soc., **114** [7] 752–58 (1967).
[74]C. B. Choudhary; Ph.D. Thesis, Indian Institute of Technology, Kanpur, 1975.
[75]J. M. Dixon, L. D. Le Grange, U. Merten, C. F. Miller, and J. T. Porter II, "Electrical Resistivity of Stabilized Zirconia at Elevated Temperatures," J. Electrochem. Soc., **110** [4] 276–80 (1963).
[76]R. E. W. Casselton, "Low Field DC Conduction in Yttria-Stabilized Zirconia," Phys. Status Solidi, **29**, 571–85 (1970).
[77]D. W. Strickler and W. G. Carlson, "Ionic Conductivity of Cubic Solid Solutions in the System CaO-Y_2O_3-ZrO_2," J. Am. Ceram. Soc., **47** [3] 122–27 (1964).
[78]W. W. Barker, "Statistical Aspects of Gross Anion Substoichiometry in Fluorite-Related Systems," Mater. Sci. Eng., **2** [4] 208–12 (1967).
[79]M. O'Keefe; pp. 609–28 in Chemistry of the Extended Defects in Non-Metallic Solids. Edited by L. Eyring and M. O'Keefe. North Holland Publishing Co., Amsterdam, 1970.
[80]E. C. Subbarao and T. V. Ramakrishnan; pp. 653–56 in Fast Ion Transport in Solids. Edited by Vashishta, Mundy, and Shenoy. Elsevier North Holland, New York, 1979.
[81]A. M. Anthony; pp. 519–26 in Ref. 66.
[82]M. Gauthier, A. Belanger, Y. Meas, and M. Kleitz; pp. 497–517 in Ref. 66.
[83]K. P. Jagannathan, H. S. Ray, A. Ghosh, S. K. Tiku, and E. C. Subbarao; pp. 201–60 in Ref. 67.
[84]V. I. Aleksandrov, V. V. Osiko, A. M. Prokhorov, and V. M. Tatarintsev; pp. 421–80 in Current Topics in Materials Science, Vol. I. Edited by E. Kaldis. North Holland Publishing Co., Amsterdam, 1978.

13

Table I. Bench marks in the history of zirconia

Event	Ref. No.
Discovery of baddeleyite in Brazil	Hussak (1892)
Identification of monoclinic symmetry of ZrO_2 and monoclinic tetragonal transformation by XRD	1
Stabilization of zirconia by RO, R_2O_3	2
and binary oxide phase diagrams	3, 4
Ionic conductivity and defect structure of stabilized ZrO_2	5–8
Existence of cubic ZrO_2	9
Martensitic nature of monoclinic-tetragonal transformation	11–13
Transformation-toughened PSZ	15–17
Crystal structure analysis	18–20
Nonrefractory uses of ZrO_2: oxygen sensor and pump, jewelry, opacifier, heating elements, extrusion dies	
Topical conference on ZrO_2	Wright-Patterson AFB (1962)

Table II. Strength and fracture toughness of zirconia

	Transverse rupture strength (MPa)	K_{IC} (MN/m$^{3/2}$)
Tetragonal + cubic ZrO_2	650	$\approx$7.1
Monoclinic + cubic ZrO_2 (overaged at 1400 °C)	250	3.7
Cubic ZrO_2 (solution-annealed at 1850 °C, 4 h)	245	2.8

Table III. Refractive Index, n, of ZrO_2-Y_2O_3 single crystals

Y_2O_3 (mol%)	n at a wavelength (nm) of				
	0.4358	0.5461	0.5779	0.5893	0.6328
9	0.22224	0.21856	0.21790	0.21768	0.21718
11	.22101	.21737	.21669	.21653	.21581
15	.21783	.21434	.21372	.21352	.21287
17	.21232	.20949	.20897	.20881	.20826

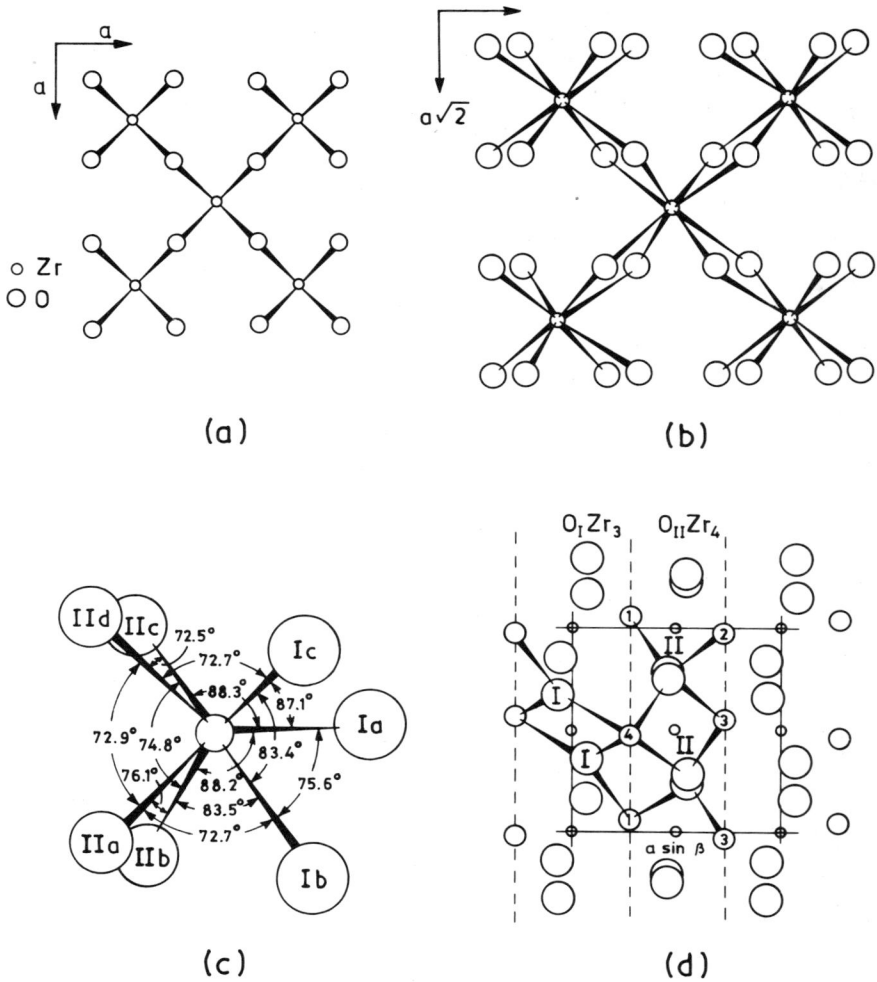

Fig. 1. Crystal structure of ZrO_2 polymorphs: (a) projection of a layer of ZrO_8 groups in the cubic ZrO_2 on the (100) plane (Ref. 19), (b) projection of a layer of ZrO_8 groups on the (110) plane of tetragonal ZrO_2 (Ref. 19), (c) angles and interatomic distances in the ZrO_7 coordination polyhedron of monoclinic ZrO_2 (Ref. 19), and (d) projection of the crystal structure of monoclinic ZrO_2 along the c_m-axis showing layers of O_IZr_3 and $O_{II}Zr_4$ polyhedra (Ref. 18).

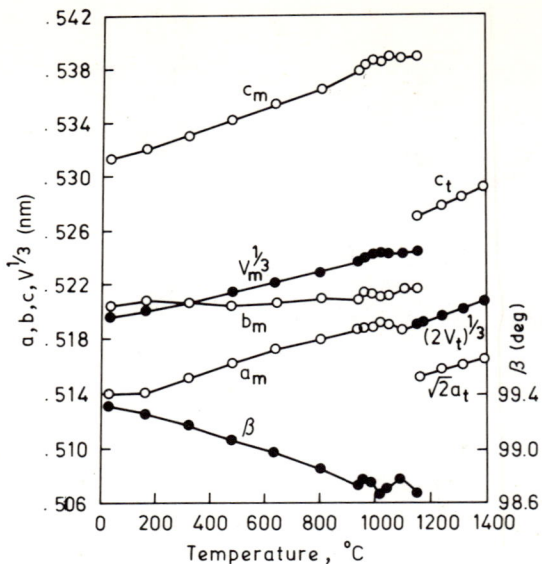

Fig. 2. Variation of lattice parameters and volume of monoclinic and tetragonal ZrO₂ with temperature (Ref. 23).

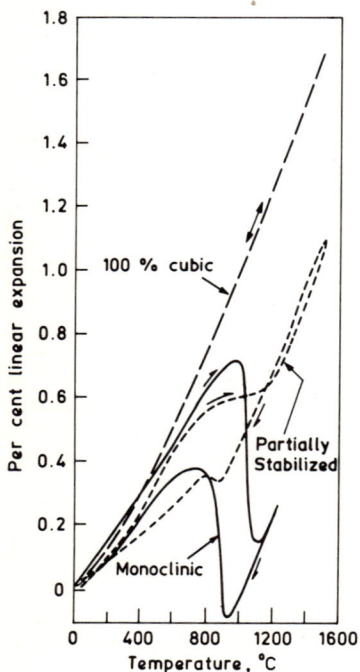

Fig. 3. Linear thermal expansion of monoclinic, partially stabilized, and fully stabilized zirconia (Refs. 25 and 26).

16

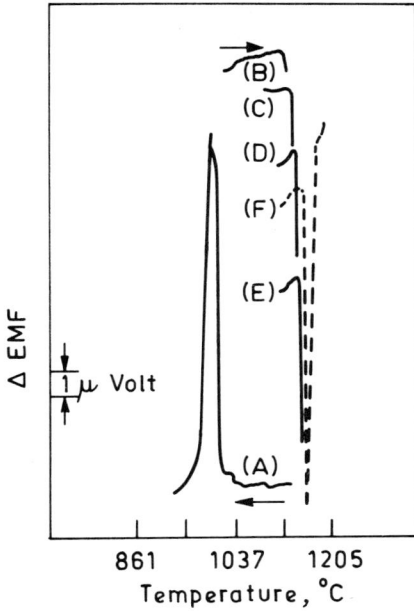

Fig. 4. Partial DTA curves illustrating athermal behavior of monoclinic-tetragonal phase change in ZrO_2 (A) total cooling, (B) to (E) allowed to partially transform, cooled to 1080 °C, then reheated, (F) as above, but transformed to completion (Ref. 27).

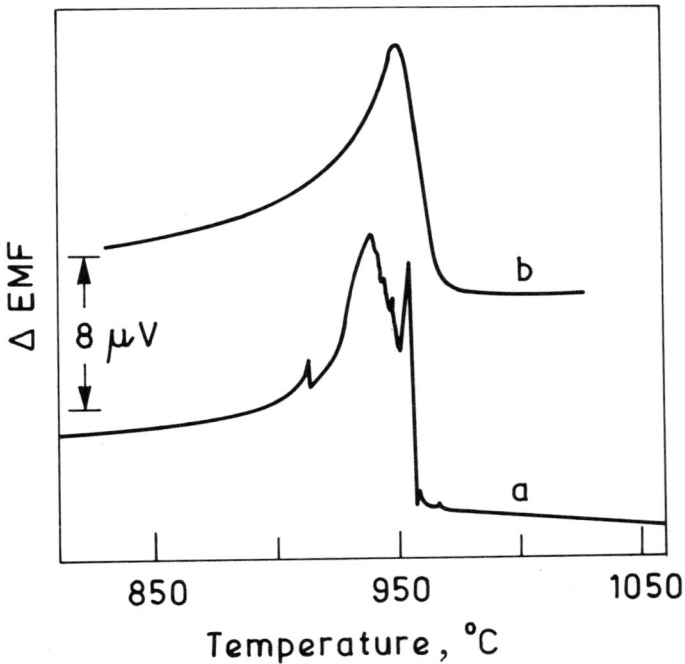

Fig. 5. Burst phenomenon in the cooling DTA trace during ZrO_2 transformation (a) peak showing a burst and (b) peak during normal cooling (Ref. 29).

17

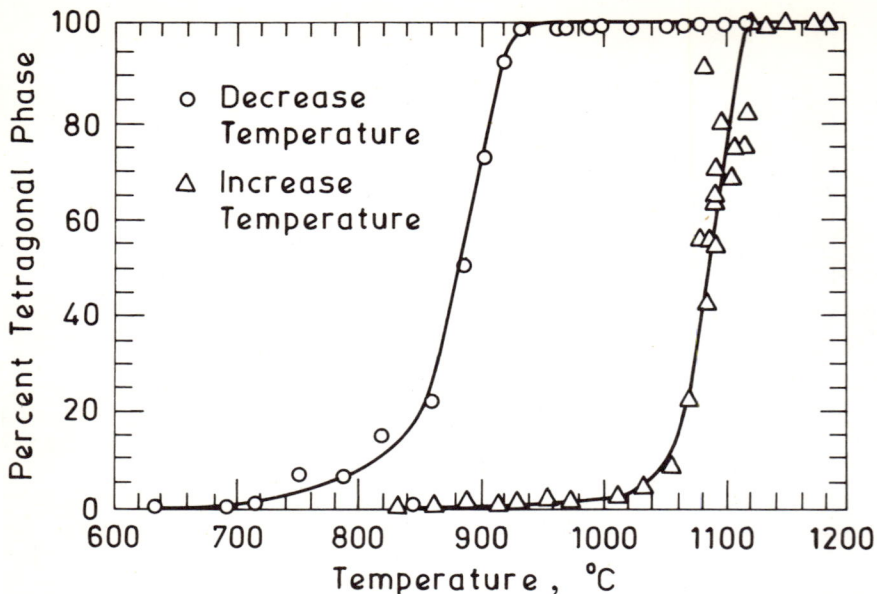

Fig. 6. Percent tetragonal phase, determined by X-ray diffraction during heating and cooling of ZrO_2 through the monoclinic-tetragonal transformation, showing thermal hysteresis (Ref. 11).

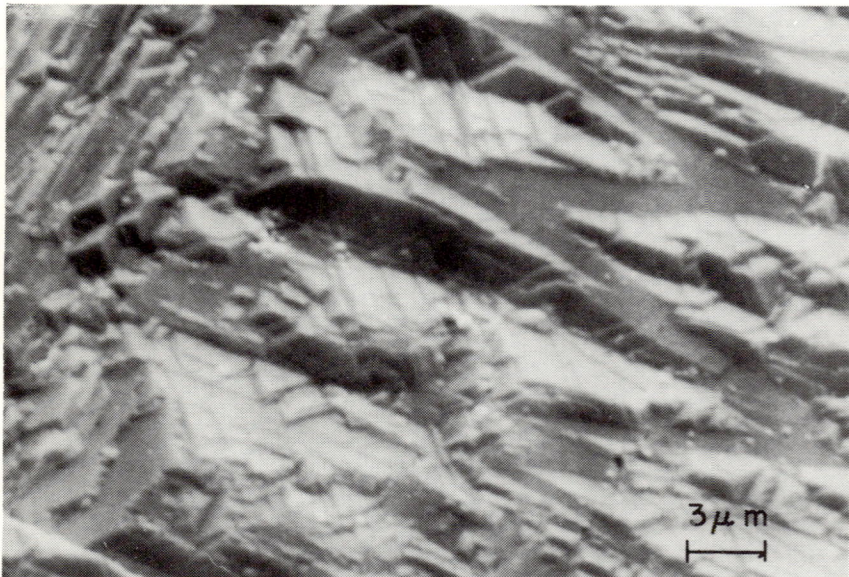

Fig. 7. Scanning electron micrograph showing surface distortion caused by martensitic phase transformation in single crystal ZrO_2 (Ref. 13).

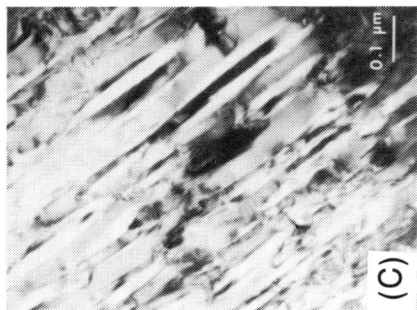

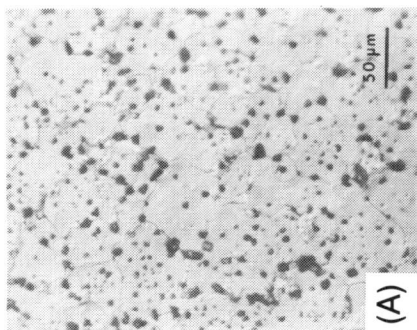

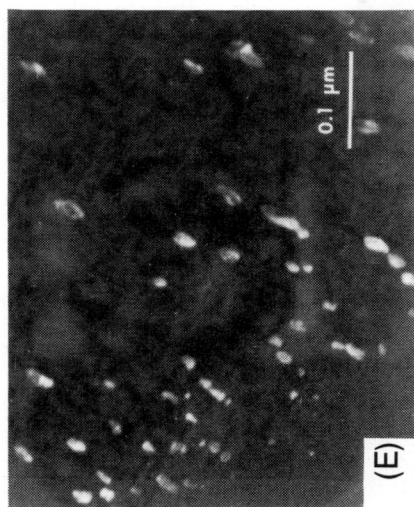

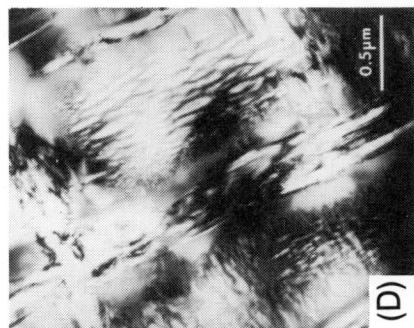

Fig. 8. See page 20 for caption.

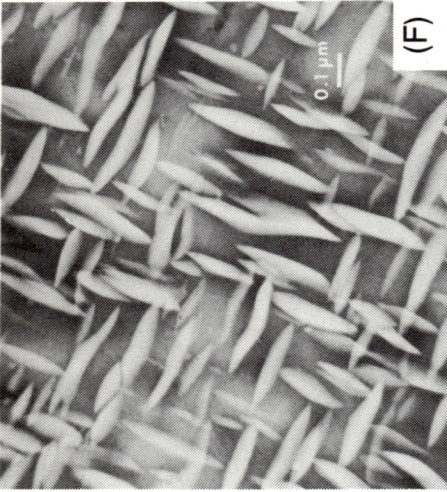

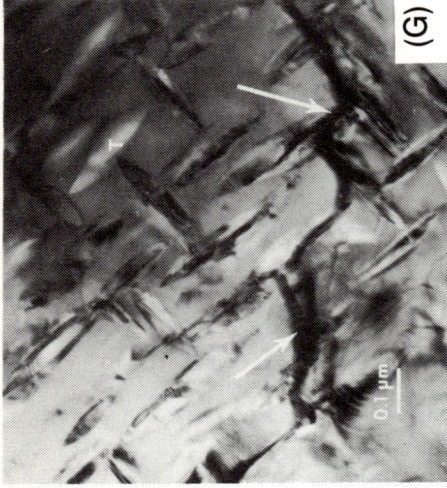

Fig. 8. Microstructural evolution of Mg-PSZ material. As-received: (*A*) optical micrograph, (*B*) electron micrograph showing large intergranular twinned monoclinic particle, (*C*) electron micrograph showing small intragranular monoclinic precipitates, (*D*) same as (*C*), but coherent tetragonal precipitate is shown, (*E*) solution-annealed (1850°C, 4 h), showing small precipitate particles, (*F*) solution-annealed (2000°C, 1–5 h in vacuum), aged at 1500°C, 1 h in air. Coherent tetragonal precipitate particles are shown. (*G*) Transformed monoclinic particles are present near a hardness indentation and untransformed tetragonal particles away from the indentation (Refs. 40, 46).

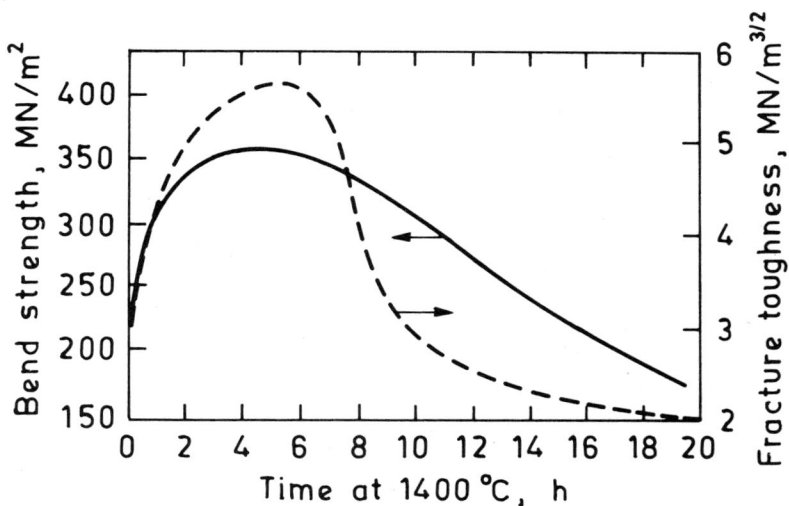

Fig. 9. Bend strength and critical stress intensity of Mg-PSZ as a function of aging time at 1400 °C (Ref. 40).

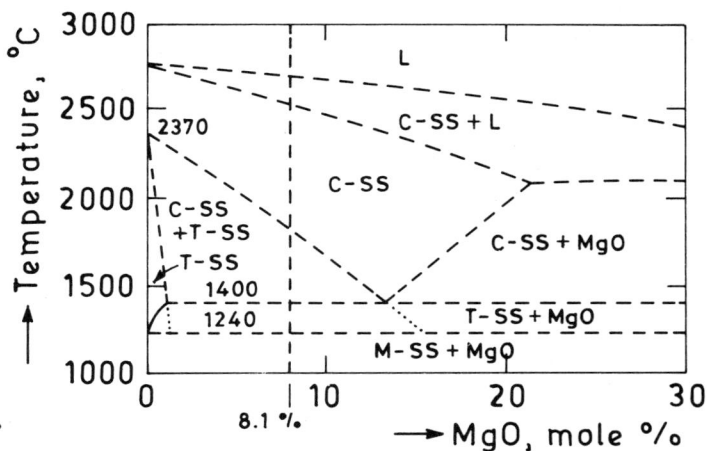

Fig. 10. Partial phase diagram of the system ZrO$_2$-MgO.

(a) Disordered

(b) Ordered

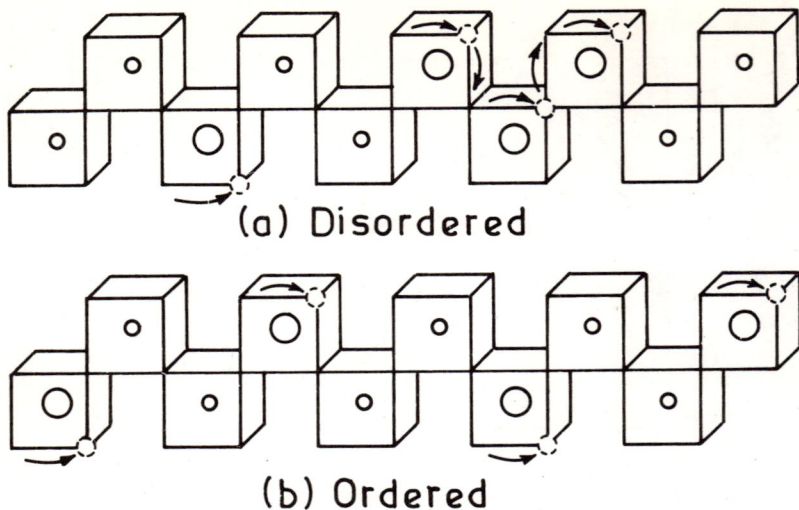

O Ca , o Zr , ⊙ Oxygen vacancy

Fig. 11. Defect structure and vacancy transport in CaO-stabilized zirconia (schematic) in disordered and ordered state (Ref. 68).

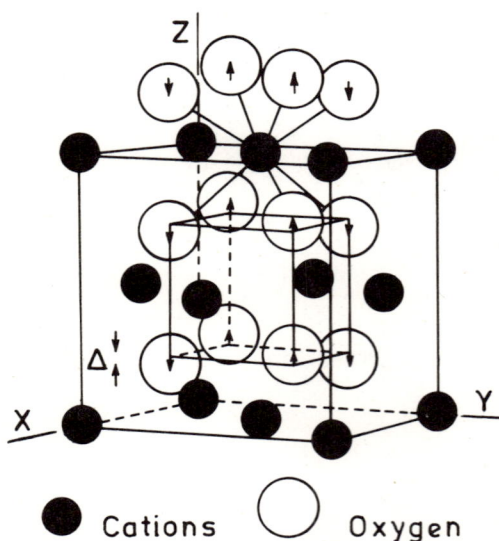

● Cations ○ Oxygen

Fig. 12. Cubic fcc fluorite unit cell. The displacement of oxygen ions from this ideal fluorite lattice site (by $\Delta/a = 4.4 \times 10^{-2}$) is shown by the arrows (not to scale) (Ref. 71).

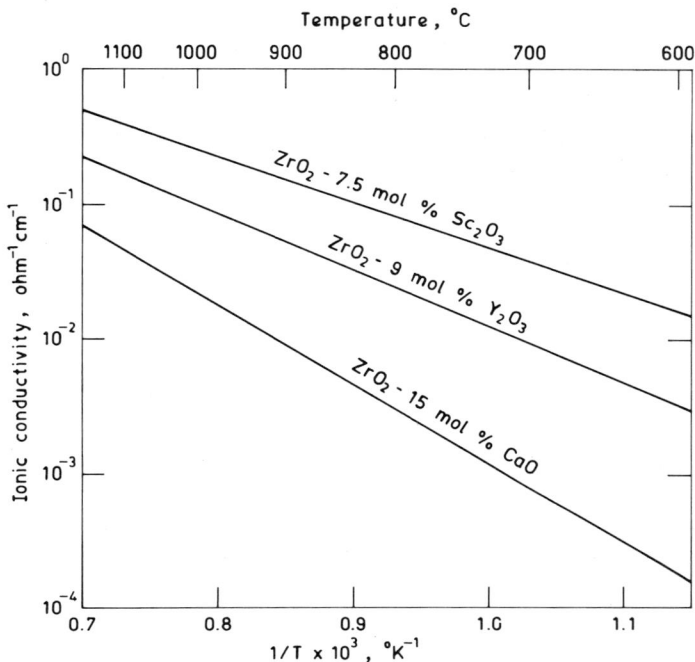

Fig. 13. Temperature dependence of ionic conductivity of zirconia doped with Sc_2O_3, CaO, and Y_2O_3.

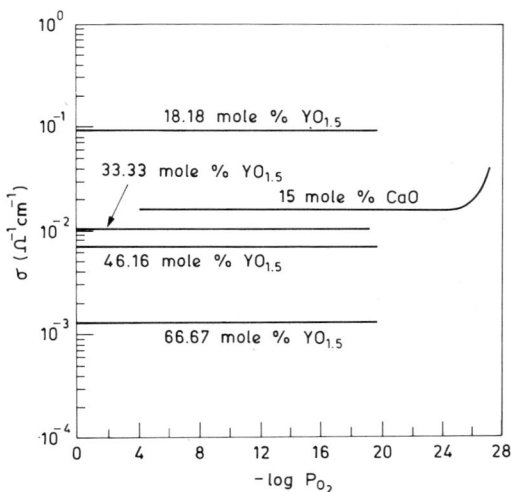

Fig. 14. Electrical conductivity (at 1000 °C) of zirconia, doped with CaO (15 mol%) (Ref. 73) and $YO_{1.5}$ (18.18, 33.33, 46.16, and 66.67 mol%), as a function of oxygen partial pressure (Ref. 74).

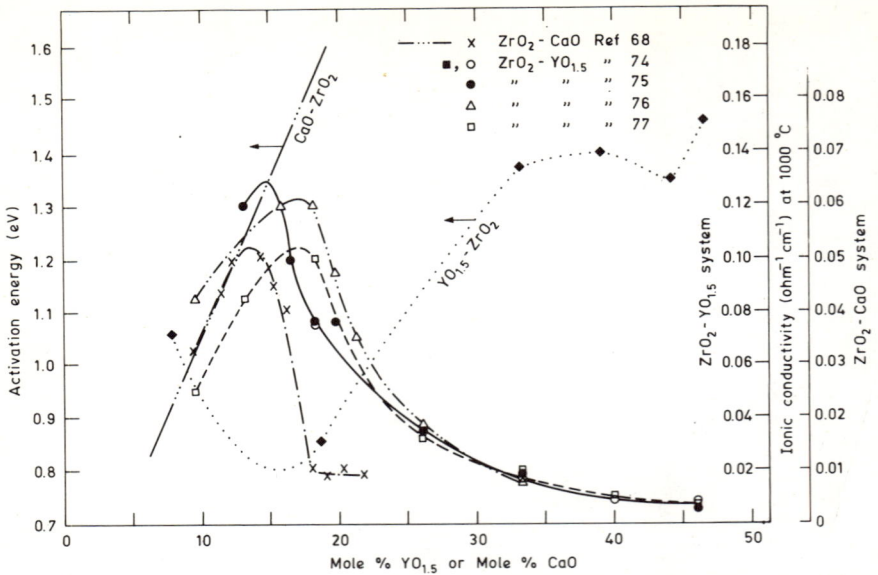

Fig. 15. Variation of electrical conductivity and activation energy for conduction with dopant concentration in zirconia-based systems.

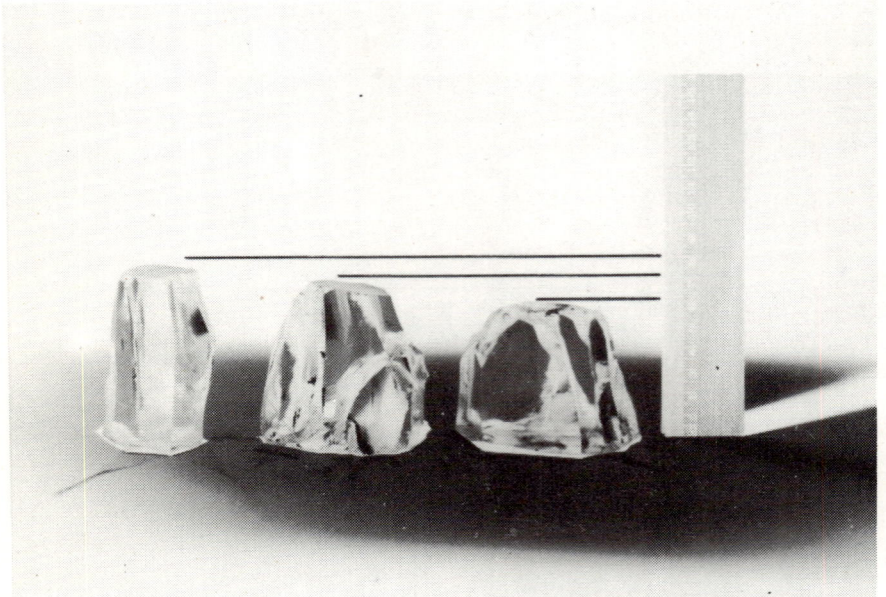

Fig. 16. Stabilized zirconia boule obtained by skull melting (Courtesy of Singh Industries).

Phase equilibria in some zirconia systems

V. S. STUBICAN AND J. R. HELLMANN

Department of Materials Science and Engineering
The Pennsylvania State University
University Park, Pa. 16802

Several phase diagrams have been proposed for each of the three most important zirconia binary systems: ZrO_2-CaO, ZrO_2-Y_2O_3, and ZrO_2-MgO. The major difficulty in establishing the correct subsolidus phase relationships arises from the fact that the reactions in these systems are relatively slow below 1400 °C. Furthermore, the establishment of metastable equilibria plays a role in zirconia systems when the cubic solid solutions are decomposed. The influence of this phenomenon on the phase boundaries is discussed for the ZrO_2-CaO system. There is definite evidence of the appearance of eutectoids in all three systems. Evidence for the short-range and long-range ordering of vacancies and impurity cations has also been obtained. Long-range ordering in ZrO_2-CaO systems results in the formation of the compound $CaZr_4O_9$, and in the systems ZrO_2-Y_2O_3 the ordered phase $Zr_3Y_4O_{12}$ is present. There is no definite evidence for the long-range ordering in the ZrO_2-MgO system. The phase relationships in two ternary systems ZrO_2-Y_2O_3-MgO and ZrO_2-CaO-MgO were determined for two temperatures.

Few materials show as much potential for advanced high temperature electrical and structural applications as zirconia-based ceramics. Stabilized zirconia ceramics, because of their superior high temperature electrical properties, have been used extensively as ionic conducting electrolytes in fuel cells, as MHD electrodes, and as oxygen gas sensors. The technological importance of zirconia-based ceramics, however, goes much further and includes applications as heat-resistant linings in furnaces and as protective coatings on alloys. The fundamental concept in zirconia ceramics is to "alloy" the pure zirconia with enough of another oxide to fully or partially stabilize high temperature polymorphs of zirconia to lower temperatures. The existence of a relatively tough, partially stabilized zirconia ceramic, consisting of a dispersion of metastable tetragonal ZrO_2 inclusions within large grains of stabilized cubic ZrO_2, seems to be derived from the stress-induced tetragonal→monoclinic transformation.

In addition to the engineering importance, zirconia and its solid solutions are of basic scientific interest. Inherent in these systems are several important reactions such as phase transitions, precipitation of metastable phases, and short and long range ordering of dopant ions and oxygen vacancies.

Knowledge of the phase relations in zirconia systems is of fundamental importance for the understanding of the physical and chemical behavior of zirconia ceramics. Establishing the correct phase diagrams, even in the simplest binary zirconia systems, is hampered by several major difficulties.

25

First, at relatively low temperatures ($<1400\,°C$), the reactions in zirconia systems are slow due to the slow diffusion. Precipitation, eutectoid decomposition, and ordering processes proceed very slowly, and equilibrium is difficult to achieve. Second, coherency strains between different phases may influence the extent of phase fields, and metastable phases may be retained at low temperatures. Finally, at very high temperatures, the presence of impurities or nonstoichiometry may influence the equilibria.

In the last 30 years, our knowledge of phase equilibria in binary and ternary zirconia systems has gradually improved. This paper briefly summarizes previous work and describes recent developments in our understanding of phase relationships in some of the most important zirconia systems.

Zirconia Binary Systems

The most commonly used oxides to form solid solutions with zirconia are CaO, MgO, Y_2O_3, Sc_2O_3, and rare earth oxides, because they exhibit a relatively high solubility in zirconia and are able to form fluorite-type phases with zirconia which are stable over wide ranges of composition and temperature.

In an early study, Duwez et al.[1] reported that the cubic field in the system ZrO_2-CaO was between 16 and 30 mol% CaO at $2000\,°C$ and that the fluorite-type cubic phase, usually called fully stabilized zirconia, was stable to room temperature. Dietzel and Tober[2] first observed a eutectoid in the composition region ZrO_2-$CaZrO_3$. This eutectoid indicated that the fluorite-type solid solution of CaO in ZrO_2 (CSZ) is thermodynamically unstable below $\approx 1220\,°C$. Several authors[3-6] who have reinvestigated the phase relationships in ZrO_2-CaO systems have obtained different results for the composition and temperature of the eutectoid in this system. Although there is agreement that the fluorite-type solid solutions are unstable at low temperature, there is no consensus on the temperature and composition of the eutectoid.

Order-disorder transition in CSZ was the subject of several studies. Tien and Subbarao[7] reported that a solid solution with 20 mol% CaO, annealed at $1000\,°C$, developed superstructure. Several workers,[5,8] using lattice parameter measurements, concluded that at 20 mol% CaO the compound $CaZr_4O_9$ is formed. The most comprehensive investigation of ordering in ZrO_2-CaO solid solutions was that of Carter and Roth[9] who used neutron diffraction and X-ray single crystal techniques. These authors indicated that there may be one or more ordered compounds in the system ZrO_2-CaO and that the ordering in ZrO_2-CaO solid solutions is associated with the growth of ordered domains of deformed oxygen polyhedra.

Recently, Michel[8] annealed single crystals of ZrO_2-20 mol% CaO for 1000 h and reported that two ordered polymorphs appeared, one in the temperature region $1050°$ to $1350°$ and the other below $1050\,°C$. Electron diffraction results of Allpress et al.[10] indicate that the $CaZr_4O_9$ compound has the same structure as $CaHf_4O_9$, which is monoclinic. It is interesting to note that the same authors found three ordered compounds, possessing distorted fluorite-type structures, in the system HfO_2-CaO. The compositions were determined to be $Ca_2Hf_7O_{16}$ (ϕ phase), $CaHf_4O_9$ (ϕ_1 phase), and $Ca_6Hf_{19}O_{44}$ (ϕ_2 phase).

It seems already well established that the ZrO_2 (tetragonal) $\rightleftharpoons ZrO_2$ (cubic) transition occurs at $\approx 2370°C$.[11]

In Fig. 1, the recent phase diagram of the system ZrO_2-CaO for the region ZrO_2-$CaZrO_3$ is shown.[11] All data $>1325\,°C$ shown on the diagram are taken from the previous work by Stubican and Ray[6] and were obtained on quenched samples using precise lattice parameter measurements. The extent of the cubic solid solution field at temperatures $<1400\,°C$ was determined by heating very reactive gels for an extended period (1 to 2 months) and quenching. The phase diagram in Fig. 1 represents the true chemical equilibria diagram.

Hudson and Moseley[12] have studied ''aging'' of the ZrO_2-CaO solid solution at $1000\,°C$ by using electron diffraction and transmission. According to these authors, the eutectoid temperature in the system ZrO_2-CaO should be lower than $1000\,°C$. However, in the process of decomposition of the cubic solid solution, tetragonal ZrO_2 (ss) and the ordered compound $CaZr_4O_9$ should form. If the formation of $CaZr_4O_9$ is prohibited by kinetic factors, such as low nucleation rate, the formation of the tetragonal ZrO_2 (ss) phase should proceed according to the metastable extension of the boundary lines.[13]

To study decomposition of the cubic solid solutions, specimens were prepared at $1900\,°C$ and then annealed at $900°$ to $1400\,°C$ for 1 to 3 months. Annealed specimens were studied by electron transmission and diffraction. The presence of diffuse features in the selected area electron diffraction patterns (Fig. 2(a)) is seen in all cubic solid solutions quenched from $1900\,°C$. It is not clear, however, if the diffuse scattering is due to the imperfect long range ordering[14] or to the appreciable short range ordering of oxygen vacancies and impurity cations.[15] Short range ordering of oxygen vacancies and doping cations is probably present in cubic solid solutions even at relatively high temperatures.

Figure 2(b) shows a $[110]_F$ zone electron diffraction pattern from a specimen containing 14 mol% CaO. The subscript F signifies a zone indexed on the basis of a cubic fluorite unit cell. The specimen, which was annealed for 2 months at $1180\,°C$, shows strong fluorite-type reflections, weaker extra reflections, and diffuse spots. Extra reflections could be satisfactorily indexed as tetragonal ZrO_2 and the diffuse spots can probably be attributed to the extensive short range ordering in the cubic phase. Similar results were obtained by Schoenlein et al.[16] A specimen containing 20 mol% CaO annealed at $1100\,°C$ shows superlattice spots and fluorite reflections (Fig. 2(c)). Large domains of the ordered compound $CaZr_4O_9$ are clearly visible in the dark field image (Fig. 2(d)). In Fig. 3 the results obtained by electron diffraction and transmission are superimposed on the equilibrium diagram shown in Fig. 1. At temperatures below the eutectoid decomposition of the cubic solid solution, precipitation of tetragonal ZrO_2 may occur according to the metastable extension of the cubic and tetragonal field, which explains the results obtained by Hudson and Moseley.[12] Metastable phase boundaries are important for the understanding of processes taking place when phase separation occurs in the cubic matrix.

One stable compound $CaZr_4O_9$ (ϕ_1) with monoclinic symmetry was found, and this compound is unstable above $1310 \pm 40\,°C$. The lower limit of stability for this compound has not yet been determined and, furthermore, the appearance of ϕ and ϕ_2 compounds in this system is still uncertain.

The system ZrO_2-MgO has been studied by various investigators. Ruff and Ebert[17] reported that the two oxides combined to form a cubic phase above $1400\,°C$. Duwez et al.[1] reported that the cubic solid solutions of MgO in ZrO_2 decompose into their constituent oxides below 1400 °C. This decomposition of the cubic phase was confirmed by several authors.[2,18,19] In 1965, Viechnicki and Stubican[19] investigated the mechanism of decomposition of the cubic solid solutions in the system ZrO_2-MgO and the influence of the amount of dopant on the tetragonal $\rightleftharpoons$ cubic transition. According to these authors, decomposition of the cubic solid solutions comprised formation of nuclei of the tetragonal phase preferentially on the grain surfaces and diffusion of the Mg^{2+} ions out of the regions of the growth of the tetragonal phase. Decomposition reactions in the system were investigated later in great detail by Bansal and Heuer[20] who used electron diffraction and transmission. The phase diagram for the system ZrO_2-MgO as given by Viechnicki and Stubican[19] is shown in Fig. 4. The more recent work on phase relationships in the ZrO_2-MgO system was done by Grain,[21] who concluded that the studies of the ZrO_2-MgO system by previous workers were too dependent on data obtained by disappearing phase methods. According to this author, the solubility of MgO in tetragonal ZrO_2 at 1300 °C is <1 mol% and the eutectoid in this system is at 1400 °C and ≈ 13 mol% MgO. The phase relations in a portion of the system ZrO-MgO, as presented by Grain,[21] are shown in Fig. 5. It is evident that this author has made investigations in a relatively narrow concentration and temperature range (1350 to 1600 °C) and that further work is necessary to clearly define this system using a much wider temperature and composition range.

It is interesting to note that no definite proof exists for the appearance of a stable compound in this system. Delamarre[22] has, however, claimed that the compound $Mg_2Zr_5O_{12}$ could be prepared in the temperature range 1850 to 2100 °C and that the same compound decomposes below 1850 °C.

The system ZrO_2-Y_2O_3 is one of the most interesting zirconia binary systems because of the relatively large cubic solid solution field. The first reported phase diagram for this system was by Duwez et al.[23] Despite many subsequent investigations, there is no general agreement among the proposed phase diagrams. Fan et al.[24] indicated the existence of a pyrochlore-type compound of the composition $Y_2Zr_2O_7$, but Smith[25] could not confirm the presence of the pyrochlore phase. Srivastava et al.[26] have incorporated the tetragonal $\rightleftharpoons$ cubic transition into the revised diagram. Recently Scott[27] proposed the existence of a miscibility gap closed below the solidus temperature in the yttria-rich solid solution region. Scott's phase diagram also indicated the formation of hexagonal yttria above ≈ 2300 °C as found by Foex and Traverse.[28]

The liquidus for the system ZrO_2-Y_2O_3 has been determined by Rouanet[29] and by Noguchi et al.[30] who found a eutectic point at 77 mol%. Skaggs et al.[31] gave an alternative liquidus with a peritectic point at ≈ 76 mol% Y_2O_3.

Recent neutron diffuse scattering experiments by Steele and Fender[32] indicate that the oxygen vacancies produced by the doping of zirconia with yttria may be associated with two Y^{3+} cations located in the nearest-neighbor positions. At the high concentration of dopant (>15 mol%) a sharing of vacancies along the body diagonals of the fluorite structure was postulated. However, no evidence for a long range ordered phase was found. Long range ordering in the ZrO_2-Y_2O_3 system, however, occurs at 40 mol% Y_2O_3 as described by Ray and Stubican[33] and Stubican et al.[34] Scott[35] also observed long range ordering at 40 mol% Y_2O_3 in a concurrent study. The ordered phase is $Zr_3Y_4O_{12}$ which has rhombohedral symmetry (space group $R\bar{3}$) and the structure of this compound could be derived from the analogous M_7O_{12} phases, $Zr_3M_4O_{12}$ (M = Sc,Er,Dy,Yb), MY_6O_{12} (M = U,Mo,W), and the binary rare earth oxides M_7O_{12} (M = Pr,Tb,Ce). In this type of structure one of the [111] cubic directions becomes the unique 3-fold inversion axis of the rhombohedral cell. Oxygen vacancies lie along this specific [111] direction resulting in infinite strings of 6-coordinated metal cations following the 3-fold axis. Structural analysis of this compound was done recently by using neutron diffraction and X rays.[36]

The recently established phase diagram by Stubican *et al.*[34] for the system ZrO_2-Y_2O_3 is shown in Fig. 6. The presence of the stable ordered compound $Zr_3Y_4O_{12}$ is now firmly established. The presence of this compound and the marked curvature in the low yttria boundary of the cubic solid solution indicate that a eutectoid decomposition reaction, cubic ZrO_2 (*ss*) → monoclinic ZrO_2 (*ss*) + $Zr_3Y_4O_{12}$, occurs at < 750 °C. The major experimental problem arises in determining the composition and temperature of the eutectoid since the slow reaction rates at low temperature make attainment of equilibrium uncertain. Results of hydrothermal investigations indicate that the eutectoid may occur at < 400 °C at a composition between 20 and 30 mol% Y_2O_3. However, the results obtained at temperatures < 700 °C must be considered as tentative.

The system ZrO_2-Sc_2O_3 shows three ordered compounds in the subsolidus region.[37,38] All three compounds ($Sc_2Zr_7O_{17}$, $Sc_2Zr_5O_{13}$, and $Sc_4Zr_3O_{12}$) have rhombohedral structure. The large rare earth ions (Sm, Gd, Nd, La) form the pyrochlore-type compounds with zirconia.

Ternary Zirconia Systems

The extent to which the stabilizing effects of calcia and magnesia or yttria and magnesia are additive can be evaluated by studying ternary phase diagrams.

Knowledge of phase boundaries in ternary systems ZrO_2-CaO-MgO and ZrO_2-Y_2O_3-MgO is important if, for example, the controlled preparation of partially stabilized zirconia with good thermal shock properties is sought. To date, the results of the phase relationships in three-component zirconia systems have been largely incomplete[39-42] because the phase relationships in binary zirconia systems were poorly understood.

Recently, in our laboratory, phase relationships in the ternary system ZrO_2-Y_2O_3-MgO and ZrO_2-CaO-MgO were studied.[43] Two isothermal sections for each system were determined by using coprecipitated gels heated at given temperatures for extended periods (1 to 2 months).

Figure 7 shows the phase diagram for the system ZrO_2-Y_2O_3-MgO at 1220 °C and Fig. 8 shows the phase diagram for the same system at 1420 °C. The 1220 °C isotherm shows three 3-phase regions, three 2-phase regions, and three single-phase regions, as well as the presence of the ordered compound $Zr_3Y_4O_{12}$. The extent of the cubic solid solution field is relatively large due to the relatively large cubic field in the binary ZrO_2-Y_2O_3 system. The binary compound $Zr_3Y_4O_{12}$ does not appear in the ternary diagram at 1420 °C because it is not stable above $\approx$ 1250 °C. Furthermore, the existence of the compound $3MgO \cdot Y_2O_3$ found by Otto[44] could not be confirmed.

In Figs. 9 and 10, isothermal sections of the ternary system ZrO_2-CaO-MgO are given. The extent of the cubic solid solution field in this system is much smaller than in the ZrO_2-Y_2O_3-MgO system, due to the fact that the cubic solution field in the binary system ZrO_2-CaO is smaller than in the ZrO_2-Y_2O_3 system. Two stable binary compounds $CaZrO_3$ and $CaZr_4O_9$ are evident on Fig. 9. However, the compound $CaZr_4O_9$ is not stable above $1310 \pm 40\,°C$ and does not appear in the ternary diagram at $1420\,°C$.

Acknowledgment

This work was sponsored by the National Science Foundation Grant No. DMR-7824360.

References

[1]P. Duwez, F. Odell, and F. H. Brown, Jr., *J. Am. Ceram. Soc.*, **35** [5] 107–13 (1952).
[2]A. Dietzel and H. Tober, *Ber. Dtsch. Keram. Ges.*, **30** [4] 71–82 (1953).
[3]R. Roy, H. Miyabe, and A. M. Diness; for abstract see *Am. Ceram. Soc. Bull.*, **43** [4] 255–56 (1964).
[4]S. Fernandes and L. Beaudin, "Subsolidus Equilibria in the CaO-ZrO$_2$ System"; presented at the 21st Annual Pittsburgh Diffraction Conference, Mellon Institute, Pittsburgh, Pa., Nov. 1963.
[5]R. C. Garvie, *J. Am. Ceram. Soc.*, **51** [10] 553–56 (1968).
[6]V. S. Stubican and S. P. Ray, *ibid.*, **60** [11-12] 534–37 (1977).
[7]T. Y. Tien and E. C. Subbarao, *J. Chem. Phys.*, **39** [4] 1041–47 (1963).
[8]D. Michel, *Mater. Res. Bull.*, **8** [8] 943–50 (1973).
[9]R. E. Carter and W. L. Roth; pp. 125–44 in Electromotive Force Measurements in High Temperature Systems. Edited by C. B. Alcock. Institution of Mining and Metallurgy, London, 1968.
[10]J. G. Allpress, H. J. Rossell, and H. G. Scott, *J. Solid State Chem.*, **14** [1] 264–73 (1975).
[11]J. R. Hellmann and V. S. Stubican; to be published.
[12]B. Hudson and P. T. Moseley, *J. Solid State Chem.*, **19** [1] 383–89 (1976).
[13]S. M. Allen and J. W. Cahn, *Acta Metall.*, **23** [9] 1017–26 (1975).
[14]J. G. Allpress and H. J. Rossell, *J. Solid State Chem.*, **15** [1] 68–78 (1975).
[15]M. Morinaga, J. B. Cohen, and J. Faber, Jr.; to be published.
[16]L. H. Schoenlein, L. W. Hobbs, and A. H. Heuer, *J. Appl. Crystallogr.*, **13**, 375–79 (1980).
[17]O. Ruff and F. Ebert, *Z. Anorg. Allg. Chem.*, **180** [1] 19–41 (1929).
[18]A. Cocco and N. Schromek, *Radex Rundsch.*, **3**, 590–99 (1961).
[19]D. Viechnicki and V. S. Stubican, *J. Am. Ceram. Soc.*, **48** [6] 292–97 (1965).
[20]G. K. Bansal and A. H. Heuer, *ibid.*, **58** [5-6] 235–38 (1975).
[21]C. F. Grain, *ibid.*, **50** [6] 288–90 (1967).
[22]C. Delamarre, *Rev. Int. Hautes Temp. Refract.*, **9**, 209–24 (1972).
[23]P. Duwez, F. H. Brown, and F. Odell, *J. Electrochem. Soc.*, **98** [9] 356–62 (1951).
[24]Fu-K'ang Fan, A. K. Kuznetsov, and E. K. Keler, *Izv. Akad. Nauk SSSR, Ser. Khim.*, **1962**, pp. 1141-46; *ibid.*, **1963**, No. 4, pp. 601–10.
[25]D. K. Smith, *J. Am. Ceram. Soc.*, **49** [11] 625–27 (1966).
[26]K. K. Srivastava, R. N. Patil, C. B. Choudhary, K. V. Gokhale, and E. C. Subbarao, *Trans. J. Br. Ceram. Soc.*, **73** [3] 85–91 (1974).
[27]H. G. Scott, *J. Mater. Sci.* **10** [9] 1527–35 (1975).
[28]M. Foex and J. Traverse, *C. R. Hebd. Seances Acad. Sci., Ser. C*, **261** [13] 2490–93 (1965).
[29]A. Rouanet, *ibid.*, **264** [23] 1581–84 (1968).
[30]Tetsuo Noguchi, Masao Mizuno, and Joyoaki Yamada, *Bull. Chem. Soc. Japan*, **43** [8] 2614–16 (1970).
[31]S. R. Skaggs, N. L. Richardson, and L. S. Nelson; for abstract see *Am. Ceram. Soc. Bull.*, **51** [4] 324 (1972).
[32]D. Steele and B. E. F. Fender, *J. Phys. C*, **7** [1] 1–11 (1974).
[33]S. P. Ray and V. S. Stubican, *Mater. Res. Bull.*, **12**, 549–56 (1977).
[34]V. S. Stubican, R. C. Hink, and S. P. Ray, *J. Am. Ceram. Soc.*, **61** [1-2] 18–21 (1978).
[35]H. G. Scott, *Acta Crystallogr.*, **B33**, 281–82 (1977).
[36]S. P. Ray, V. S. Stubican, and D. E. Cox, *Mater. Res. Bull.*, **15**, 1419 (1980).
[37]R. Collongues, F. Queyroux, M. Perez Y. Jorba, and J. C. Gilles, *Bull. Soc. Chim. Fr.*, **4**, 1141–49 (1965).

[38]R. Ruh, H. J. Garrett, R. F. Domagala, and V. A. Patel, *J. Am. Ceram. Soc.,* **60** [9] 399–403 (1977).
[39]P. A. Tikhonov, A. K. Kuznetsov, and E. K. Keler, *Izv. Akad. Nauk SSSR, Neorg. Mater.,* **7** [11] 2015–19 (1971).
[40]P. A. Tikhonov, A. K. Kuznetsov, E. K. Keler, and M. V. Kravchinskaya, *ibid.,* **11** [4] 690–94 (1975).
[41]P. A. Tikhonov, A. K. Kuznetsov, E. K. Keler, S. N. Kuznetsova, and Yu. P. Udalov, *Dokl. Akad. Nauk SSSR,* **204** [3] 661–63 (1972).
[42]V. Longo and L. Podda, *Ceramurgia,* **4**, 1 (1978).
[43]J. R. Hellmann and V. S. Stubican; to be published.
[44]E. M. Levin, C. R. Robbins, and H. F. McMurdie, Phase Diagrams for Ceramists, 1964. Edited by Margie K. Reser. American Ceramic Society, Columbus, Ohio; Fig. 263.

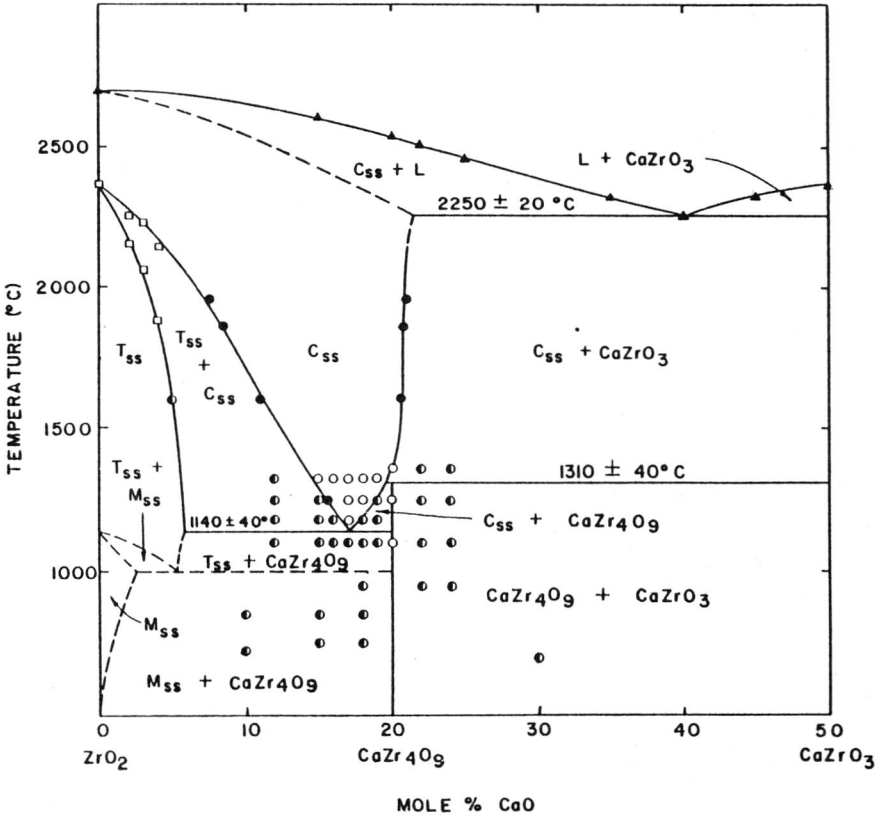

Fig. 1. Equilibrium phase diagram for the system ZrO_2-CaO.

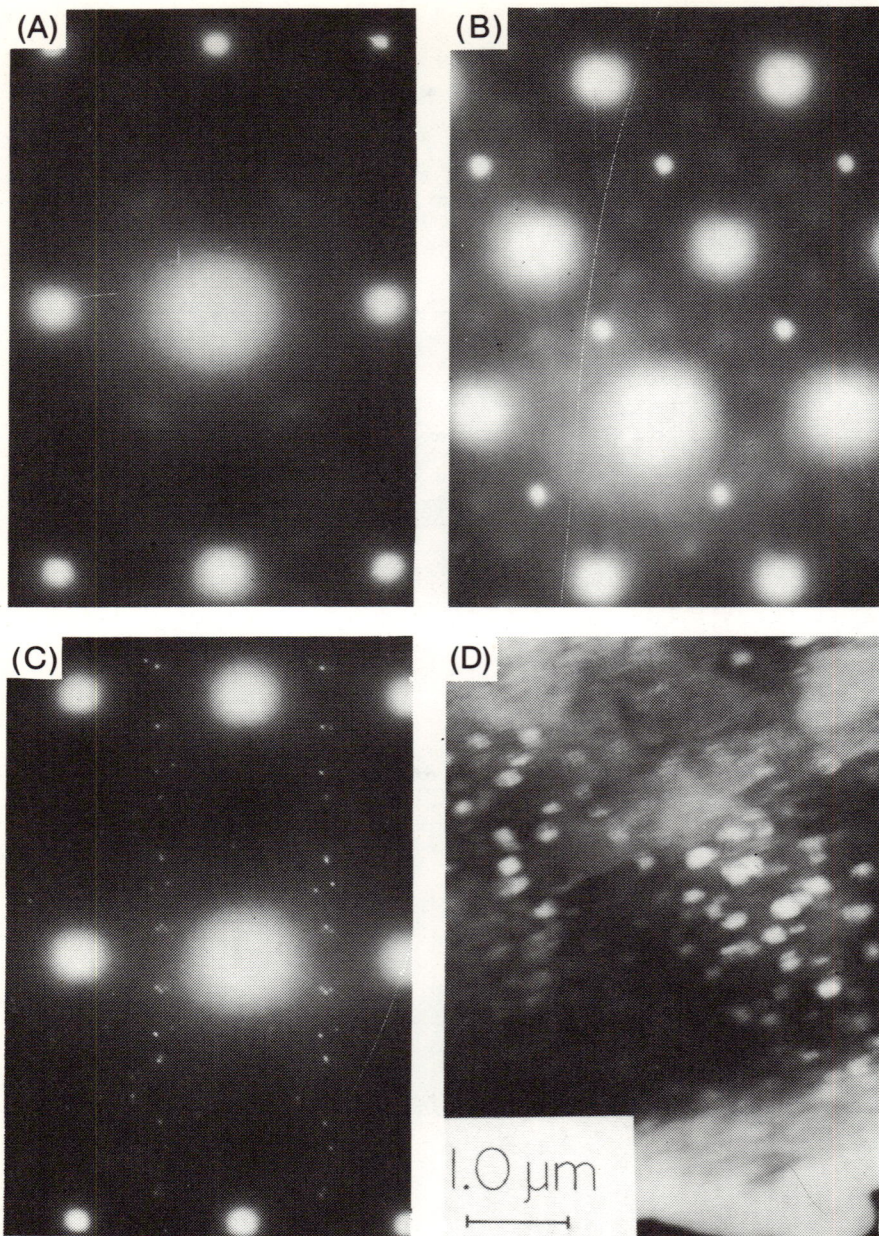

Fig. 2. (A) Selected area electron diffraction pattern (SAD) of $[112]_F$ zone of the specimen 0.83 $ZrO_2 \cdot$0.17 CaO heated at 2000ºC for 4 h and then quenched. (B) $[110]_F$ zone of the specimen 0.86 $ZrO_2 \cdot$0.14 CaO heated at 1900ºC then annealed for 2 months at 1180ºC. (C) $[112]_F$ zone of the specimen 0.80 $ZrO_2 \cdot$0.20 CaO heated at 1900ºC then annealed for 2 months at 1100ºC. (D) Dark field transmission electron micrograph obtained using a group of superlattice reflections from Fig. 2(C) as the principal beam.

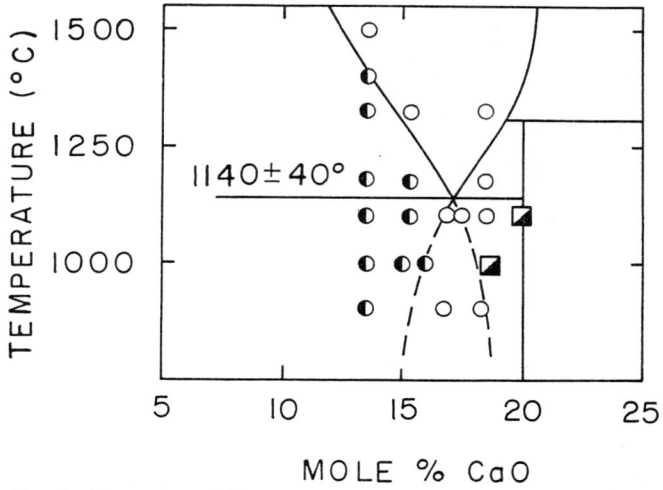

Fig. 3. Electron diffraction and transmission results superimposed on the equilibrium phase diagram given in Fig. 1. ◐ cubic and tetragonal solid solutions, ○ cubic solid solution, ◪ cubic solid solution and ordered compound.

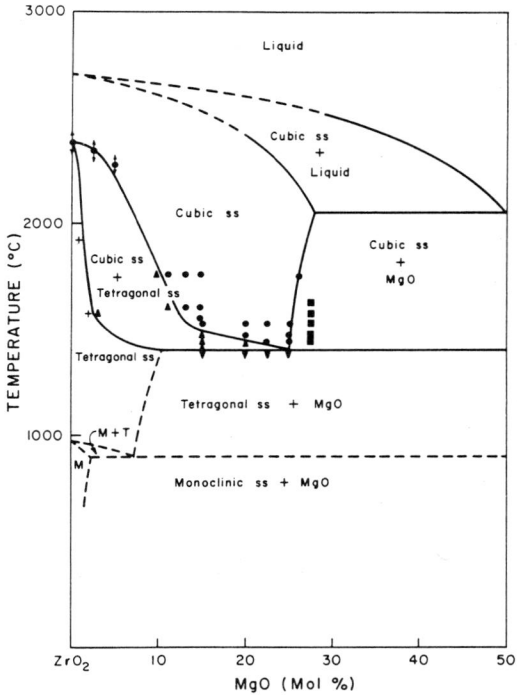

Fig. 4. Phase diagram for the system ZrO_2-MgO (Ref. 19).

33

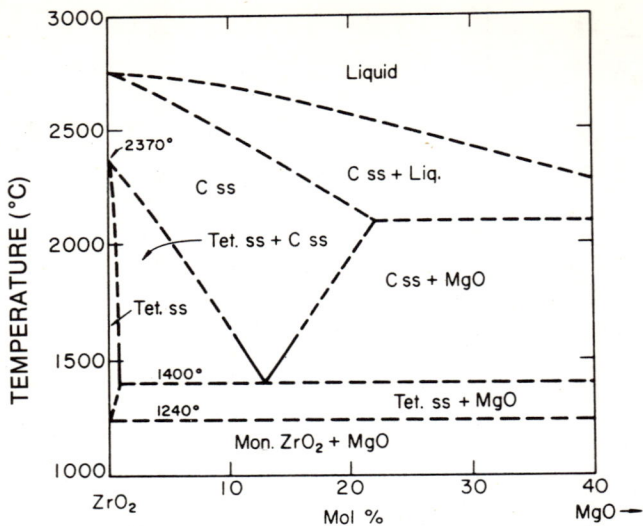

Fig. 5. Phase diagram for the system ZrO_2-MgO (Ref. 21).

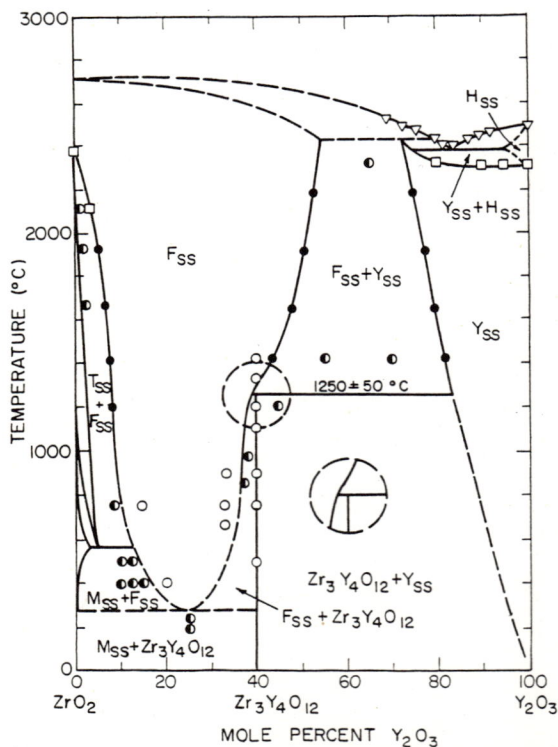

Fig. 6. Phase diagram for the system ZrO_2-Y_2O_3 (Ref. 34).

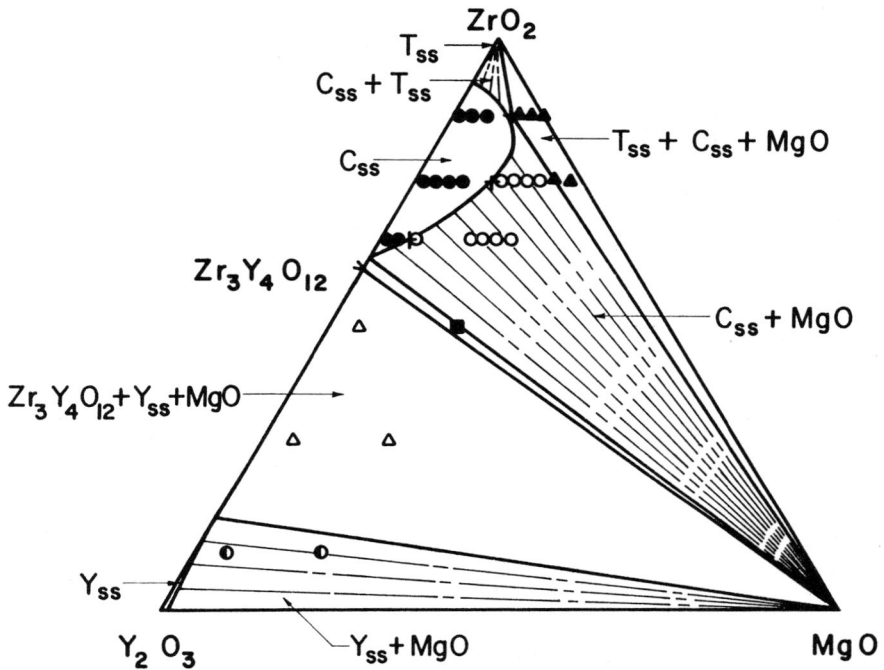

Fig. 7. Phase diagram for the system ZrO_2-Y_2O_3-MgO at $1220 \pm 5 \,°C$ (Ref. 43).

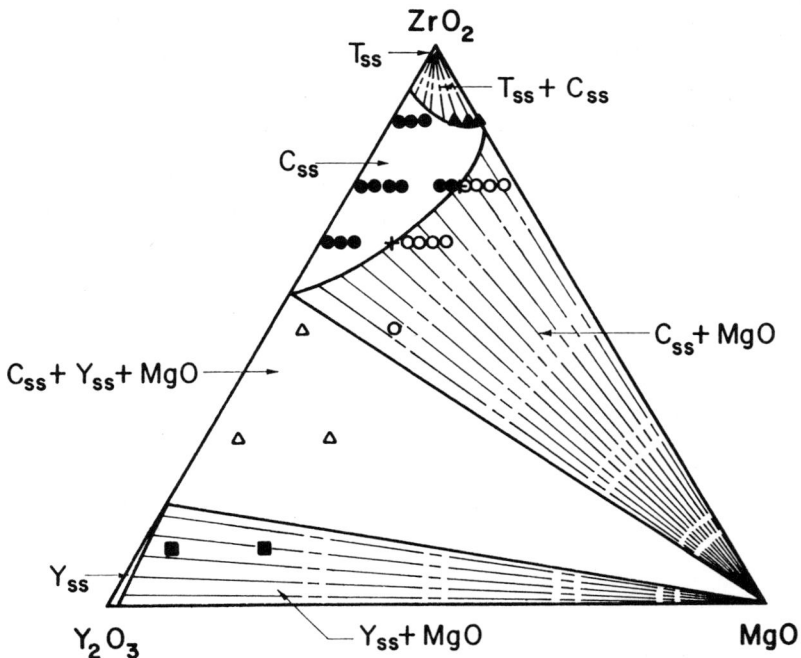

Fig. 8. Phase diagram for the system ZrO_2-Y_2O_3-MgO at $1420 \pm 5 \,°C$ (Ref. 43).

35

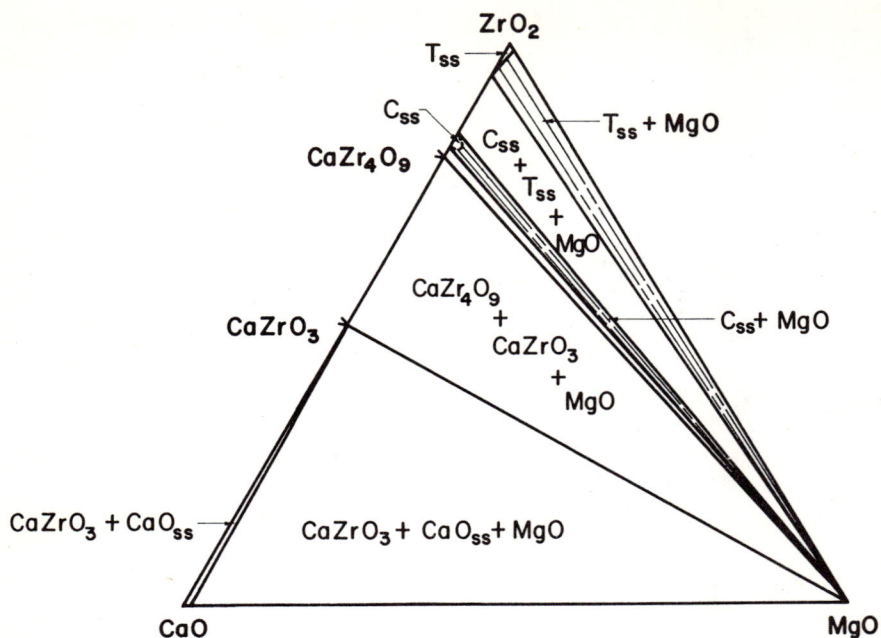

Fig. 9. Phase diagram for the system ZrO_2-CaO-MgO at $1220 \pm 5\,°C$ (Ref. 43).

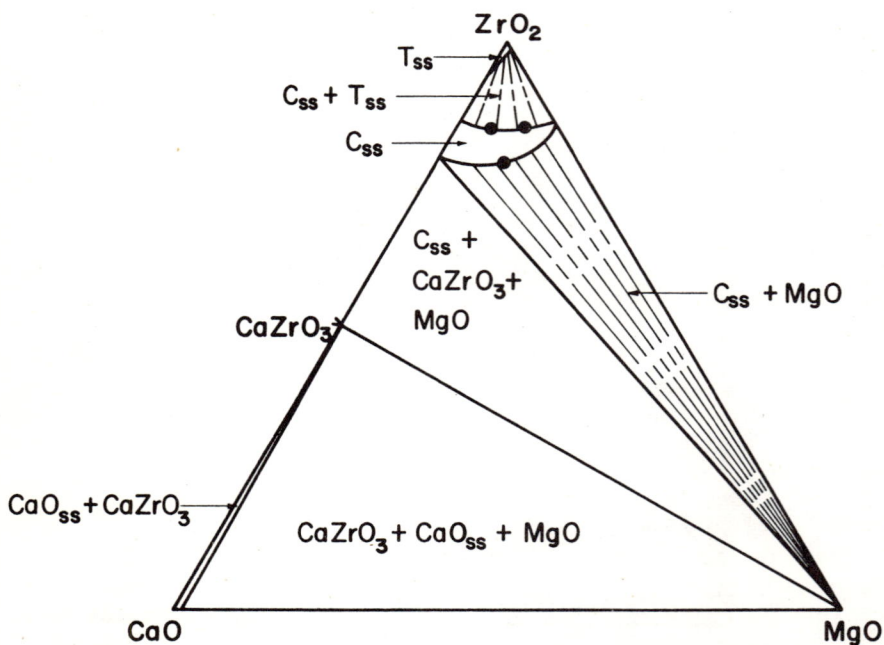

Fig. 10. Phase diagram for the system ZrO_2-CaO-MgO at $1420 \pm 5\,°C$ (Ref. 43).

The structure of (disordered) stabilized zirconias

J. B. COHEN

Department of Materials Science & Engineering
The Technological Institute
Northwestern University
Evanston, Ill. 60201

J. FABER, JR.

Materials Science Division
Argonne National Laboratory
Argonne, Ill. 60439

M. MORINAGA

Toyohashi University of Technology
Toyohashi-shi, Aichi 440, Japan

The diffuse X-ray scattering and Bragg peaks from single crystals of $Zr_{0.866}Ca_{0.134}O_{1.866}$ and $Zr_{0.786}Y_{0.214}O_{1.893}$ have been examined. The oxygen ions are displaced by $\Delta/a \approx 0.05$ in $\langle 100 \rangle$ directions. These displacements do not appear to be due to vibrational anharmonicity. In the Y-stabilized crystal there are also smaller $\langle 111 \rangle$ displacements of the cations, and not all of the oxygen ions are displaced. Ca ions are first neighbors to vacancies and these Ca ions are in rods along $\langle 332 \rangle$ directions. It is speculated that this arrangement enhances anionic conductivity and that the $\langle 100 \rangle$ displacements are a result of the formation of local tetragonal domains to reduce the electronic energy.

In this paper we summarize our recent X-ray studies of single crystals of $Zr_{0.866}Ca_{0.134}O_{1.866}$ and $Zr_{0.786}Y_{0.214}O_{1.893}$, quenched from the one-phase field. The integrated intensities of Bragg peaks were examined (with monochromatic MoKα radiation), as well as the diffuse scattering in a volume in reciprocal space (with monochromatic CuKα). Results on the average structure will be presented first and then the local atomic arrangements. Some speculation will be offered concerning the stabilization of this phase, as well as its high anionic conduction. Experimental and mathematical details can be found in Refs. 1 and 2.

The Average Structure

Only the disordered phase will be considered. Earlier investigations led to conflicting reports, one group[3,4] suggesting displacements of the oxygen ions from the ideal fluorite positions in $\langle 111 \rangle$ directions for calcia-stabilized zirconia and another group[5] favoring displacements in $\langle 100 \rangle$ directions for Y and Yb stabilization. Both studies were made with powders and were limited to only a few peaks at moderate $\sin\theta/\lambda$ where the effect of displacement on the scattering is not large. Part of the intensities measured in Ref. 1 with single crystals are shown in Fig. 1. From the structure factor F (intensity is proportional to $|F|^2$):

h, k, l all odd: $F = 4f_{cation}$
h, k, l all even, $h + k + l = 4n$: $F = 4f_{cation} + 8f_{anion}$
h, k, l all even, $h + k + l = 4n + 2$: $F = 4f_{cation} - 8f_{anion}$

The difference in these groups is due to the contribution (or lack of it) of the oxygen ions. Despite the fact that the X-ray scattering factor of oxygen is small compared to f_{cation}, the three types are clearly separated in Fig. 1. It is sometimes said that it is best to use neutron scattering to see these effects due to oxygen, in which case the scattering amplitude of oxygen is similar to that for the cations. But this is clearly not necessary.

Least-squares analysis with single crystal data like that in Fig. 1 for both Ca- and Y-stabilized crystals confirms the studies of powders by Steele and Fender[5] (on Y- and Yb-stabilized zirconias) in that there is *no* appreciable concentration of interstitial cations to account for the nonstoichiometry. This deviation *is* due to anion vacancies. Also difference electron-density maps from ($F_{measured}$-$F_{calculated \ for \ ideal \ fluorite \ phase}$) do not indicate any appreciable electron density at octahedral interstitial sites $(\frac{1}{2}, \frac{1}{2}, \frac{1}{2})$. These maps show quite clearly that the anions are displaced in $\langle 100 \rangle$ directions; an example is given in Fig. 2. Refinements based on this displacement direction, from Ref. 1, are presented in Fig. 3 where it can be seen that the unusually large Debye-Waller factors for the anions obtained by assuming the perfect fluorite structure decrease to values similar to the cations at the best refinement (see Ref. 5 for similar results). Note also that there is a significant fraction of undisplaced anions only in the Y-stabilized crystal, and this is confirmed by the difference electron densities.

For $Zr(Y)O_{2-x}$ only, in addition to these anion displacements, there are small cation displacements in $\langle 111 \rangle$ directions, about a tenth of the anion displacements. A similar effect has been reported for $Zr(Yb)O_{2-x}$.[5] Thus, there are differences between lime-stabilized zirconia and other stabilizers.

Steele and Fender[5] suggested that the anion displacement is due to the high concentration of anion vacancies. But it should also be pointed out that these displacements are similar to the movements from the cubic structure to the tetragonal phase[6] ($\Delta/a \approx 0.06$). In this regard it is interesting to note that the reported Debye-Waller factor is unusually large ($0.31 \ nm^2$) in tetragonal ZrO_2,[6] as well as in the stabilized cubic phase.

An attempt was made to see if these large anion displacements are associated with anharmonic vibrations. For third-order anharmonic effects, the displacements would cause the reflections for $h + k + l = 4n + 1$ to differ from those for $h + k + l = 4n - 1$, for the same $h^2 + k^2 + l^2$. This is not the case, as can be seen in Fig. 1. Furthermore, the displacements would be required by symmetry to be in $\langle 111 \rangle$ directions. The $\langle 100 \rangle$ type movements reported here could arise due to fourth order anharmonicity. This interpretation is possible for the data from $Zr(Ca)O_{2-x}$ but not for $Zr(Y)O_{2-x}$.[1] In view of the similarities for these two materials, such a difference seems unlikely, and we suggest that the anion displacements are static and *not* due to thermal vibrations. Perhaps they are due to soft phonon modes which are quenched during the rapid cooling to room temperature.

Local Ionic Arrangements

From internal friction measurements Wachtman and Corwin[7] suggested the near-neighbor pairing of stabilizing ions and oxygen vacancies. From a limited study of diffuse neutron scattering, Steele and Fender[5] proposed that the stabilizing ions are indeed near neighbors to oxygen vacancies, to maintain local charge neutrality, and also that there are *local* displacements of oxygen and cations which are an order of magnitude larger than those found from the Bragg peaks. But Allpress and Rossell[8] and Carter and Roth[3,4] both suggested that the diffuse scattering arises from small coherent regions of $CaZr_4O_9$ or YZr_4O_9 ≈ 3 nm in size. While the structures of these phases are not known, that for $CaHf_4O_9$ has been established.[9] In this phase, anion vacancies are in an ordered array and are *second* neighbors to a Ca ion.

Because of the unusual displacements and the conflict concerning local atomic arrangements, we examined the diffuse X-ray scattering quantitatively in a volume in reciprocal space.[2] Figure 4 is a schematic of our results, compared to the calculated intensities for $CaZr_4O_9$ or YZr_4O_9, assuming these have the same structure as $CaHf_4O_9$ and allowing for all possible variants in the orientation relationship between the precipitate and the fluorite matrix. It is clear that the calculated intensities do not agree with the measurements; some regions of diffuse scattering for the model are not observed. Until the structures of the actual precipitates are fully determined, it is unclear whether a model of coherent precipitates is really appropriate.

The diffuse X-ray intensity was carefully corrected and placed on an absolute scale.[2] A general equation in terms of local order and displacements of both cations and anions, including terms quadratic in the displacements, was then fit to the data. This was done by representing each contribution by a Fourier series, expanding each such series to a limited number of terms following Williams.[10] The coefficients of each of these series are in terms of local order and displacements between ions connected by interatomic vectors $l\bar{a}_1 + m\bar{a}_2 + n\bar{a}_3$. An examination of these equations (see Ref. 2) shows immediately that attempts to fit simple models to the diffuse scattering from such a complex material must be treated with some skepticism. In the fluorite structure there are three types of interatomic vectors, what we refer to as "type 1" (l,m,n, all $2p/4$, $l + m + n = 4q/4$, with p, q as integers) span sites on the cation fcc sublattice and the anion or tetrahedral interstitial sublattices. Contributions to *each* coefficient of *each* series are the sum of these terms. For example, the local order is described in these equations in terms of the Warren short-range order parameter α and sublattice fraction X_ν^j of component j on the νth sublattice:

$$\alpha \frac{ij}{\mu\nu}\ (lmn) \equiv 1 - \frac{P_{\mu\nu}^{ij}}{X_\nu^j} \tag{1}$$

where $P_{\mu\nu}^{ij}$ is the conditional property of finding a jth atom on the νth sublattice at the end of interatomic vector lmn, if there is an i type on the μth sublattice. For type 1 vectors, in the region of our measurement the average value $\bar{\alpha}$ can be written for $Zr(Ca)O_{2-x}$ as[2]:

$$\bar{\alpha}\ (lmn) = 0.95\alpha_{FF}^{CaZr}\ (lmn) + 0.05\alpha_{TT}^{OV_0}\ (lmn) \tag{2}$$

The first term represents the local order between cations on the fcc sublattice

39

(F), whereas the second is between anions and anion vacancies (V_0) on the tetrahedral sublattice (T). The weighting factors result because the intensity expressions involve atomic scattering factors. It is clear from Eq. (2) that $\bar{\alpha} \cong \alpha_{FF}^{CaZr}$. Proceeding in this way for type 2 vectors $[l,m,n,$ all $2p/4, l+m+n = (4q+2)/4]$ 'which belong only to the anion sublattice and type 3 vectors, which bridge the two sublattices $[l,m,n = (2p+1)/4)$, it is possible to unravel many of the pair probabilities. This is possible for the Ca-containing crystals but not the Y crystals. The diffuse scattering due to local order is proportional to the difference in the X-ray scattering factors of Zr and Y, and this difference is too small in this case; the intensity is dominated by scattering due to displacements, even at moderate scattering angles. On the other hand this is only part of the contribution for the Ca crystal. The fact that the diffuse intensities are quite similar for the two crystals[2] shows that the observed diffraction pattern can be explained qualitatively by displacements alone. A small precipitate is not required. As effects of displacement are more symmetric in electron diffraction patterns than in X-ray patterns, due to the dynamic nature of the electron scattering,[11] it is possible that the diffuse electron scattering from regions of displacement would be considered as due to small chemical regions like precipitates. This is perhaps one reason for such proposals in the past.[3,4,8]

A similar simplification is not possible for the coefficients due to displacement, except for type 2 vectors, which are entirely due to oxygen ion displacements. This overlapping of vector contributions is the principal reason that attempts to fit a simple model to the diffuse scattering have to be treated with great caution. Even if the more complete procedure described here is followed, the series expansions must be solved with various numbers of terms, to be certain that the analysis is not sensitive to this (see Ref. 2).

Our results did confirm the displacements found from the Bragg peaks. That is, the displacements of oxygen ions are appreciable only in $\langle 100 \rangle$ directions, in both Ca- and Y-stabilized phases. There is no long range correlation in the displacements (the average value for these displacements is ≈ 1 to 2% of the oxygen separation (≈ 0.002 to 0.003 nm)).

For $Zr(Ca)O_{2-x}$, it can be shown[2] that the local order parameter is:

$$\alpha_{FT}^{Ca\,V_0}\,(lmn) = -\,\frac{X_F^{Zr}}{X_F^{Ca}}\,\alpha_{FT}^{Zr\,V_0}(lmn) \tag{3}$$

It was found that $\alpha_{FT}^{Zr\,V_0}$ for the interatomic vector $\frac{1}{4}[111]$ was the largest parameter of those determined, $+0.42$. From this value, $P_{FT}^{Ca\,V_0}$ (the probability of finding a vacancy as a nearest neighbor to a Ca ion) is ≈ 0.25. If the vacancies were randomly arranged, this value would be $X_T^{V_0} = 0.075$. Clearly there is a strong tendency for Ca-ion-vacancy pairs.

To examine the local *cation* arrangements in detail, a computer simulation was used, with 32 000 cation sites in an fcc structure (20 by 20 by 20 unit cells). These cations were rearranged from an initially random array until they satisfied the measured $\alpha_{FF}^{Ca\,Zr}$ *(lmn)* for type l vectors for *(lmn)* $\frac{1}{4}[400]$ to $\frac{1}{4}[444]$, which represent shells 2–6 around a cation. (The first shell was not used because the value was uncertain.) A section of the results is shown in Fig. 5. The Ca ions are in rods in $\langle 332 \rangle$ directions on $(1\bar{1}0)$ planes. This Ca ion arrangement is similar to the arrangement in the $CaHf_4O_9$ phase (but *not* the first neighbor preference of vacant anion sites and Ca ions). Such rods are also present in a simulation of a random array[2] but their number and length are greater using the measured pair probabilities. It can also be seen in this figure that the chemical regions are much smaller than the 3 nm coherent domains that have been suggested previously to explain the diffuse scattering.

Some Speculations

Recently Morinaga et al.[12] used the cluster method to calculate the electronic structure of pure ZrO_2. Their results, some of which are shown in Fig. 6, indicate that there is a band gap between the lower (occupied) oxygen $2p$ band and the upper (empty) Zr $4d$ band, and the gap is highest in the tetragonal phase. Also, the bond order, which is a parameter showing the overlap population of electrons between the Zr and O ions, is higher in the tetragonal phase than in the cubic fluorite phase. The tetragonal phase has a stronger covalency than the cubic phase. These results imply that tetragonal ZrO_2 is more stable than the cubic phase from the point of view of electronic energy. It is not clear whether this calculation on pure ZrO_2 is directly applicable to the stabilized zirconias. But, as the local anion displacements in $\langle 100 \rangle$ directions found in stabilized zirconias are similar to those for the tetragonal ZrO_2, perhaps they arise to reduce the electronic energy in the stabilized phases. It is unlikely that they arise solely due to anion vacancies or to the presence of stabilizing ions.

The two main features of the local ionic arrangements in these materials are the rods of stabilizing cations along $\langle 332 \rangle$ directions and the fact that oxygen vacancies are near neighbors to these cations. Perhaps the high anionic conduction is due to the "guiding" of the anions by the rods of solute cations. The decrease in conductivity at higher solute concentration might then be due to the increased tendency for the formation of the compound $CaZr_4O_9$, for which the Ca ion is no longer a first neighbor to a vacancy. Studies of the diffuse scattering at higher concentrations of stabilizing ions would reveal such a change. Also (at any composition) the presence of small precipitates like $CaZr_4O_9$ could be detected by the presence of small-angle scattering, as their composition is different from that of the matrix. (The local structure reported here would yield a much smaller amount of such scattering.) It would also be of interest to repeat these studies at higher temperatures, rather than the quenched specimens used here and by others. Such investigations would reveal: (a) whether the observed diffuse intensity arises during the quench due to the beginnings of transformation, and (b) whether the large anion displacements are static at high temperatures or arise at room temperature due to the quenching of phonon modes.

Acknowledgments

We are grateful to T. H. Etsell for lending us the crystals used in this study. This research was supported at Northwestern University by the U. S. Army Research Office under Grant No. DAAG 29–76–G–0303. The X-ray work was carried out in the Long-Term X-ray Facility in Northwestern University's Materials Research Center. This facility is supported in part by the NSF-MRL through Grant No. DMR–760157. J. Faber acknowledges the support of the Department of Energy and Argonne National Laboratory.

References

[1]M. Morinaga, J. B. Cohen, and J. Faber, Jr., *Acta Crystallogr.*, **A35** [5] 789–95 (1979).

[2]M. Morinaga, J. B. Cohen, and J. Faber, Jr., *ibid.*, **A36**, 520–30 (1980).

[3]R. E. Carter and W. L. Roth, *Gen. Elec. Res. Rep.* No. 63–RL–3479M (1963).

[4]R. E. Carter and W. L. Roth; pp. 125–44 in Electromotive Force Measurements in High-Temperature Systems. Edited by C. R. Alcock. The Institute of Mining and Metallurgy, New York, 1968.

[5]D. Steele and B. E. F. Fender, *J. Phys. C*, **7**, 1–11 (1974).

[6]G. Teufer, *Acta Crystallogr.*, **15**, 1187 (1962).

[7]J. B. Wachtman, Jr. and W. C. Corwin, *J. Res. Natl. Bur. Stand. (U.S.)*, **A69**, 457–60 (1963).

[8]J. G. Allpress and H. J. Rossell, *J. Solid State Chem.*, **15**, 68–78 (1975).

[9]J. G. Allpress, H. J. Rossell, and H. G. Scott, *ibid.*, **14**, 264–73 (1975).

[10]R. O. Williams, Rep. ORNL–4828, Oak Ridge National Laboratory, Tenn.

[11]J. Cowley; Diffraction Physics, North Holland, Amsterdam, 1975; pp. 367–70.

[12]M. Morinaga, H. Adachi, and M. Tsukada; unpublished work.

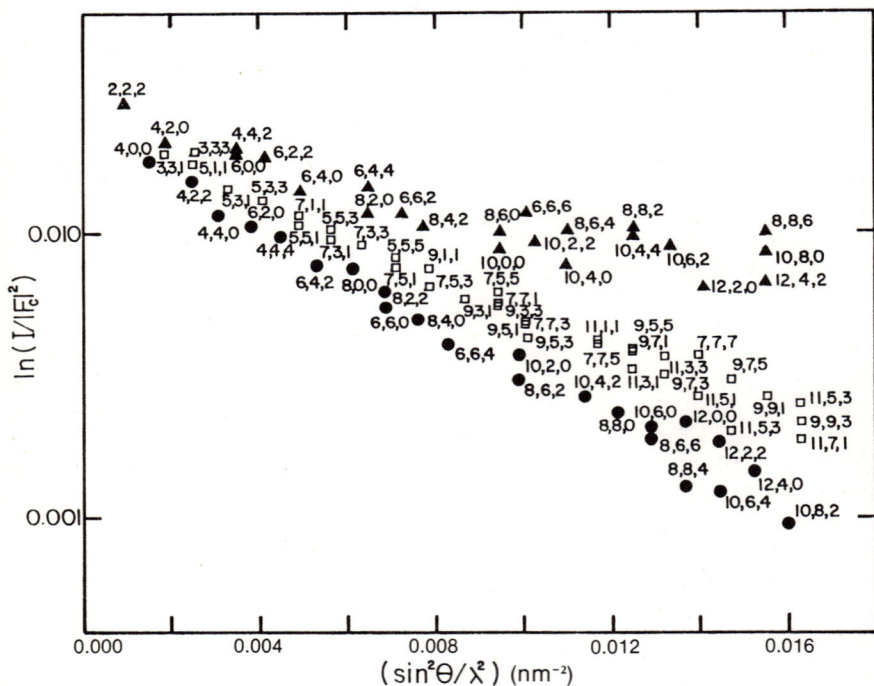

Fig. 1. Ln$(I/|F_{calc}|^2)$vs (sin^2 θ/λ^2) for Zr(Ca)O$_{2-x}$. ▲h,k,l=4n+2; □h,k,l all odd; ●h,k,l all even, $h+k+l$=4n (from Ref. 1).

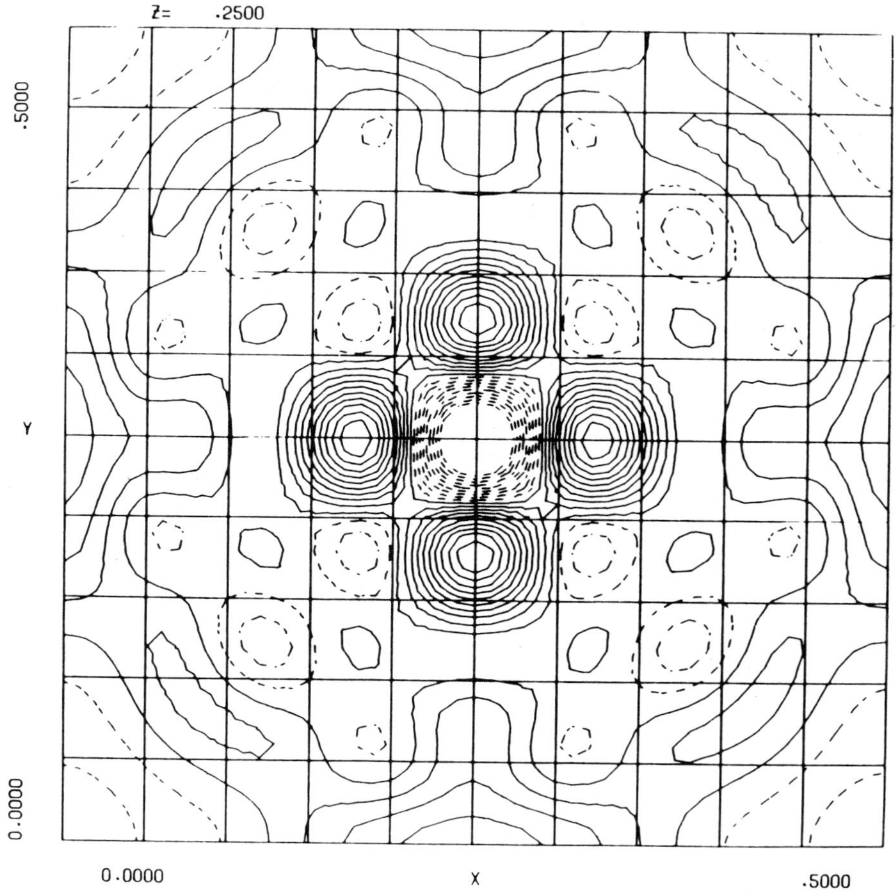

Fig. 2. Difference electron density maps at height $z = \frac{1}{4}$ of $Zr(Ca)O_{2-x}$. The contour interval is **0.00026** $e\,|\,nm^3$ (from Ref. 1).

43

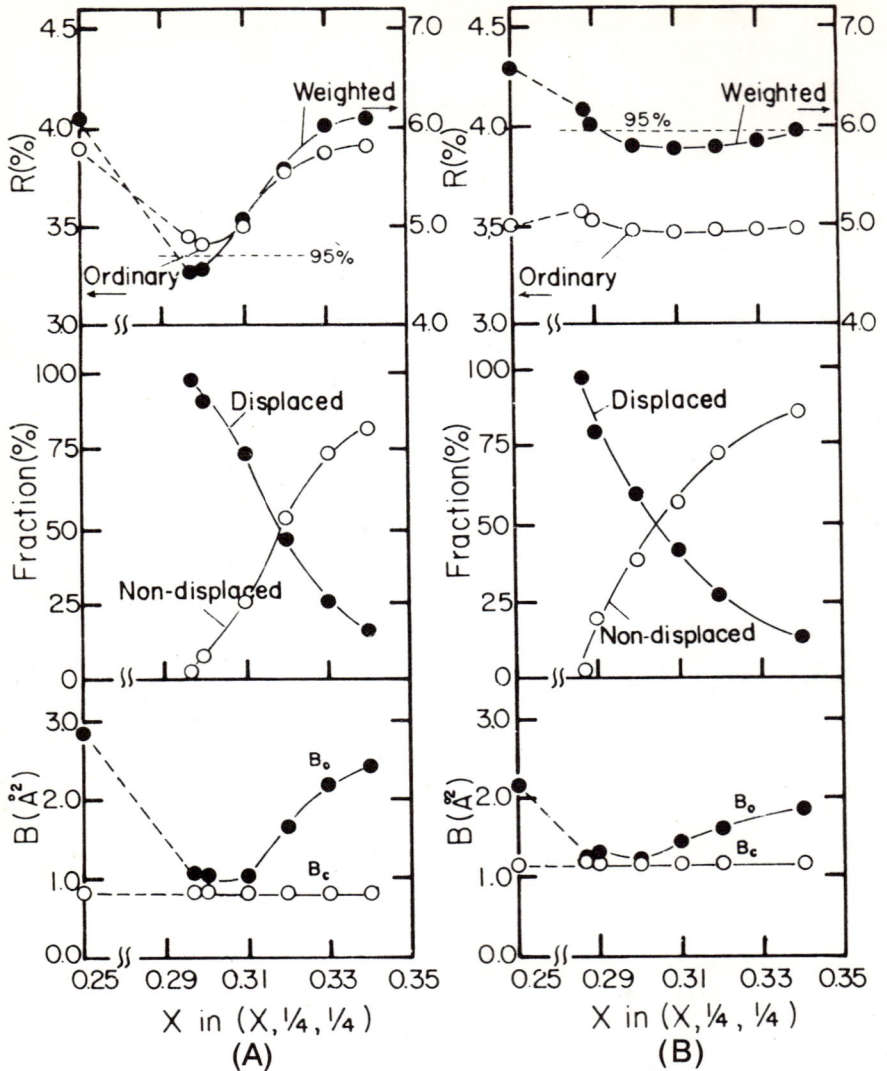

Fig. 3. Change in the residual R after least-squares refinement of the fraction of displaced and nondisplaced oxygen ions, and the temperature factors of the cations (B_c) and the anion (B_o) with the displacement of oxygen ions along $\langle 100 \rangle$ directions. The 95% confidence limits are shown. (A) $Zr(Ca)O_{2-x}$, (B) $Zr(Y)O_{2-x}$ (from Ref. 1).

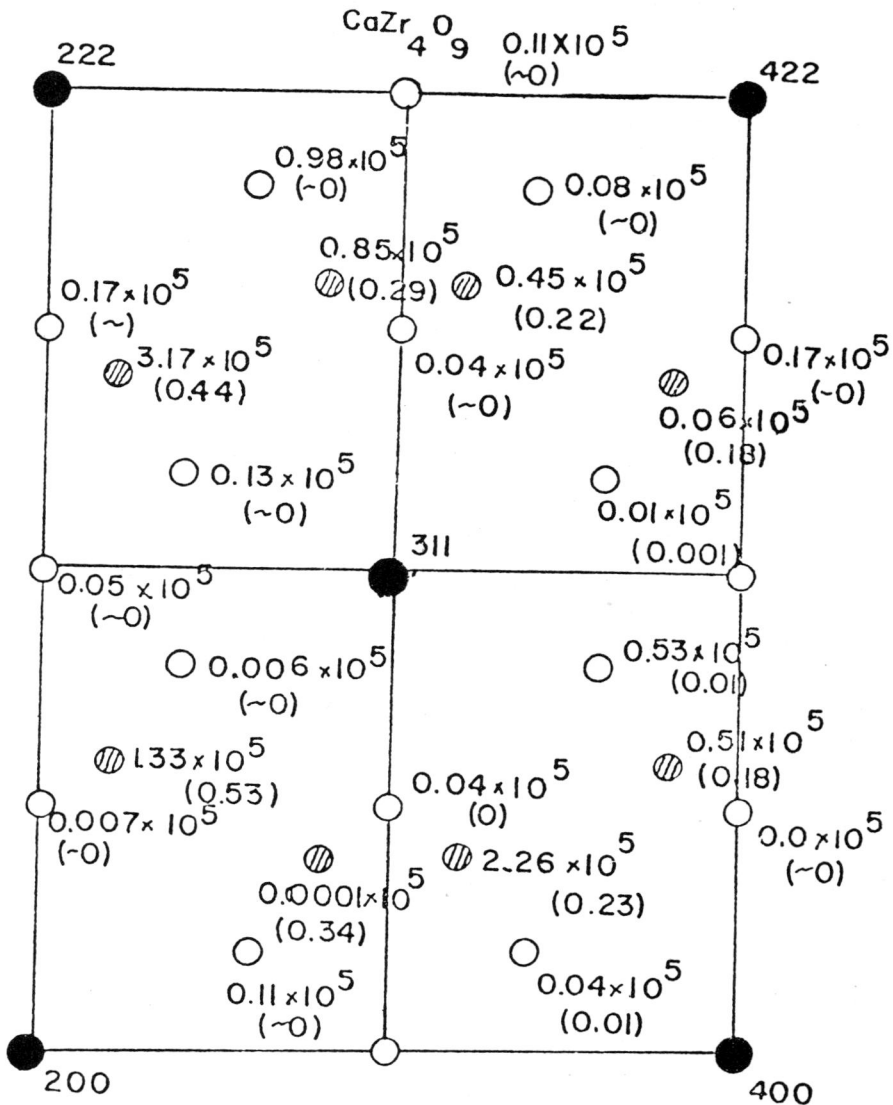

Fig. 4. Calculated intensities ($|F|^2$) due to all possible orientations of a precipitate "CaZr$_4$O$_9$." ● Fluorite peaks, ⦾ observed diffuse scattering, ○ other reflections that should arise from these precipitates. Numbers in parentheses are the calculated intensities ($\times 10^5$) without the atomic displacement in CaHf$_4$O$_9$. Other numbers include these displacements. From Ref. 2.

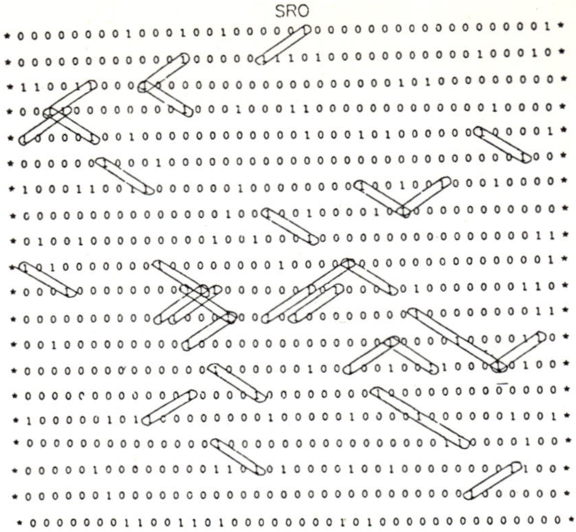

Fig. 5. Computer simulations of the local order in $Zr_{0.85}Ca_{0.15}O_{1.85}$. 1 = Ca^{2+} ions, 0 = Zr^{4+} ions (from Ref. 2).

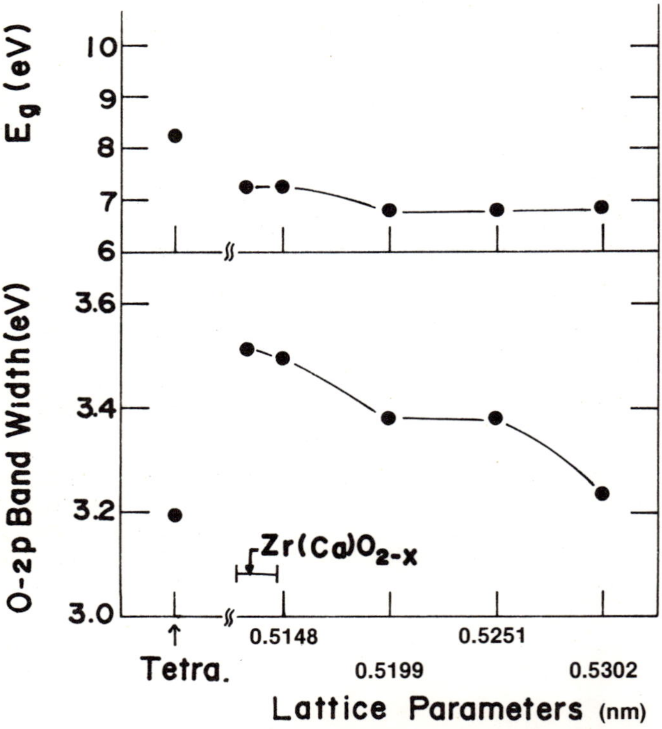

Fig. 6. Band gap, E_g, and O–$2p$ band width. Cluster calculations, from Ref. 12.

46

Ordering in anion-deficient fluorite-related oxides

H. J. ROSSELL

CSIRO Division of Materials Science
Advanced Materials Laboratory
Box 4331, GPO
Melbourne 3001, Australia

The various atomic ordering processes known to occur in the oxides of fluorite structure $MO_{2-x}\square_x$ (M = one or more cations of various valence, $\square$ = anion vacancy) are reviewed and discussed in the light of the principles of crystalline constitution found in other materials of variable composition. The power of electron-optical techniques in the structural study of all these materials is adduced. The long-range ordered fluorite-related superstructure phases exhibit complete ordering of the formal anion vacancies, together with various degrees of cation ordering. The vacancies associate in pairs separated by a $\frac{1}{2}[111]$ (fluorite) lattice vector, with a cation M' in between, which thereby becomes octahedrally coordinated by O. The arrangements of these $M'O_6$ octahedra and the cation ordering are linked, since M' is the smallest available cation type. The defect fluorite phase, exemplified by cubic stabilized zirconia, exhibits diffuse electron diffraction effects which may be explained in terms of superstructure microdomains, <5 nm diam., in all orientations coherent with a fluorite substructure. The limitations and advantages of the techniques available for the structural study of this phase are discussed.

The study of the ordering processes and structures that occur in oxides of fluorite-type (CaF_2) structure and variable composition is a relevant adjunct to the investigation of zirconia and its alloys, which are in this class of materials. The resultant information could lead to a better understanding and control of structure-dependent phenomena in zirconias, such as plasticity, ionic conductivity, precipitation, and destabilization, and may permit the design of improved engineering materials.

There is a wider aim in this study: the establishment of the structural principles underlying the accommodation of compositional variations in fluorite-related phases. The scope of the investigation may be inferred from a consideration of the structural principles that apply in other broad families of nonstoichiometric and variable-composition phases.[1,2]

It has been recognized for some time that the concept of randomly distributed point defects is inadequate for the description of phases of variable composition, i.e. nonstoichiometric compounds and heterotype solid solutions.[3-6] Subsequent experimental work has revealed the importance of extended defects, such as crystallographic shear planes[7,8] and defect clusters,[9,10] which are structurally and compositionally distinct from the parent and associated only in a formal way with point defects. These extended defects may be organized into larger schemes of crystalline order. For example, the formation of regularly spaced shear planes in rutile, TiO_2, gives rise to a "homologous series" of long-range ordered superstructure phases Ti_nO_{2n-1}, $4 \le n \le 10$.[1] Here the periodicity of the shear planes in terms of numbers of subcell units is simply related to n, and the superstructures are closely related to each other and to the substructure. Similar effects occur in MoO_{3-x} (ReO_3-type substructure)[7] and in the Nb_2O_5-based family of "block structures" (two intersecting sets of shear planes).[11] Materials that have irregularly spaced or oriented shear planes may be considered to consist of coherent intergrowths, typically 2–10 nm in extent, between different members of a homologous series of superstructures, whereas coherent intergrowth on the same scale is possible between different structures that have topological compatibility. Numerous examples of this type of crystalline constitution have been reported. Reference may be made to some reviews[12,13] and to papers presented at a recent conference.[14] These intergrowth ("Wadsley defect") structures adapt continuously to compositional variation, while their macroscopic properties simulate those of a single phase of variable composition.

The bulk of the information concerning the ordering and structural modifications that occur over small localized regions or microdomains in these materials has come from high resolution electron microscopy, which can provide images of ≈ 0.3 nm resolution that are directly interpretable in terms of structure. This technique is considerably more efficacious in the elucidation of this type of crystalline constitution than electron diffraction studies, while the use of X-ray or neutron diffraction for this purpose is futile, since such data correspond to a structure averaged over a relatively large volume.

It is not known if compositional variation is accommodated in fluorite-related materials by the formation of extended defects and structural modifications on some common plan, but there are strong indications that this may be so. For example, if anion vacancies are postulated to be point defects, then it is necessary to invoke interaction between them to explain why the ionic conductivity in many defect fluorite materials falls off with increasing concentration of vacancies,[15] while ordering effects extending over distances of 3–5 nm are implied from the widths of diffuse features which appear in diffraction patterns from single crystals of various defect fluorite phases (Fig. 1). Also, many fluorite-related systems exhibit long-range ordered phases, of narrow homogeneity range, which are superstructures based on the common fluorite substructure. Further, the formulae of the superstructures generally conform to a homologous series, such as M_nO_{2n-2}; this is particularly evident in the systems PrO_{2-x}, TbO_{2-x}, and CeO_{2-x}.

Only the results from structural investigation of anion-deficient fluorite-related systems will be considered here; this topic has been reviewed already from different standpoints.[16,17] However, it may be remarked that recently a structural feature, M_6X_{37}, which is topologically compatible with the structural elements of fluorite has been identified in two anion-excess fluorite-related superstructures.[18,19] The ordered distribution of this cluster in the fluorite structure can explain the relationship between the fluorite-related superstructures M_nX_{2n+5}, and it therefore shows promise as a common structural feature in these particular systems.

Discussion of Structures

The Fluorite Structure

The fluorite (CaF_2) structure is adopted by oxides MO_2, where M is a large tetravalent cation, e.g. Th^{4+} and Ce^{4+}, as well as by a multitude of halides, sulfides, hydrides, and intermetallic compounds of composition AB_2.[20] It is face-centered cubic, with space group *Fm3m*. For oxides MO_2, the cations M occupy the points (0,0,0) of the fcc lattice and the O anions occupy the positions $\pm(\frac{1}{4},\frac{1}{4},\frac{1}{4})$, which are tetrahedral holes in the M array and which form a primitive cubic sublattice of half the fcc cell edge. Each M is surrounded by eight oxygen ions at the vertices of a primitive cube, while every alternate cube of anions is filled by an M cation. The structure may be viewed as MO_8 cubes sharing edges in a 3-dimensional chess board arrangement.

The alloys of ZrO_2 with oxides of lower-valent metals such as CaO, MgO, or R_2O_3 (R = rare earth, Y or Sc), with overall composition MO_{2-x}, $0.0 < x < 0.4$, are said to adopt this structure. Ideally, this would require the cations to be distributed randomly on their sites so as to appear as a single cation type, and the O anions, although insufficient to fill all sites, must be distributed randomly also, i.e. there is a random distribution of vacant O sites. Under these conditions, the phase will give a diffraction pattern indistinguishable from that of perfect stoichiometric fluorite. In reality, these conditions are not met, and diffraction effects additional to those expected from fluorite occur, such as the diffuse features in Fig. 1. Such materials nonetheless are said to possess the "defect" or "disordered" fluorite structure, an expression that probably resulted from the widespread exclusive examination of materials by powder patterns, which seldom show the extra diffraction effects in easily recognizable form.

Given a material MO_{2-x} of ideal fluorite structure, states of order may be visualized in the cation array, in the anion array, or both; the ordered states may be complete or partial, long-range or short-range. Since fluorite-related oxides characteristically possess an intact and relatively immobile cation array and a mobile, defect, anion array, it may be thought that ordering could occur independently in each. However, the evidence from the long-range ordered superstructures shows that, in general, ordering in the two arrays is linked; cation order implies anion order, but not vice versa.

Long-Range Order: The Fluorite-Related Superstructures

Superstructure phases of narrow homogeneity range often appear at relatively low temperatures within a defect fluorite phase field. Large numbers of these are known from many systems, and there are many structural forms. Superstructure phases are particularly numerous in the systems Pr_{2-x}, TbO_{2-x}, and CeO_{2-x},[16] and their conformity to the "homologous series" formula M_nO_{2n-2} is clearly demonstrated. Ordered phases known to occur in ZrO_2- and HfO_2-based systems prior to about 1970 have been listed[21]; those that occur in systems involving rare earth oxides have been reviewed more recently.[22]

Attempts have been made to classify these superstructure phases,[23-26] but they contain unsatisfactory features, mainly because, at that time, structural data for these phases were very limited, and sometimes even the unit cell data were uncertain. This arose because X-ray diffraction data were limited to those obtained from powders; the materials in general form only at relatively low subsolidus temperatures and are therefore microcrystalline or multiply-twinned. Nevertheless, some structure determinations have been made from single crystal X-ray data: In_2O_3[27] and Y_2O_3[28] of C-type rare earth oxide structure, UY_6O_{12},[29] $Zr_3Yb_4O_{12}$,[30] and Ce_7O_{12}[31] of M_7O_{12} type (from the homologous series), the pyrochlore $Ca_2Sb_2O_7$[32] and the minerals braunite (Mn_7SiO_{12})[33] weberite (Na_2MgAlF_7),[34] and calzirtite $(Ca_2Zr_5Ti_2O_{16})$.[35] A few more have been determined from powder data, for example Pr_7O_{12},[36] Ce_7O_{12},[31] and $Zr_5Sc_2O_{13}$.[37]

It is extremely difficult to determine crystal structures from powder data alone because of the limitation of overlapping and coincident reflections. For the large unit cells of low symmetry that apply here, indexing of the pattern, i.e. the determination of the unit cell, is a formidable problem, the establishment of symmetry from reflection absences can be only tentative, and there are reservations concerning the application of Fourier methods to the solution of the structure. Some advances have been made recently in the use of computers for powder pattern indexing[38] and in techniques for structure solution and refinement from powder data.[39,40]

The technique used in our laboratory overcomes these difficulties in the particular case of superstructures. Finely ground material is examined in an electron microscope, and, at the magnification thus afforded, single crystal fragments can be selected and their diffraction patterns recorded. The scattering of electrons is so strong that crystallites as small as a few tens of nanometers in diameter can yield adequate diffraction patterns. By recording a set of patterns from a superstructure crystal tilted about a zone, the unit cell and its relationship to the subcell can be determined unambiguously. Also, it is possible to establish some symmetry information, but, since multiple scattering (a consequence of the strong interaction) can cause the appearance of reflections which otherwise would be absent, some care is needed in interpretation.

Diffraction patterns from superstructures characteristically have very strong reflections due to the subcell, whereas the remainder (supercell reflections) are much weaker. Thus it is reasonable to suppose that the atoms in a fluorite-related superstructure are close to the positions occupied in the ideal fluorite structure. Given the supercell/subcell relationship, it is possible to write down a set of trial atomic positions for the superstructure, and, from the composition and any known symmetry information, to construct all possible ordered atomic arrangements. The most probable trial structures can be selected from these by comparison of the calculated and observed powder X-ray diffraction intensities. Least-squares refinement procedures, based on the powder intensities, are then used to locate vacant anion sites and to refine cation positions. In most cases, refinement of anion positions is also possible. The results of the refinement are not as precise as those normally obtained from single crystal data, but they are adequate for the present purposes.

Using this procedure, structures were determined for $Ca_2Hf_7O_{16}$,[41] $CaHf_4O_9$ and $Ca_6Hf_{19}O_{44}$,[42] several compounds of M_7O_{12} structure,[43] phases Ln_3MO_7 (Ln = rare-earth or Y, M = Ta, Nb, or Sb),[44] and $CaZrTi_2O_7$ (zirconolite).[45]

It is possible to make some general observations about the types of ordering that occur in the fluorite-related superstructures from the results of the structure determinations made to date. All the superstructures have an ordered arrangement of formal anion vacancies but exhibit various degrees of cation order. In all cases, the formal anion vacancies are organized into pairs situated across the body diagonal of an MO_8 cube of the parent fluorite structure, i.e. the vacancy pair is separated by a $\frac{1}{2}[111]$ (fluorite) lattice vector. The cation involved thus becomes 6-coordinated. The anions of the resultant MO_6 group relax their positions so as to lie closer to the vertices of a regular octahedron; this involves movement of anions toward a vacant site by up to 0.05 nm. The tetrahedron of cations surrounding a vacancy is expanded radially by ≈ 0.02 nm. Both relaxational movements accord with those expected on electrostatic grounds. Formal anion vacancies separated by a vector shorter than $\frac{1}{2}[111]$ (fluorite) are known only in the C-type rare earth oxide (M_2O_3, bixbyite) structure, where, compared to fluorite, one quarter of the anion sites are vacant; even so, all the cations approach octahedral coordination.

The octahedrally coordinated cation appears to be a particularly stable configuration, i.e. it is favored energetically. The distribution of MO_6 octahedra that may be exhibited cannot be predicted, however; a great variety of arrangements has been observed, ranging from isolated octahedra through clusters of three, to chains, sheets, and 3-dimensional networks of corner-sharing octahedra.

In general, not all cation sites are associated with a pair of vacancies, so that the MO_8 cubes of fluorite may be retained or MO_7 polyhedra exhibited. An exception was found recently in zirconolite, $CaZrTi_2O_7$,[46] where one quarter of the Ti ions are coordinated by only five oxygen ions, due to the location of these Ti ions "off center" from sites of nominal 8-coordination. Cation ordering in these superstructures is such that the sites of 6-coordination are occupied by the smallest available cation type, and, as may be expected, there is frequently a correlation between cation size and occupancy of the 7- and 8-coordinated sites also. In thermodynamic terms, order of the cations can result in the optimization of M–O bonding and a decrease in enthalpy, which can overcome the effect of configurational entropy and give a net free energy decrease.

All these ordering effects may be illustrated by some specific examples.
(i) The superstructures $CaHf_4O_9(\phi_1)$, $Ca_2Hf_7O_{16}(\phi)$, and $Ca_6Hf_{19}O_{44}(\phi_2)$, have cations of different radii (Ca 0.112 nm, Hf 0.08 nm)[47] and exhibit complete cation order, with the Ca on sites of 8-fold coordination and the Hf on sites of 6- and 7-fold coordination. HfO_6 octahedra are isolated in ϕ but are corner-linked in clusters of three in ϕ_1 and helical chains in ϕ_2. The structure of ϕ_1 was determined but not refined in full; cation and vacancy order were established, but the amount of data available permitted only the cation positions to be refined. The phases ϕ_1 and ϕ_2 occur in the CaO-ZrO_2 system.[48] The structures of the Zr compounds have not been determined per se, but they have the same unit cells as the Hf compounds and may be reasonably expected to be isostructural also.
(ii) The M_7O_{12} superstructure has a rhombohedral unit cell that is 1.75 times the volume of the fluorite subcell. Cations occupying the fluorite (111) planes normal to the triad axis of the supercell are arranged so that one cation in seven is on a special site of trigonal symmetry, with octahedral coordination by O, whereas the remainder are equivalent and exist in 7-fold coordination in hexagonal arrangements around the octahedral cations. The whole struc-

51

ture contains closely spaced, but isolated, MO_6 octahedra. The cations may be ordered in several ways, with the proviso that the smallest available cation occupies the octahedron: (a) In UY_6O_{12} and $NbSc_6O_{11}F$ the cations are fully ordered. (b) In $Zr_3Er_4O_{12}$, Zr occupies the octahedral sites and the 7-coordinated sites are occupied randomly by $2Zr + 4Er$. It is thought that Pr_7O_{12} and Ce_7O_{12} are of this type, the quadrivalent cation being the smaller. (c) In $Hf_3Sc_4O_{12}$ and $Zr_3Sc_4O_{12}$ the cations occupy both sites randomly, although in the former there is a predominance of Hf on the octahedral sites. It is noteworthy that Zr^{4+}, Hf^{4+}, and Sc^{3+} are of similar radii.

All these compounds disorder at sufficiently high temperatures to the defect fluorite phase. The order-disorder process is slow and occurs at high temperatures for those compounds exhibiting cation order; e.g. disorder in $Zr_3Er_4O_{12}$ occurs at 1500 °C and takes many hours to complete, a consequence of the low diffusion rates of cations in fluorite-related structures. On the other hand, order-disorder for $Zr_3Sc_4O_{12}$ is fast (minutes) and occurs at relatively low temperatures (< 500 °C), because it involves only a rearrangement of the very mobile anion array. Order-disorder in Pr_7O_{12} is also a fast, low-temperature process; it is postulated that effective Pr^{4+}/Pr^{3+} ordering occurs through electron-hopping mechanisms. The compound $Zr_5Mg_2O_{12}$,[49] which appears in fluorite material of appropriate composition that has been retained metastably at low temperature by quenching, appears to be another example of "anion-only" ordering in this superstructure type. Its formation is also illustrative of the stability of the vacancy-pair complex, i.e. of the octahedrally coordinated cation.

The linking of cation and anion order has been demonstrated by the results of conductivity measurements on a specimen of ordered $Zr_3Er_4O_{12}$ (Fig. 2). On heating, the anionic conductivity remained low, indicating a low concentration of "free" vacancies. At 1500 °C, it is known that the whole structure disorders to defect fluorite, and this was reflected in a sharp rise in conductivity by several orders of magnitude. On cooling from this state, the fluorite phase was retained metastably because of the low ordering rate for cations and the anionic conductivity remained high, showing that anion mobility can be high in this system provided that the formal vacancies are not locked into the superstructure. The vacancy pair or octahedral cation is shown to be the more stable state, but its incorporation into a long-range ordered structure in this case, as distinct from that of $Zr_3Sc_4O_{12}$, is dependent on cation mobility. If the fluorite specimen was kept below 1500 °C for some time, the ordering process gradually occurred, being paralleled by a gradual decrease in conductivity.

(iii) The pyrochlore structure has a face-centered cubic unit cell derived from a $2 \times 2 \times 2$ array of fluorite subcells by ordering of cations and the ordered omission of one in eight anions. The formula is $A_2B_2O_7$; there are restrictions on the radii of A and B, and the ratio of these radii commonly lies in the range 1.2–1.6.[50-52] The larger A cations are in 8-fold coordination by O, while the B cations are octahedrally coordinated. The BO_6 octahedra are corner-linked into a 3-dimensional network[53] defining large connected cavities in which are located OA_4 tetrahedra.

Both A and B sites may be occupied randomly by two or more cation types with the restriction that the cation valence sum be seven. This simple isomorphous replacement contrasts sharply with gross deviations in stoichiometry which have been reported to occur in some systems, notably those containing Sb, Ta, and Nb,[54-57] e.g. materials of formulae $A_2B_2O_6$ and AB_2O_6. Nonstoichiometry is not confined to the $A^{2+}B^{5+}$ pyrochlores, since the pyrochlore phases appearing in the systems ZrO_2-Ln_2O_3 (Ln = La, Nd, Sm, or Gd) are reported to have appreciable homogeneity ranges.[58] No systematic study of nonstoichiometric pyrochlores seems to have been made.

Some pyrochlores undergo a reversible order-disorder transformation to the defect fluorite phase, which can be retained metastably on quenching. This occurs, for example, in the $Zr_2Gd_2O_7$ phase and for pyrochlores $Zr_2Ln_2O_7$; the transformation temperature increases with increasing radius of Ln, until melting intervenes, for Ln=La,Nd.

(iv) Zirconolite, $CaZrTi_2O_7$, has a C-centered monoclinic unit cell related to the fluorite subcell by $a = -a_1 - a_2 + 2a_3$; $b = a_1 - a_2$; $2c = 3a_1 + 3a_2 + 2a_3$.[59] The structure was determined from X-ray powder data[45] and later confirmed from single crystal X-ray data.[46] The cations are nominally ordered; there is a Ca site of 8-fold coordination, a 7-coordinated Zr site, two octahedral Ti sites, and one 5-coordinated Ti site. The TiO_6 octahedra are corner-linked into planar nets which have the same configuration shown by the BO_6 octahedra in {110} planes of pyrochlore. The Ca and Zr atoms lie on planes which alternate with these planar TiO_6 nets.

Exchange of Ti and Zr is possible on the Zr site and the 5-coordinated Ti site; the composition of the zirconolite phase may be expressed as $CaZr_x$-$Ti_{3-x}O_7$, $0.85 \leq x \leq 1.3$, at 1400 °C. Zirconolite also absorbs in solid solution a variety of cations such as rare earths, Th^{4+}, U^{4+}, Fe^{2+}, Fe^{3+}, and Nb^{5+},[60,61] and it retains them very strongly, a property which has been exploited in a proposal to use zirconolite as one phase in an assemblage of three synthetic minerals (SYNROC) for the immobilization of nuclear wastes.[62]

The mode of incorporation of the large foreign cations has been examined through structure determination of appropriate materials from powder X-ray data,[63] and the interesting fact has emerged that zirconolite, despite showing a tolerance for a variety of cations, nevertheless possesses no mechanism for alteration of the anion substructure, i.e. the stoichiometry is always M_4O_7. For example, small rare earths such as Yb distribute equally on the Ca and Zr sites as an isomorphous replacement of (Ca^{2+} + Zr^{4+}) by $2Yb^{3+}$. However, the larger rare-earth cations (Gd^{3+}, Nd^{3+}) and Th^{4+} enter only the Ca site, displacing calcium; and, rather than accommodating an equivalent excess of O, the system compensates by restoring the charge balance at the Zr site. This is done by replacement of Zr^{4+} by similarly sized 2+ or 3+ ions, such as Mg^{2+} or Sc^{3+}, and, somewhat remarkably, by Ti^{3+}, which is taken from TiO_2 in the reaction mixture and stabilized chemically in the zirconolite structure. Zirconolite therefore shows quite different behavior than the M_4O_7 materials, pyrochlore and Y_3TaO_7,[44] both of which apparently have mechanisms for variation of the anion substructure.

The Defect Fluorite Phase

As noted above, there is considerable evidence that the concept of isolated anion vacancies in the defect fluorite phase MO_{2-x} is inapplicable and that some vacancy interaction must occur. Indeed, this may be inferred *a priori* from a consideration of only the anion sublattice of fluorite. Each anion site has six nearest and 12 next nearest neighbors, i.e. 18 sites within

≈ 0.36 nm. Thus, for vacancy concentrations in the range 5–20% typical of the defect fluorite phase, a close approach of vacancies is inevitable.

The most direct evidence for ordering comes from diffraction studies; diffuse intensity is obtained in addition to the sharp Bragg reflections expected for an ideal fluorite structure. The diffuse intensity appears in X-ray and neutron diffraction from single crystals but is more obvious in electron diffraction. This is because the amplitude of scattered radiation in the electron case is typically 10^3–10^4 times that for X rays or neutrons, giving an enormous gain in intensity ($\propto$ amplitude^2). The large signal/noise ratio in electron diffraction means that appreciable diffraction can come from extremely small volumes of material, a property that can be exploited in the study of local atomic ordering. On the other hand, appreciable scattered X-ray or neutron radiation can be obtained only from relatively large volumes of material ($\approx 10^{12}$ unit cells), so that the corresponding structural information is averaged over this volume and resolution of detail is lost. Normally, this averaging is of little consequence, as most crystalline substances have structures that are periodic over large regions.

Typical electron diffraction patterns from three differing defect fluorite phases are shown in Fig. 1. Several points deserve notice.

(a) The patterns from a given orientation exhibit remarkably similar diffuse features, which differ mainly in size relative to the fluorite spots in different materials. For example [112] patterns always show a diffuse ring and two diffuse spots, while the diameter of the ring as a fraction of the fluorite reciprocal cell depends on the particular system studied.

(b) In any given system, the diffuse pattern does not vary if the composition is changed within broad limits; for instance, electron diffraction patterns from the fluorite phase $Ca_xZr_{1-x}O_{2-x}$ are the same for $0.08 \leq x \leq 0.20$. The intensity of the diffuse scattering may vary with composition, but this is difficult to ascertain from electron diffraction as the intensity is also a function of specimen thickness, an uncontrolled variable in our experiments, and of deviations in specimen orientation.

(c) It can be estimated from the widths of the diffuse features that they arise from structurally correlated regions of 3–7 nm extent. Dark-field images formed in the electron microscope with the beams corresponding to a diffuse feature are mottled on much this sort of scale.

(d) Electron diffraction patterns of a material will differ in appearance from the corresponding X-ray diffraction pattern, for apart from differences caused by deviations from the average structure in small volumes, electrons, unlike X rays, undergo multiple scattering. Diffracted electron beams are relatively strong and are deviated through small angles only, so that they may act as new source beams for further diffraction and thereby introduce extra intensities and symmetry elements into the pattern. In the present case, multiple electron scattering will impose on the resultant pattern the periodicity of the strong fluorite reflections and produce an intensity-averaging effect, which will be absent in the X-ray diffraction patterns.

The systems CaO-HfO$_2$ and CaO-ZrO$_2$ were studied in our laboratory.[64] It was observed that defect fluorite specimens of composition close to $Ca_{0.2}Hf_{0.8}O_{1.8}$ gave electron diffraction patterns similar to those of Fig. 1 but, on annealing the specimens for increasing periods at 1400°C, the diffuse features in the corresponding electron diffraction patterns gradually sharpened into similarly shaped groups of spots (Fig. 3). At the same time, dark field images showed a coarsening of the 3–7 nm mottling into clearly defined crystallites, which grew to 30 nm diameter. These crystallites ultimately grew to such a size that individuals could be selected and their diffraction patterns recorded; they were identified as crystals of the ordered superstructure phase ϕ_1 ($CaHf_4O_9$).

The electron diffraction patterns containing the groups of sharp spots (corresponding to the 30 nm crystallites) could be matched as regards the positions and intensities of the spots by those calculated for an assemblage of crystallites of ϕ_1 existing in equal numbers in each of the 12 orientations in which it is possible to place this superstructure so as to have the same subcell orientation. When the size of the crystallites was reduced to 3–5 nm, the spots in the calculated electron diffraction patterns broadened into diffuse features that matched those observed. A sequence of nearly identical electron diffraction patterns and images was observed for $Ca_{0.2}Zr_{0.8}O_{1.8}$ annealed at 1000 °C, and both spotty and diffuse diffraction patterns could be reproduced by calculations as above, on the assumption that $CaZr_4O_9$ and $CaHf_4O_9$ were isostructural.

Thus it is proposed that CaO-stabilized HfO_2 and ZrO_2 of ≈ 20 mol% CaO have a fluorite-type structure that exists continuously throughout a crystal. In this structure, Ca cations and anion vacancies are not randomly distributed, but rather small regions or microdomains of 3–5 nm diameter can be delineated, in which cations and formal anion vacancies are ordered as in the superstructure ϕ_1, and it is possible to identify 12 crystallographically distinct microdomain orientations. The delineation of the microdomain boundaries will be necessarily nebulous, since the microdomains on average consist of only a few unit cells of ϕ_1; however, it would be expected that the appearance of a mapped portion of a CaO-stabilized zirconia crystal would be akin to the microdomain structures portrayed for the other phases of variable composition referred to in the Introduction.

The observation that electron diffraction patterns from defect fluorite phases of HfO_2 or ZrO_2 containing < 20 mol% CaO are no different from those of the 20 mol% material may be explained if it is supposed that the microdomains of ϕ_1 are dispersed coherently in a matrix of fluorite structure (necessarily of low CaO content), and become more widely dispersed, but not otherwise altered, as the overall CaO content is reduced.

This model is compatible with the observation that anionic conductivity in CaO-stabilized ZrO_2 falls off at high CaO content. It may be supposed that the matrix, being low in CaO content, has a concentration of anion vacancies that is sufficiently low for them to behave as point defects. Then, if the resistivity of the microdomains is high, as it is in macrocrystalline superstructure phases, the greater proportion of the current will be carried in the matrix. As the overall CaO content is raised, the proportion of matrix will be reduced, and the overall conductivity will fall as the increasing concentration of microdomains causes blocking of low-resistivity pathways.

The specimens of defect fluorite phase showing the diffuse electron diffraction features were prepared by quenching from high subsolidus temperatures, or from the melt, and were examined at room temperature. In most situations involving cation order in fluorite-related oxides, it can be taken that quenching will retain high temperature structures adequately because of the extremely low cation diffusion rates, but it is conceivable that the cations can diffuse over a distance of a few nanometers during the quench, producing a microdomain structure that has little relationship to the actual structure at high temperature. If the high temperature structure of the CaO-HfO$_2$ defect fluorite phase is different from that proposed above, then the microdomains nucleate and grow during the quench when the temperature falls below $\approx 1450\,°C$, the temperature at which macrocrystalline CaHf$_4$O$_9$ disorders to the defect fluorite structure. From the value of the diffusion coefficient for Ca in CaO-ZrO$_2$,[65] it would be expected that cations could migrate ≈ 5 nm s^{-1} at $1450\,°C$, so that formation of the microdomains during the quench may be feasible in the CaO-HfO$_2$ system. The growth of microdomains must involve cation movement, since segregation and ϕ_1-type ordering of formal anion vacancies alone is scarcely credible in view of the strong cation-anion correlations imposed by size and ionic charge. On the other hand, formation of the microdomains during the quench is not as feasible in the CaO-ZrO$_2$ system; the order-disorder transformation of CaZr$_4$O$_9$ occurs at $\approx 1000\,°C$, and, at this temperature, Ca cations can be expected to move only ≈ 0.05 nm^{-1}.

Thus, microdomain structures could well exist (in a state of dynamic fluctuation) at high temperature, a condition analogous to the clustering observed in the disordered state of numerous materials showing order-disorder transformations, but there are elements of uncertainty. Hence it would be important to conduct the diffraction experiments at temperature, and, in this regard, the CaO-ZrO$_2$ system would be the more amenable because of the relatively low transformation temperature of CaZr$_4$O$_9$.

Electron diffraction patterns from specimens of ZrO$_2$-YO$_{1.5}$ in the range 15–50 mol% YO$_{1.5}$ contain diffuse features that are invariant with specimen composition. These features do not have the same geometric proportions as those from the CaO-ZrO$_2$ specimens, and there are differences in intensity distribution also. Thus, if a microdomain model is proposed, the structure of the microdomains must be different from that of ϕ_1 (a not unexpected conclusion, since it is clearly impossible for the ordering scheme of ϕ_1 to be realized in the ZrO$_2$-YO$_{1.5}$ system). The only superstructure phase reported in this system is Zr$_3$Y$_4$O$_{12}$.[66,67] Electron diffraction patterns for a material consisting of microdomains of Zr$_3$Y$_4$O$_{12}$ in all eight orientations coherent with a fluorite matrix were calculated, but the geometric proportions of the diffuse features did not match those of the observed patterns. Annealing experiments have failed to produce coarsening of the microstructure as in the CaO-HfO$_2$ case.

Thus, if the microdomain concept is adhered to, a new superstructure must be sought at compositions < 57.14 mol% YO$_{1.5}$. The fact that such a phase has not been found may reflect merely the difficulty of locating the precise conditions of temperature and composition necessary to produce it. The phase Y$_2$Hf$_7$O$_{17}$ which has been reported[21] may be of significance in this case; however, attempts to make this compound in our laboratory have been unsuccessful.

The diffuse scattering from ZrO$_2$-YO$_{1.5}$ specimens changes as the YO$_{1.5}$ content is increased beyond 50 mol%. The observed electron diffraction patterns from materials of 55–60 mol% YO$_{1.5}$ quenched from high temperature could be matched very well by patterns calculated as above for microdomains of $Zr_3Y_4O_{12}$ 3 nm in diameter. Specimens of more than 60 mol% $YO_{1.5}$ quenched from high temperature show diffuse scattering that can be readily interpreted in terms of small coherent regions of C-type rare earth oxide structure (which is that adopted by $YO_{1.5}$).

A study has been made of the eutectoid decomposition of the MgO-ZrO_2 defect fluorite phase at 13 mol% MgO and 1400 °C.[68] Electron diffraction patterns of the decomposing fluorite phase showed diffuse features similar to those in Fig. 1, which could be sharpened into spots if the specimens were suitably treated, while dark-field images indicated the growth of domains to ≈ 20 nm in diameter. The geometric proportions of both the spotty and diffuse patterns could be duplicated in patterns calculated for a fluorite matrix containing microdomains of $Mg_2Zr_5O_{12}$ in all of the eight orientations coherent with the matrix that this superstructure can adopt, but the intensity distribution did not match that observed. A more satisfactory result could be achieved if the microdomains were assumed to have the composition $MgZr_6O_{13}$ (14 mol% MgO) and to adopt a cation-disordered version of the $Yb_2Zr_5O_{13}$ structure, which has a triply primitive hexagonal unit cell derived from that of M_7O_{12} by doubling of the c axis.

Thus the concept of superstructure microdomains coherently dispersed in a common substructure matrix can be used to explain the diffraction effects observed for specimens of defect fluorite materials, with the proviso that extrapolation of this microstructure to the high temperature regime could be invalid.

An alternative concept that can be used to explain diffuse scattering effects is that of short-range order.[69] This was developed originally to describe binary metal alloys of substitutional solid solution type, in which it is supposed that successive shells of atomic positions around any given atom are not populated by the atom species at random, but instead each shell is populated by atoms in proportions which, on average, deviate from the random value and which are correlated with the preference of an atom for like or unlike nearest neighbors. A set of short-range order parameters which are related to the site occupancy probabilities in each shell may be derived from measurements of the diffuse intensity. This concept has been expanded and developed to be applicable to more complex systems.[70]

Thus, in principle, it is possible to derive order parameters describing site occupancy probabilities in the defect fluorite phase from measurements of the diffuse diffraction intensity. Since the process sets out to describe in statistical terms the average environment of all lattice points, i.e. it involves the basic assumption that the structure is homogeneous on average, it is indifferent to any structural effects (local atomic order, extended defects) that in fact may be present in the crystal. The structural correlation distances implied by the widths of the diffuse diffracted intensities are several nanometers in dimension, and this requires the specification of large and unwieldy numbers of order parameters. Since it is impossible, in general, to construct a detailed model of the original specimen from a set of order parameters, it can be held that they are descriptive rather than explanative of structure and therefore may be of value in statistical-mechanical derivation of bulk physical properties.

The measurement of diffuse diffraction intensities is practicable only with X ray or neutrons; in view of the limitations, noted above, on the applicability of these radiations and of the short-range order concept, the examination of the detailed crystalline constitution of the defect fluorite phase must be accomplished by some other means.

The most promising technique available is high-resolution electron microscopy: an account of this has been given by Clarke.[71] Electrons are scattered by the electrostatic potential distribution in the object (which need not be periodic), and an image formed from many ($\approx 10^3$) simultaneously diffracted beams is related to a projection of the actual structure. For work involving structure determination, calculation of the image contrast expected under given experimental conditions is essential: the relevant diffraction theory[72] and computational procedures[73-77] are well established.

It may be expected that for fluorite-related materials, the potential fluctuation at the site of a formal anion vacancy would be perceptible because of the associated anion and cation relaxation, but there are other factors which act so as to prevent the formation of structure images. Thus, the basic fluorite structure has few projection directions that allow the display of potential variations of resolvable scale, and the unit cell dimensions are such that only small numbers of beams are available for image formation with normal accelerating voltages and reasonable apertures. Also, if superstructure ordering occurs, repeat distances in any direction become large, so that potential variation due to the ordering may be obscured in projection, or, in extreme cases, the limits of applicability of the usual computational procedure may be exceeded. Nonetheless, the technique has been applied in studies of the homologous series of superstructures that appear in the systems PrO_{2-x} and TbO_{2-x} with encouraging success.

Calculations of image contrast as a function of various experimental parameters were made for Pr_7O_{12} and $Zr_3Sc_4O_{12}$, representatives of the only one of these ordered phases (other than M_2O_3) whose structure has been determined by conventional methods, and it was shown that for certain closely specified conditions of crystal thickness and orientation, and of instrumental parameters, the image could be interpreted intuitively as a projection of the formal anion vacancy arrangement.[78] Structures of other members of the series have been postulated from their images by a process of extrapolation from the above result, based on known unit cell data.[79] The images of the various superstructure phases are sufficiently different to be used as a means of phase identification: examples of images from fluorite-related materials have been published[16] that show local ordering at the unit cell level, coherent microdomains of an alternative phase, and twin and intergrowth structures.

The extension of this technique to the study of other fluorite-related systems may require that the conditions necessary for the formation of intuitively interpretable images be relaxed, and that more reliance be placed on the comparison of observed and calculated images; this is now essentially a matter of convenience.

References

[1] A. D. Wadsley; pp. 98–209 in Non-Stoichiometric Compounds. Edited by L. Mandelcorn. Academic Press, New York, 1964.

[2] J. S. Anderson, *J. Phys. (Orsay, Fr.),* **38** [12] Suppl., C17–C27 (1977).

[3] A. D. Wadsley, *Rev. Pure Appl. Chem.,* **5**, 165–93 (1955).

[4] S. M. Ariya and Yu. G. Popov, *J. Gen. Chem. USSR,* **32**, 2077–81 (1962).

[5] A. D. Wadsley, *Adv. Chem. Ser.,* **39**, 1–22 (1963).

[6]J. S. Anderson, *ibid.*, pp. 23–36.
[7]A. Magneli, *Acta Crystallogr.*, **6**, 495–500 (1953).
[8]S. Andersson and A. D. Wadsley, *Nature (London)*, **211**, 581–83 (1966).
[9]W. L. Roth, *Acta Crystallogr.*, **13**, 140–49 (1960).
[10]F. Koch and J. B. Cohen, *Acta Crystallogr., Sect. B*, **25**, 275–87 (1969).
[11]A. D. Wadsley and S. Andersson; pp. 1–58 in Perspectives in Structural Chemistry, Vol. 3. Edited by J. D. Dunitz and J. A. Ibers. Wiley, New York, 1970.
[12]B. G. Hyde, A. N. Bagshaw, S. Andersson, and M. O'Keefe, *Ann. Rev. Mater. Sci.*, **4**, 43–92 (1974).
[13]J. G. Allpress; pp. 87–111 in Solid State Chemistry. *NBS Spec. Publ. (U.S.)*, No. 364. Edited by R. S. Roth and S. J. Schneider, 1972.
[14]"Direct Imaging of Atoms in Crystals and Molecules," *Chem. Scr.*, **14** [1-5] (1978-79).
[15]T. H. Etsell and S. N. Flengas, *Chem. Rev.*, **70**, 339–76 (1970).
[16]L. Eyring, *Adv. Chem. Ser.*, **163**, 240-70 (1977).
[17]H. J. Rossell and H. G. Scott, *J. Phys. (Orsay, Fr.)*, **38** [12] Suppl. C28–C31 (1977).
[18]D. J. M. Bevan, O. Greis, and J. Straehle; to be published in *Acta Crystallographica.*
[19]J-P. Laval and B. Frit, *Mater. Res. Bull.*, **15**, 45–52 (1980).
[20]R. W. G. Wyckoff; pp. 239–44 in Crystal Structures, Vol. 1. Interscience Publ., N.Y., 2d ed.
[21]M. Duclot, J. Vicat, and C. Deportes, *J. Solid State Chem.*, **2**, 236–49 (1970).
[22]D. J. M. Bevan and E. Summerville; pp. 401–524 in Handbook of the Physics and Chemistry of Rare-Earths. Edited by C. A. Gschneider and L. Eyring. North Holland, Amsterdam, 1979.
[23]R. L. Martin, *J. Chem. Soc., Dalton Trans.*, pp. 1335–50 (1974).
[24]B. F. Hoskins and R. L. Martin, *ibid.*, pp. 576–88, (1975); pp. 676–85 (1976); 320–28 (1978).
[25]Yu. A. Pyatenko, *Zh. Strukt. Khim.*, **4**, 708–13 (1963); *Izv. Akad. Nauk SSSR, Neorg. Mater.*, **7**, 630–33 (1971).
[26]W. W. Barker, *Z. Kristallogr.*, **128**, 55–65 (1969).
[27]M. Marezio, *Acta Crystallogr.*, **20**, 723–28 (1966).
[28]M. G. Paton and E. N. Maslen, *ibid.*, **19**, 307–10 (1965).
[29]S. F. Bartram, *Inorg. Chem.*, **5**, 749–54 (1966).
[30]M. R. Thornber and D. J. M. Bevan, *J. Solid State Chem.*, **1**, 536–44 (1970).
[31]S. P. Ray and D. E. Cox, *ibid.*, **15**, 333–43 (1975).
[32]A. Bystrom, *Ark. Kemi, Mineral. Geol.*, **18A** [21] 1–8 (1944).
[33]P. B. Moore and T. Araki, *Am. Mineral.*, **61**, 1226–40 (1976).
[34]A. Bystrom, *Ark. Kemi, Mineral. Geol.*, **18B** [10] 1–7 (1944).
[35]Yu. A. Pyatenko and Z. V. Pudovkina, *Kristallografiya*, **6**, 196–99 (1961).
[36]R. B. Von Dreele, L. Eyring, A. L. Bowman, and J. L. Yarnell, *Acta Crystallogr., Sect. B*, **31**, 971–74 (1975).
[37]M. R. Thornber, D. J. M. Bevan, and J. Graham, *ibid.*, **24**, 1183–90 (1968).
[38]R. Shirley, *Acta Crystallogr., Sect. A*, **31**, S197 (1975) (abstract only).
[39]G. Malmros and J. O. Thomas, *J. Appl. Crystallogr.*, **10**, 7–11 (1977).
[40]P-E. Werner, S. Salome, G. Malmros, and J. O. Thomas, *ibid.*, **12**, 107–109 (1979).
[41]H. J. Rossell and H. G. Scott, *J. Solid State Chem.*, **13**, 345–50 (1975).
[42]J. G. Allpress, H. J. Rossell, and H. G. Scott, *ibid.*, **14**, 264–73 (1975).
[43]H. J. Rossell, *ibid.*, **19**, 103–11 (1976).
[44]H. J. Rossell, *ibid.*, **27**, 115–22 (1979); *ibid.*, 287–92.
[45]H. J. Rossell, *Nature (London)*, **283**, 282–83 (1980).
[46]B. M. Gatehouse, I. E. Grey, R. J. Hill, and H. J. Rossell; to be published in *Acta Crystallographica.*
[47]R. D. Shannon, *Acta Crystallogr., Sect. A*, **32**, 751–67 (1976).
[48]D. Michel, *Mater. Res. Bull.*, **8**, 943–49 (1973).
[49]O. Yovanovitch and C. Delamarre, *ibid.*, **11**, 1005–10 (1976).
[50]F. Brisse and O. Knop, *Can. J. Chem.*, **46**, 859–73 (1968).
[51]R. D. Shannon and A. W. Sleight, *Inorg. Chem.*, **7**, 1649–51 (1968).
[52]E. Aleshin and R. Roy, *J. Am. Ceram. Soc.*, **45** [1] 18–25 (1962).
[53]F. Jona, G. Shirane, and R. Pepinsky, *Phys. Rev.*, **98**, 903–909 (1955).
[54]M. Hervieu, C. Michel, and B. Raveau, *Bull. Soc. Chim. Fr.*, 1971, pp. 3939–43.
[55]G. Allais, C. Michel, and B. Raveau, *C. R. Hebd. Seances Acad. Sci.*, **274**, 1625–28 (1972).
[56]Y. Piffard and M. Tournoux, *Acta Crystallogr., Sect. B*, **35**, 1450–52 (1979).
[57]J-L. Fourquet, M. Rousseau, and R. de Page, *Mater. Res. Bull.*, **14**, 937–41 (1979).
[58]M. Perez y Jorba, *Ann. Chim. (Paris)*, **7**, 479–511 (1962).
[59]Yu. A. Pyatenko and Z. V. Pudovkina, *Kristallografiya*, **9**, 98–100 (1964).
[60]Z. V. Pudovkina and Yu. A. Pyatenko, *Tr. Mineral. Muz. Akad. Nauk SSSR*, **17**, 124–33 (1966).
[61]D. A. Wark, A. F. Reid, J. F. Lovering, and A. El Goresy; pp. 764–66 in Lunar Science IV. Edited by J. W. Chamberlain and C. Watkins. The Lunar Science Institute,

Houston, 1973.

[62]A. E. Ringwood, S. E. Kesson, N. G. Ware, W. Hibberson, and A. Major, *Nature (London)*, **278**, 219–23 (1979).

[63]H. J. Rossell; Crystal XII; 12th Meeting of Crystallographers in Austalia, Canberra, Jan. 30–Feb. 2; p. 242 (1980) (abstract only).

[64]J. G. Allpress and H. J. Rossell, *J. Solid State Chem.*, **15**, 68–78 (1975).

[65]W. H. Rhodes and R. E. Carter, *J. Am. Ceram. Soc.*, **49** [5] 244–49 (1966).

[66]H. G. Scott, *J. Mater. Sci.*, **10**, 1527–35 (1975); **12**, 311–16 (1977).

[67]H. G. Scott, *Acta Crystallogr., Sect. B*, **33**, 281-82 (1977).

[68]R. H. J. Hannink and H. J. Rossell, *Micron*, **11**, Suppl. No. 1, p. 36 (1980) (abstract only).

[69]B. E. Warren, X-ray Diffraction. Addison-Wesley, Reading, Mass., 1969; Chapter 12.

[70]M. Hayakawa and J. B. Cohen, *Acta Crystallogr., Sect. A*, **31**, 635–45 (1975).

[71]D. R. Clarke, *J. Am. Ceram. Soc.*, **62** [5-6] 236–46 (1979).

[72]J. M. Cowley and A. F. Moodie, *Acta Crystallogr.*, **10**, 609–19 (1957).

[73]J. G. Allpress, E. Hewat, A. F. Moodie, and J. V. Sanders, *Acta Crystallogr., Sect. A*, **28**, 528–36 (1972).

[74]D. F. Lynch and M. A. O'Keefe, *ibid.*, 536–48.

[75]G. R. Anstis, D. F. Lynch, A. F. Moodie, and M. A. O'Keefe, *ibid.*, **29**, 138–47 (1973).

[76]M. A. O'Keefe, *ibid.*, pp. 389–401.

[77]D. F. Lynch, A. F. Moodie, and M. A. O'Keefe, *ibid.*, **31**, 300–307 (1975).

[78]A. J. Skarnulis, E. Summerville, and L. Eyring, *J. Solid State Chem.*, **23**, 59–71 (1978).

[79]P. Kunzmann and L. Eyring, *ibid.*, **14**, 229–37 (1975).

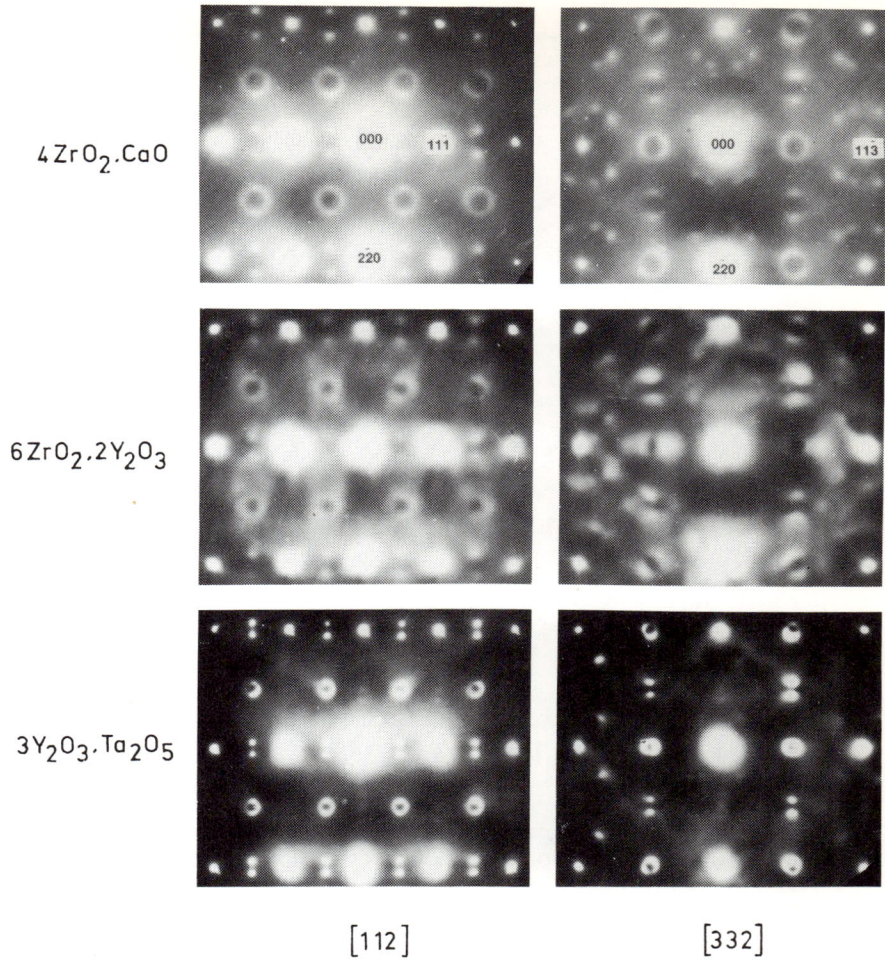

Fig. 1. Electron diffraction patterns from single crystals of the defect fluorite phase from three systems. Two representative crystal orientations are shown. Only the strong reflections (large white patches) are expected from the ideal fluorite structure; their indices are marked. The diffuse features differ mainly in relative proportions from system to system, and are indicative of additional ordering effects in the phase.

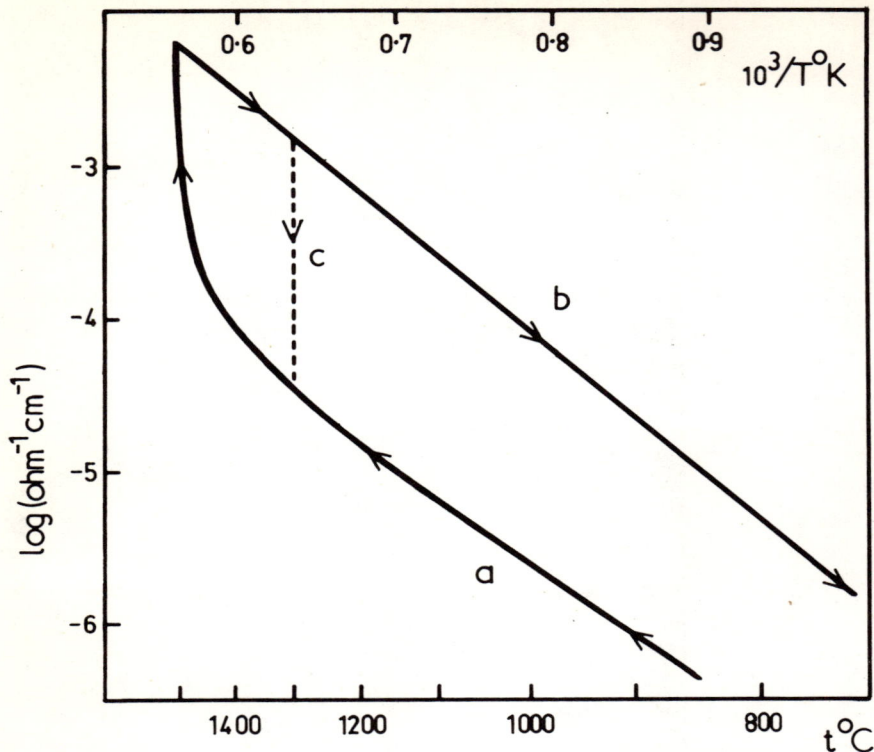

Fig. 2. Conductivity data for a specimen of $Zr_3Er_4O_{12}$: *(a)* ordered specimen, heating, *(b)* cooling relatively quickly from the 1500 °C condition, *(c)* annealing at a fixed temperature from the 1500 °C condition. (Data kindly supplied by W. G. Garrett, this laboratory.)

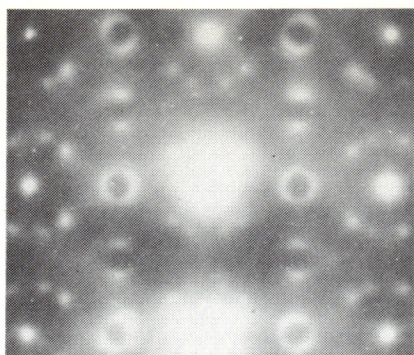
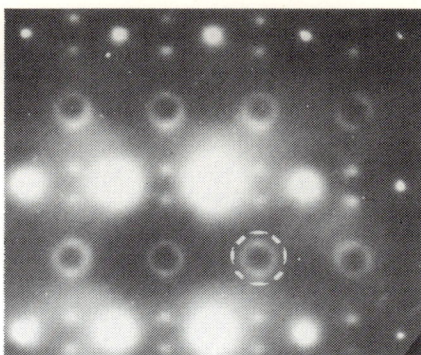

a   b

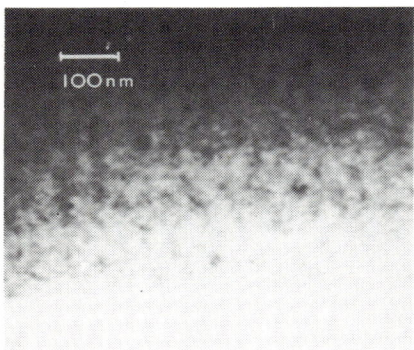
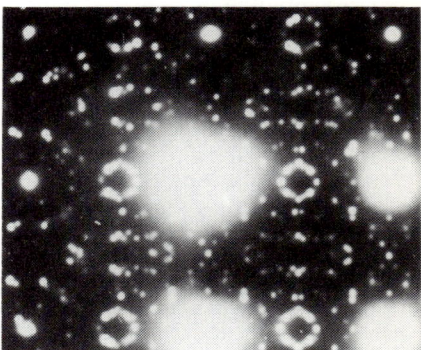

c   d

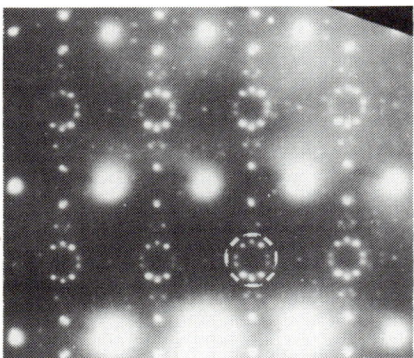
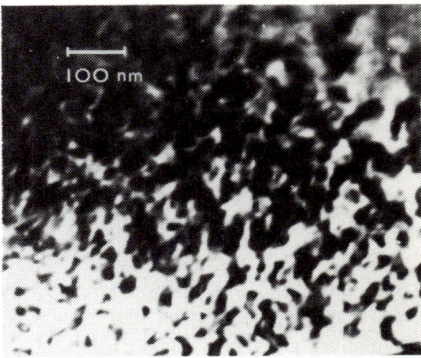

e   f

Fig. 3. *(a), (b):* Electron diffraction patterns for [332] and [112] orientations of a crystal of HfO_2-20 mol% CaO, defect fluorite phase quenched from the melt. *(c)* Dark-field image formed with the beams contributing to the diffuse features indicated in *(b)*. *(d)–(f)* Corresponding diffraction patterns and dark-field image from the same material as in *(a)–(c)* after annealing at 1400°C for 6 h.

Martensitic transformations

C. M. WAYMAN

Department of Metallurgy and Mining Engineering
and Materials Research Laboratory
University of Illinois at Urbana-Champaign
Urbana, Ill. 61801

Martensite, the name now given to the product phase of a martensitic transformation, was originally used to designate the hard microconstituent found in quenched steels. The name now far transcends its original meaning since martensitic transformations have been found in other alloys, metals, ceramics, minerals, other inorganic compounds, and polymeric materials. Quite frequently the formation of martensite leads to very desirable properties such as the conventional quench hardening of steels, age hardening, as in maraging steels, ductility and toughness enhancement (as in TRIP steels and zirconia [PSZ]), rubber-like elastic ductility, the shape memory effect, and high damping. Martensitic transformations are most readily distinguishable from other solid state phase changes on the basis of their crystallographic characteristics and "military" mode of atomic transfer. These transformations feature a coordinated structural change, involving a lattice correspondence, and a planar semicoherent parent-product interface, which, during movement (transformation), generates an invariant plane strain deformation. A fine scale inhomogeneity in the martensite, such as slip, twinning, or faulting, is usually observed at the electron microscope scale. This secondary deformation, a part of the transformation process, provides the invariant plane strain condition and maintains a semicoherent interface. Crystallographic relations between the martensite and parent, such as the habit (invariant) plane and orientation relationship, are usually nonrational.

The name martensite (in honor of the German scientist Martens) was originally used to designate the hard microconstituent found in quenched steels. Since then, materials other than steels have been found to exhibit the same type of solid state phase transformation, known as a martensitic transformation—sometimes also called a shear or displacive transformation. Indeed the name martensite, the product phase of a martensitic transformation, now far transcends its original meaning because martensitic transformations are known in other alloys, pure metals, ceramics, minerals, inorganic compounds, solidified gases, and polymeric materials.

When one considers solid state phase transformations, a variety of topics emerge for discussion, such as thermodynamics, kinetics, nucleation, growth, and crystal geometry or crystallography. Of these, martensitic transformations are most readily distinguishable from other solid state phase changes on the basis of their crystallographic characteristics, which imply a "military" (as opposed to "civilian") mode of atomic transfer from the parent to the product phase. These transformations feature a coordinated structural change involving a lattice correspondence and a planar semicoherent parent-product interface which, during movement (transformation), produces an invariant plane strain deformation. Microstructural inhomogeneity on a fine scale during a martensitic reaction—slip, twinning, or faulting—is usually observed at the electron microscope scale. This secondary deformation, an intrinsic part of the transformation process, provides the invariant plane condition at the macroscopic scale and maintains a semicoherent glissile interface between the martensite and the parent phase. Crystallographic phase relations between the martensite and the parent phase, such as the habit (invariant) plane and orientation relationship, are usually nonrational; that is, they are not expressible in terms of exact relations involving simple Miller indices.

In this paper, the various crystallographic features of martensitic transformations are discussed and illustrated with representative experimental examples. A descriptive outline of the development of the phenomenological crystallographic theory is presented, followed by an algebraic analysis. Some mechanical behavior aspects of certain martensitic transformations are also considered. Finally, some comments are made on the martensitic transformations in ZrO_2 and HfO_2. In most cases, the experimentally observed crystallographic features of martensitic transformations are in good agreement with those predicted theoretically. Some problem cases remain, however.

Experimental Observations

Figure 1 is an optical micrograph of a polycrystalline Fe-24.5% Pt alloy taken after transformation of the parent phase into martensite. Prior to transformation, and when in the parent-phase condition, the specimen was polished flat. The contrast thus observed results from a change in shape of the transformed regions. The light and dark colors correspond to regions of martensite which have been distorted in different senses with respect to the initial surface. This surface tilting or macroscopic distortion is known as the shape deformation or shape strain. If successive serial sections of the specimen were removed and the specimen observed at each stage, it would become clear that the shape of the martensite units is that of lenticular plates. In addition, an X-ray analysis would show that the plates showing different optical contrast also feature different lattice orientations with respect to the initial parent grain. Further X-ray analysis would also show that a structural change had occurred (which in this Fe-Pt case is an fcc → bcc transformation). The differently oriented plates are different crystallographic variants of the habit plane and orientation relationship.

Again considering Fig. 1, if straight scratches were intentionally abraded on the flat specimen while still in the parent-phase condition, it would be noted that, after the martensitic transformation occurred, the scratches (or interference fringes, Fig. 2) would be displaced in a characteristic manner. By analyzing several nonparallel scratches crossing a given plate and noting the initial position of the same scratches in the untransformed parent phase, a distortion matrix which describes the transformation shape deformation can be constructed. Analysis of this distortion shows that a homogeneous deformation (at the optical microscope scale) has taken place, and further consideration shows that this distortion is similar to a simple shear but, in general, is not exactly a shear. Rather, an invariant plane strain has occurred

(Fig. 3) with the plane of reference being the undistorted and unrotated habit plane. Mathematically speaking, an invariant plane strain is a homogeneous distortion such that the displacement of any point is in a common direction and proportional to the distance from a fixed plane of reference (i.e., not influenced by the strain), which is the invariant plane.[1] In most martensitic transformations a volume change also accompanies the structural change and this accounts for the normal component of the invariant plane strain. Hence, the transformation shape deformaton is not a simple shear.

The previous discussion indicates that the martensitic transformation shape deformation can be represented as an invariant plane strain, a homogeneous deformation. However, a past difficulty in the understanding of the geometry of martensitic transformations was the realization that the homogeneous shape deformation matrix, when applied to the parent structure, did not generate the known martensite crystal structure. For example, in iron alloys the martensitic transformation converts an fcc parent into a bcc (or bct) product, but the measured shape deformation matrix, when applied to the parent, will not produce a bcc structure. This apparent inconsistency will be explained later.

In general, the martensite plates in a given alloy possess a unique habit plane, as illustrated in Figs. 4 and 5. These stereographic projections show the experimentally determined habit plane poles for different martensite plates, which, as can be seen for these two particular alloys, cluster near the plane in the parent phase with Miller indices $(3\ 10\ 15)_P$ (subscripts P and M, respectively, designate the parent and martensite).

For purposes of further discussion, we consider again the fcc to bcc (or bct) martensitic transformation characteristic of iron alloys and steels. (It should be noted, however, that many other types of structure change are observed in a variety of martensitic transformations, e.g. fcc → fct in indium alloys, bcc → hcp in titanium alloys, and tetragonal → monoclinic in ZrO_2 and HfO_2.) Early work suggested that, apart from a small relative rotation of corresponding unit cells in the parent and product phases, a homogeneous, principal axis type of distortion would account for the known structural change in martensitic transformations. As early as 1924, Bain[2] suggested that the austenite (parent phase) → martensite transformation in steels could be explained by a homogeneous "upsetting" of the parent fcc lattice into the required bcc (or bct) lattice, as shown schematically in Fig. 6. This simple, intuitive lattice deformation was subsequently justified mathematically in the sense that the particular correspondence between lattices envisioned by Bain, when compared to others, involves the smallest principal strains.[3] There are many ways (correspondences) to generate a bcc product from an fcc parent by means of a homogeneous distortion and another possible correspondence is shown in Fig. 7, but analysis shows that the Bain deformation involves the smallest principal strains.[3] The lattice correspondence (which is a unique relationship between any lattice point in the initial lattice and the point it becomes in the final lattice) implied by the Bain distortion has also been verified experimentally using an ordered Fe_3Pt alloy which undergoes nominally (disregarding ordering) an fcc to bcc transformation. By observing corresponding superlattice reflections in the parent and martensitic phases by means of transmission electron diffraction, it has been deduced that the Bain correspondence actually applies.[4]

According to the lattice correspondence shown in Fig. 6, one would expect, for example, $[001]_M \| [001]_P \|$, $[010]_M \| [110]_P$, $(112)_M \| (101)_P$, $(011)_M \| (111)_P$, etc. However, this exact parallelism is not observed. For the Fe-Pt alloy shown in Fig. 1 the following orientation relationship was observed[5]

$[001]_P - [001]_M$ $9.10°$ apart
$[\bar{1}01]_P - [\bar{1}\bar{1}1]_M$ $4.42°$ apart
$(111)_P - (011)_M$ $0.86°$ apart

and this is typical of those found in iron alloys and steels. A further analysis of the above orientation relationship would show that the correspondence cell is not only distorted (upset) but also is rotated (about $10°$ from $[001]_P$ toward $[110]_P$). This has been termed a rigid body rotation. Further examination of the above orientation relationship shows that the close-packed planes $(111)_P$ and $(011)_M$ and the close-packed directions $[\bar{1}01]_P$ and $[\bar{1}\bar{1}1]_M$ are nearly parallel to each other. The mutual parallelism of close-packed planes and directions in two coexisting structures is frequently termed the Kurdjumov-Sachs type of orientation relationship.

The previous account has emphasized the following crystallographic charactistics of martensitic transformations: (1) a lattice correspondence, (2) an invariant plane strain shape deformation, (3) a habit plane, and (4) an orientation relationship. As already noted, the martensitic phase usually takes the form of lenticular plates, similar to mechanical twins. There are exceptions, however. In low carbon steels and dilute iron alloys the martensite morphology is lathlike, while still maintaining a habit plane, and in certain nonferrous alloys the martensitic transformation occurs by means of the propagation of a single interface which generates the product phase as it sweeps across the specimen.

The Crystallographic Theory

The feasibility of the Bain-type lattice correspondence and distortion has been presented as one involving minimal atomic displacements. However, the Bain distortion itself is inconsistent with the experimentally established invariant plane strain shape deformation. The reason for this is as follows. Referring to Fig. 6 again, note that the correspondence cell is contracted ($\approx 20\%$) along the z' axis and expanded the same amount ($\approx 12\%$) along the x' and y' axes. Such a homogeneous distortion will leave no plane invariant (undistorted and unrotated). But, on the other hand, suppose that the distortion along the y' axis vanished. In other words, one of the principal distortions is less than unity (along z'), one is greater than unity (along x'), and the remaining one (along y') is unity. This set of conditions is analogous to distorting an initial sphere into a triaxial ellipsoid, following which the sphere and ellipsoid can fit together along an undistorted plane of contact (habit plane) as shown in Fig. 8. However, this special set of conditions does not generally obtain in practice because the principal distortions are determined by the lattice correspondence and lattice parameters of the two phases.

This apparent dilemma can be overcome by envisioning that a lattice invariant deformation, such as slip or twinning, occurs in conjunction with the Bain distortion.[1,6] This additional deformation must be lattice invariant because the necessary structural change is completed by the Bain distortion alone. The effect of the additional deformation is effectively to shear

67

(distort) the ellipsoid resulting from the Bain distortion into tangency with the initial sphere, and then one of the principal distortions (OX in Fig. 8) becomes unity and an undistorted contact plane is provided. This secondary deformation, slip, twinning, or faulting, being a shear is known as the inhomogeneous shear or complementary shear of the crystallographic theory.

One problem still remains at this point. Even though the required structural change has been effected (by the Bain distortion) and an undistorted contact plane has been provided for (by the inhomogeneous shear) the habit plane is still not unrotated, as required from observation. Consideration of the sphere-ellipsoid analog shows that, although an undistorted plane now exists, it has been rotated from its initial unrotated position. Thus, a rigid body rotation is additionally incorporated along with the Bain distortion and inhomogeneous shear, and these are the three phenomenological steps describing the total transformation. There is no time sequence implied as to which step occurs when. The combined effect of these three operations must, of course, be equivalent to the shape deformation.

Within the framework of the theory just described, different crystallographic features such as the habit plane and orientation relationship can be predicted by supposing that the inhomogeneous shear occurs on different crystallographic planes and directions. For example, in most iron alloys (where similar lattice parameters for the parent and martensite phases are found) an inhomogeneous shear on the $(112)[\bar{1}\bar{1}1]_M$ twinning systems will predict a habit plane near $(3\ 5\ 10)_P$, but assuming the shear system to be $(011)[\bar{1}\bar{1}1]_M$ predicts a habit plane near $(111)_P$. Since the lattice parameters and correspondence are known, the Bain distortion for a given transformation is usually specified, and the flexibility in the theory comes from different suppositions concerning the plane and direction of the inhomogeneous shear. Once the inhomogeneous shear system is assumed, a shear of a certain magnitude will produce an undistorted plane, which when rigidly rotated in its original position becomes the habit plane.

Much of the previous discussion and many of the examples cited have centered around iron alloys. Because of the importance of steels there has been substantial work on them and thus an abundance of experimental data exists. However, the principles presented are quite general and apply to all martensitic transformations.

Inhomogeneous Shear and Martensite Substructure

The inhomogeneous shear was introduced in the crystallographic description of martensitic transformations to ensure that the habit plane is macroscopically undistorted. Figure 9 is a schematic representation of the appearance of internally twinned and internally slipped martensite plates. Although there are localized distortions at the interface, the serration effect because of alternating twins (called transformation twins) or slip lamellae prevents any accumulation of strain at the interface over large distances. In the case of twinned martensite, the Bain distortion is envisioned to occur along different contraction axes in the two regions, and the twinning plane in the martensite is derived from a mirror plane in the parent.[1] In the case of internally slipped martensite, the Bain distortion is the same in all regions of a plate. Figure 10 shows that the same shear γ can be accomplished by slip or twinning.

68

Because of the inhomogeneous shear, one would expect to observe some kind of substructure in the martensite. Indeed many such observations have been made since the introduction of the theory. Figure 11 is a transmission electron micrograph of a martensite plate in an Fe-Ni-C alloy.[7] The fine striations crossing the plate are transformation twins ($\{112\}_M$ twinning plane), and regions adjacent to the martensite are retained austenite. The electron diffraction pattern from the martensite shows twin related reflections, and if the twin plane is indexed specifically as $(112)_M$, the habit plane trace becomes specifically $(3\ 15\ 10)_P$. If $(112)[\bar{1}\bar{1}1]_M$ twinning is used in the habit plane calculations, the predicted habit plane is in fact $(3\ 15\ 10)_P$. Therefore the particular variant of the twin plane is observed to be consistent with the particular variant of the habit plane, and for the Fe-Ni-C alloy the experimental observations are in excellent agreement with those features which are predicted using crystallographic theory. This is also the case for many other martensitic transformations which have been studied in some detail using transmission electron microscopy and diffraction. Notable exceptions, however, are certain steels which transform to martensite with a $\{225\}_P$ habit plane. The martensite in these materials has a complex substructure, as seen in the electron microscope, and is not very well explained using the theory described above. In addition, there is as yet no adequate crystallographic description of the lath martensites found in ferrous alloys.

Algebraic Analysis

The basic equation of the crystallographic theory[1,6] just presented is

$$\mathbf{P}_1 = \mathbf{R\bar{P}B} \tag{1}$$

where $\mathbf{B}$ is the Bain distortion, $\mathbf{\bar{P}}$ is a simple shear (following, mathematically, the Bain distortion), $\mathbf{R}$ is the rigid body rotation previously mentioned, and $\mathbf{P}_1$ is the invariant plane strain shape deformation. $\mathbf{B}, \mathbf{\bar{P}}, \mathbf{R}$, and $\mathbf{P}_1$ are all (3×3) matrices. The matrix product $\mathbf{R\bar{P}B}$ is equivalent to the shape deformation $\mathbf{P}_1$ and the rotation $\mathbf{R}$ rotates the plane left undistorted by $\mathbf{\bar{P}B}$ to its original position. That is, $\mathbf{P}_1 = \mathbf{R\bar{P}B}$ is an invariant plane strain.

Although Eq. (1) shows the inhomogeneous shear $\mathbf{\bar{P}}$ following the Bain distortion, it is to be noted that the same mathematical result is obtained by "allowing" the shear to occur in the parent phase prior to the Bain distortion. By following this latter procedure, there are certain computational simplifications to be gained, and in this case the basic equation becomes

$$\mathbf{P}_1 = \mathbf{RBP} \tag{2}$$

where $\mathbf{P}$, as before, represents a simple shear.

With regard to the previous discussion, it will be noted that the invariant plane strain shape deformation can be expressed as

$$\mathbf{P}_1 = \mathbf{I} + \mathbf{mdp}'$$

$$= \begin{pmatrix} 1 & 0 & 0 \\ 0 & 1 & 0 \\ 0 & 0 & 1 \end{pmatrix} + m[d_1 d_2 d_3]\ (p_1 p_2 p_3)$$

$$= \begin{pmatrix} 1 + md_1p_1 & md_1p_2 & md_1p_3 \\ md_2p_1 & 1+ md_2p_2 & md_2p_3 \\ md_3p_1 & md_3p_2 & 1+ md_3p_3 \end{pmatrix} \tag{3}$$

where $\mathbf{p}'$ (prime meaning transpose) being a plane normal is written as a

(1×3) row matrix, in contrast to **d**, a lattice vector, which is a (3×1) column matrix. With reference to Fig. 6, the Bain distortion for the fcc → bcc (bct) case can be written as

$$\mathbf{B} = \begin{pmatrix} \sqrt{2}a/a_0 & 0 & 0 \\ 0 & \sqrt{2}a/a_0 & 0 \\ 0 & 0 & c/a_0 \end{pmatrix} \tag{4}$$

When typical values for a, a_0, and c are substituted for the fcc → bct transformation in steels, a representative **B** matrix is

$$\mathbf{B} = \begin{pmatrix} 1.12 & 0 & 0 \\ 0 & 1.12 & 0 \\ 0 & 0 & 0.8 \end{pmatrix} \tag{5}$$

where two of the principal distortions are greater than unity and the third is less than unity, and the necessary condition mentioned earlier for an invariant plane strain clearly does not exist, considering the Bain distortion per se.

Going back to Eq. (2), it is noted that **P** is a simple shear and therefore of the invariant plane strain form $(\mathbf{I} + m\mathbf{dp'})$. Further, the inverse of **P**, $\mathbf{P}^{-1} = \mathbf{I} - m\mathbf{dp'}$, corresponds to a simple shear of the same magnitude on the same plane, but in the opposite direction. Thus both **P** and $\mathbf{P}^{-1}$ are invariant plane strains. It is then convenient to rewrite Eq. (2) as

$$\mathbf{P}_1 \mathbf{P}_2 = \mathbf{RB} \tag{6}$$

where $\mathbf{P}_2 = \mathbf{P}^{-1}$. Since $\mathbf{P}_1$ and $\mathbf{P}_2$ are invariant plane strains, their product **RB** is an invariant *line* strain, **S**, defined by the *planes* (i.e., their intersection) which are invariant to $\mathbf{P}_1$ and $\mathbf{P}_2$.[1] Once the invariant line strain **S** is known, all the crystallographic features of a given martensitic transformation can be predicted.[1] It is beyond the scope of the present account to go into the details of the invariant line strain analysis, but some highlights can be mentioned. The Bain correspondence and distortion are known from the lattice parameters of the parent and martensitic phases. **R** can be determined once the plane $\mathbf{p}_2'$ and direction $\mathbf{d}_2$ of $\mathbf{P}_2$ are assumed.

The important results of the invariant line strain analysis, noting that the shape strain is $\mathbf{P}_1 = \mathbf{I} + m_1 \mathbf{d}_1 \mathbf{p}_1'$ and that the simple shear (preceding the Bain distortion) is $\mathbf{P}_2 = \mathbf{I} + m_2 \mathbf{d}_2 \mathbf{p}_2'$, are as follows[8] (where the magnitudes, directions, and planes of the component invariant plane strains are given respectively by m, **d**, and $\mathbf{p'}$).

$$\mathbf{d}_1 = [\mathbf{S}\mathbf{y}_2 - \mathbf{y}_2]/\mathbf{p}_1'\mathbf{y}_2 \tag{7}$$

$$\mathbf{p}_1 = (\mathbf{q}_2' - \mathbf{q}_2'\mathbf{S}^{-1})\mathbf{q}_2'\mathbf{S}^{-1}\mathbf{d}_1 \tag{8}$$

where $\mathbf{y}_2$ is any vector lying in $\mathbf{p}_2'$ (except the invariant line **x**) and $\mathbf{q}_2'$ is any normal (other than $\mathbf{n'}$, the row unit eigenvector of $\mathbf{S}^{-1}$, i.e., $\mathbf{n'}\mathbf{S}^{-1} = \mathbf{n'}$) to a plane containing $\mathbf{d}_2$. The normalization factor for $\mathbf{d}_1$ in Eq. (7) is $1/m_1$ and therefore $\mathbf{P}_1$, m_1, $\mathbf{d}_1$, and $\mathbf{p}_1'$ are all determinable. The matrix **R** is determined from the requirement that **x** and $\mathbf{n'}$ which are displaced by the Bain distortion must be totally invariant. **R** defines the orientation relationship within any small region of the martensite plane not involving $\mathbf{P}_2$. Thus, the assumed correspondence and lattice parameters determine **B**, the assumption of $\mathbf{p}_2'$ and $\mathbf{d}_2$ allows **R** to be determined, and $\mathbf{RB} = \mathbf{S}$ defines the elements of $\mathbf{P}_1$.

The previous description parallels the theoretical development given by Bowles and Mackenzie[1] but the treatments of Wechsler et al.[6] and Bullough and Bilby[9] are equivalent.

Some variations in the basic theory just presented include the introduction of a dilatation parameter,[1] which in effect slightly relaxes the requirements that the habit plane be undistorted, and the incorporation of two inhomogeneous shear systems[10,11] such that

$$\mathbf{P}_1 = \mathbf{RBS}_2\mathbf{S}_1 \tag{9}$$

where $\mathbf{S}_2$ and $\mathbf{S}_1$ are the two inhomogeneous shear systems involved. Neither of these modified approaches is without criticism, and the double shear approach relative to the original single shear approach loses generality.[12]

Some Mechanical Aspects of Martensitic Transformations

Since the shape deformation of martensitic transformations is essentially a shear, one would expect that an applied stress would influence a martensitic transformation. This is indeed the case. Figure 12 shows a stress-strain curve for a single crystal Cu-Zn alloy obtained above the normal transformation of M_S (martensitic start) temperature.[13] The upper plateau corresponds to the formation of stress-induced martensite, leading to an elongation of $\approx 9\%$, and the lower plateau corresponds to the reversal of the stress-induced martensite as the applied stress is released. Note that the strain due to the formation of martensite is completely recovered when the stress is released. Stress-strain curves of this type are referred to as superelastic loops, associated with the superelastic behavior which is observed. As might be expected, a higher stress is required to form martensite as the test temperature increases above the M_s temperature, as shown in Fig. 13.[13]

Another phenomenon of interest is the formation of strain-induced martensite. In this case, the plastic (irrecoverable) strain is believed to generate preferential sites for the nucleation of martensite during straining. In some steels, for example, enhanced plasticity and substantial strengthening occur as a consequence of a strain-induced martensitic transformation. These are known as TRIP (transformation-induced plasticity) steels. A somewhat similar phenomenon can be used to toughen zirconia-containing ceramics.

In some alloys, if the deformation is extensive enough, the initial martensite induced by stress is converted to another martensite with a different crystal structure. In such a case, double superelastic loops, as shown in Fig. 14, are observed.[14]

Another interesting mechanical phenomenon, which is associated with a thermoelastic martensitic transformation, is the so-called shape memory effect. This effect is related to martensite which has been deformed below the normal formation temperature range, or M_f (martensite finish) temperature. A stress-strain curve corresponding to this kind of behavior is shown in Fig. 15.[15] The region from a to b corresponds to the deformation of martensite and a residual strain of $\approx 5\%$ remains (point c) when the load is removed. However, when the specimen is heated to reverse the martensite (i.e., martensite $\rightarrow$ parent), the specimen "remembers" its initial configuration and the apparently permanent strain is completely recovered. This process is shown schematically in Fig. 16. By means of the shape memory effect, very high recovery stresses are generated as the deformed martensite reverts to the parent phase, representing the direct conversion of heat into mechanical work.

71

Many martensitic materials exhibit a very high damping capacity. This occurs because martensite-martensite boundaries and transformation twin boundaries are highly mobile and will move under stress, thus effectively attenuating an external influence.

Transformations in ZrO_2 and HfO_2

Zirconia and hafnia are structurally isomorphous and undergo a tetragonal → monoclinic martensitic transformation during cooling. The transformation in zirconia has been studied in some detail and the crystallographic documentation for its martensitic nature is convincing.[16] There is, however, one curious feature of the tetragonal → monoclinic transformation in ZrO_2. Apparently two different martensite habit planes (each with a correspondingly different lattice orientation relationship) exist,[16] one if the transformation occurs above 1000 °C, and the other if the transformation occurs below 1000 °C. The "type A" habit plane, $\sim \{110\}_{tetragonal}$, is believed to result from a $(1\bar{1}0)[001]$ inhomogeneous shear (slip) whereas the "type B" habit plane, $\sim \{100\}_{tetragonal}$, results from an inhomogeneous shear on $(1\bar{1}0)[110]$.[16] If more detailed experiments verify this, strong evidence will exist to suggest that the inhomogeneous shear (and hence habit plane and other crystallographic features) is temperature dependent.

Although studied in less detail, the tetragonal → monoclinic transformation in HfO_2 appears to be similar to that in ZrO_2. Figure 17[17] is a high temperature transmission electron micrograph showing the coexistence of the tetragonal (parent) and monoclinic (martensite) phases in hafnia. A well developed substructure in the monoclinic martensite (upper part of micrograph) can be seen.

Finally, a few words on transformation toughening of ceramics are relevant. By doping ZrO_2 with CaO, MgO, Y_2O_3, etc., it is possible to produce a room temperature microstructure consisting of a cubic matrix containing fine, coherent tetragonal precipitates. This two-phase structure is referred to as partially stabilized and can be produced in ZrO_2 by quenching from 2000 °C to room temperature and then aging for 1 h at 1500 °C.[18] The microstructure so obtained inhibits the long range propagation of cracks because the stress field of a crack acts to transform the tetragonal precipitates to monoclinic martensite and blunt the crack.[18] Hence the strain-induced martensite confers a degree of toughness not hitherto realized in ZrO_2. The detailed crystallography of the martensite transformation in the small coherent tetragonal particles remains to be investigated.

Further details on martensitic transformations can be found in Refs. 8 and 19–22.

Acknowledgments

My own work in phase transformations has been supported by the Army Research Office and by the National Science Foundation through the Materials Research Laboratory at the University of Illinois. The support of these agencies is gratefully acknowledged.

References

[1] J. S. Bowles and J. K. Mackenzie, *Acta Metall.*, **2**, 129, 138, 224 (1954).
[2] E. C. Bain, *Trans. AIME*, **70**, 25 (1924).

[3]M. A. Jaswon and J. A. Wheeler, *Acta Crystallogr.*, **1**, 216 (1948).
[4]T. Tadaki and K. Shimizu, *Trans. Jpn. Inst. Met.*, **11**, 44 (1970).
[5]E. J. Efsic and C. M. Wayman, *Trans. AIME*, **239**, 873 (1967).
[6]M. S. Wechsler, D. S. Lieberman, and T. A. Read, *ibid.*, **197**, 1503 (1953).
[7]T. Maki and C. M. Wayman, *Suppl. to Trans. Jpn. Inst. Met.*, **17**, 69 (1976).
[8]C. M. Wayman; p. 147 in the Crystallography of Martensitic Transformations in Alloys of Iron, Advances in Materials Research, Vol. III. Edited by H. Herman. Interscience Publishers, New York, 1968.
[9]R. Bullough and B. A. Bilby, *Proc. Phys. Soc.*, **B69**, 1276 (1956).
[10]A. G. Acton and M. Bevis, *Mater. Sci. Eng.*, **5**, 19 (1969/70).
[11]N. H. D. Ross and A. G. Crocker, *Acta Metall.*, **18**, 505 (1970).
[12]D. P. Dunne and C. M. Wayman, *ibid.*, **19**, 2327 (1971).
[13]T. A. Schroeder and C. M. Wayman, *ibid.*, **27**, 405 (1979).
[14]T. A. Schroeder; Ph.D. Thesis, Univ. of Illinois, Urbana, Ill., 1976.
[15]T. A. Schroeder and C. M. Wayman, *Acta Metall.* **25**, 1375 (1977).
[16]G. K. Bansal and A. H. Heuer, *ibid.*, **20**, 1281 (1972); *ibid.*, **22**, 409 (1974).
[17]B. Pieraggi, private communication.
[18]D. L. Porter and A. H. Heuer, *J. Am. Ceram. Soc.*, **60**, 183 (1977).
[19]J. W. Christian, The Theory of Transformation in Metals and Alloys. Pergamon Press, Oxford, 1965.
[20]J. W. Christian, Martensitic Transformation: A Current Assessment in the Mechanism of Phase Transformations in Crystalline Solids, Monograph 33, The Institute of Metals, London, 1969.
[21]Z. Nishiyama, in Martensitic Transformation. Edited by M. E. Fine, M. Meshii, and C. M. Wayman. Academic Press, New York, 1978.
[22]C. M. Wayman, Introduction to the Crystallography of Martensitic Transformations. Macmillan, New York, 1964.

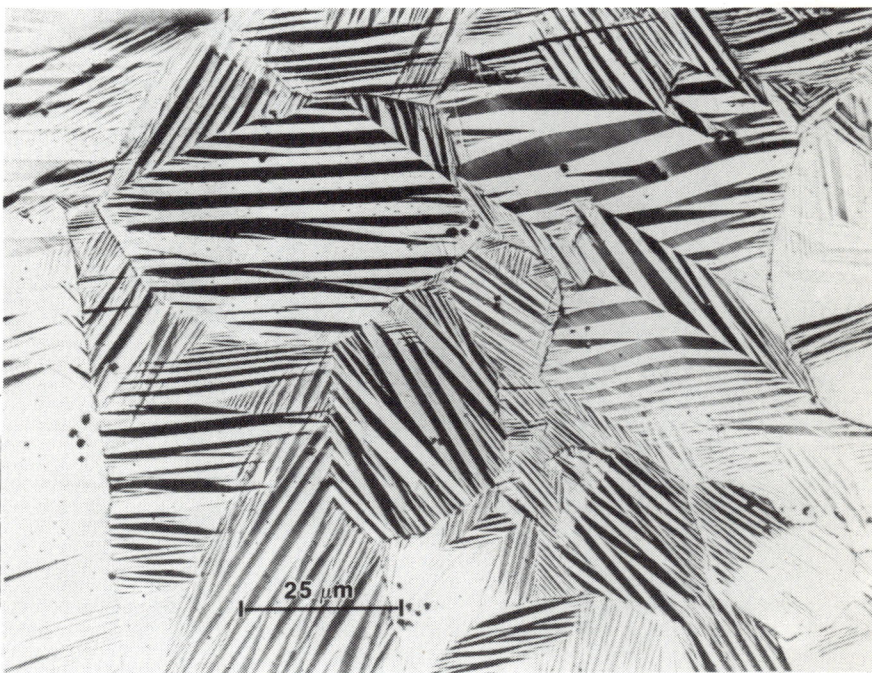

Fig. 1. Optical micrograph showing surface relief due to martensite formation in polycrystalline Fe-24.5% Pt alloy prepolished in parent-phase condition prior to transformation to martensite. The differently shaded regions correspond to martensite plates whose shape deformation produces different surface tilts.

Fig. 2. Interference micrograph showing surface relief due to martensite formation in Fe-24.5% Pt alloy.

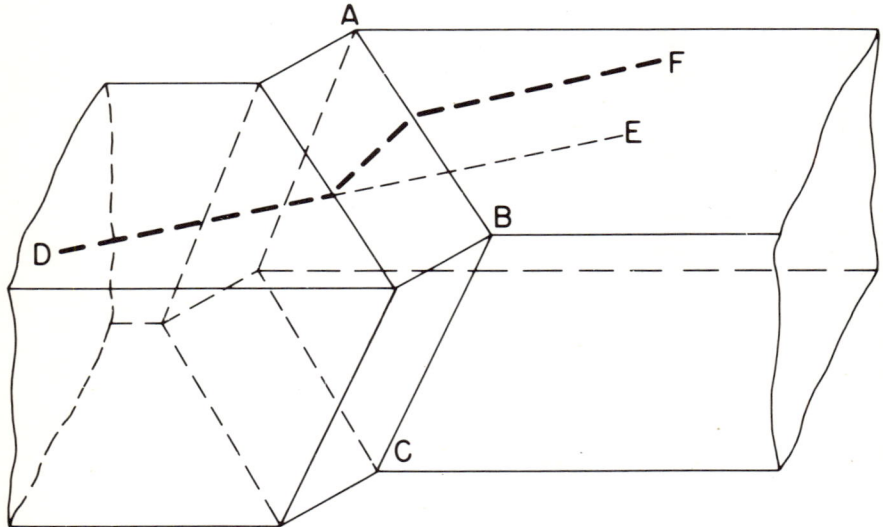

Fig. 3. Schematic representation of invariant plane strain shape deformation characteristic of the formation of martensite. The initially straight scratch DE is displaced to the DF position when the martensite plate with habit plane ABC forms. Plane ABC is both undistorted and unrotated (invariant) as a consequence of the martensitic transformation.

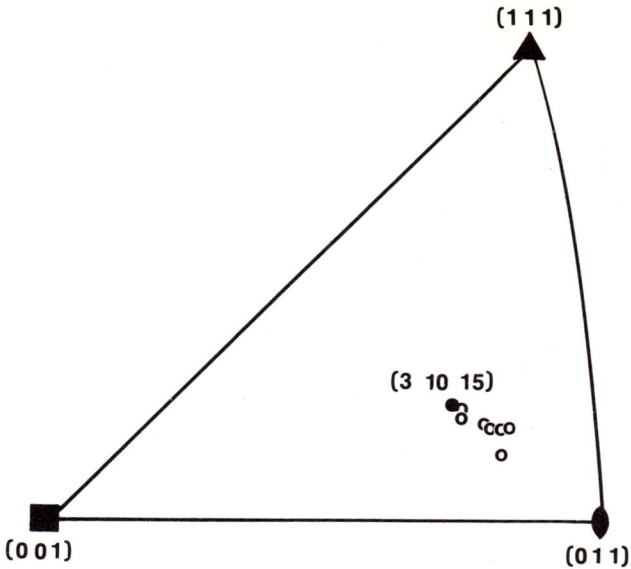

Fig. 4. Stereographic projection showing habit plane poles for seven martensite plates in an Fe-7%Al-1.5%C alloy. The mean habit plane is not one of simple Miller indices and is near the {3 10 15}$_P$ plane of the parent phase.

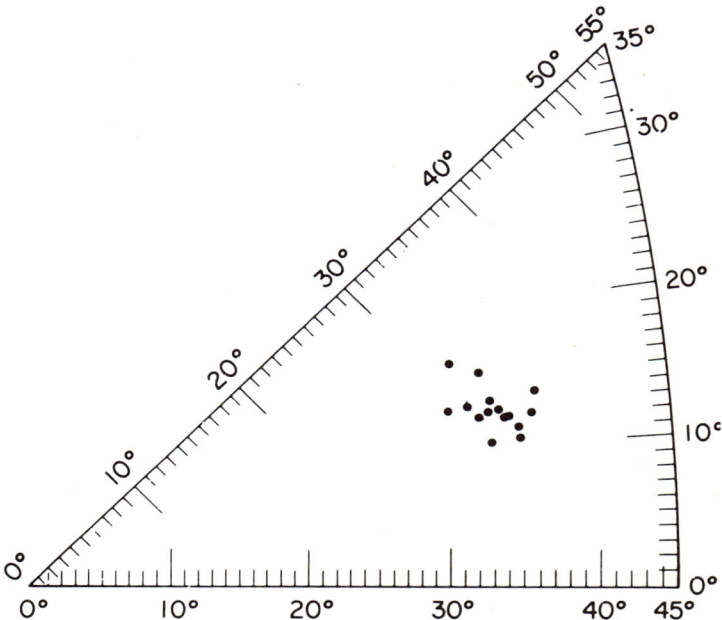

Fig. 5. Stereographic projection showing habit plane poles for different martensite plates in an Fe-31%Ni alloy. Courtesy R. P. Reed.

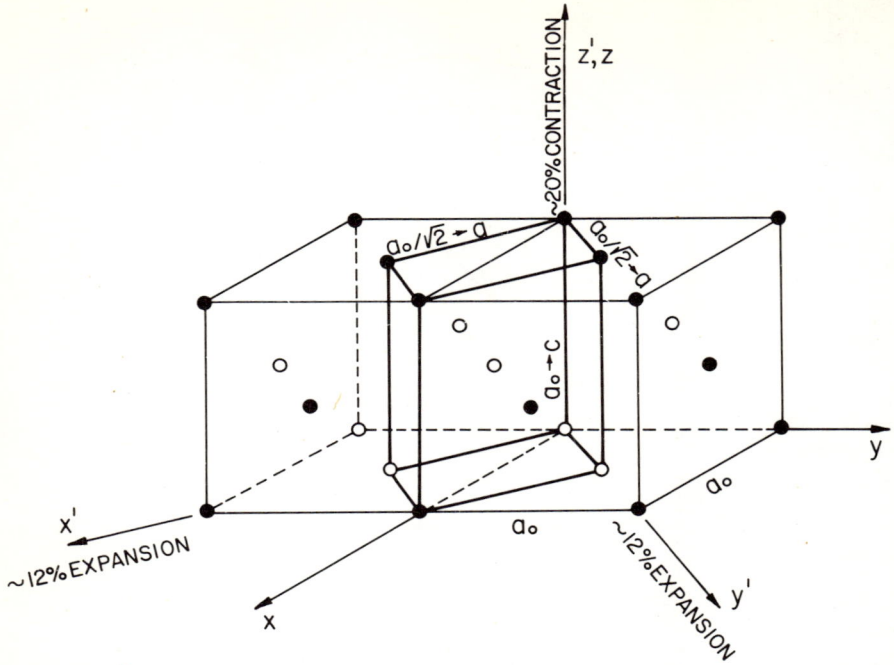

Fig. 6. Lattice correspondence and distortion proposed by Bain for the fcc→bcc (bct) transformation in iron alloys. The delineated (by heavier lines) correspondence cell in the parent becomes a unit cell in the martensitic phase after a homogeneous contraction (upsetting) with respect to the z' axis. The principal distortions along the x', y', and z' are indicated.

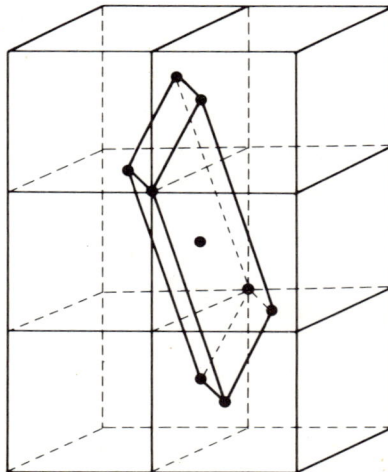

Fig. 7. Alternative lattice correspondence for fcc → bct transformation involving larger principal distortions than the Bain correspondence.

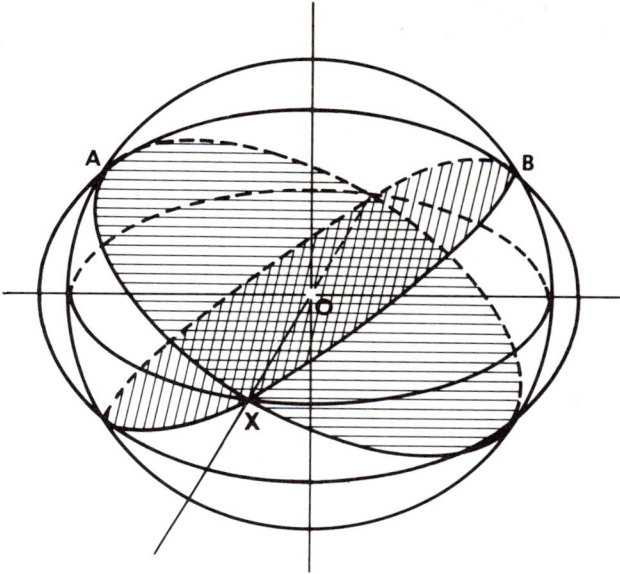

Fig. 8. Sphere-ellipsoid analog of the Bain distortion.
See text for discussion.

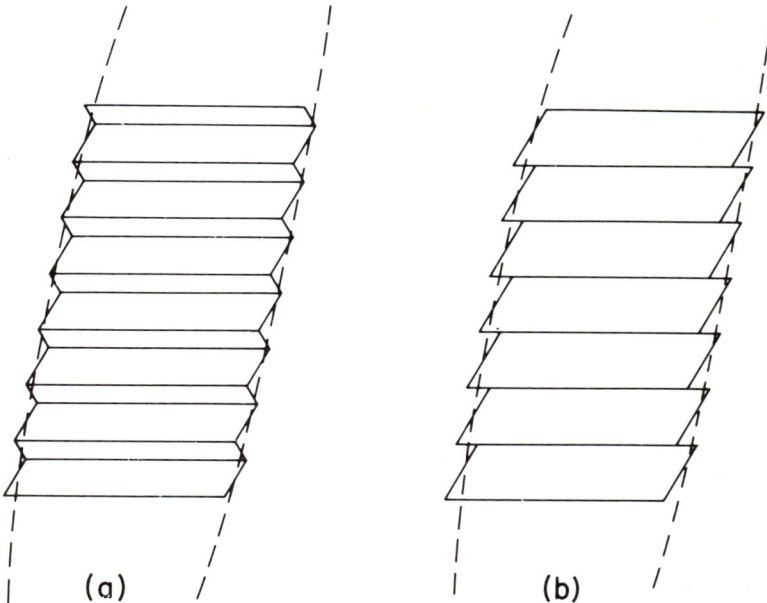

(a)   (b)

Fig. 9. Schematic representation of the inhomogeneous shear in
martensitic transformations showing (a) internally twinned and (b)
internally slipped martensite plates. Because of the inhomogeneous deformation, the "sawtooth" effect at the interface
prevents the widespread accumulation of strain, and the interface
remains macroscopically undistorted.

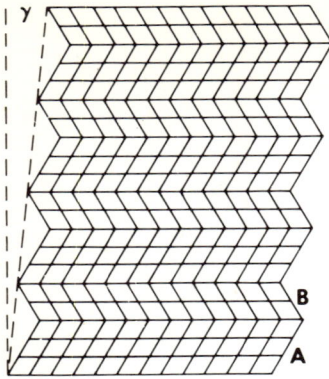
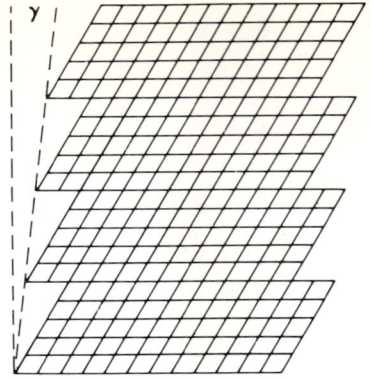

(a)   (b)

Fig. 10. Schematic representation of the slip and twinning showing that the same magnitude of inhomogeneous shear (angle γ) can be accomplished by either.

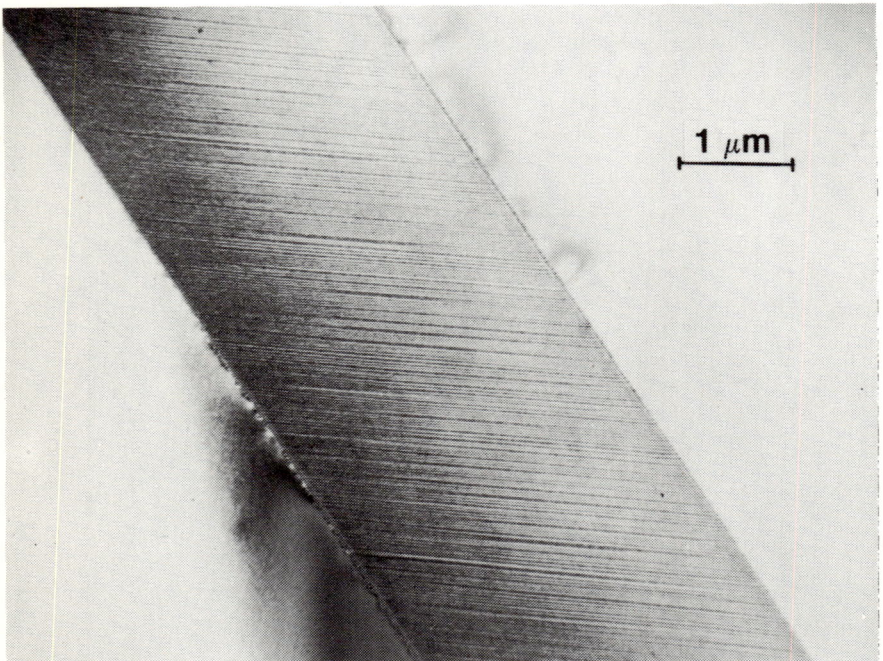

Fig. 11. Transmission electron micrograph showing martensite plate in an Fe-30%Ni-0.39%C alloy. The lateral striations within the plate are transformation twins as idealized in Fig. 9(a), and the regions adjacent to the martensite are untransformed parent phase (retained austenite).

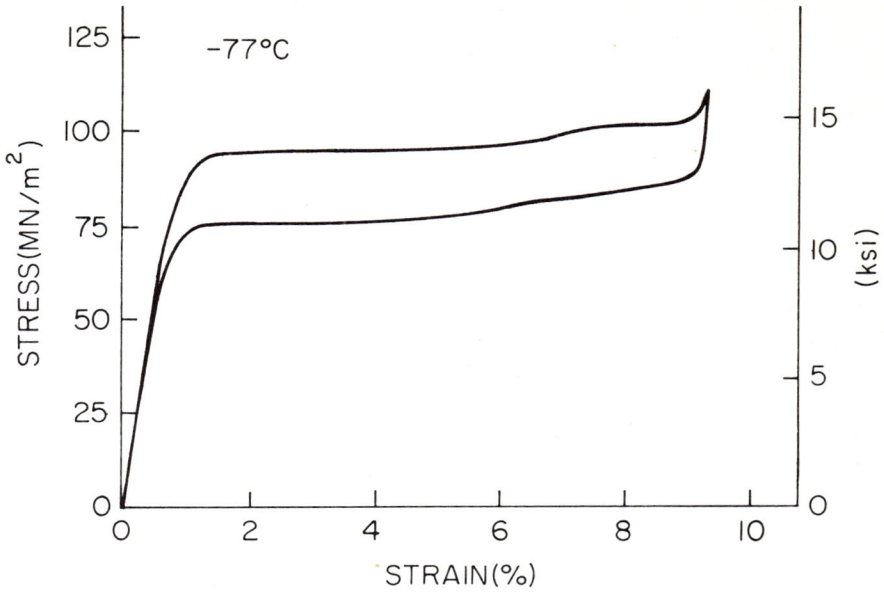

Fig. 12. Stress-strain curve for Cu-39.8%Zn single crystal specimen showing superelastic loop due to the formation and reversion of stress-induced martensite.

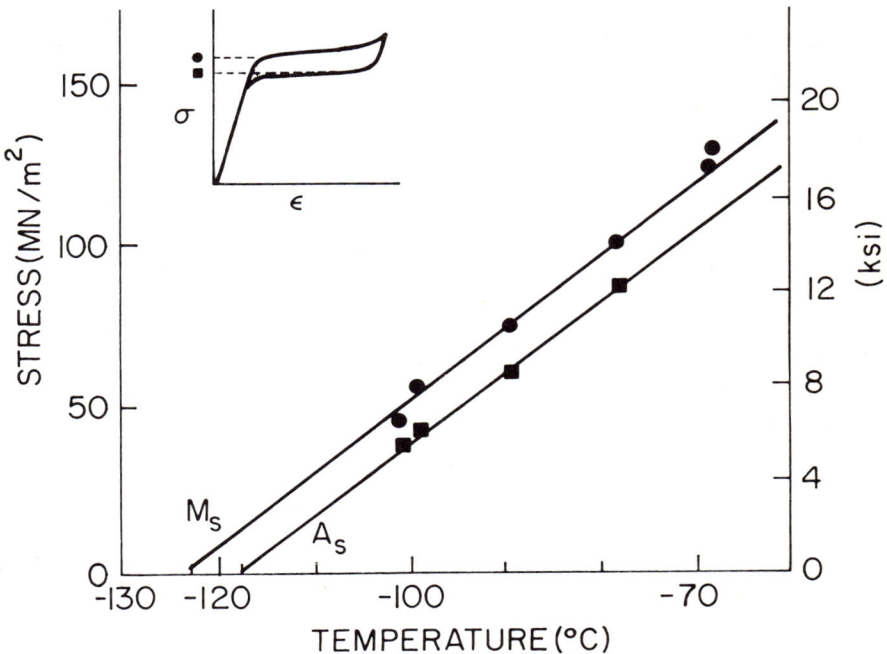

Fig. 13. Temperature dependence of the stress required to form stress-induced martensite in Cu-39.8%Zn alloy.

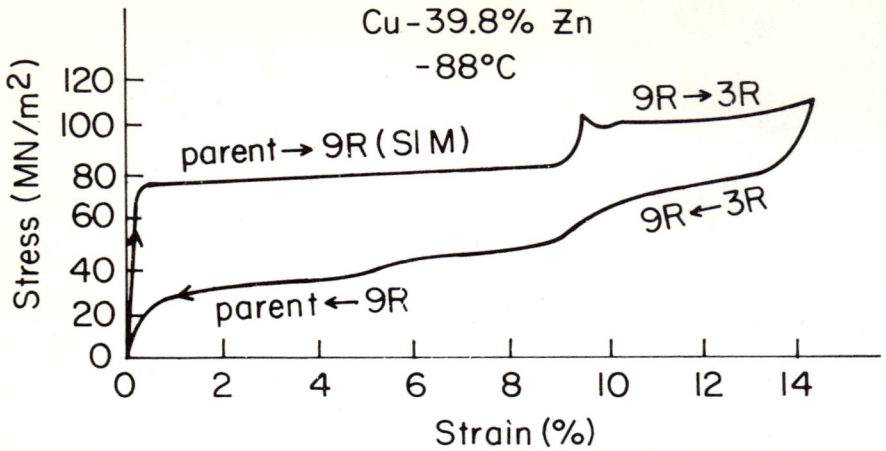

Fig. 14. Double superelastic loops due to martensite → martensite transformation in Cu-39.8% Zn alloy.

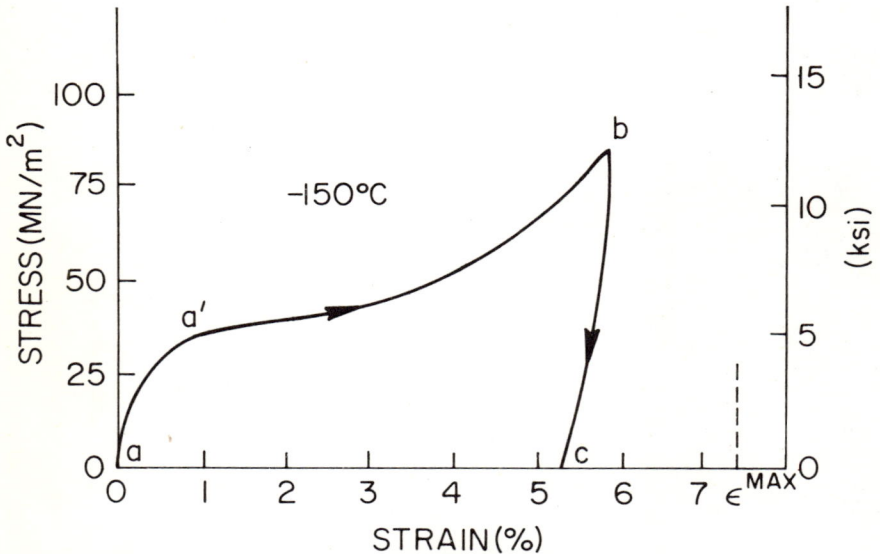

Fig. 15. Stress-strain curve corresponding to the shape memory effect. The remaining strain shown at c is completely recovered when the specimen is heated.

WHAT IS THE SHAPE MEMORY EFFECT (SME)?

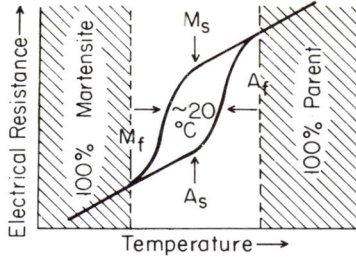

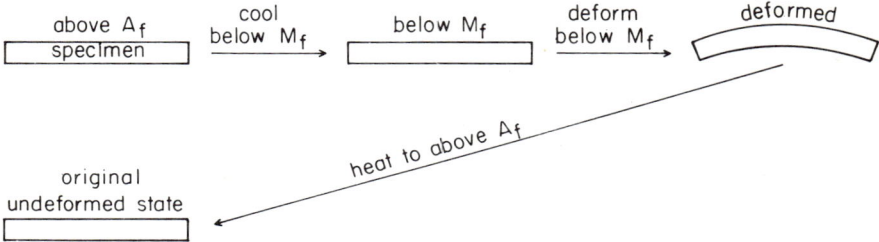

Recoverable Strains Typically ~7%

Fig. 16. Schematic description of shape memory effect.

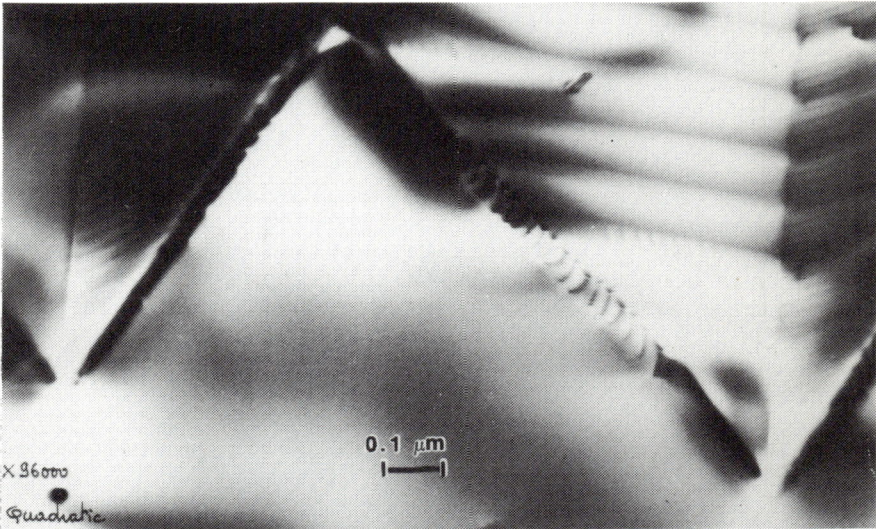

Fig. 17. High temperature transmission electron micrograph showing coexisting tetragonal and monoclinic (martensite) phases in HfO_2. Courtesy of B. Pieraggi.

The martensite crystallography of tetragonal zirconia*

W. M. Kriven,† W. L. Fraser,‡ and S. W. Kennedy

Department of Physical and Inorganic Chemistry
University of Adelaide
Adelaide, S.A., 5001
Australia

Crystallographic analyses of possible martensitic transformation mechanisms were conducted on the tetragonal-to-monoclinic transformation in zirconia. Such analyses are purely geometrical and based on the Invariant Plane Strain model of a martensitic transformation in a bulk crystal. Calculations were done for lattice invariant shear (LIS) slip and twin systems for all three reported lattice correspondences. The computations predict habit planes, shape changes, and orientation relations for each variant in terms of the parent tetragonal structure. Most of the slip LIS's had shape strains of $\approx 11\%$ to 17%, while one twin system had a 5% shape strain. It was also shown that deformation twinning on $(001)_m$ or $(100)_m$ may also act as an LIS system.

It is now well established that zirconia is among the most refractory thermal shock-resistant and corrosion-resistant oxides. At atmospheric pressure, zirconia exists in three crystallograpic modifications[1]:

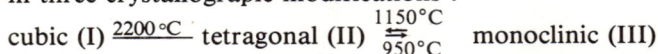

$$\text{cubic (I)} \xrightarrow{\text{2200 °C}} \text{tetragonal (II)} \underset{950°C}{\overset{1150°C}{\rightleftharpoons}} \text{monoclinic (III)}$$

The tetragonal-to-monoclinic transformation is martensitic and accompanied by a 3% volume increase, which is physically deleterious to the ceramic. The transformation may be suppressed, however, by small additions of other oxides, notably CaO, Y_2O_3, and MgO.[2] The cubic, fluorite-type solid solution then persists to room temperature, as fully stabilized zirconia. The most useful mechanical properties of zirconia arise from composite microstructures of monoclinic[3] or tetragonal[4] particles of pure zirconia dispersed in a matrix of cubic zirconia, which is known as partially stabilized zirconia (PSZ).

The crystal structures of tetragonal and monoclinic ZrO_2 are shown in Fig. 1. Three lattice correspondences may arise[5] between the tetragonal and monoclinic polymorphs as illustrated in Fig. 2. If a right-hand screw convention is adopted for unit cell axes, lattice correspondences (LC) A, B, or C may be defined, depending on which monoclinic axis, a_m, b_m, or c_m, is parallel to c_t.

Bailey[5] studied the II and III transformation by transmission electron microscopy and found twinning on (110), $(1\bar{1}0)$, and (100) monoclinic planes, as well as direct evidence for an orientation relationship consistent with LC C and indirect evidence for orientation relations derived from all three lattice correspondences.

A martensitic mechanism for the tetragonal-to-monoclinic transformation was first reported by Wolten,[6] who found symmetry options and variants for the orientation relation derived from lattice correspondence C. He concluded that the interface had a monoclinic habit plane (101) which becomes (101), (110), and (011) of the tetragonal cell. His X-ray orientation relations imply all three lattice correspondences. As reviewed by Subbarao *et al.*,[1] several other determinations of orientation relations essentially arise from LC B or LC C.

Bansal and Heuer[7,8] examined single crystal II and III transformations by transmission electron microscopy and X-ray precession experiments. They found that an orientation relation derived from LC B occurred for an A_s temperature above 1000 °C. A_s is the temperature at which the reverse (monoclinic-to-tetragonal) temperature starts. An LC C (i.e., type 3) orientation resulted from an M_s below 1000 °C. Habit planes of the type $(671)_m$ or $(761)_m$, approximately 7° off $(110)_m$, were termed type A and usually occurred inside the crystal, whereas surface regions showed type B plates, which were lenticular and twinned along a midrib close to $(100)_m$ or $(010)_m$. Coarse $(100)_m$ twins were also observed in plate-free regions.

Martensite calculations were performed by Bansal and Heuer,[8] using the algebraic method of Bowles and MacKenzie,[9] for eight slip lattice invariant shear (LIS) systems. The slip systems and lattice correspondences for which calculations were performed are listed below:

LC B $(1\bar{1}0)[001]$ $(111)[1\bar{1}0]$ $(100)[001]$ $(100)[011]$
LC C $(1\bar{1}0)[001]$ $(1\bar{1}0)[110]$ $(10\bar{1})[010]$ $(111)[1\bar{1}0]$

Martensite analyses on LC C by $(1\bar{1}0)[001]$ slip system predicted $\{671\}_m$ habit planes, while $(100)_m$ habit planes were predicted by slip on $(1\bar{1}0)[110]$. No martensite analyses using LIS twinning systems have yet been reported.

In the details of the Bansal and Heuer[8] calculations, a numerical error was noted. A check on the volume increase in the transformation using the algebraically calculated magnitudes of the principal distortions indicated values of 1.8360% and 1.8356% for the lattice correspondences B and C, respectively. The volume change from the unit cell dimensions at 1000 °C, as used by Bansal and Heuer,[8] showed that the volume increase is actually 1.9030%. Errors in these initial stages of the principal strain calculations could result in misleading conclusions.

In this work, we present calculations of possible crystallographic transformation mechanisms in bulk, unconstrained zirconia crystals. Possible slip systems for all three lattice correspondences, as well as one twin LIS system, are analyzed. Habit planes, shape changes, and orientation relations for each variant are predicted in terms of the tetragonal structure. The shape changes associated with each mechanism and its variants are compared.

Analysis

Lattice parameters of both tetragonal and monoclinic phases corresponded to a transformation temperature of 950 °C. They were calculated from experimental parameters and thermal expansion coefficients determined by Patil and Subbarao[10,11] and gave a II to III volume expansion of ≈ 3.0%. Thermal expansion was shown to be anisotropic. The present analysis at 950° gave a volume expansion of 3.047%, which was consistent with the volume increase calculated from lattice parameters. Buljan,

McKinstry, and Stubican determined lattice parameters at 1000 °C which were used by Bansal and Heuer[8] and also in this analysis to compare the effect of lattice parameter differences.

The martensite analysis formulated by Bowles and MacKenzie[9] was programmed into Fortran IV computer language by Ledbetter and Wayman[12] in PRØGRAM MRTNST. Calculations were done in an orthonormal (cubic) frame of reference in which input data also needed to be expressed. In zirconia, part of the lattice deformation may be visualized as a simple contraction or expansion in the direction corresponding to b_m, which was here designated as the η_3 direction. The remainder of the lattice deformation occurred in the plane normal to this direction. The two principal axes and distortions in this plane were readily determined by standard geometrical theory for a homogeneous strain in two dimensions[13] (Appendix I). These calculations of principal strain magnitudes and directions were confirmed by the general method of Bowles and MacKenzie.[9] Table I presents lattice parameters of the two phases at 950°C; calculated values of the deformations are given in Table II.

Important features arising from the above are that (1) two of the principal deformation directions are irrational, and (2) the magnitude of η_3 varies only by $\approx 1\%$ from unity for all three lattice correspondences. Finally, the values of $\eta_3 - 3$, which are usually taken as measures of the strain energy, favor lattice correspondence C, but the differences are small between all calculations.

Physically, lattice invariant shears occur in the product (monoclinic) phase, but LIS systems were specified in the tetragonal phase, to which they were referred from the monoclinic cell, by inspection of lattice correspondences. The LIS systems investigated were derived mainly from slip and twinning shears found in the monoclinic phase after transformation in previous experimental studies. For lattice correspondence B, the twin system $(010)[100]_t$ was analyzed, where the direction of twinning shear was specified in the input data and treated as a slip system. (Mathematically, computations are identical for slip and twinning.) This approach was necessary as PRØGRAM MRTNST was unable to calculate twin directions for nonorthogonal systems.

Results

It was sometimes difficult to assess whether or not a predicted magnitude of LIS was physically reasonable. LIS by twinning posed no problem, as magnitudes of shear determined relative proportions of each twin in the product. The LIS angle must therefore be less than the angle of twinning shear. For slip or stacking fault shears, the shear angle associated with an individual dislocation or stacking fault was readily determined and, in this work, any LIS angle greater than half the maximum value was taken to indicate an unreasonable defect density. Thus, an LIS of 20° was taken as an upper limit for "reasonable" solutions. Shape strains (m_1) $>20\%$ were considered unlikely and hence not reported.

Table III compares some of the calculations of Bansal and Heuer[8] with those computed by the present method, using the same lattice parameters at 1000 °C. It is seen that, although there is general agreement, there is a scatter of habit planes and variation in the magnitude of shape strain by $\approx 10\%$. Table IV summarizes habit planes and shape strains (m_1, d_1) predicted at 950 °C for each mechanism. The combination of all three lattice correspondences with LIS systems yielded 32 distinct reasonable solutions.

Orientation relations were determined by multiplying vectors and plane normals by the total strain matrix.[12] A vector $\mathbf{V}$ was rotated to become a vector $\mathbf{V}'$, both in the same cubic basis. The mutual angle of rotation was then calculated for that crystal system. With reference to the operating lattice correspondence, the product vectors and plane normals were then relabelled according to the monoclinic axes. The details of predicted orientation relations are presented in Table V.

Twinning on $\{010\}\langle\bar{1}00\rangle$ as an LIS resulted in four symmetry-equivalent solutions for each twin plane. The shear angle associated with twinning was $8.78°$. The proportion of each twin in the product was predicted to be $0.409{:}0.591$ and $0.617{:}0.383$ (for an LIS angle $= 3.61°$ and $5.44°$, respectively). These ratios were close to 2:3 and 3:2. Habit planes approximated to $\{121\}_t$ and the total shape change was significantly less than the shape change of the lattice. The four different (but equivalent) orientation relations for each twinning LIS system could each be described by:

$$[001]_t \ {}^{\Lambda}[010]_m \ = \ 1.18°, \quad \{100\}_t \ {}^{\Lambda}(100)_m \ = \ 1.14°$$

Discussion

The calculations presented above are quantitative analyses of possible crystallographic transformation mechanisms operating at $950°C$. Comparison of results based on different lattice parameters and values of principal strains indicates that such predictions are dependent on lattice parameters of both parent and product phases and that their values at the same transformation temperature should be used. Since lattice parameters are affected by solute content,[14] which also affects the M_s temperature, different mechanisms may thus occur under different conditions.

For all three lattice correspondences, the outstanding features of the results obtained with stacking fault or slip modes (apart from $\{010\}\langle101\rangle$ is the small ($<1°$) amount of LIS shear required. The predicted crystallography is therefore very similar for a number of possible LIS systems. Individual orientation relations therefore might only be resolved with precise transmission electron microscopy, since particular reflections from one variant would occur within 2 to 3° of equivalent reflections from other variants.

The shape strain may be resolved into components parallel and perpendicular to habit planes (Fig. 3).[15] From Table III it is seen that most of the shape strain (m_1) is resolved parallel to the habit plane. Table IV shows that slip LIS mechanisms produce shape strains of 11% to 17%, while the $(010)_{tet}$ twin system produced strains of 5% per variant.

As mentioned in the Introduction, there is some experimental evidence for the occurrence of both lattice correspondences B and C. There is no apparent reason why lattice correspondence A should be less probable. From Table V it is seen that the rotational component of the total mechanism is small in all instances. Hence $[axis]_t \parallel [axis]_m$ does imply that $[axis]_t$ becomes $[axis]_m$ in the transformation.

The $(010)_t$ twin system, by lattice correspondence B (Fig. 2) becomes $(001)_m$ or $(100)_m$ when equivalent a_t and b_t axes are interchanged. $(100)_m$ and $\{110\}_m$ deformation twinning was found by Bailey[5] in thin TEM specimens. Kriven[16] also found the same twin systems in small included ZrO_2 particles in Al_2O_3-50% pure ZrO_2 (42% volume fraction tetragonal) ceramics.

The calculations presented here, however, imply that the $(100)_m$ and $(001)_m$ can also act as lattice invariant shear systems in a martensitic mechanism, giving rise to a macroscopic shape change and habit plane. This finding may be relevant to understanding how stress-induced martensitic transformations in included zirconia particles are able to toughen composite ceramics.

The fact that $(100)_m$ may act as an LIS twin system further suggests that martensite calculations be performed for all the $(100)_m$ and $\{110\}_m$ twin systems in combination with the three lattice correspondences. Such calculations would require modification of the method used here.

Summary and Conclusion

We have quantitatively analyzed some martensitic transformation mechanisms in zirconia. The analyses presented were purely geometrical and were based on the invariant plane strain criterion of a martensitic transformation in a bulk crystal. Lattice parameters were used for both tetragonal and monoclinic phases to the same transformation temperature of 950 °C. Three principal strain axes were determined in the monoclinic unit cell, and the different tetragonal LIS systems were obtained from the monoclinic cell by inspection, with reference to three lattice correspondences. The computed martensite analyses then yielded solutions which predicted habit planes, shape strains related to the habit planes, and orientation relations for each variant. Most of the LIS slip systems had shape strains of the order of 11% to 17%, while one twin LIS system had 5% shape strains, resolved essentially parallel to the habit plane.

Thus, the work presented here has shown that the $(100)_m$ or $(001)_m$ twin systems may act as LIS systems, giving a martensitic solution with macroscopic shape changes and habit planes, etc., and it illustrates the different shape changes associated with each mechanism and variant.

Acknowledgments

The authors thank H. M. Ledbetter and C. M. Wayman for a copy of PRØGRAM MRTNST. S. W. Kennedy acknowledges a calculations grant from the Australian Research Grants Committee. W. M. Kriven thanks D. M. Huang and A. G. Evans, of the University of California at Berkeley, for valuable discussions during the preparation of this manuscript; it was prepared with the support of a U.S. Navy Grant, Contract No. 842456–25989.

Appendix I

The Determination of Directions and Magnitudes of Principal Distortions for the Tetragonal → Monoclinic Phase Transformation in Zirconia

Directions and magnitudes of the principal distortions associated with the three possible lattice correspondences between tetragonal and monoclinic zirconia have been determined by analytical geometry.

The problem is two-dimensional, as one principal axis is obtained directly by inspection (Fig. 2, $\eta_3//b_m$). The remaining two principal distortions must

86

lie in the plane of which the η_3 direction is the normal. The theory of finite homogeneous strain in two dimensions (e.g. Jaeger[13]) gives a simple method for the location of these two vectors:

The changes in the relevant tetragonal plane can be factorized into *(a)* a change of dimensions (expansion/contraction) and *(b)* a change of shape (simple shear). This is illustrated in Fig. 4.

Consider a point *(x,y)* in the tetragonal plane:

(i) $x \rightarrow \dfrac{x_m}{x_t} x$ $\qquad\qquad$ $y \rightarrow \dfrac{y_m \sin \beta}{y_t} y$ (by expansion)

$\quad (x \rightarrow ax)$ $\qquad\qquad\qquad$ $(y \rightarrow dy)$

(ii) $ax \rightarrow ax + \tan(90-\beta)dy$ $\qquad$ dy unchanged (by shear)

$\quad (ax \rightarrow ax + by)$ $\qquad\qquad\quad$ $(dy \rightarrow dy)$

Hence

$$x \rightarrow \frac{x_m}{x_t} x + \tan(90 - \beta)\frac{y_m \sin \beta}{y_t} y$$

$$(x \rightarrow ax + by)$$

and

$$y \rightarrow \frac{y_m \sin \beta}{y_t} y$$

$$(y \rightarrow dy)$$

From two-dimensional strain theory, for $x \rightarrow ax + by$ and $y \rightarrow cx + dy$, directions of the principal distortion axes are given by:

$$\tan 2\alpha = \frac{2(ab + cd)}{a^2 + c^2 - b^2 - d^2} = \frac{2ab}{a^2 - b^2 - d^2}$$

(since $c = 0$ in this case).

(α is the angle made with the x axis, and the required angles are α, $\alpha + \Pi/2$).

Magnitudes (A, B) are given by:

$(A + B)^2 = (a + d)^2 + (b - c)^2 = (a + d)^2 + b^2$

$(A - B)^2 = (a - d)^2 + (b + c)^2 = (a - d)^2 + b^2$

Directions of the principal distortion axes in the monoclinic phases can also be determined:

$$\tan 2\alpha' = \frac{2(ac + bd)}{a^2 + b^2 - c^2 - d^2} = \frac{2bd}{a^2 + b^2 - d^2}$$

(α', $\alpha' + \Pi/2$ are the angles formed by the principal distortion axes and the x axis).

Rotation of the principal axes during the deformation is therefore ($\alpha' - \alpha$).

These formulae can be applied to tetragonal $\rightleftharpoons$ monoclinic, orthorhombic $\rightleftharpoons$ monoclinic, and cubic $\rightleftharpoons$ monoclinic changes.

References

[1]E. C. Subbarao, H. S. Maiti, and K. K. Srivastava, *Phys. Status Solidi A*, **21**, 9–40 (1974).
[2]B. C. Weber, H. J. Garrett, F. A. Mauer, and M. A. Schwartz, *J. Am. Ceram. Soc.*, **39**

[6] 197–207 (1956).
  [3]R. C. Garvie and P. S. Nicholson, *ibid.*, **55** [3] 152–57 (1972).
  [4]R. C. Garvie, R. H. Hannink, and R. T. Pascoe, *Nature*, **258**, 703–704 (1975).
  [5]J. E. Bailey, *Proc. R. Soc. A*, **279**, 359–412 (1964).
  [6]G. M. Wolten, *Acta Crystallogr.*, **17**, 763–65 (1964).
  [7]G. K. Bansal and A. H. Heuer, *Acta Metall.*, **20**, 1281–89 (1972).
  [8]G. K. Bansal and A. H. Heuer, *ibid.*, **22**, 409–17 (1974).
  [9]J. S. Bowles and J. K. MacKenzie, I, *ibid.*, **2**, 129–37 (1954); II, *ibid.*, pp. 138–47; III,
ibid., pp. 224–34.
  [10]R. N. Patil and E. C. Subbarao, *J. Appl. Crystallogr.*, **2**, 281–88 (1969).
  [11]R. N. Patil and E. C. Subbarao, *Acta Crystallogr.*, **26**, 535–42 (1970).
  [12]H. M. Ledbetter and C. M. Wayman, *Mater. Sci. Eng.*, **7**, 151–57 (1971).
  [13]J. C. Jaeger, Elasticity, Fracture, and Flow. Methuen & Co., Ltd., London, 1964;
pp. 23–29.
  [14]R. H. J. Hannink, *J. Mater. Sci.*, **13**, 2487–96 (1978).
  [15]Z. Nishiyama; pp. 372–73 in Martensitic Transformation. Edited by M. E. Fine, M.
Meshii, and C. M. Wayman. Academic Press, New York, 1978.
  [16]W. M. Kriven; this issue, pp. 168–83.

  *This work was done as partial fulfillment for Ph.D. degrees by W. L. Fraser, in 1974, and W. M. Kriven, in 1976, at the University of Adelaide.
  †Now with the Department of Materials Science and Mineral Engineering, University of California, Berkeley, Calif. 94720. Currently on leave at the Max-Planck-Institut fur Metallforschung, Institut fur Werkstoffwissenschaften, Stuttgart, Germany.
  ‡Now with Kodak Research Laboratories, Coburg, 3058, Victoria, Australia.

Table I. Lattice parameters

Temp. (°C)	a_m	b_m	c_m	β	Temp. (°C)	a_t	c_t
956 (meas.)	0.51882	0.52142	0.53836	81.217	1152 (meas.)	0.51518	0.52724
950 (calc.)	.51881	.52142	.53835	81.22	950 (calc.)	.51485	.52692
Thermal expansion coeff. (nm × 10⁻⁶ C°⁻¹)	1.031	0.135	1.468			1.160	1.608

Table II. Principal distortions*

	LC A	LC B	LC C
η_1: Direction	[0, 0.8267, 0.5627]	[0, 0.7383, −0.6745]	[0.7860, −0.6183, 0]
Magnitude	1.0956	0.9337	0.9428
η_2: Direction	[0, −0.5627, 0.8267]	[0, 0.6745, 0.7383]	[0.6183, 0.7860, 0]
Magnitude	0.9287	1.0897	1.1045
η_3: Direction	[100]	[100]	[001]
Magnitude	1.0128	1.0128	0.9896
$\sum \eta_i^2 - 3 =$	0.0885	0.0850	0.0880

ΔV(increase) = $100 - |1 - \eta_1\eta_2\eta_3| = 3.047\%$.

*Indices referred to an orthonormal basis with axes parallel to the base axes of the tetragonal lattice.

Table III. Comparison of computed method of calculations with those of Bansal and Heuer (Ref. 8), using uncorrected lattice parameters at 1000°C. The shape strain is resolved into components parallel and perpendicular to habit planes (Ref. 15)

Lattice Correspondence	Shear System	Habit Plane	Direction d_1	Magnitude M_1	% Strain x	% Strain y
BANSAL & HEUER (SLIP):						
LC C	$(1\bar{1}0)[001]$	$\begin{pmatrix}-.62070\\-.775527\\-.115215\end{pmatrix}$ ∿$(\bar{6}71)$	$\begin{bmatrix}-.0345ᵗ\\-.01366\\-.99931\end{bmatrix}$	.12363	1.8	12.2
LC C	$(1\bar{1}0)[1\bar{1}0]$	$\begin{pmatrix}.996597\\-.014290\\.081068\end{pmatrix}$ ∿(100)	$\begin{bmatrix}.02766\\.10149\\.99445\end{bmatrix}$	.16248	1.73	16.1
FRASER, KRIVEN & KENNEDY (SLIP)						
LC C	$(1\bar{1}0)[001]$ (a)	$\begin{pmatrix}-.802\\-.586\\-.118\end{pmatrix}$ ∿$(\bar{7}51)$	$\begin{bmatrix}-.168\\.123\\-.955\end{bmatrix}$ ∿$[\bar{1}18]$	.110	1.9	10.5
	(b)	$\begin{pmatrix}-.207\\.090\\-.998\end{pmatrix}$ ∿$(\bar{1}05)$	$\begin{bmatrix}-.799\\.598\\-.061\end{bmatrix}$ ∿$[1\bar{1}0]$	.110	1.9	10.5
LC C	$(1\bar{1}0)[1\bar{1}0]$ (a)	$\begin{pmatrix}-.014\\.997\\.084\end{pmatrix}$ ∿(010)	$\begin{bmatrix}.002\\.043\\.975\end{bmatrix}$ ∿$[001]$	.153	1.9	14.8
	(b)	$\begin{pmatrix}.001\\.113\\1.017\end{pmatrix}$ ∿(019)	$\begin{bmatrix}.015\\1.000\\.007\end{bmatrix}$ ∿$[010]$	.153	1.87	15.1

Lattice Correspondence	Shear System	Habit Plane	Direction d_1	Magnitude M_1	% Strain X	Y
FRASER, KENNEDY & KRIVEN (TWINNING):						
LC $\mathcal{B}$ at 1000°C	(010)[$\overline{1}$00]	$\begin{pmatrix} \overline{.}483 \\ -.721 \\ -.509 \end{pmatrix}$ $\sim(3\overline{2}\overline{2})$	$\begin{bmatrix} .478 \\ -.714 \\ .499 \end{bmatrix}$ $\sim[\overline{3}22]$	.039	1.9	3.37
at 950°C	(010)[$\overline{1}$00]	$\begin{pmatrix} -.383 \\ -.810 \\ .455 \end{pmatrix}$ $\sim(\overline{1}\overline{2}1)$	$\begin{bmatrix} -.379 \\ -.801 \\ -.452 \end{bmatrix}$ $\sim[\overline{1}\overline{2}1]$	.052		

Table IV. Predictions of habit planes and shape strains

LATTICE INVARIANT SHEAR SYSTEM	HABIT PLANE INDICES (P1)	DIRECTION OF SHAPE STRAIN D1	MAGNITUDE OF SHAPE STRAIN M1

Lattice Correspondence: A

LATTICE INVARIANT SHEAR SYSTEM	HABIT PLANE INDICES (P1)	DIRECTION OF SHAPE STRAIN D1	MAGNITUDE OF SHAPE STRAIN M1
(010) $[\bar{1}0\bar{1}]$	$\begin{pmatrix} 0.162 \\ -0.971 \\ 0.181 \end{pmatrix}$ $\sim(1\bar{6}1)$	$\begin{matrix} 0.638 \\ -0.281 \\ 0.701 \end{matrix}$ $\sim[2\bar{1}2]$	0.122
	$\begin{pmatrix} 0.637 \\ -0.335 \\ 0.711 \end{pmatrix}$ $\sim(2\bar{1}2)$	$\begin{matrix} 0.126 \\ -0.967 \\ 0.217 \end{matrix}$ $\sim[\bar{1}8\bar{2}]$	0.122
(110) $[\bar{1}1\bar{0}]$	$\begin{pmatrix} -0.012 \\ -0.996 \\ 0.094 \end{pmatrix}$ $\sim(0\bar{1}0)$	$\begin{matrix} -0.002 \\ -0.266 \\ -0.942 \end{matrix}$ $\sim[0\bar{2}\bar{7}]$	0.173
	$\begin{pmatrix} -0.003 \\ -0.346 \\ -0.960 \end{pmatrix}$ $\sim(0\bar{1}3)$	$\begin{matrix} -0.012 \\ -0.984 \\ 0.171 \end{matrix}$ $\sim[0\bar{6}\bar{1}]$	0.173
(101) $[\bar{1}0\bar{1}]$	$\begin{pmatrix} 0.012 \\ 0.280 \\ 0.982 \end{pmatrix}$ $\sim(027)$	$\begin{matrix} 0.000 \\ 0.995 \\ -0.094 \end{matrix}$ $\sim[01\bar{0}]$	0.164
	$\begin{pmatrix} -0.001 \\ -1.000 \\ 0.018 \end{pmatrix}$ $\sim(0\bar{1}0)$	$\begin{matrix} -0.012 \\ -0.203 \\ 0.957 \end{matrix}$ $\sim[0\bar{1}5]$	0.164

Lattice Invariant Shear System	Habit Plane Indices (P1)	Direction of Shape Strain D1	Magnitude of Shape Strain M1
$(\bar{1}11)$ $[\bar{1}10]$	$\begin{pmatrix} -0.057 \\ -0.996 \\ 0.075 \end{pmatrix}$ $\sim(0\bar{1}0)$	$\begin{array}{c} 0.031 \\ -0.238 \\ 0.948 \end{array}$ $\sim[0\bar{1}\bar{4}]$	0.185
	$\begin{pmatrix} 0.025 \\ -0.324 \\ -0.968 \end{pmatrix}$ $\sim(0\bar{1}\bar{3})$	$\begin{array}{c} -0.061 \\ -0.985 \\ 0.159 \end{array}$ $\sim[0\bar{6}\bar{1}]$	0.185
	$\begin{pmatrix} 0.661 \\ 0.354 \\ -0.678 \end{pmatrix}$ $\sim(21\bar{2})$	$\begin{array}{c} 0.796 \\ -0.557 \\ 0.232 \end{array}$ $\sim[8\bar{5}2]$	0.178
	$\begin{pmatrix} 0.839 \\ -0.515 \\ 0.180 \end{pmatrix}$ $\sim(5\bar{3}1)$	$\begin{array}{c} 0.598 \\ 0.407 \\ -0.674 \end{array}$ $\sim[32\bar{2}]$	0.178

Lattice Correspondence: B

Lattice Invariant Shear System	Habit Plane Indices (P1)	Direction of Shape Strain D1	Magnitude of Shape Strain M1
(010) $[\bar{1}0\bar{1}]$	$\begin{pmatrix} -0.615 \\ -0.334 \\ 0.730 \end{pmatrix}$ $\sim(\bar{2}\bar{1}2)$	$\begin{bmatrix} -0.067 \\ -0.984 \\ -0.160 \end{bmatrix}$ $\sim[0\bar{6}\bar{1}]$	0.120
	$\begin{pmatrix} -0.103 \\ -0.988 \\ 0.122 \end{pmatrix}$ $\sim(\bar{1}9 1)$	$\begin{bmatrix} -0.620 \\ -0.281 \\ 0.716 \end{bmatrix}$ $\sim[\bar{6}\bar{3}7]$	0.120
(011) $[01\bar{1}]$	$\begin{pmatrix} -0.049 \\ -0.999 \\ 0.011 \end{pmatrix}$ $\sim(0\bar{1}0)$	$\begin{bmatrix} -0.990 \\ -0.144 \\ -0.000 \end{bmatrix}$ $\sim[\bar{7}\bar{1}0]$	0.159
	$\begin{pmatrix} -0.976 \\ -0.219 \\ 0.001 \end{pmatrix}$ $\sim(\bar{5}\bar{1}0)$	$\begin{bmatrix} 0.028 \\ -1.000 \\ 0.010 \end{bmatrix}$ $\sim[0\bar{1}0]$	0.159

Lattice Invariant Shear System	Habit Plane Indices (P1)	Direction of Shape Strain D1	Magnitude of Shape Strain M1
(101) $[\bar{1}01]$	$\begin{pmatrix} 0.018 \\ -1.000 \\ -0.002 \end{pmatrix}$ $\sim (0\bar{1}0)$	$\begin{bmatrix} -0.979 \\ -0.203 \\ 0.011 \end{bmatrix}$ $\sim [\bar{5}\bar{1}0]$	0.164
	$\begin{pmatrix} 0.960 \\ 0.280 \\ -0.011 \end{pmatrix}$ $\sim (720)$	$\begin{bmatrix} -0.097 \\ 0.995 \\ 0.003 \end{bmatrix}$ $\sim [010]$	0.164
($\bar{1}$11) $[01\bar{1}]$	$\begin{pmatrix} 0.056 \\ 0.996 \\ -0.075 \end{pmatrix}$ $\sim (010)$	$\begin{bmatrix} 0.992 \\ 0.126 \\ 0.018 \end{bmatrix}$ $\sim [810]$	0.170
	$\begin{pmatrix} -0.978 \\ -0.206 \\ -0.013 \end{pmatrix}$ $\sim (\bar{5}\bar{1}0)$	$\begin{bmatrix} 0.026 \\ -0.997 \\ 0.074 \end{bmatrix}$ $\sim [0\bar{1}0]$	0.170
(010) $[\bar{1}00]$ Twin System	$\begin{pmatrix} -0.383 \\ -0.810 \\ 0.455 \end{pmatrix}$ $\sim (\bar{1}2\bar{1})$	$\begin{bmatrix} -0.379 \\ -0.801 \\ -0.452 \end{bmatrix}$ $\sim [\bar{1}2\bar{1}]$	0.052
	$\begin{pmatrix} -0.383 \\ -0.810 \\ -0.455 \end{pmatrix}$ $\sim (\bar{1}2\bar{1})$	$\begin{bmatrix} -0.379 \\ -0.801 \\ -0.452 \end{bmatrix}$ $\sim [\bar{1}2\bar{1}]$	0.052

Lattice Correspondence: C

Lattice Invariant Shear System	Habit Plane Indices (P1)	Direction of Shape Strain D1	Magnitude of Shape Strain M1
(110) $[00\bar{1}]$	$\begin{pmatrix} -0.178 \\ -0.047 \\ 1.006 \end{pmatrix}$ $\sim (\bar{1}06)$	$\begin{bmatrix} -0.859 \\ 0.494 \\ 0.134 \end{bmatrix}$ $\sim [\bar{7}41]$	0.115
	$\begin{pmatrix} -0.855 \\ 0.483 \\ 0.195 \end{pmatrix}$ $\sim (\bar{9}52)$	$\begin{bmatrix} -0.132 \\ -0.076 \\ 0.966 \end{bmatrix}$ $\sim [\bar{1}07]$	0.115

Lattice Invariant Shear System	Habit Plane Indices (P1)	Direction of Shape Strain D1	Magnitude of Shape Strain M1
(010) [$\bar{1}$00] (cont.)	$\begin{pmatrix} 0.383 \\ -0.810 \\ -0.455 \end{pmatrix}$ $\sim(1\bar{2}\bar{1})$	$\begin{matrix} 0.379 \\ -0.801 \\ 0.452 \end{matrix}$ $\sim[1\bar{2}1]$	0.052
	$\begin{pmatrix} 0.383 \\ -0.810 \\ 0.455 \end{pmatrix}$ $\sim(1\bar{2}1)$	$\begin{matrix} 0.379 \\ -0.801 \\ -0.452 \end{matrix}$ $\sim[1\bar{2}\bar{1}]$	0.052
(110) [$\bar{1}$10]	$\begin{pmatrix} 0.000 \\ -0.132 \\ -1.015 \end{pmatrix}$ $\sim(0\bar{1}\bar{8})$	$\begin{bmatrix} -0.013 \\ -0.998 \\ 0.063 \end{bmatrix}$ $\sim[0\bar{1}\underline{0}]$	0.156
	$\begin{pmatrix} -0.013 \\ -0.990 \\ -0.142 \end{pmatrix}$ $\sim(0\bar{7}\bar{1})$	$\begin{bmatrix} 0.000 \\ -0.057 \\ -0.976 \end{bmatrix}$ $\sim[00\bar{1}]$	0.156
(101) [$\bar{1}$0$\bar{1}$]	$\begin{pmatrix} -0.013 \\ -0.049 \\ 1.022 \end{pmatrix}$ $\sim(001)$	$\begin{bmatrix} -0.004 \\ -0.990 \\ -0.140 \end{bmatrix}$ $\sim[0\bar{7}\bar{1}]$	0.159
	$\begin{pmatrix} -0.005 \\ -0.976 \\ -0.224 \end{pmatrix}$ $\sim(0\bar{9}\bar{2})$	$\begin{bmatrix} -0.013 \\ 0.028 \\ -0.977 \end{bmatrix}$ $\sim[00\bar{1}]$	0.159
($\bar{1}$11) [110]	$\begin{pmatrix} -0.062 \\ -0.988 \\ -0.144 \end{pmatrix}$ $\sim(0\bar{7}\bar{1})$	$\begin{bmatrix} 0.014 \\ -0.041 \\ -0.976 \end{bmatrix}$ $\sim[00\bar{1}]$	0.169
	$\begin{pmatrix} 0.009 \\ -0.122 \\ -1.016 \end{pmatrix}$ $\sim(0\bar{1}\bar{8})$	$\begin{bmatrix} -0.064 \\ -0.996 \\ -0.058 \end{bmatrix}$ $\sim[0\bar{1}0]$	0.169

Table V. Details of predicted orientation relations (angles are between planes and directions marked X)

System	Solution	Tet. Planes {100}	Tet. Planes (001)	Mon. Planes (100)	Mon. Planes (010)	Mon. Planes (001)	Angle (°)	Tet. Dir. <100>	Tet. Dir. [001̄]	Mon. Dir. [100]	Mon. Dir. [010]	Mon. Dir. [001̄]	Angle (°)
Lattice Correspondence A													
(010) [101̄]	1 (a)	X				X	0.47	X				X	0.86
	1 (b)	X				X	6.40	X				X	4.37
(110) [11̄0]	1 (a)	X				X	0.76	X				X	0.72
	1 (b)	X				X	0.57	X				X	0.74
(101) [1̄01̄]	1 (a)		X	X			0.77	X				X	0.73
	1 (b)		X	X			0.00	X				X	0.72
(1̄11) [11̄0]	1 (a)	X				X	1.32	X				X	1.16
	1 (b)	X		X			0.53	X				X	0.51
	2 (a)	X				X	11.10	X				X	7.88
	2 (b)	X				X	5.47	X				X	5.78
Lattice Correspondence B													
(010) [1̄00] (Twin Shear)	1 (a)	X				X	1.40		X		X		1.18
	1 (b)	X				X	1.40		X		X		1.18
	2 (a)	X				X	1.40		X		X		1.18
	2 (b)	X				X	1.40		X		X		1.18
(010) [101̄]	1 (a)	X		X			4.61		X		X		4.91
	1 (b)	X		X			0.31		X		X		0.56
(011) [011̄]	1 (a)	X			X		0.60		X		X		0.59
	1 (b)	X			X		0.05		X		X		0.61
(101) [1̄01̄]	1 (a)	X		X			0.03		X		X		0.61
	1 (b)	X		X			0.64		X		X		0.59
(1̄11) [011̄]	1 (a)	X			X		0.74		X		X		0.79
	1 (b)	X			X		0.05		X		X		0.73
Lattice Correspondence C													
(110) [001]	1 (a)	X		X			3.15	X			X		5.50
	1 (b)	X		X			0.43	X			X		0.45
(110) [11̄0]	1 (a)		X			X	0.07	X			X		0.74
	1 (b)		X	X			0.72	X			X		0.72
(101) [1̄01]	1 (a)	X		X			0.74	X			X		0.73
	1 (b)		X			X	0.05	X			X		0.72
(1̄11) [110]	1 (a)	X		X			1.00	X			X		1.48
	1 (b)	X				X	0.07	X			X		0.65

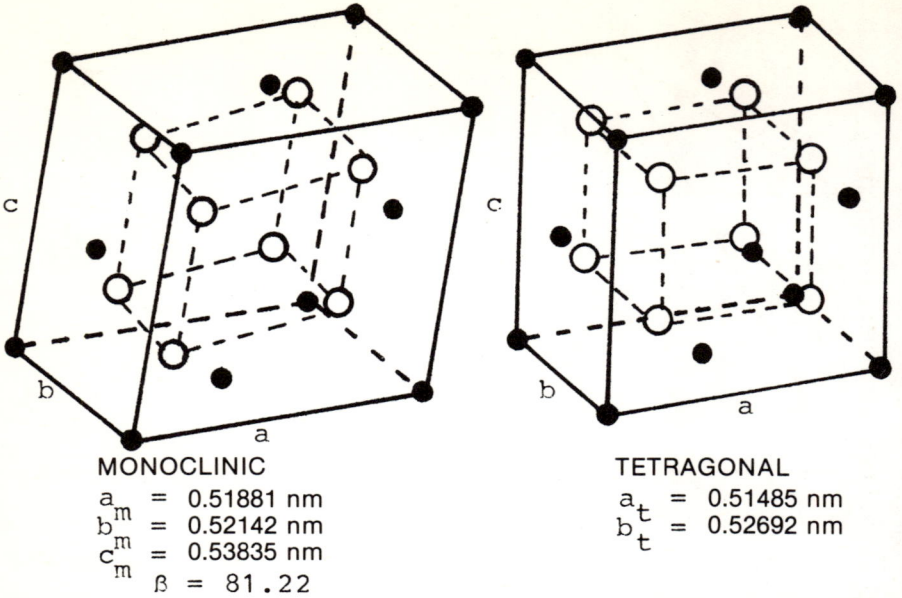

Fig. 1. Crystal structures of monoclinic and tetragonal zirconia with lattice parameters at 950 °C for both phases. Monoclinic twin planes (Ref. 5) are (100), (110), and (1$\bar{1}$0); slip occurs on (010).

MONOCLINIC

a_m = 0.51881 nm
b_m = 0.52142 nm
c_m = 0.53835 nm
β = 81.22

TETRAGONAL

a_t = 0.51485 nm
b_t = 0.52692 nm

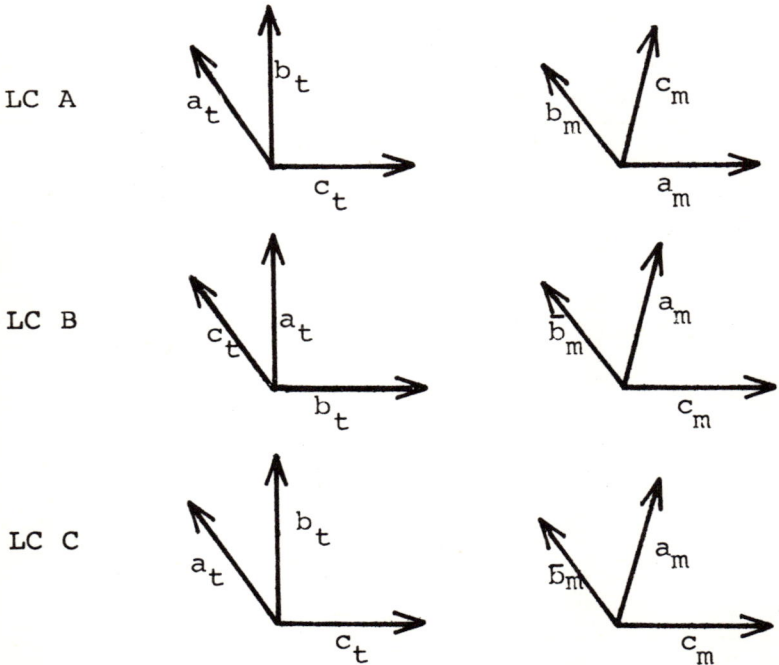

Fig. 2. Lattice correspondences for zirconia.

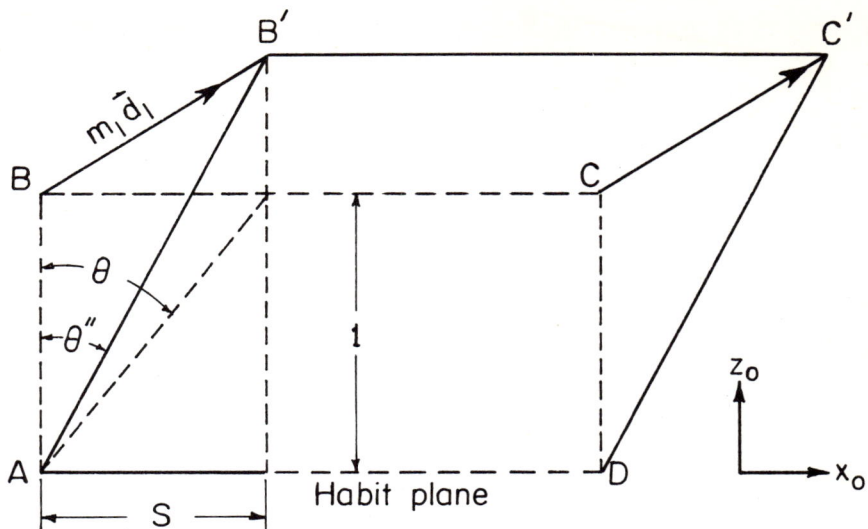

Fig. 3. The relation between shape deformation, angle of shear, and habit plane.

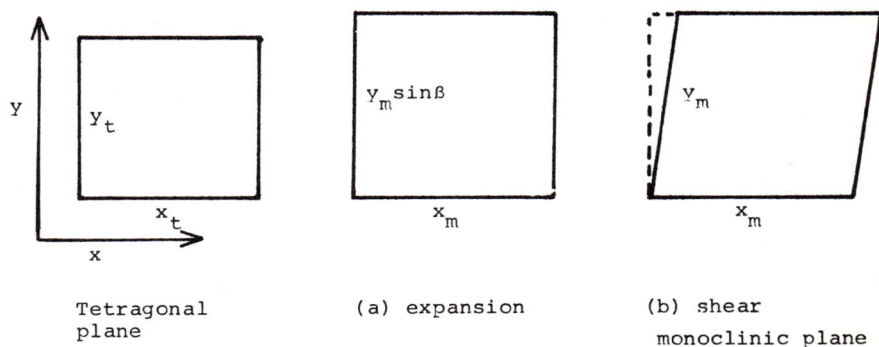

Tetragonal plane

(a) expansion

(b) shear

monoclinic plane

Fig. 4. The determination of magnitudes and directions of principal strains by a coordinate geometry method.

Alloy design in partially stabilized zirconia

A. H. HEUER

Department of Metallurgy and Materials Science
Case Institute of Technology
Case Western Reserve University
Cleveland, Ohio 44106

The principles of alloy design in *transformation-toughened* partially stabilized ZrO_2 are reviewed. Optimum microstructures are those that contain a homogeneous distribution of coherent tetragonal ZrO_2 particles in a cubic ZrO_2 matrix. The transformation toughening arises from a stress-induced martensitic transformation in these particles to monoclinic symmetry in crack tip stress fields. The energetics of such constrained transformations are discussed in the framework of absolute reaction rate theory.

Interest in strong and tough ZrO_2-containing ceramics has increased markedly since it was discovered[1-4] that ceramics containing fine tetragonal ZrO_2 particles are tougher and stronger than either single-phase ceramics or ceramics containing monoclinic ZrO_2 particles. The mechanism responsible for the superior properties, at least in the case of optimally-fabricated partially-stabilized zirconias (PSZ's), is the stress-induced martensitic transformation of these tetragonal particles to monoclinic symmetry in the stress field of a propagating crack.[2,3,5] Optimal PSZ's are two-phase ceramics in which the major phase is a *matrix* of solute-rich cubic stabilized zirconia which contains fine *precipitates* of a lower-solute-content tetragonal zirconia phase. In this paper, we confine our attention to Mg-PSZ because (1) the principles of alloy design necessary to optimize both the microstructure and mechanical properties emerge particularly clearly in the system, and (2) other authors in these Proceedings discuss Ca-PSZ[6] and Y-PSZ.[7-9]

The second section of this paper briefly reviews the evidence that the tetragonal → monoclinic transformation in ZrO_2 occurs martensitically and indicates how this martensitic transformation can be controlled and exploited to improve the strength of ZrO_2-containing ceramics. The third section reviews the studies on Mg-PSZ, particularly how optimum microstructures can be obtained in this system. The fourth section contains a thermodynamic formalism by which the several important transformation characteristics—particularly the dependence of transformation temperature on particle size—can be rationalized. The fifth section discusses the physics of the *stress-induced* transformation in zirconia-based ceramics which controls the transformation toughening and which can also be related to the thermodynamic formalism.

Martensitic Transformation in Zirconia

The tetragonal → monoclinic reversible transformation in ZrO_2 was first reported by Ruff and Ebert in 1929,[10] and the phase transformation has often

98

been referred to as "disruptive" in the ceramic literature because of the cracking that generally accompanies the $\approx 3\%$ volume increase on transformation to monoclinic symmetry. Murray and Allison[11] (among others) studied the large hysteresis in the transformation temperature on heating and cooling. However, the seminal work in this field occurred in the period 1963-1965, when Wolton[12] suggested that the transformation was martensitic, Bailey[13] studied the transformation *in-situ* in the electron microscope using foils prepared by oxidizing zirconium metal and recognized the shear-like nature of the transformation, and Fehrenbacher and Jacobson[14] noted surface striations by hot stage metallography for the reverse (monoclinic → tetragonal) transformation. The most persuasive experimental evidence for the martensitic nature of the forward (tetragonal → monoclinic) transformation was provided by Bansal and Heuer,[15] who obtained single crystals of monoclinic symmetry (which had been grown from a fluxed-melt system at temperatures within the monoclinic stability field[16]) and cycled these crystals through the transformation. The vicinal surfaces on these crystals after one cycle of transformation (heating and cooling) showed extensive surface rumpling, as shown in Fig. 1. Such surface upheaval is the best experimental evidence of the occurrence of a martensitic transformation.[17] Electron microscopic study by these authors[15] showed that two habit plane orientations occurred. In principle, the two habit planes could correspond to two crystallographically-identical variants arising from a single martensitic transformation mechanism or to two different transformation mechanisms. In addition, surface cracking due to the anisotropic volume expansion on transformation can be seen in Fig. 1.

Further study by Bansal and Heuer[15] involving high temperature single crystal X-ray diffraction showed that two different transformation mechanisms are operating, the two products being shown in Fig. 2. The lenticular plates arising from one of these mechanisms, Fig. 2(*a*), have $(671)_m$ or $(761)_m$ habit planes (the subscripts m and t hereafter refer to the monoclinic and tetragonal phases, respectively), and application of the crystallographic theory of Wechsler *et al.*[18] by Bansal and Heuer[15] showed that this martensite

product could have arisen from the $\begin{bmatrix} 100 \\ 010 \\ 001 \end{bmatrix}$ Bain correspondence matrix and

a $(1\bar{1}0)_t$ $[001]_t$ lattice invariant shear deformation. The second type of martensite product is also in the form of plates and is shown in Fig. 2(*b*). The habit plane of the midribs of these plates is close to $(100)_m$.[15] Although the plates in Fig. 2(*b*) are finely twinned on $(100)_m$, similar untwinned plates of this orientation have been observed (see Fig. 4(*b*) of Ref. 15 (Part I)). Application of the crystallographic theory showed that the plates of Fig. 2(*b*) could have arisen from the same Bain correspondence matrix but with $(1\bar{1}0)_t$ $[110]_t$ slip as the lattice invariant deformation.

The cracking already mentioned due to the anisotropic volume expansion has prevented pure zirconia ceramics from being used in other than powder form. As one solution to this problem, it has been known for more than 50 years[10] that additions of MgO, CaO, Y_2O_3, or any of the rare earth oxides will "stabilize" the high temperature fluorite structure of cubic zirconia. As shown by the MgO-rich portion of the MgO-ZrO_2 phase diagram of Fig. 3,[19] the solutes actually lower the temperature of the cubic→tetragonal transformation and most of these "stabilized" zirconias are metastable at room temperature—eutectoid decomposition should occur on cooling from high temperatures. In the majority of cases, however, rapid cooling can suppress the sluggish eutectoid decomposition reaction and such "fully stabilized" zirconias find wide application in electrolytic devices.

While the electrolytic properties of such stabilized zirconias are quite good, and the ceramics can be used over a wide temperature range where eutectoid decomposition is not a problem, the mechanical properties of these stabilized zirconias leave much to be desired. In particular, the thermal conductivity is low and the thermal expansion coefficient high, and thus these materials have exceedingly poor thermal shock resistance. "Partially-stabilized" zirconias—two-phase materials containing insufficient solute to yield the single phase cubic solid solution—have found wide acceptance. Those PSZ's available prior to 1975 were a mixture of cubic and monoclinic ZrO_2 at room temperature (they probably also contained a minor amount of tetragonal ZrO_2). As shown in Fig. 4, a considerable portion of the improved thermal shock resistance of these PSZ's arises from their lower overall thermal expansion coefficient, a result of the expansion attendant on the transformation to monoclinic symmetry in the tetragonal phase of these materials.

However, the very best PSZ's are those in which the martensitic tetragonal→monoclinic transformation of the tetragonal component has been suppressed altogether and is achieved by keeping this component as very small particles. A typical microstructure is shown in Fig. 5 (the microstructural evolution leading to such materials is discussed in the next section). In these materials, the improved properties arise from an increase in the fracture toughness by a factor of 2 or more.[3] This increase is due to the occurrence of a stress-induced martensitic transformation in regions adjacent to the crack tip, due to the high stress fields existing in these regions. Experimental evidence for this phenomenon is shown in Fig. 6, where it is seen that a transformation "zone" of ≈ 0.5 μm is formed adjacent to the propagating crack.[21] As discussed by Porter et al.,[3] the transformed inclusion theory of Eshelby[22] can provide good qualitative agreement on the magnitude of the increased fracture toughness due to transformation toughening.

If the particles in these two-phase ceramics are larger than a certain critical size, which in Mg-PSZ is ≈ 0.2 μm (and in Ca-PSZ ≈ 0.1 μm[6]), they transform to monoclinic symmetry on cooling through a critical temperature, typically between 500 and 1000 °C. These two-phase (monoclinic + cubic) materials are still stronger and tougher than the single-phase solid solution,[23] although in this case it appears that the improved properties (relative to the single-phase solid solution) arise from microcracking adjacent to the particles in a zone ahead of propagating cracks, as has been discussed by Rice[24] and Porter and Heuer.[2,25]

Microstructural Evolution in Magnesia Partially-Stabilized Zirconia

Commercial Mg-PSZ's have a complex microstructure and contain cubic zirconia matrix grains, very fine intragranular tetragonal particles, and monoclinic zirconia particles with two morphologies—large, coarsely twinned grain boundary particles and finer but still twinned intragranular precipitates. A typical microstructural characterization of one such material, containing 8.1 mol% MgO and taken from the work of Porter and Heuer,[23] is shown in Fig. 7. The "optimal" microstructure, shown in Fig. 5, can be produced in this material by heating in the single-phase cubic field, $\approx 1850\,°C$ for the composition of Fig. 7 (see Fig. 3), for a long enough time to dissolve the minor phases in the cubic matrix. The homogeneous solid solution so formed is then cooled as rapidly as possible (consistent with avoiding thermal stress induced fracture) through the two-phase field and below the eutectoid reaction temperature. Because of the low thermal conductivity of zirconia, the quenching rate must be limited, and it is not possible to suppress precipitation completely; rather, fine, homogeneously-nucleated precipitates form during quenching, as shown in Fig. 8. These materials, when heated in the cubic (ss)+tetragonal (ss) two-phase field, for example, at $1400\,°C$ for 2-4 h, develop the microstructure shown in Fig. 5, which has been termed "optimally-aged" by Porter and Heuer.[23]

Well-developed tetragonal precipitates in this system have the shape of oblate spheroids[21] and are shown in good diffracting conditions in Fig. 9(a). The elliptical particles present have been sectioned in the process of making the thin foil; the c axis indicated in the figure is the minor axis of revolution of the spheroids, and the particle shapes visible are cross sections perpendicular to the particle diameters. One elliptical precipitate variant is readily visible in Fig. 9(a), but the other two precipitate variants are out of contrast (the foil plane in this figure is nearly parallel to a [100] zone axis orientation); the arrowed features are believed to be intragranular pores.[21] Figure 9(b) is a convergent beam electron diffraction pattern from one of the elliptical particles; the lack of four-fold symmetry shows that the c axis lies in the foil plane.

As shown by the conventional selected area diffraction pattern of Fig. 9(c), the c axis of tetragonal ZrO_2 is considerably larger than the a axis of the cubic matrix ($c_{tet} = 1.018\ a_{tet}$), whereas a_{tet} is almost identical to a_{cubic}. The precipitate shapes—thin oblate spheroids, with the c axis of the precipitates parallel to the shortest dimension of the particle—arise from minimization of lattice strain. The critical particle size, at which particles are still tetragonal on cooling to room temperature, is $\approx 0.2\ \mu m$.[2,21,23]

For longer aging times at $1400\,°C$, during which precipitate coarsening occurs, the particles have monoclinic symmetry after cooling to room temperature and are twinned on at least two systems: "midrib" twins parallel to $(100)_m$ and "cross" twins, parallel to $(110)_m$ (A and B respectively in Fig. 10). The factors which determine which type of twins form on cooling from elevated temperatures have not yet been determined and are the subject of continuing study.

As will be already apparent, whether the low-solute tetragonal ZrO_2 precipitate particles retain their tetragonal symmetry on cooling to room temperature or transform to monoclinic symmetry at some temperature $\leq 1000\,°C$ is crucial in determining mechanical behavior. In metallurgical jargon, the temperature at which a martensite reaction first occurs is known as M_s, and the particle size dependence of M_s is of paramount importance.

Characteristics and Thermodynamics of Martensitic Transformations in Constrained Particles

For present purposes, the three important characteristics of martensitic transformations which dominate the transformation behavior are that they are (1) diffusionless, (2) athermal, and (3) involve a shape deformation. The diffusionless character of the transformation means that atoms in the parent and product phases have the same near neighbors, albeit differently arranged. As no long-range diffusion or other mass transport is involved, such transformations can occur at quite low temperatures.

An athermal transformation is one in which the extent of transformation at any one temperature depends only on that temperature and is independent of time. Typically, athermal transformations are those in which the onset of transformation causes an increase in strain energy, which therefore opposes the reduction in volume free energy causing the transformation in the first place. At constant temperature, therefore, the reaction ceases when it is partially complete, and further transformation requires a further increase in driving force, i.e. a greater supercooling.

Finally, and most importantly, the transformation involves a shape deformation, the strain of which cannot be relieved by long-range diffusion (see (1) above). The surface upheaval shown in Fig. 1 is very good evidence that the transformation involves an invariant plane strain, and the theory worked out for such invariant plane strain martensitic transformations yields the important result that the shape deformation of a macroscopic crystal is different from that of the unit cell. The thermodynamics of such transformations can be usefully illustrated with the free energy-reaction coordinate diagrams introduced by Eyring and his colleagues in their development of absolute reaction rate theory.[26] While the latter theory mostly involves thermally activated reactions, no loss of generality accrues by using this formalism in the present context.

We consider the Helmholtz free energy F and, in the first instance, an isolated particle. As shown in Fig. 11, the change in the Helmholtz free energy on transformation can be represented by Eqs. (1)–(3) in the inset, and the difference between the Gibbs and Helmholtz free energies for an isolated particle at atmospheric pressure is essentially nil. At the temperature for which Fig. 11 is appropriate (e.g. room temperature), the monoclinic phase is stable, and the net decrease in Helmholtz free energy of transformation from tetragonal to monoclinic symmetry is given by ΔF. As in most phase transformations, an activation barrier, of magnitude ΔF^*, exists between the parent (tetragonal) and product (monoclinic) states. (Strictly, ΔF^* is the energy of the activated state the tetragonal phase must pass through during the phase transformation.) If classical nucleation theory applies, ΔF^* will be proportional to $(1/\Delta F)^2$ and the transformation will occur at appreciable rates only at that temperature T where $\Delta F^* \approx kT$, where k is the Boltzmann constant.

The transformation energetics for this same particle at the same temperature T but now contained inside a rigid matrix are shown in Fig. 12. At this temperature, no transformation will occur, because T is greater than M_s. Although there is still a large chemical driving force for the transformation, which is represented by ΔF_{chem} in Fig. 12, the free energy of the monoclinic particle formed by transformation inside a rigid matrix is in fact greater than that of the untransformed tetragonal particle, and there is no net driving force for the transformation. The difference in energy between an isolated monoclinic particle and one that has formed by transformation in a matrix is given by $\frac{1}{2}P_{ij}^I e_{ij}^T$, where e_{ij}^T is the transformation strain tensor of an unconstrained particle[22] and P^I_{ij} is the stress tensor within the particle due to the constrained transformation.[22]

It must also be recognized that the free energy of a constrained tetragonal particle will not necessarily be the same at temperature T as that of an isolated particle because of coherency strains that may occur upon precipitation, thermal expansion mismatch between particle and matrix, etc. As T is reduced, spontaneous transformation will first occur at a particular supercooling which now corresponds to the M_s temperature. At this temperature, ΔF_{chem} is sufficiently larger than the strain energy caused by transformation, $\frac{1}{2}P^I_{ij}e^T_{ij}$, that the martensitic reaction can occur spontaneously (Fig. 13).

As already mentioned, there exists in the PSZ's studied to date a critical size below which particles will not transform martensitically and above which they will. This critical size problem can be addressed using the free energy diagrams. Consider Fig. 13, which can now be interpreted as describing the situation where small particles will not transform to monoclinic symmetry but large particles will. For the large particles to transform, changes have had to occur in either the strain energy term $\frac{1}{2}P^I_{ij}e^T_{ij}$, the nucleation barrier ΔF^*, or the driving force for transformation, ΔF_{chem}. Evans et al.[27] have argued that it is the P^I_{ij} term which is important. Specifically, they suggest that the martensitic product is always subdivided into twins or crystallographically equivalent martensite product variants in such a way that the shear strain of neighborin gregions is of opposite sign and thus no long-range shear strain fields exist in the matrix. However, there will be a much shorter range strain field at the interface between particle and matrix, which is localized at the site where the twins intercept the interface. This strain energy is much smaller than the strain energy that would result if the transformed particle were a single crystal and furthermore scales with the size of the particle. On this argument, it is only large particles—those in which the chemical free energy on transformation is larger than the strain energy caused by the transformation—that will transform.

Lange and Green[28] argued that the size dependence of the transformation temperature may reside in three surface area/surface energy terms:

(1) The volume change on transformation requires that the surface area change; furthermore, tetragonal and monoclinic ZrO_2 can have different interfacial energies with the matrix in which they are included.

(2) Appreciable twin boundary energy results from the inevitable twinning that accompanies transformation (Fig. 10).

(3) Considerable surface area will be associated with the (postulated) microcracks that accompany transformation.

There is no direct evidence for the exact cancellation of long-range stresses required by the Evans et al.[27] model, and evidence is available in at least one case* that the model does not apply. Likewise, calculations[29] based on Lange's model yielded a critical size for room temperature transformation a factor of 20 smaller than experiment. It thus appears that neither surface energy terms, which would alter the magnitude of ΔF_{chem}, or the Evans et al. concept of minimizing the strain energy term, $\frac{1}{2}P^I_{ij}e^T_{ij}$, controls the size dependence of the transformation temperature, and attention must therefore be focused on other effects, e.g. on changes in either of the terms ΔF^* or ΔF_{chem} to explain the particle size dependence.

The δ fringes present in Fig. 9(a) show that tetragonal zirconia particles in Mg-PSZ are coherent with the cubic matrix. As is well known from the metallurgical literature, precipitate growth will eventually cause loss of coherency at a critical particle size, the interface then becoming either semicoherent or incoherent. It is likely that martensite nucleation in the coherent particles shown in Figs. 5 and 9 will occur with difficulty, and it is only the dislocations introduced when the interface becomes semicoherent or incoherent that permit the martensite nucleus to form at M_s. In other words, the dislocations introduced by the loss of coherency cause a reduction in ΔF^*, the activation free energy of nucleation.

The coherency of the particles shown in Fig. 9 may also cause a size-dependent ΔF_{chem}. The phase diagram of Fig. 3 is an incoherent diagram, and no lattice strain should be present between the tetragonal and cubic phases. The coherent diagram (appropriate for coherent precipitates) will have solvuses displaced from those shown in Fig. 3, and the solute content of coherent particles will be different from that of incoherent particles. This could give rise to a chemical driving force, ΔF_{chem}, which is different for different sized particles, and therefore to a size-dependent M_s. We have tried to measure the magnesia content of both tetragonal and monoclinic (i.e. large and small) particles in magnesia-stabilized zirconia using energy-dispersive X-ray spectrometry (EDS) in a TEM/STEM microscope, but this has not been possible[21] because of the difficulty in detecting magnesium X rays at low concentration levels in the small precipitates. However, a size-dependent solute content apparently occurs in Al_2O_3/ZrO_2 composites containing dispersed particles in tetragonal ZrO_2, as is discussed elsewhere.[29]

Stress-Induced Transformations

Absolute reaction-rate diagrams can also be used to consider the physics of the stress-induced transformations. The applied stress (P_{ij}^A) can do useful work if particles transform and is indicated by the $P_{ij}^A e_{ij}^T$ term in Fig. 14. If such stress-induced transformation occurs, then crack propagation requires an additional expenditure of energy, $\tilde{\Delta}F$, which provides the transformation toughening. If there are N_p particles per unit length of crack in the transformation zone, the change in fracture surface energy, $\Delta\gamma_f$, is just $N_p\tilde{\Delta}F$. Inasmuch as $P_{ij}^A = (1/2\pi r)^{1/2}K_I f_{ij}(\theta)$,[31] $\tilde{\Delta}F$ for a single particle is proportional to K_I, and its maximum value is set by K_{1C}.

If more-resistant particles are present, that is particles where the difference between $F_{mono(free)}$ and $F_{mono(constrained)}$ is greater, a larger $\tilde{\Delta}F$ and a smaller transformation zone occur, and N_p will be reduced. The maximum improvement in $\Delta\gamma_f$, and hence the maximum transformation toughening, will thus be given for particles whose transformation characteristics are such as to give a large N_p but a still appreciable $\tilde{\Delta}F$. It is clear that optimizing transformation toughening in ZrO_2-containing ceramics requires detailed knowledge of these several energy terms.

Acknowledgment

It is a pleasure to acknowledge my former students G. K. Bansal, D. L. Porter, and L. S. Schoenlein, who have contributed so much to my understanding of ZrO_2-containing ceramics, and the National Science Foundation, who have supported my research on transformation toughening in PSZ.

References

[1]R. C. Garvie, R. H. Hannink, and R. T. Pascoe, "Ceramic Steel?," *Nature (London)*, **258** [5537] 703–704 (1975).

[2]D. L. Porter and A. H. Heuer, "Mechanisms of Toughening Partially Stabilized Zirconia (PSZ)," *J. Am. Ceram. Soc.*, **60** [3-4] 183–84 (1977).

[3]D. L. Porter, A. G. Evans, and A. H. Heuer, "Transformation Toughening in Partially Stabilized Zirconia (PSZ)," *Acta Metall.*, **27** [2] 1649–54 (1979).

[4]N. Claussen, "Stress-Induced Transformation of Tetragonal ZrO_2 Particles in Ceramic Matrices," *J. Am. Ceram. Soc.*, **61** [1-2] 85–86 (1978).

[5]A. G. Evans and A. H. Heuer, "Review—Transformation Toughening in Ceramics: Martensitic Transformations in Crack-Tip Stress Fields," *ibid.*, **63** [5-6] 241–48 (1980).

[6]R. H. Hannink, K. A. Johnston, R. T. Pascoe, and R. C. Garvie, "Microstructural Changes During Isothermal Aging of a Calcia Partially Stabilized Zirconia Alloy"; this volume, pp. 116–35.

[7]C. A. Anderson and T. K. Gupta, "Phase Stability and Transformation Toughening in Zirconia"; this volume, pp. 184–201.

[8]R. A. Miller, J. L. Smialek, and R. G. Garlick, "Phase Stability in Plasma-Sprayed Partially Stabilized Zirconia-Yttria"; this volume, pp. 241–53.

[9]F. K. Moghadam, T. Yamashita, and D. A. Stevenson, "Characterization of Yttria-Stabilized Zirconia Oxygen Solid Electrolytes"; this volume, pp. 364–79.

[10]O. Ruff and F. Ebert, "Ceramics of Highly Refractory Materials: I," *Z. Anorg. Allg. Chem.*, **180** [1] 19–41 (1929).

[11]P. Murray and E. B. Allison, "A Study of the Monoclinic⇌Tetragonal Phase Transformation in Zirconia," *Trans. Br. Ceram. Soc.*, **53**, 335–61 (1954).

[12]G. M. Wolten, "Diffusionless Phase Transformation in Zirconia and Hafnia," *J. Am. Ceram. Soc.*, **46** [9] 418–22 (1963).

[13]J. E. Bailey, "The Monoclinic-Tetragonal Transformation and Associated Twinning in Thin Films of Zirconia," *Proc. R. Soc., London*, **279**, 395–412 (1964).

[14]L. L. Fehrenbacher and L. A. Jacobson, "Metallographic Observation of the Monoclinic-Tetragonal Phase Transformation in ZrO_2," *J. Am. Ceram. Soc.*, **48** [3] 157–61 (1965).

[15]G. K. Bansal and A. H. Heuer, "Martensitic Phase Transformation in Zirconia (ZrO_2): I," *Acta Metall.*, **20** [11] 1281–89 (1972); "II," *ibid.*, **22** [4] 409–17 (1974).

[16]A. S. Chase and J. A. Osmer, "Growth of Single Crystals of ZrO_2 and HfO_2 from PbF_2," *Am. Mineral.*, **51**, 1808–11 (1966).

[17]J. W. Christian, The Theory of Transformations in Metals and Alloys, Vol. 1, 2nd ed. Pergamon Press, Elmsford, N.Y., 1975.

[18]M. S. Wechsler, D. S. Lieberman, and T. A. Read, "On the Theory of the Formation of Martensite," *Trans. AIME*, **197** [11] 1503–15 (1953).

[19]C. F. Grain, "Phase Relations in the ZrO_2-MgO System," *J. Am. Ceram. Soc.*, **50** [6] 288–90 (1967).

[20]O. J. Whittemore, Jr., and N. N. Ault, "Thermal Expansion of Various Ceramic Materials to 1500 °C," *ibid.*, **39** [12] 443–44 (1956).

[21]L. S. Schoenlein and A. H. Heuer; unpublished work.

[22](a) J. D. Eshelby, "Determination of the Elastic Field of an Ellipsoidal Inclusion and Related Problems," *Proc. R. Soc., London*, **241A**, 376–96 (1957).
(b) J. D. Eshelby, pp. 89–140 in Progress in Solid Mechanics, Vol. 2. Edited by I. N. Sneddon and R. Hill. Wiley-Interscience, New York, 1961.

[23]D. L. Porter and A. H. Heuer, "Microstructural Development in MgO-Partially Stabilized Zirconia (Mg-PSZ)," *J. Am. Ceram. Soc.*, **62** [5-6] 298–305 (1979).

[24]R. Rice, "Further Discussion of 'Precipitation in Partially Stabilized Zirconia,'" *ibid.*, **60** [5-6] 280 (1977).

[25]D. L. Porter and A. H. Heuer, "Reply to "Further Discussion of 'Precipitation in Partially Stabilized Zirconia,'"" *ibid.*, pp. 280–81.

[26]S. Glasstone, K. J. Laidler, and H. Eyring, The Theory of Rate Processes. McGraw-Hill, New York, 1941.

[27]A. G. Evans, N. Burlingame, M. Drory, and W. M. Kriven, "Martensitic Transformations in Zirconia—Particle Size Effects and Toughening," *Acta Metall.*, **29** [2] 447–56 (1981).

[28]F. F. Lange and D. J. Green, "Effect of Inclusion Size on the Retention of Tetragonal ZrO_2: Theory and Experiments"; this volume.

[29]A. H. Heuer, N. Claussen, W. Kriven, and M. Rühle, "The Stability of Tetragonal ZrO_2 Particles in Ceramic Matrices"; unpublished work.

[30] M. Rühle; unpublished work.
[31] B. R. Lawn and T. R. Wilshaw, Fracture of Brittle Solids. Cambridge University Press, London, 1975.

*The experiment in question was done by Rühle (Ref. 30) on Al_2O_3-ZrO_2 composites containing 20 vol% spherical intragranular precipitates of tetragonal ZrO_2. Approximately 10-20% of the particles remained untransformed on cooling to 20 K. However, the propensity for transformation was diminished on high temperature annealing (30-40% remained untransformed at 20 K), even though the annealing did not change the size, size distribution, shape, or location of the particles. It is presumed that the annealing eliminated some nucleation sites in these particles or alternatively altered the solute content so as to stabilize the tetragonal form (Ref. 29).

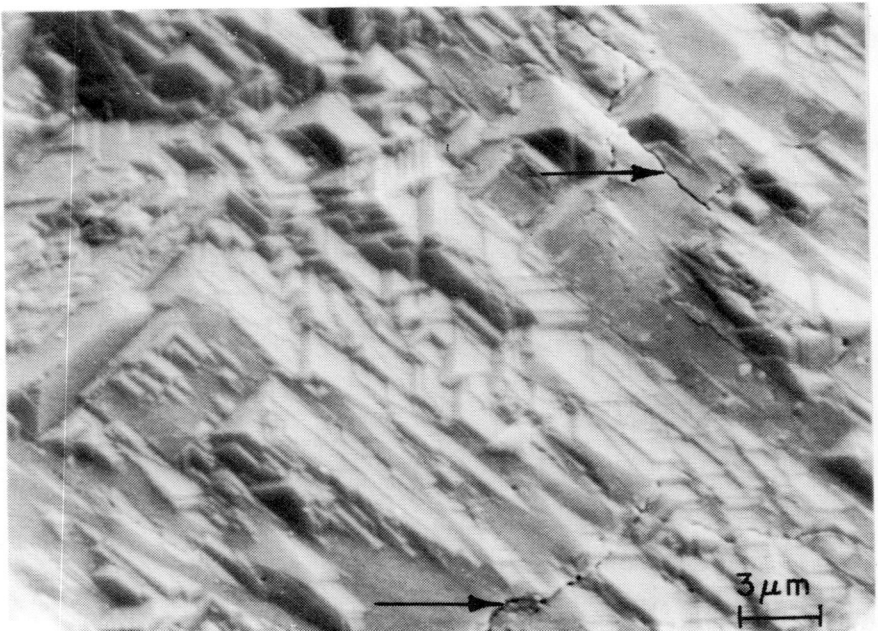

Fig. 1. Optical micrograph showing surface relief ("upheaval") on vicinal surface of ZrO_2 single crystal after one cycle of transformation (heating and cooling). This surface relief is firm experimental evidence that the tetragonal→monoclinic transformation is martensitic. The surface cracks (arrowed) are due to the anisotropic volume increase on transformation (Ref. 15).

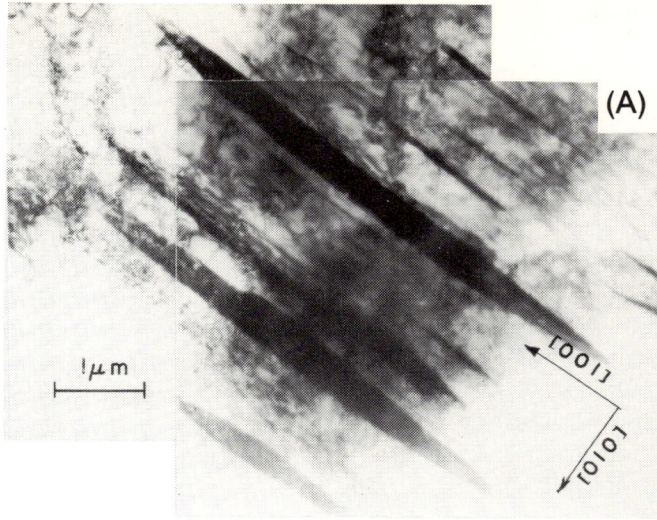

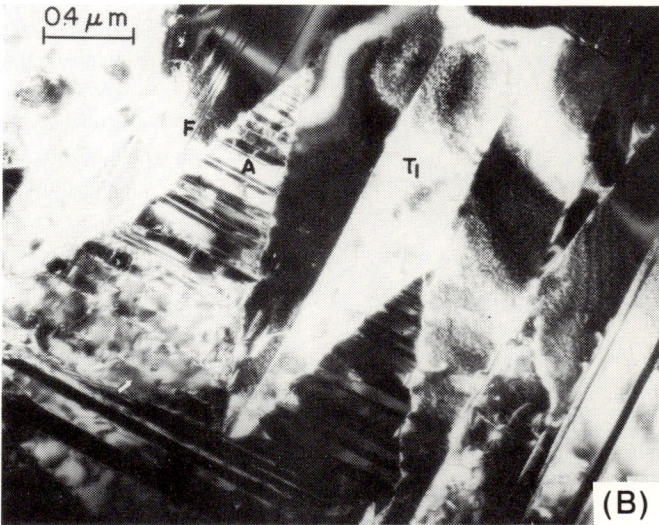

Fig. 2. Transmission electron micrographs of martensite plates formed in ZrO_2 single crystals during the tetragonal→monoclinic transformation. (A) is a bright field micrograph shoeing heavily dislocated plates with $\sim(671)_m$ and $\sim(761)_m$ habit planes. (B) is a dark field micrograph of internally twinned plates formed by a second martensite mechanism. The twinning plane is $(100)_m$; the coarse twins between plates are also of this type, but they are rotated by 90° from the fine twins (Ref. 15).

107

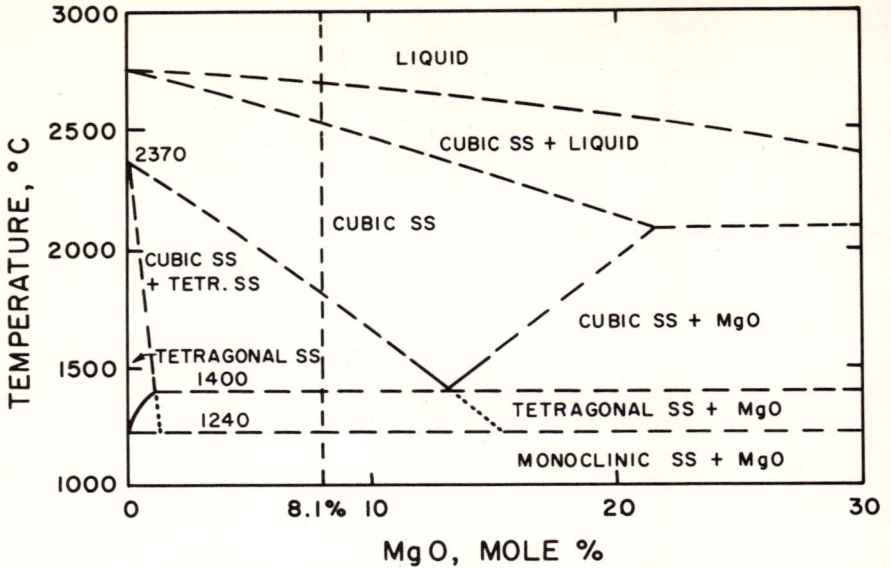

Fig. 3. Phase diagram of the high-ZrO_2 portion of the MgO-ZrO_2 binary phase diagrams (after Ref. 19). (The dotted lines are metastable extensions of the solvuses separating the two-phase (tetragonal solid solution (ss) + cubic (ss) field from the single-phase fields adjacent to it.) The vertical line at 8.1% MgO refers to the material characterized in Fig. 7.

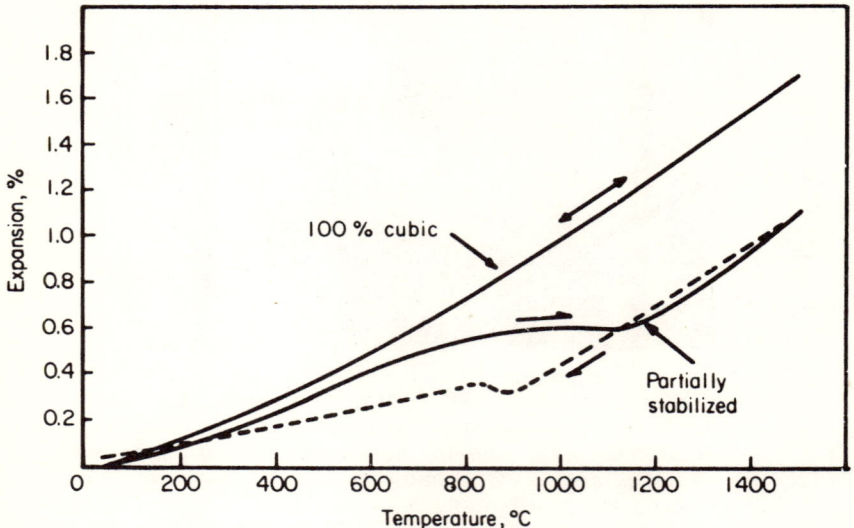

Fig. 4. Thermal expansion curves for a fully stabilized and partially stabilized zirconia ceramic (Ref. 20).

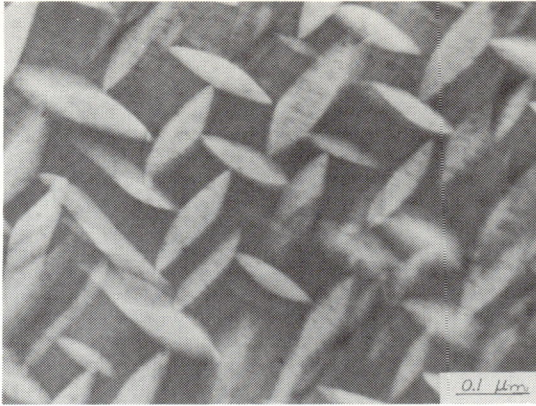

Fig. 5. Transmission electron micrograph of optimally aged MgO-partially stabilized ZrO_2 (Mg-PSZ). The matrix is an MgO-rich cubic solid solution and contains oblate spheroid particles (seen here in cross section) of an MgO-lean tetragonal solid solution (Ref. 21).

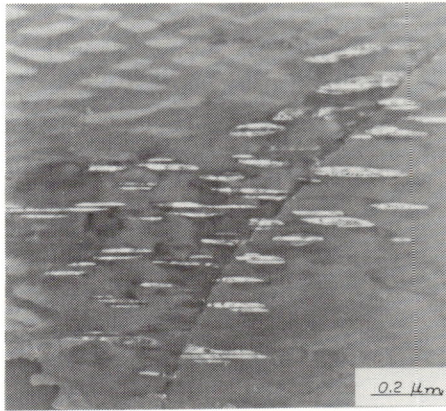

Fig. 6. Dark field transmission electron micrograph of crack tip transformation zone in optimally aged Mg-PSZ. The cracks were introduced into a 50 μm thin section of Mg-PSZ by overloading with a Vickers indenter. Only particles within ≈ 0.5 μm of the crack tip have transformed; they are the bright internally twinned particles and have monoclinic symmetry (this micrograph was taken with a monoclinic reflection). Numerous untransformed particles are visible in the upper left (Ref. 21).

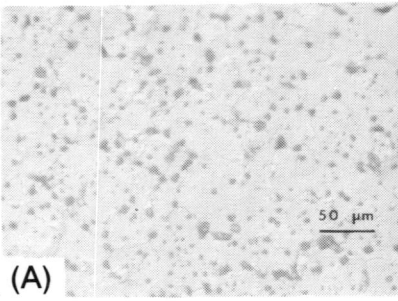

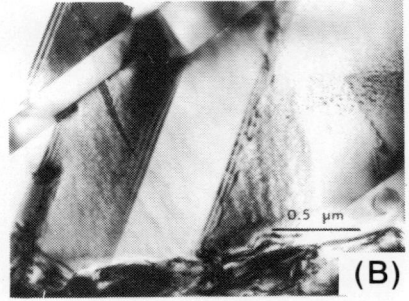

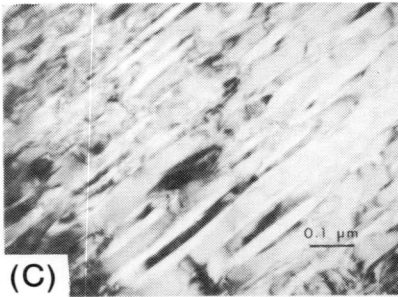

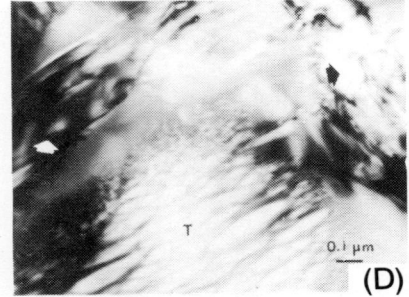

Fig. 7. Microstructural characterization of commercial Mg-PSZ. (A) is an optical micrograph and shows ≈5% porosity residual in a matrix with an average grain size of ≈50 μm. Monoclinic ZrO_2 is present both as coarsely twinned grain boundary particles (B) or as finer but still twinned intragranular particles (C). Still finer intragranular tetragonal ZrO_2 particles are also present (D) and are separated from the larger intragranular monoclinic particles (arrowed in (D)) by zones "denuded" of particles. ((B), (C), and (D) are transmission electron micrographs.) See Ref. 23 for further discussion of the microstructural evolution implied by these micrographs.

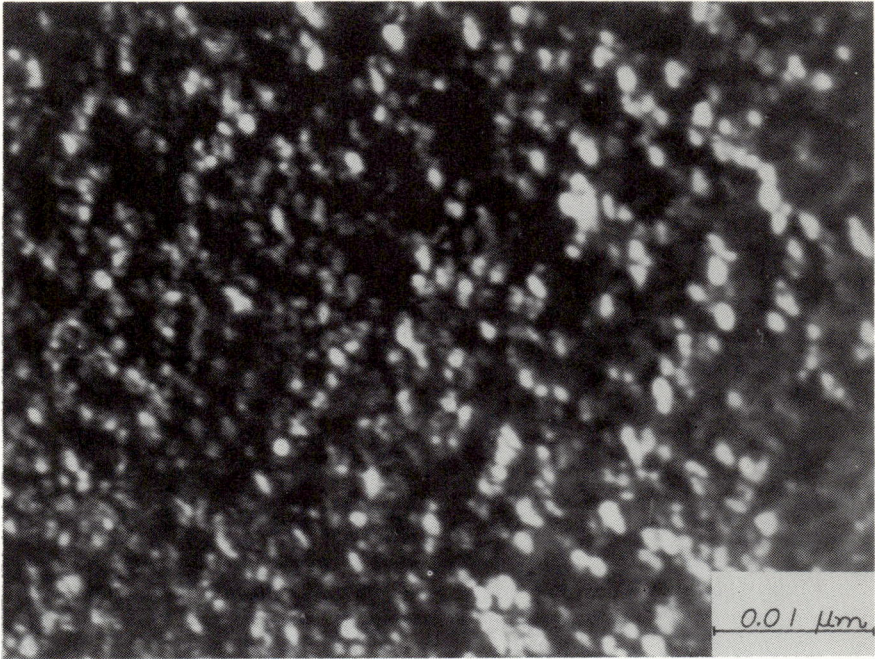

Fig. 8. Dark field transmission electron micrograph of a quenched and solution-annealed Mg-PSZ. This micrograph was taken with a $\underline{g} = 112$ fluorite-forbidden reflection; the bright features in the micrograph are due to one of the three variants of the homogeneously nucleated tetragonal ZrO_2 precipitates, which are ≈ 2 nm in diameter.

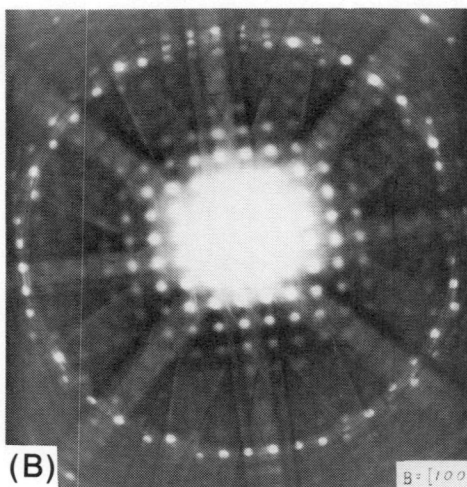

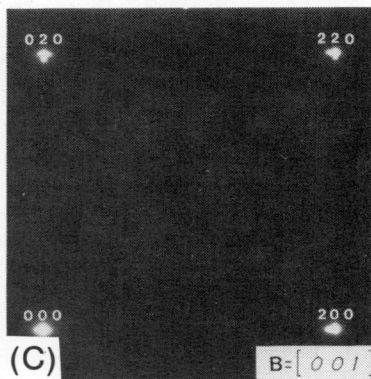

Fig. 9. Transmission electron micrograph and diffraction patterns of tetragonal ZrO_2 precipitates in optimally aged Mg-PSZ. (A) is a bright field micrograph taken in good dynamical contrast; two of the three precipitate variants have an elliptical cross section in this foil orientation.(B) is a convergent beam electron diffraction pattern from one of the elliptical particles present in (A); the tetragonal c axis lies in the foil plane, and thus the pattern lacks four-fold symmetry. (C) is a conventional selected area diffraction pattern. The second reflections shown at the 020 and 200 positions are due to the 002 tetragonal reflections of two of the three variants; reflections from the third circular variant superimpose those of the cubic matrix (Ref. 21).

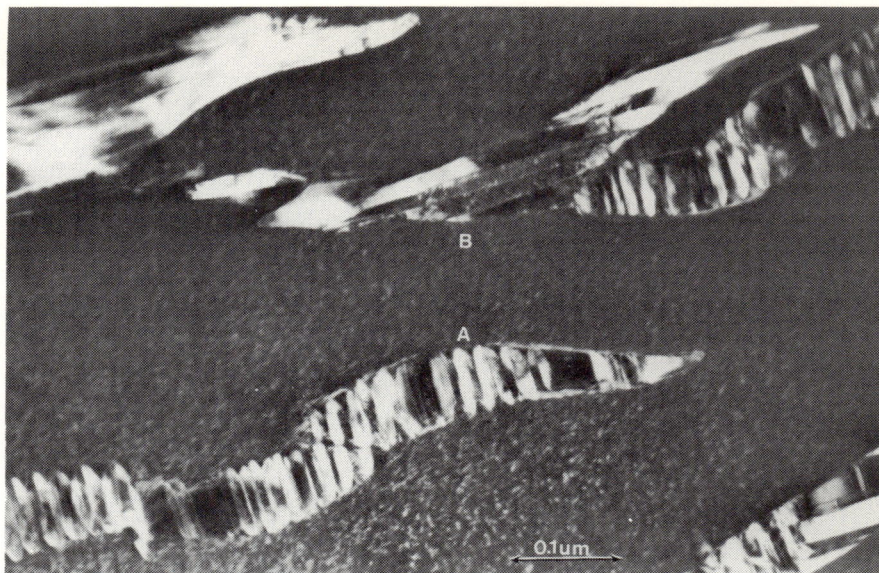

Fig. 10. Dark field transmission electron micrograph of twinned monoclinic ZrO_2 precipitates in overaged Mg-PSZ. A and B are twinned on different systems, as described in the text (Ref. 23).

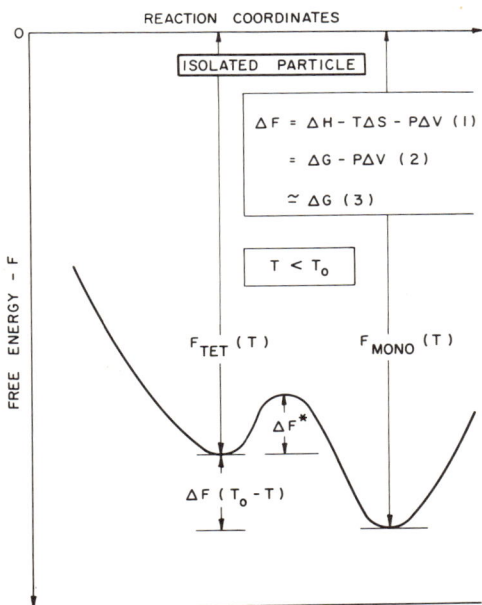

REACTION COORDINATES

ISOLATED PARTICLE

$\Delta F = \Delta H - T\Delta S - P\Delta V$ (1)

$= \Delta G - P\Delta V$ (2)

$\simeq \Delta G$ (3)

$T < T_0$

FREE ENERGY – F

$F_{TET}(T)$ $F_{MONO}(T)$

ΔF^*

$\Delta F(T_0 - T)$

Fig. 11. Absolute reaction rate diagram for isolated ZrO_2 particle at room temperature. See text for details.

113

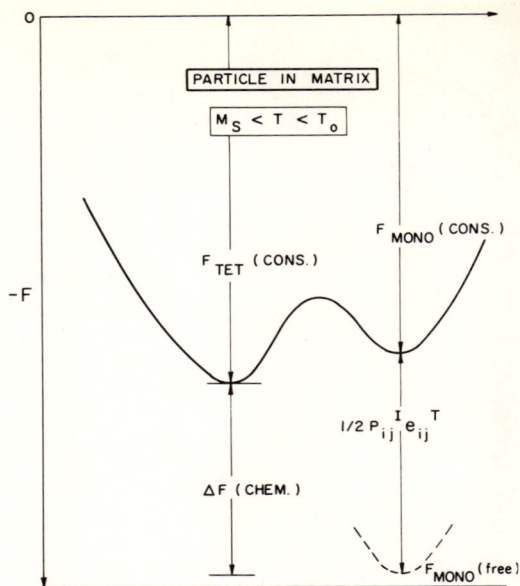

Fig. 12. Absolute reaction rate diagram for the particle of Fig. 11 but now in a rigid matrix.

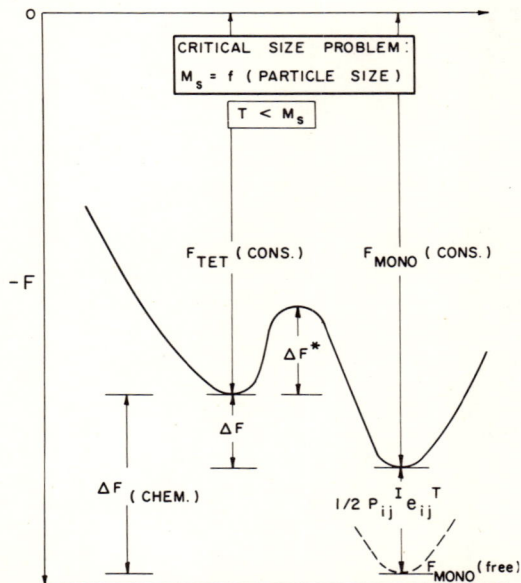

Fig. 13. Absolute reaction rate diagram for the particle of Fig.12 but now supercooled sufficiently for transformation to occur.

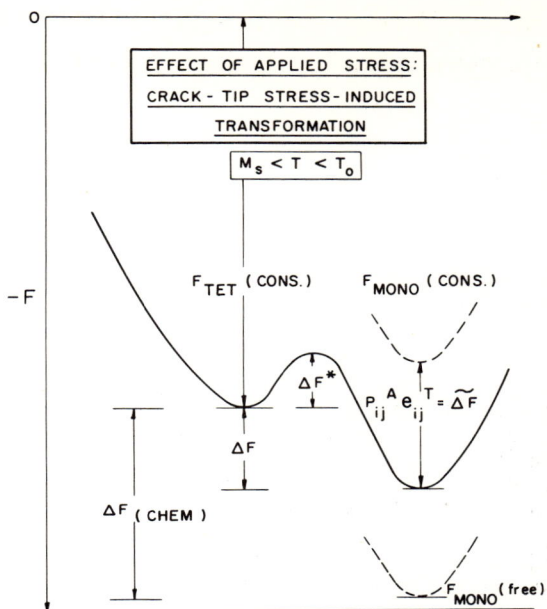

Fig. 14. Absolute reaction rate diagram for the particle of Fig. 12 but in a crack tip stress field.

Microstructural changes during isothermal aging of a calcia partially stabilized zirconia alloy

R. H. J. Hannink, K. A. Johnston,* R. T. Pascoe,* and R. C. Garvie

CSIRO
Division of Materials Science
Advanced Materials Laboratory
Melbourne, Australia

Lime-stabilized zirconia of a limited composition range can be strengthened by an isothermal aging treatment in which essentially pure tetragonal zirconia is precipitated from a cubic zirconia matrix. The structural changes during the aging of an 8.4 mol% CaO alloy at 1300 °C are described here. Nucleation and growth of the equilibrium tetragonal phase are very rapid and are completed in a short time compared with the time required to reach the highest strengths. The principal structural change during the remainder of the aging process is particle coarsening. It is found that small particles maintain their tetragonal structure on cooling to room temperature, but particles larger than ≈ 90 nm transform to the stable monoclinic structure. The existence of the tetragonal phase at room temperature is explained by a balance among chemical, interfacial, and strain energies.

$\mathbf{P}$ure zirconia exists as three polymorphs. The high temperature form, stable above ≈ 2400 °C, is cubic and has the fluorite-type structure. On cooling, the cubic polymorph transforms first to tetragonal symmetry and then, at ≈ 1200 °C, to the monoclinic form. The tetragonal phase can be retained below the standard transformation temperature when the particle size is very small or when a constraint opposes either or both of the volume and shape changes which accompany the phase change.[1]

Although zirconia has the potentially useful properties of high melting temperature and chemical inertness, the transformation from tetragonal to monoclinic symmetry produces cracking and even disintegration, which renders the material useless as a structural ceramic. The transition is fast and diffusionless and cannot normally be suppressed in pure zirconia but can be avoided by adding other oxides (such as CaO, MgO, and $YO_{1.5}$) which stabilize the high temperature phase by forming solid solutions having the fluorite structure.[2]

The best mechanical properties (e.g. fracture strength, resistance to thermal shock) in zirconia-based systems are found in two-phase alloys known as partially stabilized zirconia (PSZ). The strongest and toughest PSZ alloys contain a dispersion of metastable tetragonal precipitates in a cubic fluorite matrix.[3,4]

The proposed toughening/strengthening mechanism is that the tetragonal precipitates within a critical size range are transformed by stress to the stable monoclinic structure in a process zone at the crack tip. The region within the process zone is placed in self-compression because of the 4 vol% expansion accompanying the transformation. The stress at the crack tip is redistributed by this mechanism so that the material is tougher/stronger.

In this paper, the effects of isothermal aging on the microstructure of a specific CaO-PSZ alloy are examined and are related to the strength of the alloy and to the stability of the tetragonal phase. The strengthening process, being associated with the precipitation and growth of a second phase, shows close parallels with age hardening in metals. The microstructures developed are sensitive to composition and heat treatment. Similar processes occur, generally at different rates, in other PSZ alloys and with other heat treatments.[4,5]

Experimental Procedure

Sample Preparation

The samples all contained 8.4 mol% (4.0 wt%) CaO and were prepared from zirconium dioxide† and calcium acetate.‡ The starting materials were mixed, calcined at 1050 °C, ball-milled, and pressed either into disks (10 mm in diam. and 2 mm thick) or into bend test bars (3.5 by 3.5 by 40 mm). The samples were then sintered and solution-treated in the cubic phase field at 1800°C (Fig. 1) using a gas-fired furnace. Additional details of sample preparation are given elsewhere.[5,6]

For aging studies, the material should have a known microstructure; ideally this should be the single phase high temperature structure retained to low temperature by rapid cooling. Disk samples, quenched into water from the solution temperature using a special quenching furnace, gave a structure close to the ideal but were, of course, useless for mechanical property measurements.[7] For the bar samples, on which the progress of aging was monitored by changes in strength, a standard cooling procedure was adopted to avoid thermal shock damage associated with quenching.[8] The cooling times from 1800 °C were: 15 min to reach 1450 °C, then 15 min to 1300 °C, and a further 5 h to 500 °C.

After cooling to room temperature, the samples were aged at 1300° in air in electrically heated tube furnaces. The samples were steadily pushed into the hot zone over a 10-min period to avoid thermal shock and, after aging, were cooled in the furnace at its natural rate.

Physical Examination

Fracture Strength: Fracture strengths (modulus of rupture) were measured at room temperature in four-point bending on unnotched bars, all faces of which were ground with 150-grit diamond.

X-Ray Diffraction: X-ray powder photographs were obtained, using a Guinier-Hagg camera, from thin foils prepared for electron microscopy and from crushed disk samples. From the photographs the phases present were determined and also accurate lattice parameters were measured by reference to a thoria standard deposited on the foils or mixed with the powder.

X-ray diffractometry was used to estimate the proportions of phases present at the surface of test bars in the as-fired, as-polished, and as-ground conditions. Approximate lattice parameters could also be determined. The phase analysis was carried out by comparing the areas of selected diffractometer peaks. {111} reflections were used to determine the ratio of monoclinic-to-cubic plus tetragonal and the tetragonal-to-cubic ratio was determined from {400} reflections.[9]

Electron Microscopy: Transmission electron microscopy was used to reveal phase morphology and crystallographic relationships. This technique, used in conjunction with X-ray diffraction, is particularly useful when the three phases with cubic, tetragonal, and monoclinic symmetry coexist.

Foils were prepared by diamond-sawing thin slices, grinding these to a thickness of ≈ 80 μm, and finally thinning by bombardment with a beam of argon ions at 5 kV. The foils were examined in both bright and dark field at 100 or 200 kV.

Precipitate sizes were measured from electron micrographs and corrected for orientation, assuming a {110} habit plane.[10]

Dilatometry: A dilatometer was used to measure thermal expansion of selected samples from room temperature to 1200 °C, heating at 5 °C min^{-1}. Dilatometry is particularly useful for the detection of phase changes in these materials.

Results

Fracture Stress

Figure 2, a typical isothermal aging curve for the 8.4 mol% CaO alloy, shows strength as a function of aging time at 1300 °C. The as-fired strength here was 200 MPa. This is higher than that of a fully stabilized zirconia, i.e. ≈ 170 MPa. From this as-fired value the strength rose to a peak of 620 MPa at 64 h, then fell rapidly to 200 MPa, at which level it remained for at least 300 h.

The specimens will be described using the nomenclature of age hardening. The highest strength material will be referred to as peak-aged, that aged for longer times as overaged, and that for shorter times as underaged.

X-Ray Diffraction

The X-ray diffraction techniques showed the presence of the expected cubic phase in all specimens. The room-temperature-stable monoclinic phase was found in some specimens as was the tetragonal form which was apparently metastably retained from high temperature.

Examination of surfaces before and after grinding indicated that the grinding process could transform near-surface tetragonal material to monoclinic symmetry. For this reason specimens for phase analysis by diffractometry were always examined in two surface conditions: the original sintered surface of the specimen and the surface ground by 150-grit diamond after completion of all heat treatments. Inserting the grinding operation between firing and aging treatments gave inconsistent results because widely differing values of phase content could be obtained from two surfaces of the same specimen.

Occasionally as-fired surfaces suffered damage before examination and as a result gave anomalous values for the proportion of phases. It was found possible to remove such damaged surfaces by polishing with 1 μm diamond paste until the diffractometer profile was unchanged by further polishing. The peak profiles so obtained were shown to be identical to those from undamaged as-fired surfaces.[11]

Diffractometer peak profiles from specimens covering a representative range of the heat treatments studied are presented in Fig. 3. Profiles of the {111} group of peaks from as-fired and ground surfaces and of the {400} peaks from as-fired surfaces are shown. These are described in detail in the following sections. Table I lists the lattice parameters determined from Guinier-Hagg films of samples in various aged conditions.

Underaged Material: The {111} peak profiles showed little variation with heat treatment for the three underaged conditions presented in Fig. 3. For the {400} peaks, however, there was a considerable evolution. The quenched disk specimen showed a sharp pattern which indexed as cubic. The α_1, α_2 doublet was clearly resolved at large diffraction angles.

For a quenched disk after 5 h aging at 1300 °C, the higher index peaks of the diffraction pattern showed probable strain broadening. The {400} reflection profile (Fig. 3) was consistent with the presence of both tetragonal and cubic phases, but the resolution was inadequate for unequivocal indexing. The X-ray patterns from quenched material aged for 10 h and from carefully cooled bar samples were similar and, although their actual thermal histories were quite different, in terms of the evolution of sample microstructure, as revealed by X rays, they were nearly equivalent.

The subsequent underaged materials described here were all bar samples, carefully cooled and then aged at 1300 °C for various times. After 2 h aging, the cubic and the two tetragonal peaks in the {400} group were just resolved, and after 10 h (Fig. 3) the peaks were easily resolved, and the proportion of tetragonal phase could be estimated as 55 ± 10 wt%. Additional aging caused only a small improvement in resolution of the {400} peaks and no significant change in the proportion of tetragonal phase.

Guinier photographs of underaged material, whether powdered samples, quenched disk samples, or thin foils cut from bars, correlated well with diffractometer patterns from as-fired surfaces.

The first signs of change due to grinding occurred in the 10-h aged specimen when a small monoclinic satellite appeared beside the {111} cubic reflection (Fig. 3). Longer aging increased the amount of monoclinic phase produced by grinding (Fig. 4) and both $(11\bar{1})$ and (111) monoclinic peaks became clearly defined. Interestingly, the ratio of the areas under these diffractometer peaks, $(11\bar{1})_m:(111)_m$, in underaged specimens, was $> \approx 2:1$ compared to a theoretically calculated ratio of 1.3:1. This preferred orientation effect will be discussed elsewhere (R. Hannink, unpublished work).

The {400} peak profiles from ground specimens, aged for longer than 10 h, were broadened, probably due to strain, and the proportion of tetragonal phase could not be determined accurately.

Peak-Aged Material: In terms of the diffractometer measurement, the peak-aged specimens were indistinguishable from underaged material. However, a change was noted in Guinier photographs. Instead of correlating well with diffractometer data from the as-fired surface, there was a close relation to the ground surface data; cubic, tetragonal, and monoclinic phases were all present, as shown in Fig. 5 (the tetragonal phase is indexed as a distorted fluorite structure). Lattice parameters derived from this photograph are included in Table I. Guinier photographs, made from several peak-aged foils, showed variations in the proportion of monoclinic phase, as

119

indicated by change in monoclinic line intensities. The variations depended on foil preparation technique, specifically whether the foils were ion-thinned from both surfaces or whether the ground surface was left intact.

Overaged Material: Aging beyond the strength peak produced a marked change in the {400} diffractometer profile. The distinct tetragonal peaks were absent and a broad asymmetric cubic peak remained, which could have masked both monoclinic and residual tetragonal reflections. The {111} profile showed the presence of monoclinic phase in as-fired surfaces. The intensity ratio of the monoclinic peaks was close to the theoretical 1.3:1. The Guinier photographs clearly showed cubic and monoclinic phases to be present; the lines were, however, somewhat diffuse.

Electron Microscopy

Underaged Material: A microstructure of small tetragonal precipitates in a cubic matrix was observed in all underaged specimens. Included were fragments of the water-quenched disk which, from the X-ray diffraction examination, appeared to be solely cubic.

The size of the precipitate particles increased with aging time but the shape remained reasonably equiaxed. Some of the larger particles showed faceting but no particular crystal habit was identified. The distribution of the precipitates was distinctly modulated into bands with a spacing which increased with aging time. Impingement between precipitate particles was infrequent except in specimens approaching the peak-aged condition.

A typical dark-field image of underaged material is shown in Fig. 6. The aging treatment for this specimen was 20 h at 1300 °C and the precipitate particles were mostly in the size range 30 to 70 nm. Considering the tetragonal cell as a slightly distorted cube, the cubic-to-tetragonal orientation relationship, shown by the electron diffraction photograph (inset, Fig. 6), was a simple cube-to-cube. The direction of the distortion, the tetragonal *c*-axis, was random with respect to the three cube cell axes.

Peak-Aged Material: The tetragonal precipitates observed in the electron microscope had reached an average size of ≈ 90 nm and some impingement was evident (Fig. 7). Both tetragonal and monoclinic particles could be found within a single grain but unambiguous identification was not always possible because of the close structural relationship between the phases and because the twinning which usually identified the monoclinic appeared to be absent. A significant observation was that occasionally, with fixed diffraction conditions, an individual particle which was initially tetragonal was seen to change contrast, indicating that it had transformed to monoclinic symmetry.

In peak-aged foils, electron diffraction patterns from grains containing monoclinic particles were different from those obtained from slightly overaged foils in that the β-angle splitting characteristic of monoclinic symmetry was absent. Tetragonal and monoclinic could only be distinguished by viewing in the $\langle 100 \rangle_c$ orientations where the extra monoclinic spots occur but extra tetragonal spots are forbidden. This preferred orientation effect will be discussed elsewhere (R. Hannink, unpublished work).

Overaged Material: The electron microstructures showed an almost continuous network of highly twinned monoclinic particles in a cubic matrix.

120

Some tetragonal particles may also have been present but their unambiguous identification was not possible due to overlapping diffraction spots. The average precipitate size increased with aging at 1300 °C from ≈ 100 nm in slightly overaged material to > 250 nm after 300 h, the longest period studied. A 70-h aged specimen is shown in Fig. 8. The average precipitate size in this specimen was near 100 nm. The diffraction pattern from the same specimen (inset to Fig. 8) showed the monoclinic spot splitting typical of all overaged specimens. It was clear from this and similar patterns that the tetragonal-to-monoclinic transformation occurred without appreciable lattice rotation and thus the simple cube-to-cube relationships with, of course, the added distortion due to the tetragonal-to-monoclinic transition, were preserved.

Dilatometry

Thermal expansion measurements were made to determine temperature at which discontinuities, corresponding to phase transitions, occurred. Some dilatometer traces are shown in Fig. 9. The monoclinic-tetragonal transition shows in all overaged alloys as a marked contraction on heating and expansion on cooling. The temperatures of the start and finish of the forward and reverse transition were determined from these traces and are shown in Fig. 10 as a function of precipitate size. In samples where transformation occurred, the four temperatures all increased with increasing precipitate size. In the slightly overaged alloy, with 95 nm mean particle size, the transformation did not go to completion on cooling. No transition could be detected in underaged alloys, nor could one be induced by immersion in liquid nitrogen.

Discussion

Precipitate Coarsening

After precipitation is complete the principal structural change which occurs at the aging temperature is particle coarsening, in which large precipitates grow at the expense of small ones, to reduce the total interfacial energy. It is believed that, during growth of the coherent tetragonal particles, a critical size is exceeded and coherency is lost. At this point the tetragonal structure can no longer be preserved at room temperature where the stable phase has monoclinic symmetry.

The general features of the aging process are shown in Fig. 4, which is a plot of the monoclinic phase content as measured by X-ray diffraction of specimens progressively aged at 1300 °C. Two curves are presented, one obtained from as-fired samples, the other from samples whose surfaces had been ground. The "ground" curve has a discontinuity at ≈ 64 h which is not understood.

It would be expected that the "as-fired" and "ground" curves would coincide for overaged materials. Figure 4 shows that the as-fired curve is considerably above the ground curve after an aging time of ≈ 100 h. The discrepancy is explained by the fact that stresses due to grinding cause a *decrease* in the amount of monoclinic phase in samples that are overaged, which is the reverse of the effect of grinding stresses on samples which have been under- or peak-aged. This was shown by measuring the monoclinic content of the surface of a sample (aged for 120 h at 1300 °C) which first was in the aged condition, then ground, and then annealed for various times at

1000 °C. The data are given in Table II. The as-fired value of the monoclinic content of 41% fell to 29% when the surface was ground. When the sample was annealed for 2.5 h the monoclinic content was restored to 37%. Further annealing for ≈ 52 h nearly restored the monoclinic content to its original value. The monoclinic content correlated inversely with the half-peak width (a measure of strain in the surface) of the cubic/tetragonal (111) diffraction profile.

Presumably grinding the surface of overaged samples caused the transformation of some monoclinic precipitates to the tetragonal structure. With a high concentration of monoclinic precipitates, a given precipitate would be within the stress fields of its neighbors. The interparticle distance in overaged materials is about the same as the particle diameter. Grinding the surface raises the stress level here to an even higher value. The transformation of some monoclinic precipitates to tetragonal symmetry would reduce the mechanical strain energy of the system probably with an overall net reduction in the free energy of the system.

The coarsening process, that is, particle growth where a constant volume fraction of precipitate is present, has been analyzed by Greenwood,[12] Lifshitz and Slyozov,[13] and Wagner.[14] Their theories lead to a relationship of the type:

$$r = kt^{1/3} \tag{1}$$

where r is the particle radius, t is the time, and k is the material constant.

The data measured for precipitate growth at 1300°C are plotted on logarithmic axes in Fig. 11. It was found that the precipitate size of a sample cooled at the standard rate was approximately the same as that of a quenched sample after aging for 10 h. Accordingly, the aging times plotted in Fig. 11 were adjusted by adding 10 h to the time for all standard cooled samples. For the aging time interval, 0–74 h, the curve is reasonably linear with a slope $\approx 1/3$. At about the time required to attain peak strength (74 h) there is a break in the slope which is consistent with the suggestion that at this time the tetragonal precipitates lose coherency. The constant k, in Eq. (1), is defined as follows, using Wagner's formulation:

$$k = \left(\frac{8\gamma D\, C_e V_m^2}{9RT} \right)^{1/3} \tag{2}$$

where γ is the precipitate/matrix interfacial energy, D is the diffusion coefficient of the transported species in the matrix, C_e is the solubility in the matrix of a precipitate of ''infinite'' radius at the aging temperature, V_m is the molar volume of the precipitate, and R and T have their usual meanings. Our interpretation is that the break in slope reflects a change in k, which, in turn, is the result of an increase in γ, caused by the originally coherent precipitate/matrix interface becoming semicoherent.

An attempt was made to calculate γ for each type of interface because all the quantities in Eq. (2) are known. The coherent and semicoherent values of γ amounted to 0.01 and 0.05 Jm^{-2}, respectively. These values may be too low, in light of the fact that the lowest known interfacial energy for metals is the twin boundary in copper, which amounts to 0.025 Jm^{-2}.[15] The difficulty could be due to the value used for D, the least accurately known parameter in Eq. (2), which was estimated by extrapolating the data of Rhodes and Carter[16] determined in the range 1700–2150 °C. Equation (2) effectively estimates γ at 1300°C. This may be an important factor in reducing the values considerably below those expected at room temperature because of the known temperature dependence of surface energy.

It has been observed elsewhere that, in a CaO-PSZ alloy of similar composition, the rate of precipitate growth, as indicated by the time to reach peak strength, increases with increasing temperature.[5] Using these observations and correcting for the differences in C_e and T, an apparent activation energy is obtained for the precipitate growth of $\approx 330 \pm 30$ kJ/mol. This value agrees reasonably well with the activation energies for the diffusion of zirconium and calcium ions in fully stabilized zirconia (12–16 mol% CaO) of 388 and 420 kJ/mol, respectively, measured between 1700 and 2150°C.[16] The cations are the least mobile species in calcia-stabilized zirconia and therefore they would be expected to be the rate-controlling species.

The Strength-Aging Curve and The Critical Precipitate Size Range

The X-ray diffraction and dilatometer data, transmission electron micrographs, and kinetics of coarsening are all consistent with and support the proposed toughening mechanism. These results can be used also to interpret the strength-aging curve (Fig. 2).

The as-fired material at 200 MPa is already stronger than cubic stabilized zirconia (170 MPa) because of the presence of precipitates. They behave as any second phase distributed in a brittle matrix in inhibiting the propagation of a crack as discussed by Evans and Langdon.[17] At this stage of microstructural development there would be little or no contribution to the strength from transformation toughening.

We established experimentally that tetragonal precipitates less than ≈ 60 nm in diameter are stable even in the presence of a crack, while precipitates larger than ≈ 90 nm lose coherency and transform to the stable monoclinic structure during cooling of the material to room temperature (Figs. 4 and 10). The inference is that a critical size range exists such that 60 nm $< d <$ 90 nm (where d is the particle diameter), within which tetragonal precipitates are transformable by stress to the monoclinic phase and thereby contribute to the toughening of the material.

During aging in the interval 0–64 h (underaged material), the coherent tetragonal precipitates coarsen with an increasing fraction of the population falling within the critical size range. The increasing value of this "active fraction," as a function of aging time, causes the strength to rise smoothly from the as-fired value of 200 MPa to the peak value of 620 MPa. This description is supported by the X-ray diffraction and dilatometer studies which show that the monoclinic phase only occurs on ground surfaces during this aging interval. Also the amount of monoclinic phase formed by the grinding stresses increases with aging time, as does the strength of the material. At the time required to attain peak strength, the active fraction is at its maximum value.

With further aging, the population in the active fraction rapidly declines. The precipitates in the aging interval 64–"∞" h (overaged material) lose coherency and transform to the stable monoclinic structure when the material is cooled to room temperature. During this aging interval, the strength rapidly falls to a low plateau value of ≈ 200 MPa because there are few precipitates which can be transformed by stress. This view is supported by the kinetics of coarsening data. The Wagner-Lifshitz plot shows a break in the slope at about the time required to attain peak strength, which is consistent with a change in the precipitate/matrix interface from coherent to semicoherent. The X-ray diffraction and dilatometer studies show the presence of the monoclinic phase in the overaged material which has not been ground.

A critical size *range* rather than a single critical size exists within which precipitates can be transformed by stress, because the value of the applied stress field driving the crack varies inversely with distance from the crack tip. At the tip, the stress has its maximum value and the large tensile strains at this point allow the transformations of precipitates as small as 60 nm. Precipitates this size further away from the tip would not transform but larger precipitates (up to 90 nm) would do so. This argument is quantified in the following section.

Stabilization of the Tetragonal Phase

The equilibrium tetragonal-to-monoclinic transition temperature for a single crystal ZrO_2 is 1073 °C, and coarsely crystalline pure zirconia is always monoclinic below 800 °C even with the widest temperature hysteresis reported (218 °C). The tetragonal phase can be retained to room temperature without deliberate addition of stabilizer by two reported methods: high pressure and reducing the material to very small particle sizes.

Whitney[18] demonstrated that hydrostatic pressure lowered the monoclinic-to-tetragonal transformation temperature of pressed zirconia disks, in accord with Le Chatelier's principle. His temperature-pressure data extrapolate to give 3.9 GPa (39 kbars) for the pressure required to retain the tetragonal form at room temperature, in agreement with his earlier thermodynamic predictions.[19]

Microcrystals prepared by decomposing zirconia-containing salts at low temperatures are usually tetragonal. Garvie[1] explained this particle-size effect in terms of the lower surface energy of the tetragonal phase compensating for the differences in chemical free energy in powders of high specific surface; he estimated the critical diameter of a microcrystal to stabilize the tetragonal form to be ≈ 10 nm. This explanation is not universally accepted. For example, Porter *et al.*[20] doubt the existence of a crystallite size effect in pure, unrestrained material because they observed a monoclinic microcrystal only 5 nm in diameter.

Our observed critical size of ≈ 90 nm for thermally transformed monoclinic precipitates is much larger than found experimentally or predicted theoretically for pure zirconia so that the simple particle size effect above will not account for the existence of metastable tetragonal ZrO_2. If the microcrystals dissolved significant amounts of CaO, then the transformation temperature would be suppressed. However Green *et al.*[21] showed there was little or no solubility of CaO in ZrO_2 microcrystals. Also the lattice parameter data of tetragonal and monoclinic precipitates shown in Table I are consistent with the suggestion that there is no appreciable solubility of CaO in zirconia. Both tetragonal precipitates aged from 10 to 64 h and monoclinic precipitates aged from 64 to 130 h showed no significant change in lattice parameters during these aging intervals.

The remaining factors which can influence the transformation temperature are strain and interfacial energy changes. When a constrained tetragonal precipitate transforms it experiences a large hydrostatic compressive stress due to the 4% volume increase accompanying the transforma-

124

tion. This stress reduces the transformation temperature, as shown by Whitney.[18,19] The precipitate/matrix interface in underaged material would change from coherent to incoherent when precipitates are transformed by stress, with a substantial increase in interfacial energy. In overaged material cooled to room temperature, the increase in interfacial energy would be caused by the interface changing from semicoherent to incoherent.

The influence of the various factors on the transformation temperature can be seen by considering the free energy of the transformation (ΔG) of a spherical microcrystal of radius r as follows:

$$\Delta G = \Delta \psi \frac{4}{3}\pi r^3 + \Delta \epsilon \frac{4}{3}\pi r^3 + \Delta \gamma \, 4\pi r^2 \tag{3}$$

where $\Delta \psi$ is the change in chemical free energy/unit volume, $\Delta \epsilon$ is the change in the bulk strain energy/unit volume and $\Delta \gamma$ is the change in interfacial energy.

By setting the total free energy change to zero, and expanding the chemical free energy in a Taylor's series about the transformation temperature, T, and with a little algebra, Eq. (2) can be expressed in the following more useful form[1]:

$$\frac{1}{r} = \left(\frac{q}{3\Delta \gamma T_b} \right) T - \left(\frac{q + \Delta \epsilon}{3\Delta \gamma} \right) \tag{4}$$

where T_b is the transformation temperature of a crystal of "infinite" radius and q is the heat of transformation.[22,23] According to Eq. (3), a plot of the reciprocal radius of a microcrystal as a function of its transformation temperature is linear, with slope = $q/3\Delta \gamma T_b$ and intercept = $-(q + \Delta \epsilon/3\Delta \gamma)$.

Figure 10 is a plot of the transformation temperature as a function of precipitate size. It was constructed by combining data from Figs. 9 and 11. The interesting features of Fig. 10 are the pronounced hysteresis of the transformation temperature with respect to heating and cooling of the sample, the existence of transformation temperature bands rather than a sharp transformation temperature, and the clear indication of a size effect. The bands could be the result of the precipitate size distribution and also the existence of anisotropic stresses. Figure 10 can be used to test Eq. (4) by replotting the reciprocal crystallite size as a function of the transformation temperature, on the assumption that the bands can be replaced by mean values (Fig. 12).

The slopes and intercepts of the linear plots (Fig. 12) were used to estimate the changes in the strain and interfacial energies. The results are given in Table III along with the average transformation temperatures for heating and cooling and also the enthalpy of the phase transformation, taken from the literature. The chemists' convention (negative enthalpy for an exothermic reaction) was used to fix the sign of the later quantity, which then determined the signs of the other energy terms.

During cooling, the mean transformation temperature was $\approx 200\,°C$. The experimental value of the change in bulk strain of the system was 111 MPa. Using the approach of Davidge and Green,[24] the hydrostatic stress generated by the transformation was estimated to be 53 MPa. The difference between the experimental and calculated values is likely due to the presence of a shear component in the former; the twinning of the monoclinic precipitates does not relieve all of the large shear stresses accompanying the transformation. The change in interfacial energy was $\approx 1.8\ Jm^{-2}$ and would

125

be associated with a change of the precipitate/matrix interface from semicoherent to incoherent, as we are concerned here with overaged materials. This value is higher than the range of values reported in studies on metals. The reason could be due to the transformation occurring at low temperatures in a brittle material with a high Tamman temperature. Also the twinning process would introduce many sharp angles in the interface of the monoclinic precipitate which would be regions of enhanced interfacial energy. The figures in parentheses after each term in Table III indicate what amount of the total energy change is associated with that particular term. The chemical term accounts for about one-half the energy change, and the interfacial and strain terms account for about one-fourth each.

It is the change in strain energy which is involved in redistributing the stress at the crack tip and thereby toughens/strengthens the material. Inside the process zone at the crack tip where the stress-induced transformation of the precipitate occurs, the compressive component of the strain term subtracts from the tensile stress driving the crack. To maintain propagation of the crack, extra energy must be supplied to the tip which appears as an enhanced fracture energy. This approach to the toughening mechanism has been discussed by Evans et al.[25] and by Hornbogen.[26]

During heating, the mean transformation temperature is $\approx 724\,°C$. The changes in the strain and interfacial energies are reduced to 45 MPa and 1.1 Jm^{-2}, respectively. These reduced values can be expected because of the temperature dependence of the elastic moduli and the Poisson ratios as well as the effect of the different coefficients of thermal expansion on the lattice parameters of the various phases. This is shown by the fact that the hydrostatic component of strain calculated at 724 °C was only 30 MPa. The proportion of the total energy change due to the chemical term is increased now to 71% because its temperature dependence is low compared to those of the strain and interfacial terms.

The data in Table III can be used to explain the enhanced hysteresis associated with the constrained transformation. During cooling, a transforming precipitate expands against the matrix, placing itself in compression which lowers the transformation temperature. During heating, transforming precipitate pulls away from the matrix and experiences a tensile stress which raises the transformation temperature. Further, the sign of the change in interfacial energy causes this term to augment the effect of the strain terms on the transformation temperature. Thus, for a given size of crystallite, the strain and interfacial terms are double-valued with different signs. The particular value and sign operating depends on whether the material is being heated or cooled. This situation operating in conjunction with Eq. (4) guarantees the existence of two transformation temperatures for a given size of precipitate. The width of the hysteresis loop is not as great as would occur in zirconia polycrystalline powders because the surface-to-volume ratio is much greater and so the constraints on a transforming crystallite by its neighbors are reduced.

Equation (4) can be used to quantify the discussion from the previous sections on the critical size range of precipitates which can be transformed by stress. The upper bound for the size range is found by inserting the experimental values (during cooling) for the strain and interfacial terms from Table III in Eq. (4) and setting the transformation temperature to be room temperature. The calculated value for a precipitate to be transformed thermally under these conditions is ≈ 99 nm, in good agreement with the experimental value of the upper limit of the critical size range. To estimate the lower bound of the range, consider a precipitate transforming near a crack tip. The strain term in Eq. (4) is compensated by the strains resulting from the applied stress driving the crack and is effectively zero. Therefore, by neglecting the strain term in Eq. (4), the lower bound of the critical size range was estimated to be ≈ 51 nm, in fair agreement with the observed value.

Conclusions

1. Zirconia stabilized with lime can be effectively strengthened by isothermal aging.

2. Nucleation of the precipitates is very rapid.

3. The principal structural change during the aging process after furnace-cooling is the growth of second phase precipitates of equilibrium composition.

4. Small precipitates retain their tetragonal structure to room temperature and are coherent.

5. The highest strength occurs with a mean precipitate size of 90 nm.

6. Precipitates larger than ≈ 120 nm have a monoclinic structure. The appearance of the monoclinic phase is accompanied by a sharp decrease in strength.

7. The stability of the tetragonal phase can be explained by a balance between chemical, strain, and interfacial energies.

8. Particles near to the critical size transform to monoclinic symmetry on removing bulk constraints, such as during thinning for electron microscopy.

Acknowledgments

We thank our colleagues of the Engineering Ceramics and Refractories Laboratory for their assistance with this work. We particularly thank M. J. Bannister, N. A. McKinnon, and M. V. Swain for criticism and discussion.

References

[1]R. C. Garvie, *J. Phys. Chem.,* **82** [2] 218–24 (1978).
[2]E. C. Subbarao, H. S. Maiti, and K. K. Srivastava, *Phys. Status Solidi A,* **21** [9] 9–40 (1974).
[3]R. C. Garvie, R. H. J. Hannink, and R. T. Pascoe, *Nature (London),* **258** [5537] 703–704 (1975).
[4]D. L. Porter and A. H. Heuer, *J. Am. Ceram. Soc.,* **60** [3-4] 183–84 (1977).
[5]R. T. Pascoe, R. H. J. Hannink, and R. C. Garvie; pp. 447–54 in Science of Ceramics, Vol. 9. Edited by K. J. de Vries. The Nederlandse Keramische Vereniging, 1977.
[6]R. C. Garvie, R. H. J. Hannink, and R. T. Pascoe, Aust. Pat. Application No. 85680/75 (1979).
[7]R. K. Stringer and R. T. Pascoe, *Proc. Aust. Ceram. Conf., 7th,* 1976.
[8]R. C. Garvie, R. R. Hughan, and R. T. Pascoe, *Mater. Sci. Res.,* **11,** 263–74 (1978).
[9]R. C. Garvie and P. S. Nicholson, *J. Am. Ceram. Soc.,* **55** [6] 303–305 (1972).
[10]R. H. J. Hannink, *J. Mater. Sci.,* **13,** 2487–96 (1978).
[11]R. T. Pascoe and R. C. Garvie; pp. 774–84 in Ceramic Microstructures '76. Edited by R. M. Fulrath and J. A. Pask. Westview, Boulder, Colo., 1977.
[12]G. W. Greenwood, "The Mechanism of Phase Transformation in Crystalline Solids," *Monogr. and Rep. Series Inst. Met. (London),* No. 33, 103–10 (1966).
[13]I. M. Lifshitz and V. V. Slyozov, *J. Phys. Chem. Solids,* **19** [1-2] 35–50 (1961).
[14]C. Wagner, *Z. Elektrochem.,* **65** [7-8] 581–91 (1961).
[15]J. Burke; pp. 140–41 in The Kinetics of Phase Transformations in Metals.

Pergamon, Oxford, 1965.

[16]W. H. Rhodes and R. E. Carter, *J. Am. Ceram. Soc.*, **49** [5] 244–49 (1966).

[17]A. G. Evans and T. G. Langdon, *Prog. Mater. Sci.*, **21** [3-4] 171–441 (1976).

[18]E. D. Whitney, *J. Am. Ceram. Soc.*, **45** [12] 612–13 (1962).

[19]E. D. Whitney, *J. Electrochem. Soc.*, **112** [1] 91–94 (1965).

[20]D. L. Porter, A. G. Evans, and A. H. Heuer, *Acta Metall.*, **27** [10] 1649–54 (1979).

[21]D. J. Green, D. R. Maki, and P. S. Nicholson, *J. Am. Ceram. Soc.*, **57** [3] 136–39 (1974).

[22]T. Mitsuhashi and Y. Fujiki, *ibid.*, **56** [9] 493 (1973).

[23]J. P. Coughlin and E. G. King, *J. Am. Chem. Soc.*, **72** [5] 2262–65 (1950).

[24]R. W. Davidge and T. J. Green, *J. Mater. Sci.*, **3**, 629–34 (1968).

[25]A. G. Evans, A. H. Heuer, and D. L. Porter, *Proc. 4th Int. Conf. on Fracture*, **1**, 529–55 (1977).

[26]E. Hornbogen, *Acta Metall.*, **26**, 147–52 (1978).

*Deceased.

†99.9% ZrO_2, "glass quality," Ugine Aciers, 15 rue du Rocher, Paris, France.

‡Analytical reagent grade, British Drug Houses, Chemicals Division, Poole, England.

Table I. Lattice parameter and cell volumes

Sample condition	Cubic (nm)	Cell vol (nm³)	Tetragonal (nm)	Cell vol (nm³)	Monoclinic (nm)	Cell vol (nm³)
Solution-treated	0.51238 ± 0.00003	0.1345				
Under-aged	0.5132 ± 0.0001	0.1352	0.5092 ± 0.0001 0.5177 ± 0.0001	0.1342		
Peak-aged	0.51325 ± 0.00010	0.1352	0.5094 ± 0.0001 0.5177 ± 0.0003	0.1343	0.5156 ± 0.0004 0.5191 ± 0.0005 0.5304 ± 0.0004 98.8 ± 0.9°	0.1403
Over-aged	0.5135 ± 0.0002	0.1354			0.5153 ± 0.0003 0.5194 ± 0.0003 0.5294 ± 0.0004 98.9 ± 0.7°	0.1400

Table II. Phase analysis of surfaces with various treatments

Specimen history	% monoclinic	Half-peak width (°2θ)
As-aged	41.1	0.46
Ground	29.0	.54
2.3 h/1000 °C	37.3	.47
3.6 h/1000 °C	37.5	.47
51.6 h/1000 °C	40.4	.46

Table III. Chemical, strain, and interfacial energies of the transformation

	Heating	Cooling
Mean temp. (°C)	724	203
$\Delta\epsilon$ (expt.) (MPa)	-45	$+111$ (22%)
$\Delta\epsilon$ (calc.) (MPa)	30	53
$\Delta\gamma$ Jm^{-2}	-1.1 (18%)	$+1.8$ (23%)
q (MPa)	$+284$ (71%)	-284 (55%)

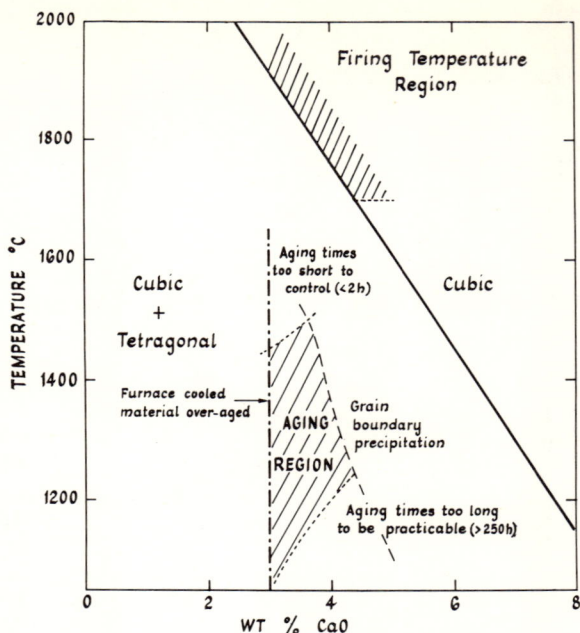

Fig. 1. "Working" phase diagram for the ZrO_2-CaO system. The solubility of CaO in monoclinic and tetragonal zirconia is unknown but small. The tetragonal-to-monoclinic transition is subject to considerable hysteresis and an "average" transition temperature is shown.

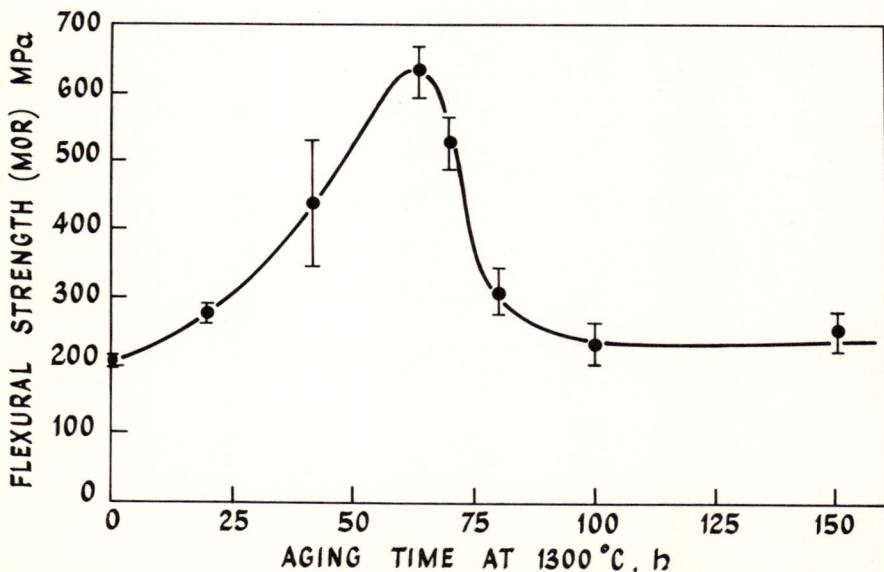

Fig. 2. Changes in flexural strength during aging of a ZrO_2-8.4 mol% CaO alloy at 1300 °C.

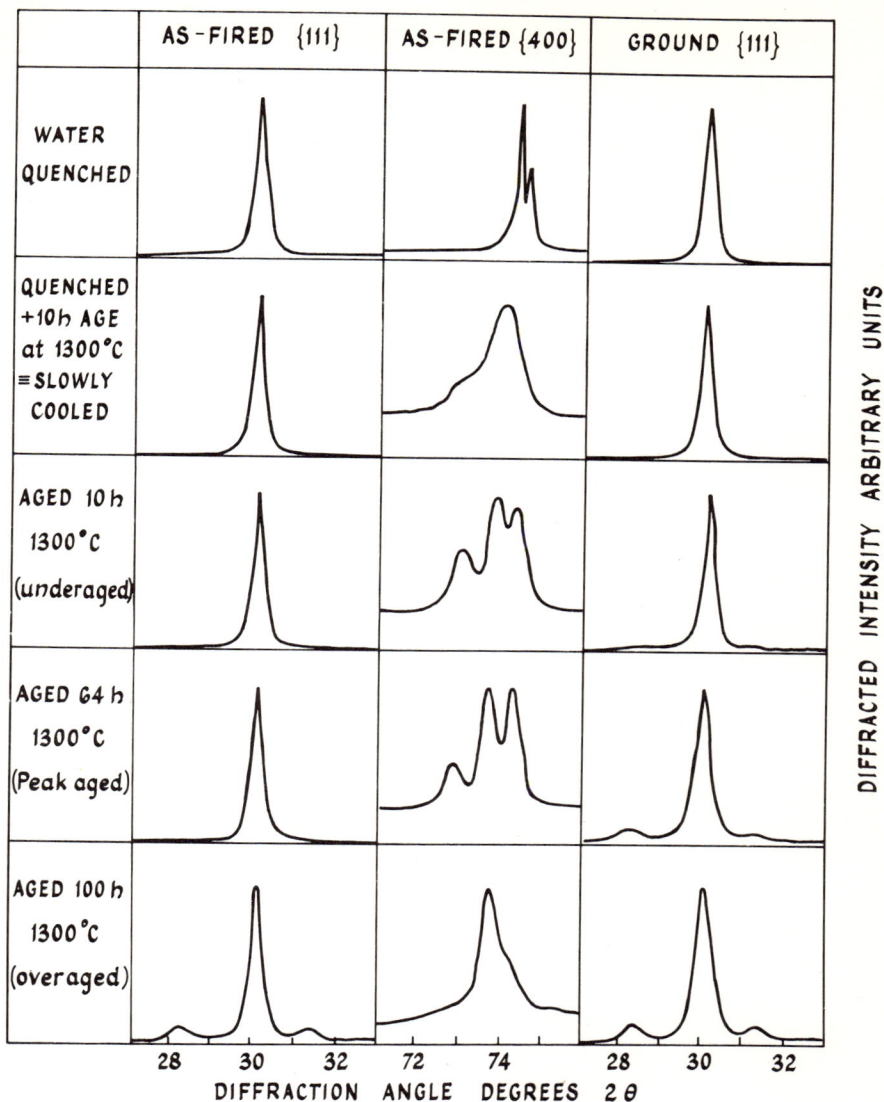

Fig. 3. X-ray diffraction profiles showing the structural development during aging. The central {111} peak is due to cubic and tetragonal phases and the side peaks to the monoclinic phase. The {400} peaks show the decomposition of an initially well-resolved cubic phase into a mixture of cubic and tetragonal phases.

131

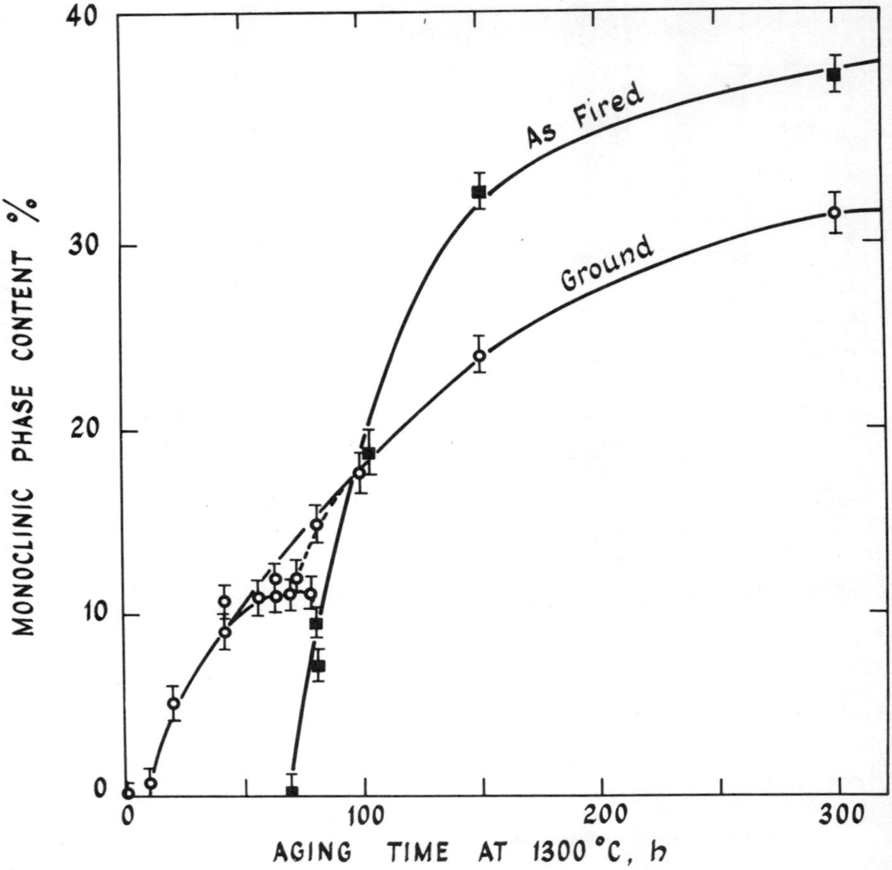

40
30
20
10
0
0
100
200
300
MONOCLINIC PHASE CONTENT %
AGING TIME AT 1300 °C, h
As Fired
Ground

Fig. 4. Change in monoclinic content measured on ground and as-fired surfaces during aging of a ZrO_2-8.4 mol% CaO alloy at 1300 °C.

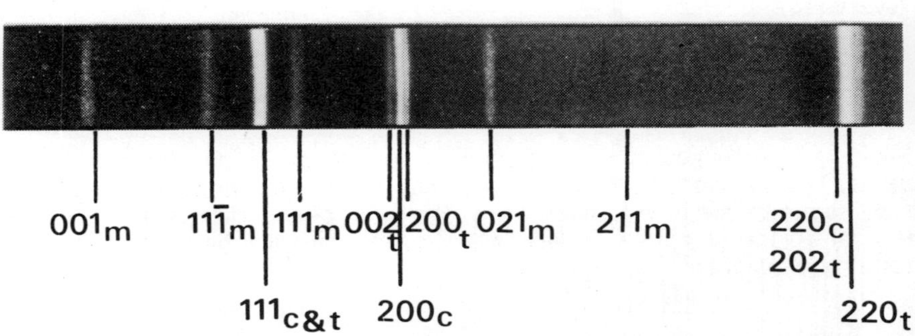

001m
111̄m
111m
002t
200t
021m
211m
220c
202t
111c&t
200c
220t

Fig. 5. Guinier X-ray film, obtained from peak-aged ion beam thinned sample, illustrating the coexistence of the cubic, tetragonal, and monoclinic phases.

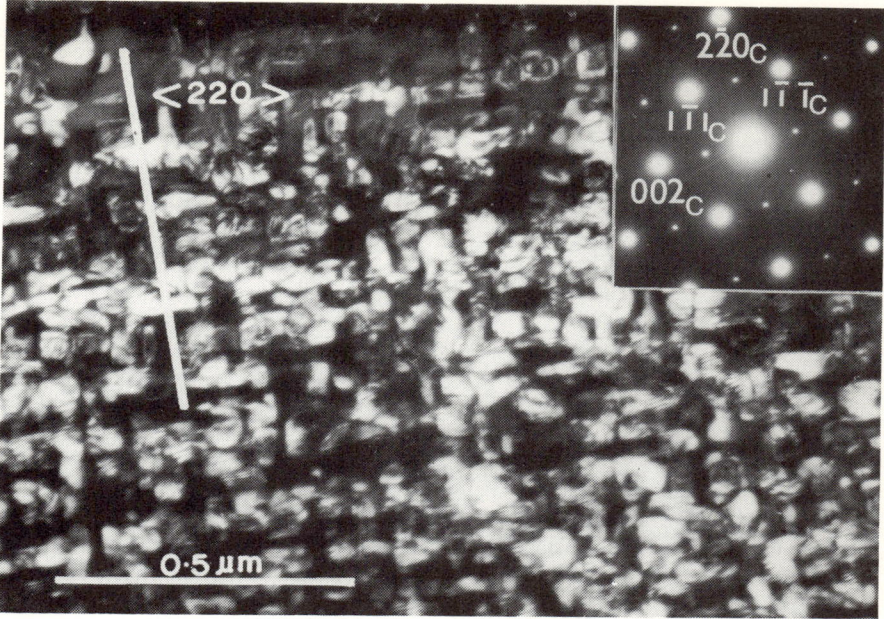

Fig. 6. Tetragonal precipitate distribution in the underaged sample. The inset, a {110} cubic selected area diffraction pattern, shows the cubic-tetragonal orientation relationship. Dark-field electron microscope image, inset not corrected for rotation.

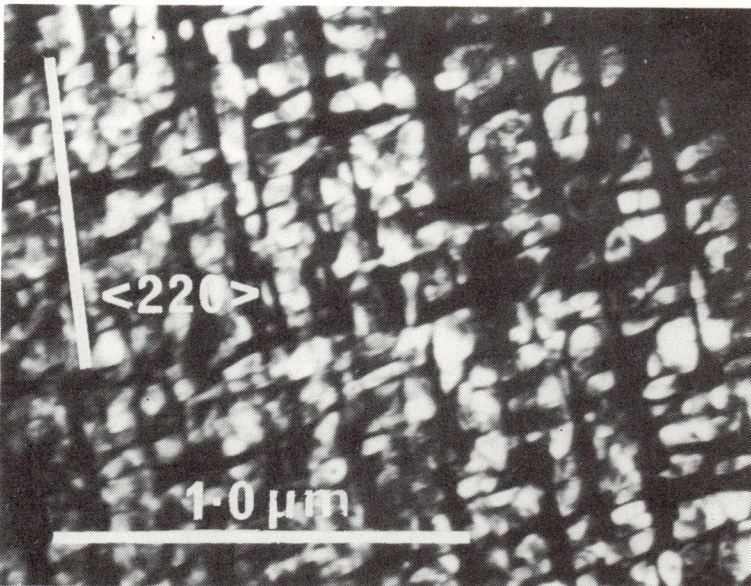

Fig. 7. Tetragonal precipitate morphology in peak-aged material. Dark-field image.

Fig. 8. Dark-field image of overaged material showing the twinned monoclinic particles. The tetragonal-monoclinic transformation has occurred in the bulk sample. Inset is a {110} selected area diffraction pattern and shows variants and twin spots of the monoclinic phase; it is not corrected for rotation.

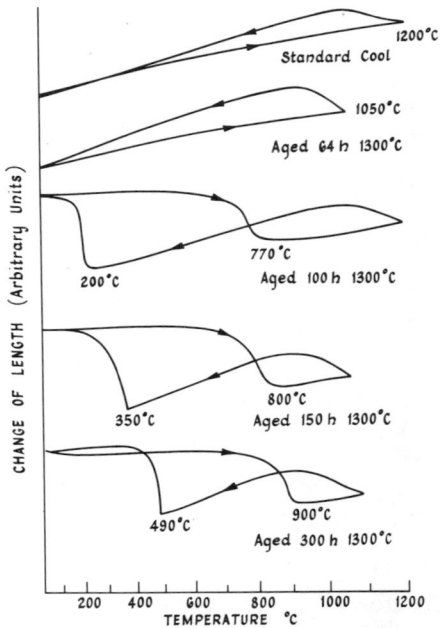

Fig. 9. Typical dilatometer traces after various aging treatments.

134

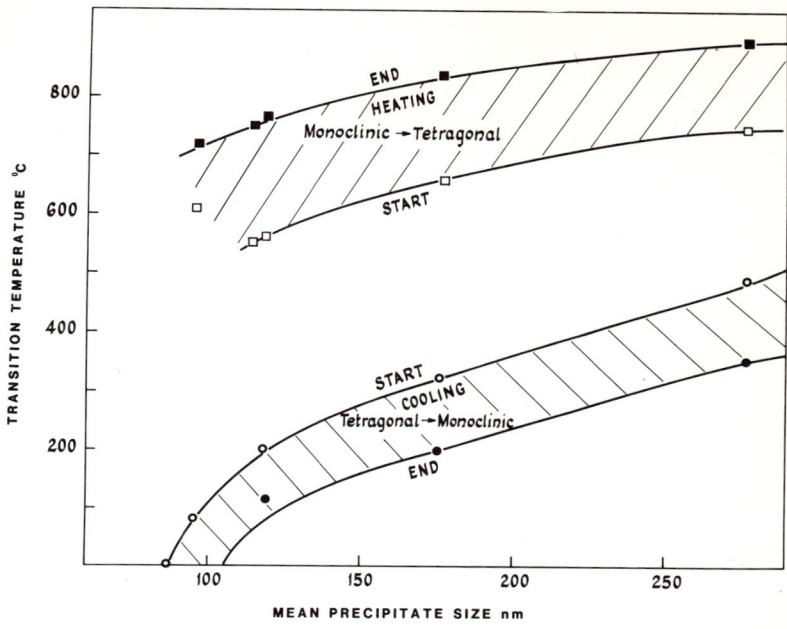

Fig. 10. Temperature for the start and end of the monoclinic⇄tetragonal transition as determined by dilatometer measurements.

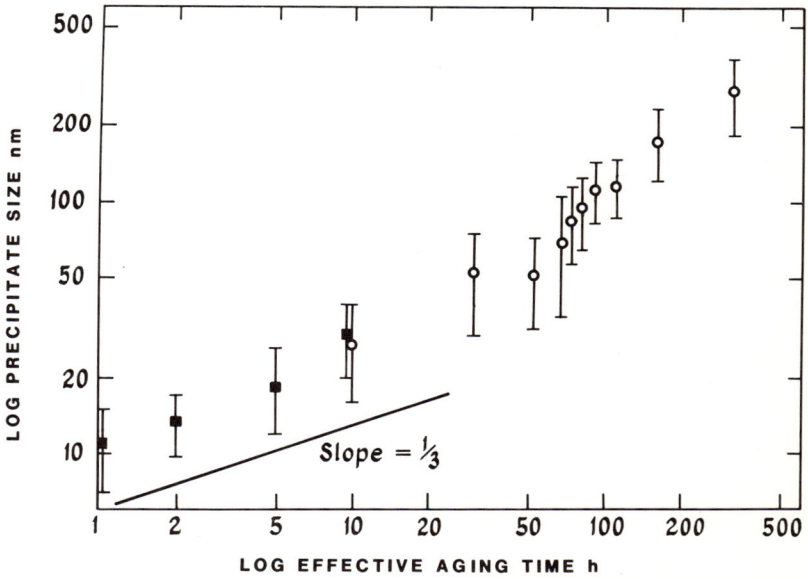

Fig. 11. Precipitate growth data plotted on logarithmic axes; 10 h has been added to the aging times of the slow cooled samples to allow for growth during cooling.

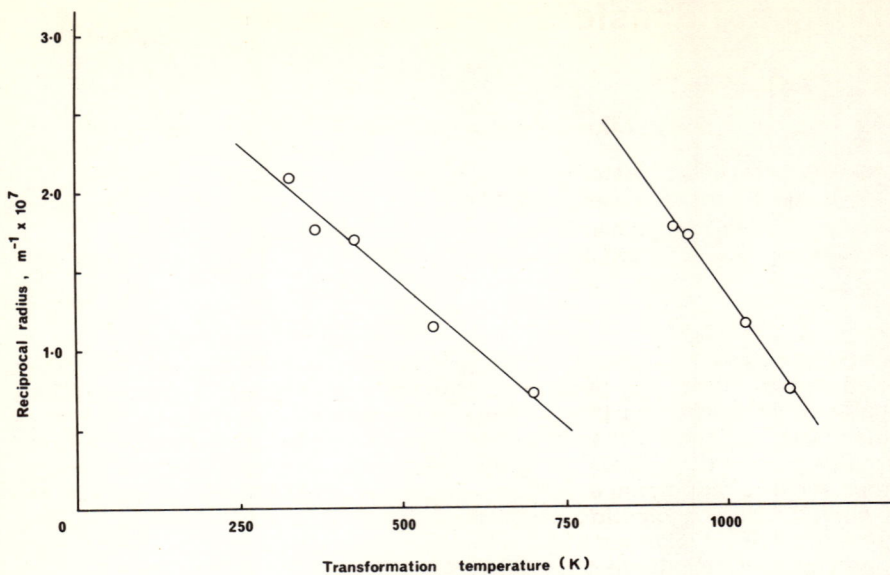

Fig. 12. Reciprocal precipitate size plotted as a function of the transformation temperature.

Design of transformation-toughened ceramics

NILS CLAUSSEN AND MANFRED RÜHLE

Max-Planck-Institut für Metallforschung
Institut für Werkstoffwissenschaften
Pulvermetallurgisches Laboratorium
Heisenbergstr. 5, D-7000 Stuttgart, Germany

Ceramics can be toughened by utilizing the tetragonal-to-monoclinic phase transformation of ZrO_2 particles dispersed in the ceramic matrix. Various parameters controlling this martensitic transformation (especially M_s) are discussed, together with in situ-TEM observations and the respective stress analysis around the ZrO_2 particles. The possible toughening mechanisms that allow for additional energy dissipation at crack tips are stress-induced transformation, nucleation of microcracks, and crack branching. Further strengthening is achieved by introducing steady-state compressive stresses. The type and magnitude of toughening is controlled by the microstructural design parameters such as volume fraction, chemical composition, size, and size distribution of ZrO_2 particles, which again determine the M_s temperature. The homogeneous dispersion of ZrO_2 is the most important step in the technology of ZrO_2-toughened ceramics. Mechanical mixing, sol-gel, and in-situ dispersion techniques are compared and possibilities of simulating the "parent material," PSZ, are shown. Special measures have to be taken to densify nonequilibrium systems, such as SiC-ZrO_2, to prevent destructive reactions. Experimental examples for the toughening effect are given for the ceramic matrices Al_2O_3, spinel, mullite, Si_3N_4, and SiC. The influence of MgO and HfO_2 on the transformation behavior of ZrO_2 in Al_2O_3 is indicated.

The tetragonal→monoclinic transformation of ZrO_2 and/or HfO_2 particles and the resulting volume increase of the particles can be utilized to enhance the toughness and strength of ceramic materials.[1-10] Even though this fact has been well recognized, the exact mechanisms of the toughness increase still remain under active discussion. This is especially due to the fact that at least two mechanisms have been found which result in this enhancement of toughness; in one case, the martensitic transformation of ZrO_2 particles near the advancing crack tip is directly involved in the energy absorption (stress-induced transformation) and, in the other case, nucleation of matrix microcracks and residual stresses, which are due to particles transformed on cooling before the specimen is loaded, are responsible for increased energy absorption during crack propagation (microcrack nucleation).

"Transformation toughening," which was previously thought to be a phenomenon inherent only in partially stabilized ZrO_2 (PSZ) ceramics, has been shown to be also applicable to Al_2O_3 and other ceramic matrices[2,4] in which ZrO_2 can be incorporated. Another important aspect of transformation toughening is the generation of compressive surface stresses (surface toughening) resulting in considerable strength increases.

The objective of the present paper is to review the experimental results which show that ZrO_2 dispersions increase the strength through enhanced toughness and introduction of residual compressive surface stresses. The influence of the ZrO_2 transformation on the thermal shock behavior has been treated in a separate paper.[11] Special consideration will be given to a most important microstructural design parameter, the ZrO_2 particle size, because it controls the temperature for spontaneous and stress-induced transformation and it determines whether microfracture can take place spontaneously or under an external stress. Since the critical particle sizes lie between 0.1 and 1.0 μm, results of TEM analysis will form an essential background for the understanding of toughening mechanisms and for further technological developments. The theoretical background will be touched on only as it helps in understanding experimental findings.

The Transformation of ZrO_2 Particles in a Ceramic Matrix

The martensitic transformation of single crystals and polycrystalline bulk ZrO_2 has been presented in detail (see for example Subbarao et al.[12] and Bansal and Heuer[13]). The M_s temperature (start of transformation on cooling) occurs between 950° and 850°C. ZrO_2 particles embedded in other ceramic matrices are usually retained to lower temperatures[7,8,14] due to the constraint on the transformational volume expansion and shape change by the surrounding matrix. Relief of this constraint, arising from matrix microcracking[15] and/or externally applied stress, triggers the transformation to the stable monoclinic symmetry.

A convenient way of representing and characterizing the transformation behavior of ZrO_2 particles in a dispersion composite is by a dilatation-temperature curve, as shown schematically in Fig. 1. One important quantity, as related to the toughening mechanisms, is the M_s temperature, because it determines which type of toughening may be dominant. M_s depends in a very complex way on various parameters such as particle size, chemical composition, shape, site (within the grains or along grain boundaries), elastic properties, thermal expansion coefficients, etc. In composites with a ZrO_2 particle size distribution, M_s relates only to the martensite start temperature of a few, usually the largest, particles in a composite. Correspondingly, M_f pertains to the temperature of the completion of the transformation of the smallest particles. Thus, since most ZrO_2 composites studied so far contain some particles which are small enough to be retained in tetragonal symmetry to absolute zero, a correct M_f cannot be defined. However, for the present purposes, M_f will be taken as the temperature at which the dilatometric cooling curve approaches the heating curve to 90% of the transformational height, H. The difference between M_s and M_f, which more or less represents the slope of the forward transformation, is controlled by the particle size distribution and, to some extent, by the homogeneity of the spatial distribution. The reverse transformation starting at A_s has been found to be much less influenced by the various parameters, although a tendency similar to that for M_s exists; e.g. A_s decreases slightly with smaller particle size. Additions like HfO_2, however, that increase the chemical driving force can drastically increase A_s. A_s refers to only one group of particles, probably the smallest particles, i.e. the same remarks true for M_s apply to A_s.

Conditions for the Tetragonal→Monoclinic Transformation

The transformation takes place when the particle/matrix system can reduce its total free energy, G. The change in free energy between the two lattice modifications per unit transformed volume is given by[16]

$$\Delta G = \Delta G_{chem} + \Delta U_T + \Delta U_a \tag{1}$$

where ΔG_{chem} is the chemical free energy difference between the transformed and untransformed states. ΔU_T is the transformational strain energy* which is modified by an applied stress

$$\sigma_a = -\Delta U_a/\epsilon_T \tag{2}$$

where ϵ_T is the transformational strain and ΔU_a the applied strain energy. A more detailed analysis of the energies involved in the transformation is presented in Ref. 14. A surface energy term has been neglected in Eq. (1). The conditions required for the transformation to occur are

$$-\Delta U_a \geq \Delta U_T + \Delta G_{chem} \tag{3}$$

Thus, the applied stress necessary to cause the transformation should be

$$\sigma_a > \frac{\Delta U_T + \Delta G_{chem}}{\epsilon_t} \tag{4}$$

Useful Parameters for Controlling M$_s$

The most convenient and effective parameters for controlling M_s are the particle size[7,18] and the chemical composition[1,3,8]; the parameters mentioned previously are more or less fixed for a given particle-matrix system. The exact origin of the size dependency of M_s is yet unclear. Possible reasons are the difficulty of martensite nucleation, a size-dependent activation energy, and the possibility that the shear strain energy is proportional to the surface area of the particles.[17]

Aging heat treatments are usually applied to PSZ ceramics to optimize the precipitate size and, hence, the transformation behavior. Another way of controlling M_s through the particle size is by milling the composite powder mixture for various times. Dilatometer curves for $Al_2O_3 + 15$ vol% ZrO_2 attrition milled between 10 min and 16 h are shown in Figs. 3 and 4 of Ref. 7. While the M_s of the 10 min composite is 650 °C (corresponding to a critical size $d_c \approx 1.0\ \mu$m), M_s of the 6 h composite is 350 °C ($d_c \approx 0.7\ \mu$m), and M_s of the 16 h composite is below RT ($d_c \approx 0.4\ \mu$m). The last case is thus suitable for stress-induced transformation toughening.

A second example for the M_s size dependence is given in Fig. 2 for an Al_2O_3 matrix in which ZrO_2 has been dispersed by an oxychloride decomposition technique.[19] The as-prepared composite contained 100% tetragonal ZrO_2 at room temperature. After annealing at high temperatures for various times, to permit grain growth, dilatometer results were correlated to linear grain size analysis by SEM, as depicted in the upper left of the figure. In the special case where the ZrO_2 particles are located predominantly at the Al_2O_3 grain boundaries, the average critical size for spontaneous transformation at RT is 0.52 μm.

Some critical particle sizes for M_s = room temperature, determined by different methods, are listed in Table I. The differences in critical size are not

139

so much due to a thermoelastic mismatch but rather to variations in chemical composition of the ZrO_2 particles in the different matrices. This can also be deduced from the extensive work on the influence of stabilizing additives. All compounds except HfO_2, soluble in ZrO_2, have been shown to reduce M_s to a lesser or greater extent. Even such oxides as Al_2O_3 or SiO_2, not actually considered as stabilizers, act in lowering M_s.[26,27] Up to $\approx 10\%$ Al_2O_3 has been detected by EELS in ZrO_2 particles dispersed in a mullite matrix,[22] probably causing the relatively large critical size. Only the addition of HfO_2[20] increases ΔG_{chem} and therefore M_s.

The utilization of compositional changes to control M_s is somewhat limited because a simultaneous effect of the matrix properties must be expected; for example, in Al_2O_3-ZrO_2 composites, additions of CaO will preferentially migrate to the Al_2O_3 grain boundaries and hence deteriorate the inherent properties of the matrix rather than reduce the M_s temperature of the ZrO_2 particles. Similar results have been obtained with MgO additions exceeding the usual 0.25–0.5% added to enhance sintering and to control the Al_2O_3 microstructure. Even the use of fully MgO-stabilized ZrO_2 powder yields M_s temperatures in the Al_2O_3 matrix almost identical to those when pure ZrO_2 powder of the same particle size is used. During sintering, the MgO-doped ZrO_2 is destabilized when MgO diffuses along the Al_2O_3 grain boundaries during the sintering process; spinel formation is the result. A similar effect is seen when spinel and ZrO_2 powders are sintered; transformation toughening still occurs.[23] Y_2O_3 and most rare earth oxides are more efficient in reducing M_s because their solid solutions with ZrO_2 are more stable at the sintering temperatures; less Y_2O_3 is required to lower M_s than is required of CaO or MgO. But, as will be discussed later, in order to keep toughening potential as high as possible, it seems to be more advisable to control M_s by changing only the particle size.

In-Situ TEM Observations of the Transformation

The martensitic transformation from the tetragonal to the monoclinic structure of small ZrO_2 particles embedded in a ceramic matrix can be observed directly in the TEM. Specimens were selected in which the major fraction of the ZrO_2 particles was present at room temperature in the tetragonal phase. The TEM specimens were prepared by grinding, polishing, and ion thinning with a standardized technique.[28,29] Since most of the tetragonal particles were transformed during grinding and ion thinning of the specimen, the thin specimens were annealed at 1200°C prior to TEM observations.

Tetragonal particles could easily be identified in TEM. They showed, as demonstrated in Fig. 3(A), a slightly stronger absorption contrast of the electrons than the adjacent Al_2O_3 grains. Particles embedded in the Al_2O_3 grains were round (Fig. 4), and ZrO_2 particles at the grain boundaries were irregularly shaped. For the latter case it has been determined[30] that, at room temperature, nearly all particles with diameters <0.5 μm had tetragonal symmetry, while the larger particles were predominantly monoclinic.

The martensitic transformation was observed during in-situ experiments using HVEM by either (A) cooling the specimen to a temperature below M_s, (B) by straining the specimen with an in-situ straining stage,[31] or (C) by buckling a specimen due to heating of the specimen with the electron beam.

140

Figure 3*(B)* shows the same area of the specimen shown in Fig. 3*(A)* after cooling to 183 K. It can be seen that the two particles arrowed in Fig. 3 have transformed. Twins introduced during the transformation can easily be detected. It was observed that $M_s \geq$ 183 K for a large fraction of the particles. The same specimen was cooled to 20 K. It was found that 16% of the round particles remained tetragonal at that temperature, while nearly all irregularly shaped particles transformed martensitically.

Specimens of a special geometry were developed for a double tilting deformation stage[31] which fitted to an HVEM (operated at 1 MeV). Figure 4 shows a micrograph of Al_2O_3 containing 4% ZrO_2. The specimen was prepared with a special sol-gel technique so that nearly all ZrO_2 particles were enclosed in the Al_2O_3 grains. It can be seen from Fig. 4 that these particles were regularly shaped and spherical. Most of the particles in the ≈ 4 μm thick foil were tetragonal. The ZrO_2 particles transformed when a strain was applied, as shown in Fig. 5, and the transformation was recorded on video tape. A small fraction of ZrO_2 particles transformed in discrete steps with an overall transformation time up to several seconds. Post-transformation rearrangements of the twinned structure were frequently observed.[30]

Details of the crystallography of the transformation and the difference in the transformation behavior after cooling and straining will be reported elsewhere.[30]

The formation of cracks could also be observed in the in-situ deformation and indentation experiments. In fine-grained Al_2O_3 (≈ 2 μm) containing 8 vol% intercrystalline ZrO_2 a transformation zone of ≈ 3 μm was determined; all particles adjacent to the crack transformed.

Stress Analysis

The stresses in the Al_2O_3 matrix surrounding a ZrO_2 particle were analyzed by evaluation of TEM micrographs taken under well defined diffraction conditions. This can, of course, reliably be done only if the particle is embedded completely in an Al_2O_3 grain. For the analysis, the specimen is imaged under different diffraction conditions (with respect to the Al_2O_3 matrix). The observed strain contrast profiles are then compared to calculated contrast profiles obtained from different strain field models. It was observed that even tetragonal particles embedded in Al_2O_3 show strain contrast (Fig. 6*(A)*), which arises from the thermal mismatch between Al_2O_3 and ZrO_2. From a qualitative analysis it can be concluded that the Al_2O_3 matrix is under radial tension, and a quantitative analysis yields the expected volume mismatch of $\approx 2\%$. Furthermore, it was found that the strain around the tetragonal particle is approximately isotropic. For ZrO_2 particles embedded in spinel, very little strain contrast could be observed (Fig. 6*(B)*). This is expected since almost no thermal mismatch exists between spinel and tetragonal ZrO_2.

After transformation, the sign of the strain field changed; thus the Al_2O matrix was under radial compression. From a qualitative inspection (Fig. 7*(A)* *(B)*) it can be concluded that the strain field is at least locally larger (and of opposite sign) than before the transformation. The TEM studies showed, by imaging with different diffraction conditions, that the strain around a monoclinic particle is very anisotropic (Fig. 8) due to the anisotropic transformation behavior. Evaluations of the contrast showed that strains may reach up to 0.1.[32] The decrease of strain with distance, r, from the particle can also be determined. Depending on the orientation of the transformed particles, the

strain is proportional to r^{-n}, where n varies between 3 and 6.

Toughening Mechanisms

Increasing the fracture toughness of ceramics implies the activation of mechanisms that reduce the stress concentration ahead of the tip of a loaded crack. This "crack blunting" allows further loading until the critical stress intensity factor, K_{Ic}, is reached. During this process, elastic energy is converted into other forms of energy. The increase in fracture toughness can therefore be considered a measure of the maximum energy, U_c, absorbed ahead of the macrocrack tip when the crack is loaded from zero to a critical value where crack extension occurs. U_c will be absorbed with an energy density, n_c, in a volume which is characteristic for the participating absorption mechanisms, V_c. Assuming a cylindrical shape for V_c of radius ϱ_c located as shown in Fig. 9, the absorbed energy per unit crack front length is[33]:

$$U_c = n_c \pi \varrho_c^2 \tag{5}$$

In the dissipation of U_c, the material within ϱ_c is "used up," i.e. the critical process zone has developed. This state corresponds to the point when the crack extends. Hence U_c is equivalent to the critical energy rate, G_c, released along the path $\pi/2 \cdot \varrho_c$. Therefore

$$K_{Ic} = \sqrt{2E n_c \varrho_c} \tag{6}$$

where E is the Young's modulus of the material. When different mechanisms are involved in the energy absorption, U_c is composed of a combination of the different energy absorption mechanisms, i.e. that inherent to the matrix, U_0, and to stress-induced transformation and microcracking nucleation, U_T and U_M, respectively. The total fracture toughness is then

$$K_{Ic} = K_0 + \sqrt{2E n_T \varrho_T} + \sqrt{2E n_M \varrho_M} \tag{7}$$

where K_0 is the toughness of the matrix and $n_{T,M}$ and $\varrho_{T,M}$ are characteristic magnitudes of the respective mechanisms.

Transformation toughening addresses both mechanisms, since improvements in toughness have been observed in bodies containing both tetragonal and monoclinic ZrO_2 particles. It is yet unclear whether the mechanisms are additive or exclude each other. Furthermore, an interaction may be possible which is not considered in Eq. (7).

Another kind of transformation toughening which also involves the reduction of stress concentration at the crack tip is by superimposing compressive stresses. This type of "toughening," commonly used in glass technology, is especially applicable to surface cracks and therefore may be termed "surface toughening."

Stress-Induced Transformation

This type of toughening is possible for matrix/particle systems in which the M_s of a large fraction of ZrO_2 particles is below the test temperature, e.g. for $T = RT$ the forward transformation curve 1 in Fig. 1 would apply. Rewriting Eq. (7) yields the fracture toughness

$$K_{Ic} = K_0 + \sqrt{2E n_T \varrho_T} \tag{8}$$

where $n_T = v(-\Delta U_a) = v(\Delta U_T + \Delta G_{chem})$, v being the volume fraction of transformable ZrO_2 particles.

The transformation zone size can be estimated from[4,37]

$$\sigma_a = K_{1c} / \sqrt{2\pi \varrho_I} \tag{9}$$

With Eq. (4), the zone size is

$$\varrho T = \frac{K^2_{1c}\epsilon^2_T}{2\pi(\Delta U_T + \Delta G_{chem})} \tag{10}$$

Setting Eq. (10) into Eq. (8) the toughness can be expressed by

$$K_{1c} = K_0 \left[1 - \sqrt{\frac{Ev\epsilon_T}{\pi(\Delta U_T + \Delta G_{chem})}} \right]^{-1} \tag{11}$$

The relative volume expansion $\Delta V/V$ must be corrected for a thermal mismatch:

$$\epsilon_T = \frac{\Delta V}{V} - 3\Delta\alpha\Delta T \tag{12}$$

where $\Delta\alpha$ is the thermal expansion mismatch between the tetragonal form of ZrO_2 and the matrix, ΔT is the difference between preparation and test temperature (e.g., ϵ_T in PSZ ≈ 0.045, in Al_2O_3-$ZrO_2 \approx 0.025$). Since it is more difficult to measure ΔG_{chem} than ϱ_T a more useful expression for the fracture toughness is obtained by rewriting Eq. (8) using $\eta_T = v\epsilon_T^2 E$:

$$K_{1c} = K_0 + \epsilon_I E \sqrt{2v\varrho_I} \tag{13}$$

The effect of ZrO_2 particle density (number per volume) within ϱ_I, which is not contained in Eqs. (11) and (13), can best be indicated by comparing the two most important systems, $Al_2O_3 + ZrO_2$ and PSZ, with different ZrO_2 particle size but the same transformation zone and volume fraction, e.g. 15 vol% and $\varrho = 2\ \mu m$. In this case the transformed zone (see Fig. 9) would just contain one particle of 0.5 μm (typical for $Al_2O_3 + ZrO_2$). However, for PSZ ceramics ≈ 70 precipitate particles (largest axis 0.2 μm)³ would be present in the process zone. It is obvious that, for the PSZ-type composite, a quasi-hydrostatic state of compression will exist. This suggests that U_T can be considered as an average strain energy within the process zone volume, rather than just U_T associated with a single transformed particle. On the other hand, the stress state around the single particle in the Al_2O_3 matrix is highly localized and may even attract the main crack, i.e. uniform energy density distribution is not available. Therefore, it should be realized that, although identical energy quantities are dissipated, process zones with inhomogeneous energy density distribution are less effective preventing crack extension than zones with homogeneous energy distribution. Thus, a large number of small particles within a given ϱ_T would be preferred to a few large particles.

Microcrack Nucleation

Stress-induced transformation toughening cannot, by definition, take place when exclusively monoclinic particles are present at the test temperature. In this case, a toughness increase is termed "microcrack toughening," because it is hypothesized that additional energy is absorbed by microcracking[2,4,35] in the stress field around the transformed particles. Microcracks, especially between ZrO_2 particles, have been observed in Al_2O_3-ZrO_2 composites by TEM.[15,36] However, recent TEM studies utilizing an in-situ tensile straining stage have not detected microcracking in thin foils.[37] On the other hand, extensive microcracking has been detected during in-situ SEM investigation of Al_2O_3-ZrO_2 composites.[38] Further indirect evidence for microcracking being responsible for energy absorption may be concluded from K_{Ic}-vs-grain size results of ZnO,[33] where K_{Ic} increases with decreasing grain size and with a simultaneously increasing fraction of intergranular fracture surface. In spite of these indications of microcracking, other physical interpretations of monoclinic ZrO_2 toughening, involving interaction of residual stresses with propagating cracks, may be possible.

ZrO_2 particles which have spontaneously transformed, i.e. where $M_s > T$ (for $T = RT$, curve 2 in Fig. 1 would be representative) are thought to contribute to the toughness by forming an additional source for microcracking. In the nucleation of microcracks, elastic energy is mainly converted into surface energy. In analogy to Eq. (8)

$$K_{Ic} = K_0 + \sqrt{2E\eta_M\varrho_M} \tag{14}$$

where

$$\eta_M = \gamma m \tag{15}$$

γ is the specific energy of the cracked surfaces, usually the grain boundaries, and m is the microcrack area density created by the externally applied stress[4]:

$$m = (v/d)\,F \tag{16}$$

where F is a function $T(\propto 1/d)$ which takes into account the microcrack area fraction used to release the particle transformational strain energy and a particle/matrix grain size relation.[33]

In order for the ZrO_2 particles to induce microfracturing, the particle size must be $d \leq d_c'$,[39] where $d_c' \geq c/\sigma^2$. The hoop stress σ results from a superposition of transformational stress and crack tip stresses due to the externally applied load. Since from Eqs. (14) to (16) the toughness increases with $1/d$, the transformational stresses should be as high as possible to allow the use of the smallest possible particle size. Furthermore, d should be below d_c'', the critical size for spontaneous microcracking (for ZrO_2 particles[2] in Al_2O_3 $d_c'' \approx 3$ μm) so that little of the available microcrack area is used to release transformational strain energy. A stress analysis,[30,32] however, suggests that the stresses are highly localized at the interface, decreasing with distance r from the interface proportional to $1/r^3$ to $1/r^6$. This indicates that the ZrO_2 transformation is highly suitable to just nucleate microcracks without causing excessive extension of these microcracks.

A rough estimate of the process zone ϱ_M can be made by a surface flaw evaluation[40] such that

$$\varrho_M = \frac{1}{4}\left(\frac{K_{Ic}}{\sigma_F}\right)^2 \tag{17}$$

where σ_F is the flexural strength of the unnotched specimens. According to Eq. (17), ϱ_M values for Si_3N_4-ZrO_2 and Al_2O_3-ZrO_2 range from 20 to 100 μm. However, no experimental evidence for microcrack process zones has yet been found in these materials.

Surface Toughening

The strength of ceramic materials can be increased by introducing compressive surface stresses, as is state of the art in glass technology.[41] In

144

ceramics containing ZrO_2, these stresses are produced by transforming tetragonal particles in the surface layer to monoclinic symmetry and maintaining those in the bulk with tetragonal symmetry. This localized volume expansion can be stress-induced, i.e. by grinding,[7,18,42] by sand blasting,[19] or by a low-temperature treatment when tetragonal ZrO_2 is present.

Another method of producing these surface stresses is by chemically destablizing tetragonal or cubic particles near the surface and allowing them to transform. Transient compressive stresses are generated on cooling composites containing monoclinic ZrO_2 to room temperature. Some techniques for the introduction of compressive stresses are listed in Table II.

The compressive stresses at $x=0$ (x = distance from the surface) can be estimated for a flat plate by[44]

$$\sigma_{c[0]} = \left(\frac{\Delta V}{3V} - \Delta\alpha\,\Delta T \right) \frac{v \cdot E}{1 - v} \tag{18}$$

assuming that the transformational depth s is small with respect to the thickness of the plate, that the Young's moduli of matrix and particles are similar, and that the volume fraction of transformable particles is modest. These conditions usually apply to surface-toughened ZrO_2 ceramics. The monoclinic fraction of particles transformed by grinding has been found to decrease with depth according to[19,42]

$$X_{(x)} = X_0 e^{-bx}$$

where X_0 is the monoclinic fraction at the surface after grinding. An example is shown in Fig. 10 for Al_2O_3-ZrO_2 composites with various volume fractions of tetragonal ZrO_2 particles which were surface ground under different pressures (s ≈ 40 μm). In a first approximation, the exponential decrease can be replaced by a linear function. Then the stress distribution between the surface and s is given by[45]

$$\sigma_{(x)} = -\sigma_{c(0)}(1 - x/s) \qquad 0 < x < s \tag{19}$$

The surface toughening is expressed by a critical stress intensity factor

$$K_{ST} = K_{Ic} + K_c \tag{20}$$

where K_{Ic} is the fracture toughness of the composite and K_c the stress intensity factor due to the compressive field[46]

$$K_c = y\sigma_{c(0)}\sqrt{\pi a \cdot M} \tag{21}$$

where y is a geometrical factor. The correction factor M, which depends on the ratio of both the flaw size, a, and s to the thickness of the plate, $2t$, is given by

$$M = 1 - (2a/\pi s) \qquad 0 \leq a \leq s \tag{22}$$

The fracture strength of a surface-toughened ceramic can now be predicted from Eqs. (18) to (21):

$$\sigma_F = \frac{K_{Ic}}{y\sqrt{\pi a}} + \frac{MvE}{1 - v}\left(\frac{\Delta V}{3V} - \Delta\alpha\,\Delta T \right) \tag{23}$$

Summary of Design Criteria for Potential Toughening

Some design criteria for optimum toughening given in the preceding sections can be summarized:

145

Toughening by Stress-Induced Transformation

The ZrO_2 particles dispersed in the ceramic matrix must be retained with tetragonal symmetry such that M_s of the composite is below the application (test) temperature. Within the frame of this basic requirement, the following parameters, which are partly interdependent and connected, can be optimized by:

Minimization of: *(a)* particle size, *(b)* stabilizing solute concentration (CaO, MgO, etc.), *(c)* particle size distribution, *(d)* particle-matrix thermal expansion mismatch.

Maximization of: *(e)* chemical driving force, ΔG_{chem}, *(f)* volume fraction of ZrO_2 particles, *(g)* destabilizing solute content (HfO_2), *(h)* inherent toughness and E of matrix.

The variation of these parameters, except for the particle size and size distribution, is usually rather limited for a given ZrO_2/matrix system. However, extra toughness may be obtained by adding high-modulus additions and the highest possible HfO_2 solute content.

Toughening by Microcrack Nucleation

Toughening by nucleation of microcracks (or interaction of residual stresses) may be optimized, as suggested from Eqs. (14) to (16), by:

(A) Minimizing the ZrO_2 particle size, d. However, $d_c' \le d \le d_c''$, where d_c' is the critical size for spontaneous or stress-induced transformation and d_c'' the critical size for spontaneous microfracture.

(B) Minimizing the thermal mismatch $(\alpha_p - \alpha_M)$ to allow maximum possible transformational stresses to build up; hence, a small particle size $(d \ge c/\sigma^2)$ can nucleate microcracks.

(C) "Controlling" the particle chemistry so that peak stress results from shear stresses rather than from volume expansion.

(D) Uniform particle spacing.

(E) Minimizing particle size distribution.

(F) Optimizing volume fraction; v is limited by agglomeration and too close spacing causing spontaneous microcracking.

(G) Arranging ZrO_2 particles at the matrix grain boundaries, while keeping the grain size of the matrix small, i.e. close to the ZrO_2 particle size (≈ 0.5 to ≈ 2 μm).

(H) Maximizing the inherent fracture toughness, the grain boundary energy, and Young's modulus.

Surface Toughening

The effectiveness of surface toughening depends on the technique by which transformation in the surface layer is achieved. The most effective technique so far investigated is by grinding such that the requirements listed in the section on Stress-Induced Transformation apply. In addition, Eq. (22) suggests that optimum strengthening is obtained by minimizing the ratio of flaw size to transformation depth. At the same time, the ratio of the transformed zone to the thickness of the body should be small ($\ll 1$).

Experimental Evidence of ZrO_2 Toughening

Effective toughening has been observed in PSZ ceramics; e.g. the strength of Ca-PSZ[42] was increased from 200 to 640 MPa and the fracture toughness of an Mg-PSZ[3] from 2.8 to 6 MPa $\sqrt{m}$ by optimally aging. In these cases, it is evident that only stress-induced transformation toughening takes place. In other ZrO_2 composites, however, toughening by microcrack nucleation may be dominant, especially when the monoclinic fraction of ZrO_2 is high at the test temperature. In the following, some new results obtained with composites other than PSZ are presented.

Al_2O_3-ZrO_2

An experiment, set up to decide whether microcracking or stress-induced transformation is the more powerful mechanism, is shown in the dilatometer curve of Fig. 11. An Al_2O_3-18 vol% ZrO_2 composite with 84% of the particles monoclinic at RT and an M_s of $\approx 600°C$ was investigated in an SENB test at 630°C. In one case, the SENB specimens were heated to 1200°C, a point at which all of the particles were tetragonal, then cooled to only 630°C, at which all of the particles remained 100% tetragonal. In the other case, the body was heated to only 630°C, a point at wich 84% of the particles were monoclinic, and tested. Identical values of the fracture toughness resulted from both procedures. One possible interpretation is that in both cases microcracking is responsible for the toughening. As predicted from Eq. (11), K_{Ic} at 630°C is reduced compared to the room temperature K_{Ic}, since ΔG_{chem} decreases with increasing temperature.

The dilatometer curve and the retained strength data of an Al_2O_3 material, in which the ZrO_2 was introduced by attrition wear of the ZrO_2 milling media, is shown in Fig. 12. Even though the ZrO_2 balls were fully stabilized (≈ 13 mol% MgO), the composite contained 80% monoclinic particles at room temperature after sintering at 1530°C for 2 h. This also indicates that the activity of MgO is lower in Al_2O_3 than in ZrO_2. The strength of annealed specimens was ≈ 600 MPa, double the matrix strength. In confirmation of the results by Becher,[10] the strength after quenching in boiling water[47] was retained for all temperature differences of thermal shock tested. However, ΔT_c, the critical temperature difference for drastic loss of strength for the 20 °C water quench was only slightly improved over the value for the pure Al_2O_3 matrix, $\Delta T_c \approx 200$ °C. It is felt that, in this material, the strength increase is due mainly to microcracking.

The flexural strength of composites containing >95% tetragonal ZrO_2,[19] surface toughened by grinding, is plotted versus the monoclinic fraction (true local fraction, see Fig. 10) and the resulting compressive stresses are computed from Eq. (18) in Fig. 13. The degree of transformation, i.e. the monoclinic fraction, was changed by variation of the grinding pressure. It is interesting to note that the "roughest" grinding, which produced the largest surface flaws, resulted in the maximum strength. Surface polishing after grinding (8% ZrO_2 composite) reduced the monoclinic fraction, hence also the compressive stresses; however, at the same transformed ZrO_2 level, the strength was higher due to decreased surface flaw size. Annealing at various temperatures ($T > A_f$) caused the particles to retransform, the surface stresses to heal out, and, consequently, the strength to be reduced to the as-fabricated value (Fig. 14).

A similar result is presented in Fig. 15 for composites in which the ZrO_2 particle size was varied by varying the milling time.[7,11] At a milling time of 10 h an optimum fraction of transformable particles (>90% tetragonal in the annealed condition at room temperature) was achieved. The apparent as-notched fracture toughness was ≈ 16 MPa $\sqrt{m}$, which, by annealing at

1350°C for 10 min, dropped to the value of the matrix. This indicates that stress-induced transformation ahead of a crack is ineffective, possibly due to a small transformation zone. Another explanation may be an "R-curve effect," i.e. a crack, when annealed, starts to propagate at a lower K, reaching K_{Ic} after complete development of the process zone ($\Delta a \approx 3 \cdot \varrho_T$).[48] It remains surprising that grinding and notching are especially effective in these mechanically mixed composites. It must furthermore be pointed out that the "annealed K_{Ic}" of the 10 min composite, which contains >90% monoclinic particles, is higher than that of the matrix. Hence, toughening by microcracking should be the prevailing mechanism.

The influence of small amounts of MgO on the fracture toughness of a 7.5 vol% ZrO_2-Al_2O_3 composite is demonstrated in Fig. 16. At $\approx 0.12\%$ MgO a maximum of ≈ 7 MPa $\sqrt{m}$ is obtained. It is interesting to note that these composites exhibit identical fracture toughness in the as-notched and in the annealed state; contrary to the material of Fig. 15, toughening by stress-induced transformation seems to be very effective in spite of the relatively low ZrO_2 content. K_{Ic} of the Al_2O_3 matrix with the same MgO content is ≈ 4.5 MPa $\sqrt{m}$. The composites were prepared by decomposition of Mg and Zr acetates which were mixed with Al_2O_3 powder. Even though the material of Fig. 15 (milled for 10 h contains similar ZrO_2 fraction and particle size ($\approx 90\%$ tetragonal; ≈ 0.5 μm) at room temperature, transformation is obviously enhanced. This may be due to slight differences in solute content, and thus ΔG_{chem} and transformation behavior.

A final example of the difficulty in interpreting the toughening origins and deciding the efficiency of the various mechanisms in Al_2O_3-ZrO_2 systems is shown by sintered 15 wt% ZrO_2-Al_2O_3 cutting tools†; materials with equal amounts of monoclinic and tetragonal ZrO_2 exhibit optimum resistance to fracture during high-speed turning of steel. Most likely, all three mechanisms contribute to the toughening in this case.

Spinel-ZrO₂

The strength increase of spinel ($MgO \cdot Al_2O_3$) from ≈ 200 MPa to ≈ 500 MPa when 17.5 vol% ZrO_2 is incorporated (Fig. 17)[23] exemplifies the fact that ceramic systems having high MgO content are toughenable under suitable conditions.‡ The composites were prepared by mixing Al_2O_3 and MgO powder with Zr acetate which, during heat treatment, decomposed before spinel formation took place (≈ 900 °C). After hot-pressing at 1350 °C for 20 min or sintering at 1670 °C for 1 h, the ZrO_2 particle content was 60% and 20% tetragonal, respectively, at room temperature. No cubic ZrO_2 could be detected, indicating that all MgO was used up in the spinel formation. As demonstrated in Fig. 11, however, no improvement in the critical thermal shock temperature difference was obtained when using the boiling water quench test[47]; this is probably due to the high inherent thermal expansion coefficient of spinel. It should be pointed out, however, that these materials have in no way been optimized.

Mullite-ZrO₂

This system has been prepared by in-situ reaction of presintered Al_2O_3-zircon mixtures.[21,22] Since pure mullite cannot be obtained by this technique, a direct evaluation of the improvement afforded by ZrO_2 inclusions is not possible. However, the fact that the sintered composite attained ≈ 400 MPa and the best value of pure hot-pressed mullite found in the literature is only 269 MPa[50] suggests that considerable toughening in this system is possible.

Si$_3$N$_4$-ZrO$_2$

Toughness and strength of sintered and hot-pressed composites have been shown to increase with increasing monoclinic ZrO$_2$ content.[24] Due to the high thermal mismatch, retention of tetragonal ZrO$_2$ was not possible; the increase in toughness should be ascribed to microcracking. Because of the high fabrication temperatures necessary to densify the material, reactions can hardly be avoided; some ZrN is formed and a fraction of the ZrO$_2$ particles ($\approx 30\%$) is stabilized in the cubic phases by nitrogen.[51] This fraction, however, is readily reoxidized to monoclinic ZrO$_2$, leading to surface toughening.[43,52] This is demonstrated by the thermal shock results[52] (Fig. 18). Hot-pressed sialon 1 (Si$_3$N$_4$ containing 1 eq.% Al) with 10 vol% ZrO$_2$ exhibited a strength increase from 600 to ≈ 700 MPa after a 50 h anneal at 800 °C in air. The same retained strength was measured when the unannealed composite was quenched from above ≈ 1000 °C. In this case, transient "compressive stress,"[24] developing from the tetragonal$\rightarrow$monoclinic inversion, oppose the tensile stresses due to the thermal gradient in the 20° water quench. This phenomenon is not observed with the ZrO$_2$-free sialon 1.

The improvement in strength properties of ZrO$_2$-containing Si$_3$N$_4$ ceramics, which is even more pronounced at high volume fractions of ZrO$_2$, can, however, only be utilized for moderate application temperatures since the ZrO$_2$ additions enhance grain-boundary oxidation and, thus, result in a deterioration of the material properties. Only volume fractions of less than 10% ZrO$_2$ result in oxidation-resistant composites.[52]

SiC-ZrO$_2$

The very high preparation temperatures involved in SiC densification make special precautions necessary to prevent gross reactions with ZrO$_2$. One way is by reducing the temperature for liquid-phase formation, e.g. by adding considerable amounts of Al$_2$O$_3$.[53] For such composites toughening has been observed for both ZrO$_2$ and HfO$_2$ additions. However, at present, no high-temperature applications seem to be advisable.

Other Systems

Improvements have been demonstrated also in systems with ZnO[33] and ThO$_2$.[54] The concept of ZrO$_2$ toughening seems to be applicable to many more systems (especially oxides); however, the technically most interesting systems are already under intensive investigation.

Technological Aspects

Effective ZrO$_2$ toughening is achieved only when the processing conditions for manufacturing the composite bodies are closely controlled, especially with respect to ZrO$_2$ particle size and size distribution. Since, however, the inherent properties of the ceramic matrix are also to be retained as much as possible (e.g. high density, small grain size), compromises, which are extremely difficult to optimize, must be made. The technological aspect of ZrO$_2$ toughening seems to be by far the most unexplored area; a possible exception is the fabrication of PSZ ceramics (extrusion and calibrating dies) and some cutting tool aluminas.

In the following, some of the ZrO_2-dispersion techniques, which have so far been applied, will be summarized. In some cases combinations of these techniques yield useful results. Optimum conditions would certainly be those which result in the same particle distribution as that in PSZ ceramics—evenly spaced particles of almost the same size:

Mechanical Mixing

Wet mechanical mixing of the powder components in combination with spray drying is being widely applied. A very intensive and homogeneous dispersion is achieved by attrition milling.[7,18,24] A disadvantage, however, is that considerable wear of the milling media contaminates the powder. The composition of these media must therefore be selected such that they are compatible with the composites, e.g. Si_3N_4 media have been used for sialon systems.[52]

Wear of ZrO_2 Milling Media

A combination of intensive milling and fine particle dispersion can be obtained by using ZrO_2 milling media. The specific surface area of the resulting ZrO_2 powder has been determined to be > 100 m^2/g. In this technique the ZrO_2 volume fraction is controlled by the milling time.

Sol-Gel Techniques

A great variety of sol-gel techniques is available. When all components are mixed in the form of sols (e.g., Al nitrates, Zr oxychlorides, Zr acetates) good ZrO_2 dispersions are obtained upon calcination; however, large quantitites are usually difficult and expensive to prepare and the matrix properties are usually inferior.[19] The use of high-quality matrix powder (e.g., high-purity Al_2O_3) together with sols of Zr compounds (Zr acetate) yield even particle distribution without significantly changing the matrix properties.[23]

Chemical In-Situ Reactions

This technique has so far been applied only to the Al_2O_3-zircon system.[22] However, a series of similar reactions with zircon or other Zr compounds may be possible.

Acknowledgment

The authors thank G. Petzow for helpful discussions. Thanks are also due to the DFG (German Research Foundation) for supporting parts of this work.

References

[1]R. C. Garvie, R. H. Hannink, and R. T. Pascoe, "Ceramic Steel?" *Nature (London),* **258** [12] 703–704 (1975).

[2]N. Claussen, "Fracture Toughness of Al_2O_3 with an Unstabilized ZrO_2 Dispersed Phase," *J. Am. Ceram. Soc.,* **59** [1-2] 49–51 (1976); *Am. Ceram. Soc. Bull.,* **54** [4] 403 (1975) (abstract only).

[3]D. L. Porter and A. H. Heuer, "Mechanisms of Toughening Partially Stabilized Zirconia (PSZ)," *J. Am. Ceram. Soc.,* **60** [3-4] 183–84 (1977).

[4]N. Claussen, J. Steeb, and R. F. Pabst, "Effect of Induced Microcracking on the Fracture Toughness of Ceramics," *Am. Ceram. Soc. Bull.,* **56** [6] 559–62 (1977).

[5]T. K. Gupta, J. H. Bechtold, R. C. Kuznicki, L. H. Cadoff, and B. R. Rossing, "Stabilization of Tetragonal Phase in Polycrystalline Zirconia," *J. Mater. Sci.,* **12** [12] 2421–26 (1977).

⁶U. Dworak, H. Olapinski, and G. Thamerus, "Strength Increases in the Polyphase Ceramic Systems ZrO_2-ZrO_2, Al_2O_3-ZrO_2, and Al_2O_3-TiC," *Ber. Dtsch. Keram. Ges.,* **55** [2] 98–101 (1978).

⁷N. Claussen and J. Jahn, "Transformation of ZrO_2 Particles in a Ceramic Matrix," *ibid.,* **55** [11] 487–91 (1978).

⁸F. F. Lange, "Transformation Toughening in the Al_2O_3/ZrO_2 Composite System," Tech. Rept. No. 7, Rockwell Int., 1979.

⁹D. L. Porter and A. H. Heuer, "Microstructural Development in MgO-Partially Stabilized Zirconia (Mg-PSZ)," *J. Am. Ceram. Soc.,* **62** [5-6] 298–305 (1979).

¹⁰P. F. Becher, "Transient Thermal Stress Behavior in ZrO_2-Toughened Al_2O_3," *J. Am. Ceram. Soc.,* **64** [1] 37–39 (1981).

¹¹N. Claussen and D. P. H. Hasselman; pp. 381–95 in Thermal Stresses in Severe Environments. Edited by D. P. H. Hasselman and A. Heller. Plenum Press, New York, 1980.

¹²E. C. Subbarao, H. S. Maiti, and K. K. Srivastava, "Martensitic Transformation in Zirconia," *Phys. Status Solidi A,* **21** [1] 9–40 (1974).

¹³G. K. Bansal and A. H. Heuer, "Martensitic Phase Transformation in Zirconia (ZrO_2): I," *Acta Metall.,* **20** [11] 1281–89 (1972); "II," *ibid.,* **22** [4] 409–17 (1974).

¹⁴A. G. Evans and A. H. Heuer, "Transformation Toughening in Ceramics: Martensitic Transformations in Crack-Tip Stress Fields," *J. Am. Ceram. Soc.,* **63** [5-6] 241–48 (1980).

¹⁵M. Rühle, C. Springer, N. Claussen, and H. Strunk, "TEM Studies of Al_2O_3-ZrO_2 Composites," *Proc. Int. Conf. on HVEM, 5th,* Kyoto, pp. 633–36 (1977).

¹⁶J. D. Eshelby; pp. 89–140 in Progress in Solid Mechanics, Vol. 2. Edited by I. N. Sneddon and R. Hill. Wiley-Interscience, New York, 1961.

¹⁷A. C. Evans, N. Burlingame, M. Drory, and W. M. Kriven, "Martensitic Transformations in Zirconia: Particle Size Effects and Toughening"; submitted to *Acta Metallurgica.*

¹⁸N. Claussen, "Stress-Induced Transformation of Tetragonal ZrO_2 Particles in Ceramic Matrices," *ibid.,* **61** [1-2] 85–86 (1978).

¹⁹R. Wagner; Ph. D. work at MPI Stuttgart.

²⁰N. Claussen, F. Sigulinski, and M. Rüle, "Phase Transformation of Solid Solutions of ZrO_2 and HfO_2 in an Al_2O_3 Matrix"; this volume, pp. 164–67.

²¹N. Claussen and J. Jahn, "Mechanical Properties of Sintered, In Situ-Reacted Mullite-Zirconia Composites," *J. Am. Ceram. Soc.,* **63** [3-4] 228–29 (1980).

²²J. Wallace, M. Rühle, G. Petzow, and N. Claussen, "Development of Phases in In Situ-Reacted Mullite-Zirconia Systems," *Proc. Int. Conf. on Surface and Interfaces,* Berkeley, 1980.

²³T. Kosmac; Ph. D. work at MPI Stuttgart.

²⁴N. Claussen and J. Jahn, "Mechanical Properties of Sintered and Hot-Pressed Si_3N_4-ZrO_2 Composites," *J. Am. Ceram. Soc.,* **61** [1-2] 94–95 (1978).

²⁵T. K. Gupta, F. F. Lange, and J. H. Bechtold, "Effect of Stress-Induced Phase Transformation on the Properties of Polycrystalline Zirconia Containing Metastable Tetragonal Phase," *J. Mater. Sci.,* **13** [7] 1464–70 (1978).

²⁶A. Krauth and H. Meyer, "Rapid Quenching Modifications and Crystal Growth in Systems Containing ZrO_2," *Ber. Dtsch. Keram. Ges.,* **42** [2] 61–72 (1965).

²⁷C. A. Sorrell and C. C. Sorrell, "Subsolidus Equilibria and Stabilization of Tetragonal ZrO_2 in the System ZrO_2-Al_2O_3-SiO_2," *J. Am. Ceram. Soc.,* **60** [11-12] 495–99 (1977).

²⁸C. Springer and M. Rühle, "Preparation of Ceramic Raw Materials for Radiation Electron Microscopy," *Sonderb. Prakt. Metallogr.,* **9,** 223–36 (1978).

²⁹B. Kraus, A. Strecker, and M. Rühle, "A Special Grinding Device for Ceramic Specimens"; unpublished work.

³⁰M. Rühle, B. Kraus, and N. Claussen, "In-Situ TEM Observations of Transformation Zones in ZrO_2-Containing Ceramics"; unpublished work.

³¹B. Kraus, W. Mader, and M. Rühle, "An In-Situ HVEM Straining Device for Ceramic Specimens"; unpublished work.

³²N. Claussen, W. Mader, and M. Rühle, "In Situ TEM Studies of the Martensitic Transformation of Small Zirconia Particles"; for abstract see *Am. Ceram. Soc. Bull.,* **59** [3] 363 (1980).

³³H. Ruf; Ph.D. work at MPI Stuttgart.

³⁴B. R. Lawn and T. R. Wilshaw, Fracture in Brittle Solids. Cambridge University Press, 1975.

³⁵W. Pompe, H.-A. Bahr, G. Gille, and W. Kreher, "Increased Fracture Toughness of Brittle Materials by Microcracking in an Energy Dissipative Zone at the Crack Tip," *J. Mater. Sci.,* **13,** 2720–23 (1978).

³⁶D. J. Green, "Critical Microstructures for Microcracking in Al_2O_3-ZrO_2 Composites"; for abstract see *Am. Ceram. Soc. Bull.,* **60** [3] 363 (1980).

³⁷M. Rühle; unpublished research.

[38]C. Cm. Wu, R. W. Rice, and P. F. Becher, "The Character of Cracks in Fracture"; unpublished work.

[39]R. W. Davidge and T. J. Green, "Strength of Two-Phase Ceramic/Glass Materials," *J. Mater. Sci.,* **3** [6] 629–34 (1968).

[40]R. F. Pabst, J. Steeb, and N. Claussen, pp. 821–33 in Fracture Mechanics of Ceramics, Vol. 4. Edited by R. C. Bradt *et al.* Plenum Press, New York, 1978.

[41]M. E. Nordberg, E. L. Mochel, H. M. Garfinkel, and J. S. Olcott, "Strengthening by Ion Exchange," *J. Am. Ceram. Soc.,* **47** [5] 215–19 (1964).

[42]R. T. Pascoe, R. R. Hughan, and R. C. Garvie, "Strong and Tough Zirconia Ceramics," *Sci. Sintering,"* **11** [3] 185–92 (1979).

[43]F. F. Lange, "Compressive Surface Stresses Developed in Ceramics by an Oxidation-Induced Phase Change," *J. Am. Ceram. Soc.,* **63** [1-2] 38–40 (1980).

[44]O. Richmond, W. C. Leslie, and H. A. Wriedt, "Theory of Residual Stresses Due to Chemical Concentration Gradients," *ASM Trans. Q,* **57** [1] 294–300 (1964).

[45]M. V. Swain, "Grinding-Induced Tempering of Ceramics Containing Metastable Tetragonal Zirconia"; unpublished work.

[46]B. R. Lawn and D. B. Marshall, "Contact Fracture Resistance of Physically and Chemically Tempered Glass Plates," *Phys. Chem. Glasses,* **18** [1] 7–18 (1977).

[47]P. F. Becher, D. L. Lewis III, K. R. Carman, and A. C. Gonzalez, "Thermal Shock Resistance of Ceramics: Size and Geometry Effects in Quench Tests," *Am. Ceram. Soc. Bull.,* **59** [5] 542–45 (1980).

[48]A. G. Evans; personal communication.

[49]G. Popp; Ph. D. work at MPI Stuttgart.

[50]K. S. Mazdiyasni and L. M. Brown, "Synthesis and Mechanical Properties of Stoichiometric Aluminum Silicate (Mullite)," *J. Am. Ceram. Soc.,* **55** [11] 548–52 (1972).

[51]N. Claussen, R. Wagner, L. J. Gauckler, and G. Petzow, "Nitride-Stabilized Cubic Zirconia," *ibid.,* **61** [7-8] 369–70 (1978).

[52]R. Lupold; Ph. D. work at MPI Stuttgart.

[53]J. Lorenz, L. J. Gauckler, and G. Petzow, "Improved Fracture Toughness of SiC-Based Ceramics"; for abstract see *Am. Ceram. Soc. Bull.,* **58** [3] 338 (1979).

[54]R. M. Cannon and T. O. Ketchum, "ThO$_2$ Toughened by ZrO$_2$"; unpublished work.

---

*ΔU_T is due to unconstrained volume change of 3 to 4.5% (Ref. 8) and constrained shear strains of up to 17% (Ref. 17).

†Type SN 80 Steel Master, Feldmühle AG, Plochingen, Germany.

‡Unstabilized ZrO$_2$ can even be obtained in pure MgO by controlled heat treatment.

Table I. Critical sizes, d_c, for $M_s \cong$ room temperature obtained with different ZrO_2-containing composites

	Al_2O_3	Al_2O_3	Mullite	Spinel	Si_3N_4	ZrO_2 (Mg-PSZ)	ZrO_2 (Y-PSZ)
$d_c(\mu m)$	0.52	0.3	1	0.8–1.0	<0.1	0.1–0.2	0.32
ZrO_2*	16	15	22	17.5	15	8.1 mol% MgO	≈2 mol%† Y_2O_3
		$Zr_{0.5}Hf_{0.5}O_2$					
Ref.	19	20	21,22	23	24	9,13	25

*Vol% except as otherwise indicated. †Exact amount not revealed.

Table II. Possible techniques for introduction of residual compressive stresses

Technique	Transformation induced	Example
Grinding, sand blasting, quenching	By stresses	Al_2O_3-ZrO_2, PSZ
Heat treatment in HfO_2 powder in air in vacuum	Chemically	Al_2O_3-ZrO_2 Si_3N_4-ZrO_2 Mg-PSZ, Mg-CSZ
Cooling in liquid He or N_2 for short periods	Low temp. $(T_c < M_s)$	Al_2O_3-ZrO_2
Coating with (a) higher vol fraction, or (b) larger particle size	M_s of coating $> T$ M_s of bulk $< T$, T: applic. temp.	Al_2O_3-ZrO_2

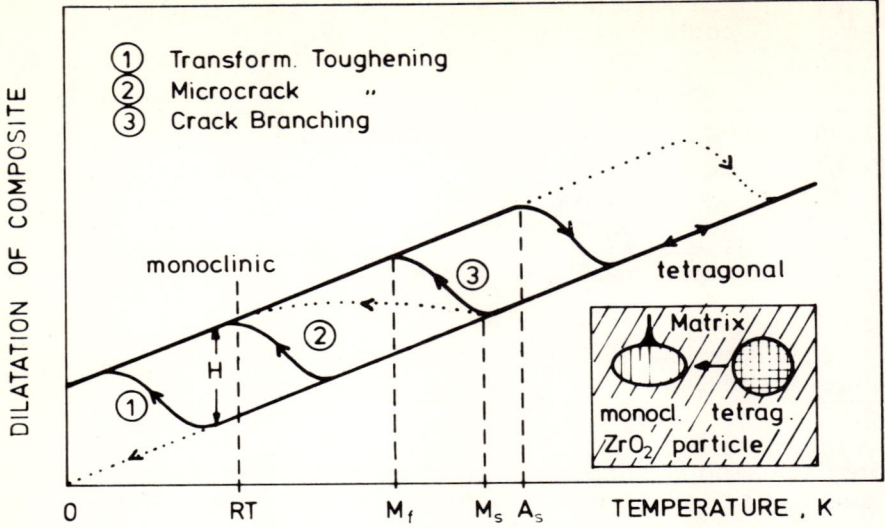

Fig. 1. Expansion-temperature curve of a ceramic containing ZrO_2 particles. Transformation in temperature ranges 1, 2, and 3 indicates possible toughening mechanisms.

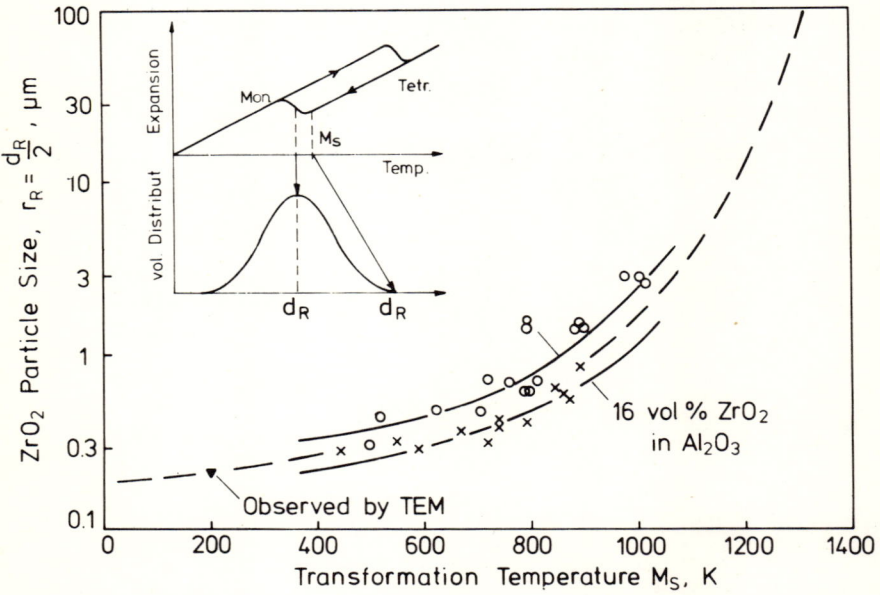

Fig. 2. Size dependency of M_s temperature of $Al_2O_3 + ZrO_2$ composites as determined by correlation of dilatometer results with linear SEM analysis (indicated in the upper left).

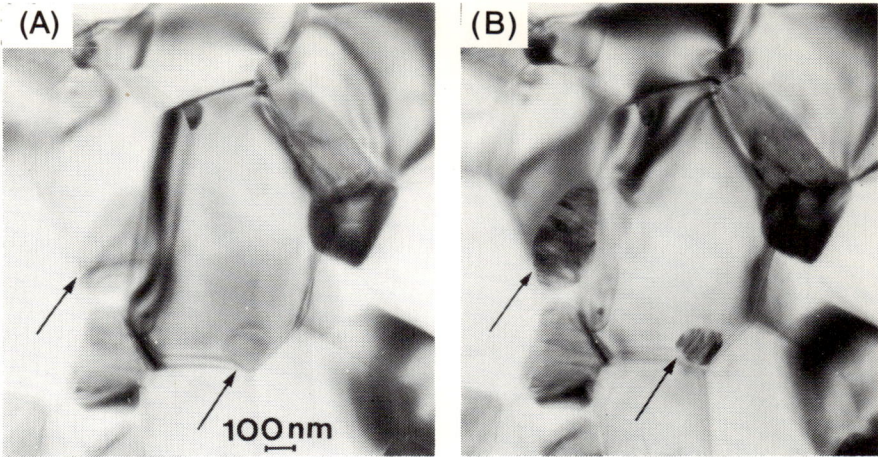

Fig. 3. Al$_2$O$_3$-ZrO$_2$ dispersion ceramics. *(A)* Tetragonal particles of ZrO$_2$ are visible at the grain boundary of Al$_2$O$_3$. *(B)* The particles transformed on cooling in the HVEM to $T < 183$ K; twins are clearly visible.

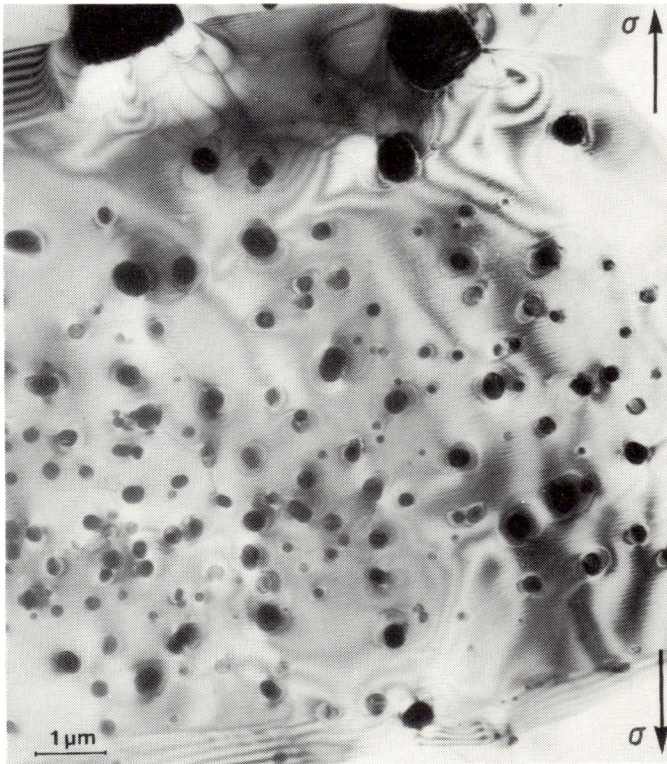

Fig. 4. Low magnification TEM of a strained specimen. Nearly all particles are included in the Al$_2$O$_3$ matrix.

155

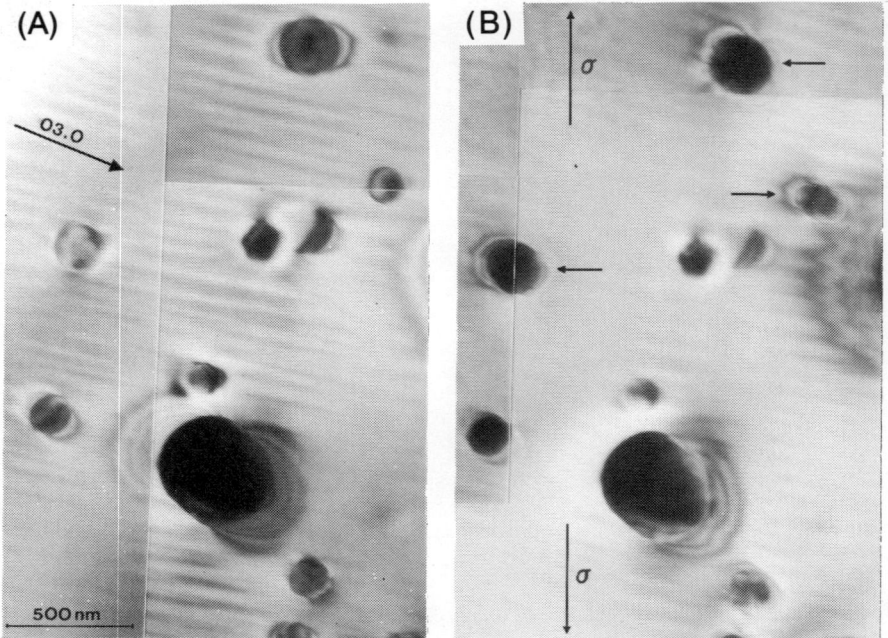

Fig. 5. In-situ deformation experiment. *(A)* Before straining, and *(B)* after straining. Transformed particles are indicated by an arrow.

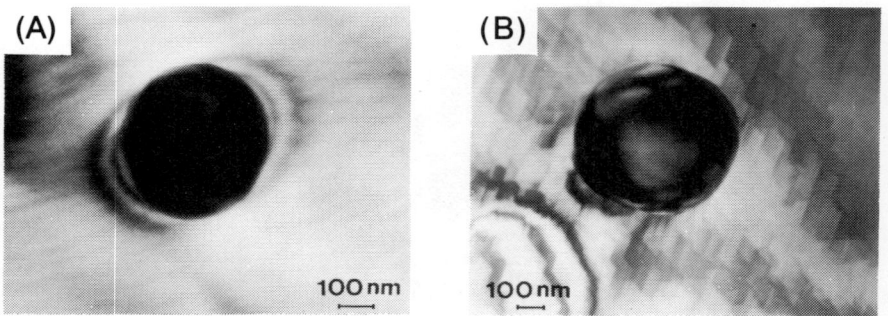

Fig. 6. Tetragonal particles in different matrices. *(A)* In Al_2O_3; *(B)* in spinel. A strain contrast is observed in *(A)* due to the thermal mismatch. The fine structure in *(B)* is due to surface roughness.

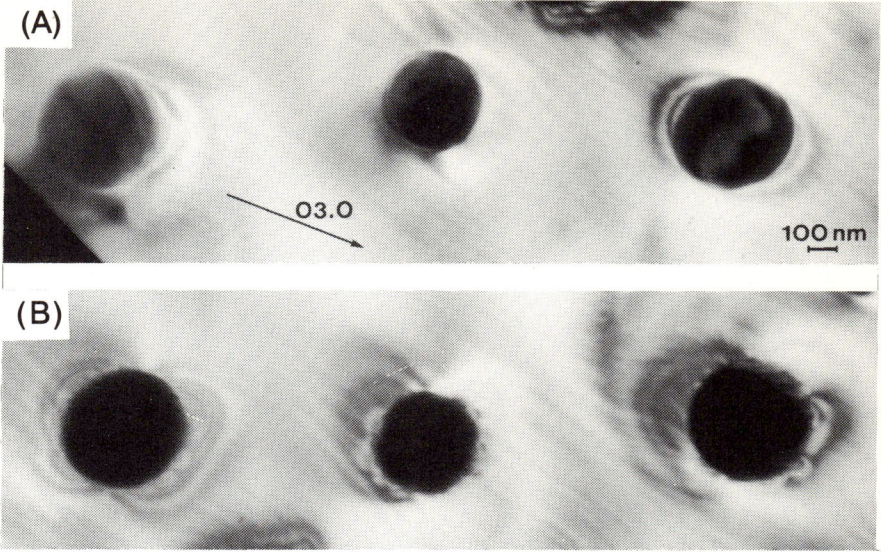

Fig. 7. Strain contrast of tetragonal *(A)* and monoclinic particles *(B)*. The transformation was caused by straining in the HVEM.

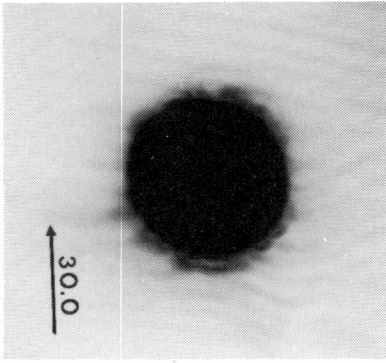

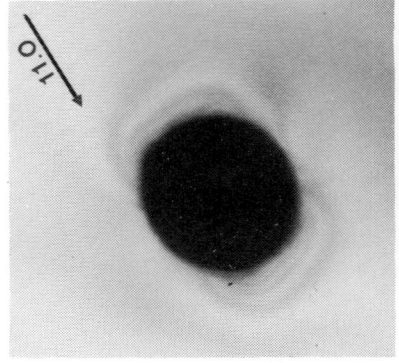

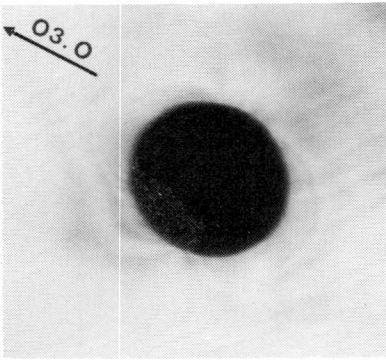

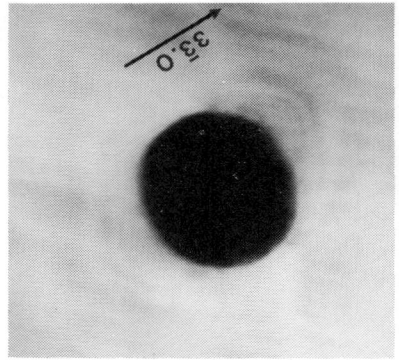

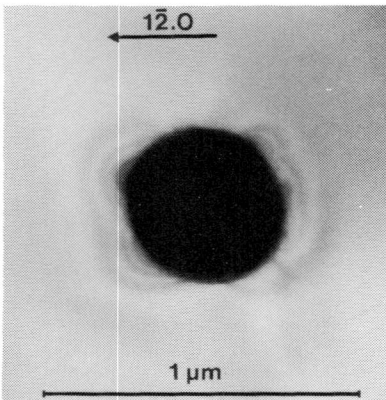

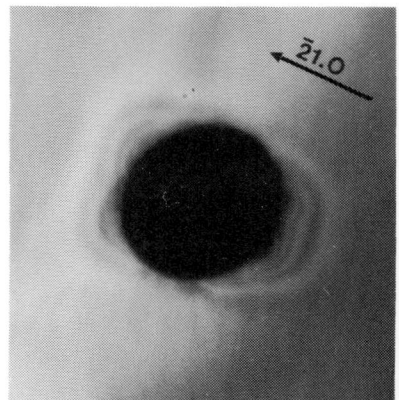

Fig. 8. Dynamical DF imaging of the same particle with different diffraction vectors (indicated on the micrographs). The observable strain contrast depends strongly on the selected diffraction vector which indicates that the strain is anisotropic.

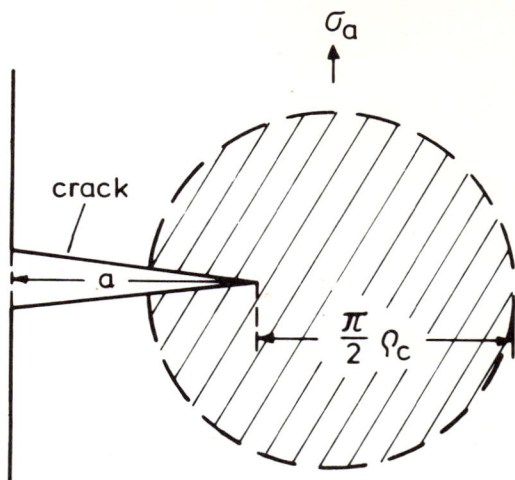

Fig. 9. Process zone ahead of a crack tip.

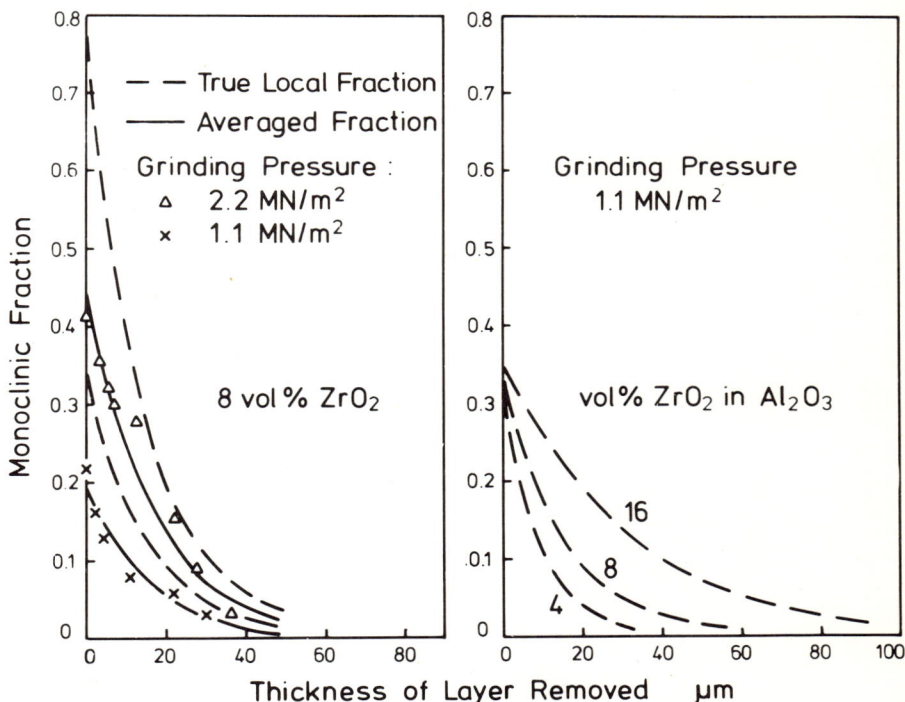

Fig. 10. Fraction of monoclinic ZrO_2, which was transformed by grinding, as function of the depth beneath the surface. X-ray data were obtained after successive, careful removing of the surface. (A) $Al_2O_3 + 8$ vol% ZrO_2, different grinding pressures. (B) Al_2O_3 with different volume fraction ZrO_2 at constant grinding pressure of 1.1 MPa.

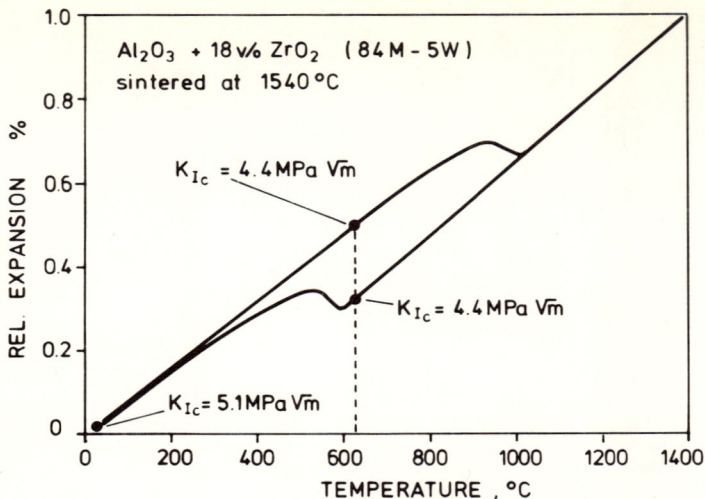

Fig. 11. Expansion-temperature curve of a sintered Al_2O_3 + 18 vol% ZrO_2 composite, containing 84% monoclinic ZrO_2 at room temperature. Fracture toughnesses measured after different temperature treatments (see text) are marked in the plot.

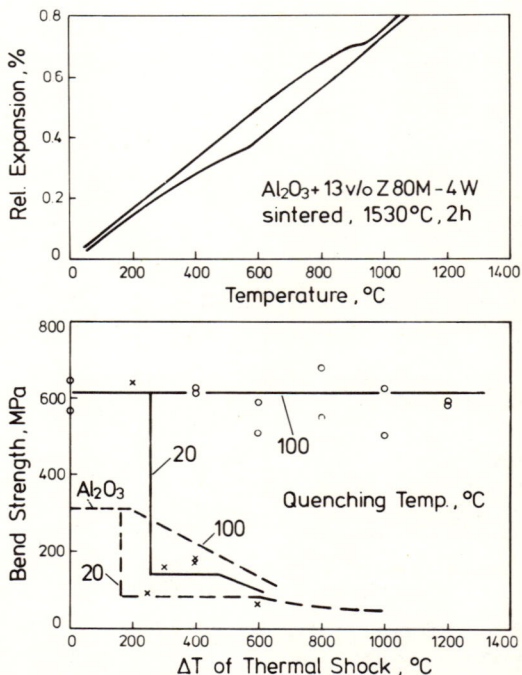

Fig. 12. Dilatometer curve and retained flexural strength of Al_2O_3-ZrO_2 composites containing 80% monoclinic ZrO_2 at room temperature.

160

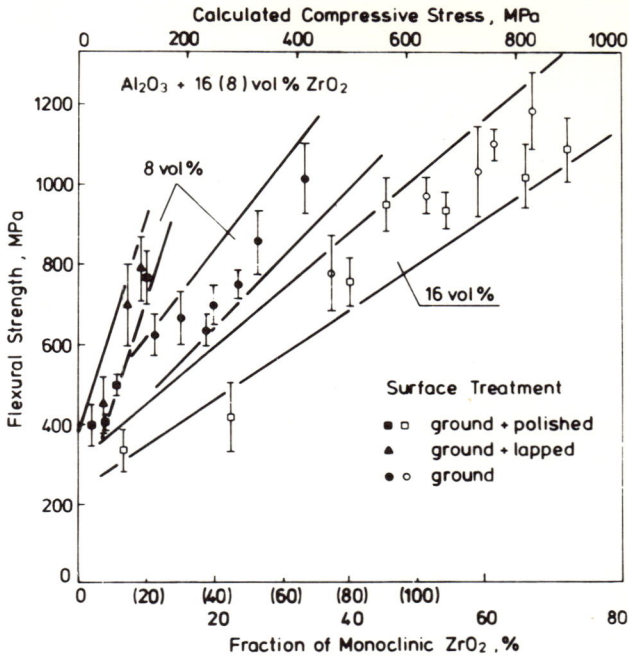

Fig. 13. Flexural strength of Al_2O_3–ZrO_2 composites vs fraction of transformed ZrO_2 in the surface layer (true local, see Fig. 10). Monoclinic fraction was changed, applying different grinding pressures. The compressive stresses were calculated from Eq. (18).

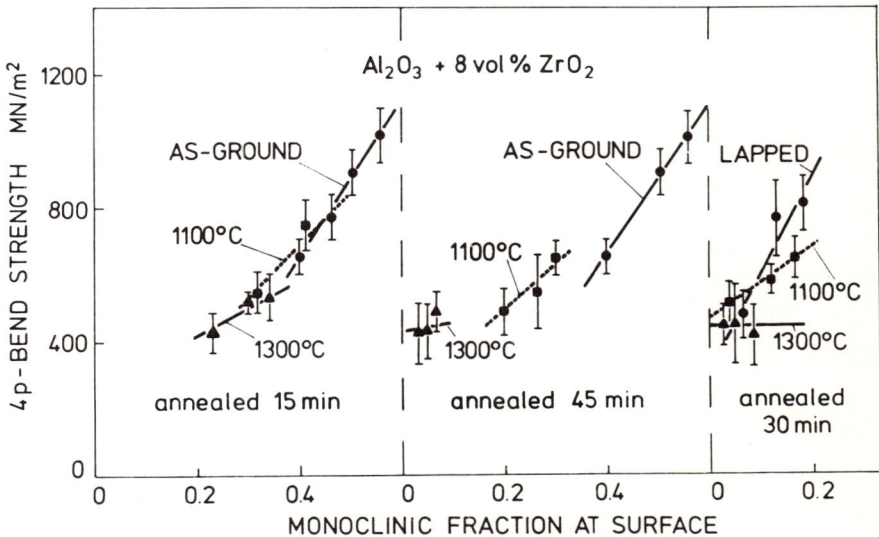

Fig. 14. Effect of annealing on flexural strength of ground Al_2O_3-ZrO_2 composites.

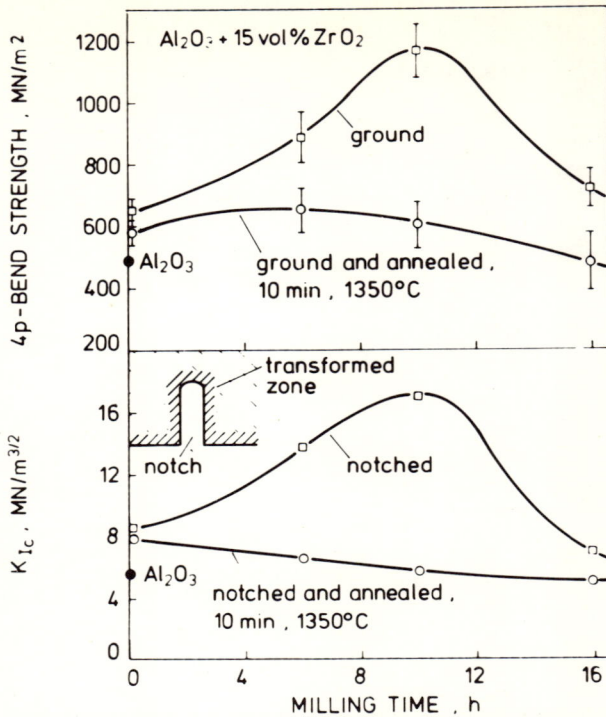

Fig. 15. Flexural strength and fracture toughness of $Al_2O_3 + 15$ vol% ZrO_2 composites attrition milled for various times. Grinding-induced transformation is "healed out" by the indicated annealing process.

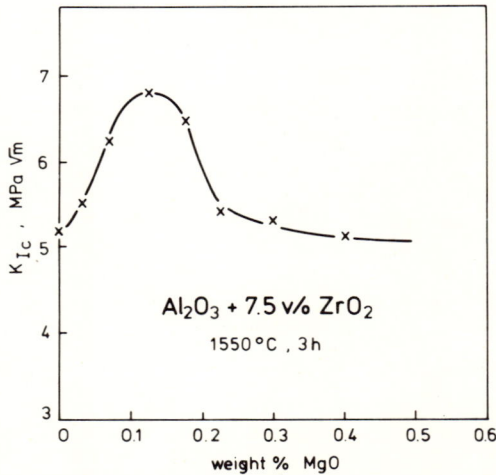

Fig. 16. Fracture toughness of sintered $Al_2O_3 + 7.5$ vol% ZrO_2 as a function of MgO content.

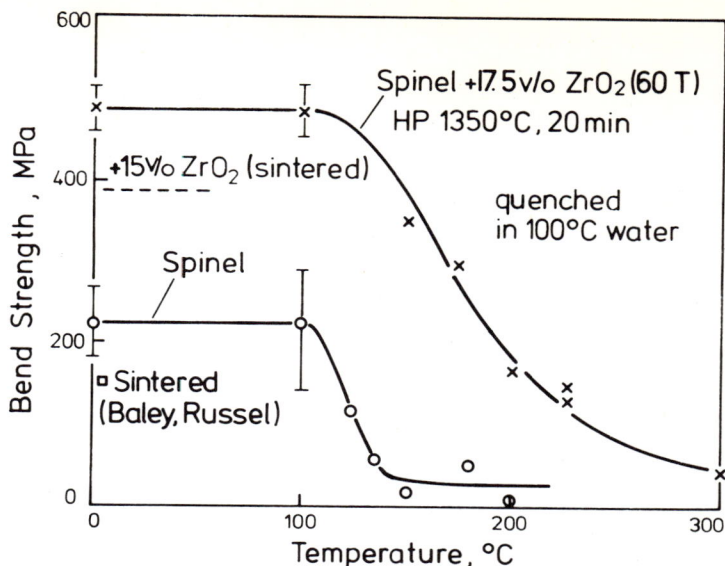

Fig. 17. Flexural strength of sintered and hot-pressed spinel-ZrO_2 obtained from mixtures of MgO, Al_2O_3, and ZrO_2 powders. The hot-pressed material contains 60% tetragonal ZrO_2 at room temperature.

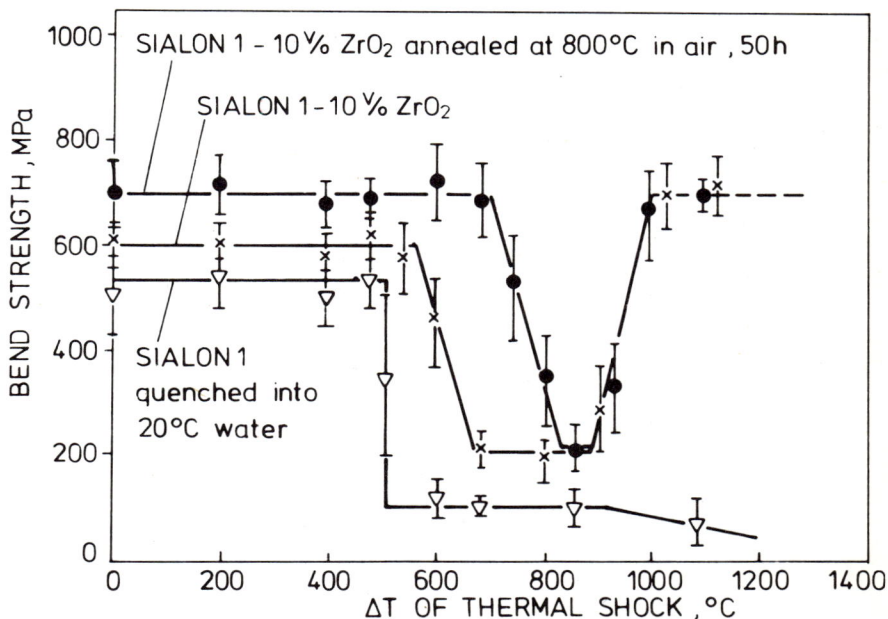

Fig. 18. Retained flexural strength of hot-pressed sialon-ZrO_2. Annealing in air causes the oxidation of N-stabilized ZrO_2 in the surface region and consequently causes compressive surface stresses (sialon 1 = Si_3N_4 with 1 eq.% Al).

163

Phase transformation of solid solutions of ZrO$_2$ and HfO$_2$ in an Al$_2$O$_3$ matrix

Nils Claussen, Feodor Sigulinski, and Manfred Rühle

Max-Planck-Institut für Metallforschung
Institut für Werkstoffwissenchaften
Pulvermetallurgisches Laboratorium
D-7000 Stuttgart-80, Federal Republic of Germany

HfO$_2$ was alloyed with ZrO$_2$ particles in order to increase the driving force for the tetragonal-to-monoclinic transformation and to shift M_s and A_s to higher temperatures. Dilatometer curves of Zr$_{0.5}$Hf$_{0.5}$O$_2$-Al$_2$O$_3$ composites exhibit unusual expansion when heated from 800 to 1460°C (A_s); this is accompanied by transformation of some ($\approx$30%) particles which are tetragonal at room temperature to the monoclinic symmetry. Simultaneously, twin coarsening occurs in the monoclinic particles. A tentative explanation is presented.

Transformation toughening has been dealt with in various papers of this volume. In some of them,[1-3] it has been stated that particles with an increased chemical driving force (ΔG_{chem}) for the tetragonal→monoclinic transformation would yield more powerful toughening as long as the tetragonal form can be retained at the test temperature. HfO$_2$, which is isomorphous with ZrO$_2$ and has a phase transition at $M_s > 1600$°C,[4] is the only known solute which increases the chemical driving force.

The purpose of this note is to report first results obtained with sintered Al$_2$O$_3$* + 15 vol% (Hf,Zr)O$_2$ composites. Both prealloyed ZrO$_2$†-HfO$_2$‡ solutions as well as the single components were attrition mixed with Al$_2$O$_3$, spray dried, isostatically pressed at 600 MPa and sintered at 1550°C for 1 h in air. HfO$_2$-to-ZrO$_2$ volume ratios of 3:7, 1:1, and 6:4§ were used. The density of the composite bodies was >99% theoretical in all cases.

Samples which were diamond-cut into 30 by 7 by 3 mm bars were used for dilatometer and SENB tests. The expansion in air of a composite containing an HfO$_2$/ZrO$_2$ ratio of 1:1, recorded with a heating/cooling rate of 10°C/min, is given in Fig. 1. Similar dilatometer curves were obtained for composites with 3:7 and 6:4 HfO$_2$/ZrO$_2$ ratios. However, for these composites, A_s and M_s were shifted to lower and higher temperatures, respectively. No difference in the shape of the curve could be detected between samples for which prealloyed (Zr,Hf)O$_2$ mixtures or HfO$_2$ and ZrO$_2$ powders were used in the composite preparation. This demonstrates that the sintering time was sufficient to allow complete solid solution through diffusion along Al$_2$O$_3$ grain boundaries. Furthermore, since no changes occurred in the expansion curves during repeated cycling at temperatures between S and A_f, it is felt that chemical stability had occurred. The A_s-A_f-M_s-S curve in Fig. 1 resembles that of Al$_2$O$_3$-ZrO$_2$ alloys, but considerable deviations are observed in the S-M_f''-A_s curve.

The martensitic transformation A_s-A_f occurs in the same temperature range as is observed in bulk $Hf_{0.5}Zr_{0.5}O_2$[4] which may indicate (a) that complete formation of solid solution had occurred, (b) that the transformation takes place under stress conditions equivalent to those encountered in the bulk material, and (c) that there is no influence of a possible Al_2O_3 solute content on A_s. An accidental agreement cannot yet, however, be excluded. As expected, M_s is reduced, mainly due to the matrix constraint and the small particle size,[5] as compared to M_s^* and M_f^* of the respective bulk composition. At the M_s temperature, the transformation from tetragonal to monoclinic symmetry begins. However, under the cooling conditions used, this transformation is not completed, i.e. $M_f < 20\,°C$ and $\approx 30\%$ of the $(Hf,Zr)O_2$ particles remain tetragonal at room temperature. Completion of the transformation occurs only on reheating to M_f'' at $1140\,°C$. Samples cooled from M_f'' to S follow the dashed curve to S' indicating that no transformational changes occur; these samples contain 100% monoclinic particles. Thus, transformation from tetragonal (at S) to monoclinic (at M_f'') took place on heating. Further heating above M_f'' to A_s initiates a transformation to the tetragonal form for some of the particles.

The unusual transformation behavior along M_s-S-M_f'' suggests that the athermal forward transformation from M_s to $900\,°C$ is followed by a "thermally activated process" which is not possible without the athermal start at M_s. Two explanations are possible: (a) After some of the monoclinic has formed, the matrix begins to relax by creep,§ thereby releasing constraint of the $(Hf,Zr)O_2$ particles and allowing them to continue to transform athermally. However, if this explanation would apply, it is surprising that the shape of the curve is retained after repeated cycling from S to A_f. This may be considered as a type of "shape-memory effect."[6] (b) The transformation starts athermally but the continuation is thermally activated as also found in metal alloys.[7,8] Evidence for such "thermal activation" by creep or diffusion is shown by an isothermal treatment at $900\,°C$ where the reaction proceeds along the dashed line to M_f'.

The M_f''-A_s part of the curve is also difficult to interpret. A small fraction of the material transforms to tetragonal. Since holding the temperature at $1300\,°C$ leads to A_s, a diffusional process may be active.[9] This path, however, involves a much larger contraction than can be explained by the respective transformational volume change. TEM observations show that the fine twin structure of particles at S' (Fig. 2(A)) has considerably coarsened after heating to A_s (Fig. 2(B)), which, however, does not account for the extensive contraction between M_f'' and A_s. The twin coarsening is also expressed by sharpening of the monoclinic (200) and (020) peaks which preferentially result from oxygen ions. Cooling from A_s to room temperature is shown with the dashed line in Fig. 1, i.e. the contraction is not reversible.

The A_f-M_s cooling curve is parallel to the Al_2O_3 reference line, i.e. the thermal expansion coefficient of the tetragonal form of $(Zr_{0.5},Hf_{0.5})O_2$ is nearly equivalent to that of Al_2O_3. This enables the full transformational strain to be utilized in toughening. Furthermore, it is interesting to note that the fine particles retained in tetragonal symmetry at S ($\approx 0.3\ \mu m$) readily transform to monoclinic when the electron beam is concentrated on the particle in TEM. Critically sized ZrO_2 particles ($\approx 0.5\ \mu m$) in the same matrix are much more difficult to transform under identical TEM conditions, even though a relatively large thermal mismatch is enhancing the transformation.[1]

Hence, the chemical driving force of the $(Zr,Hf)O_2$ particles is definitely larger than that of pure ZrO_2 particles.

Acknowledgment

The authors thank G. Petzow and W. M. Kriven for critical discussions. F. Sigulinski is especially indebted to the International Office of KFA, Jülich, Germany.

References

[1]N. Claussen and M. Rühle, "Design of Transformation-Toughened Ceramics"; this volume, pp. 137–63.

[2]F. F. Lange and D. J. Green, "Stress-Induced Transformation-Toughened ZrO_2 Materials, Phase Retention, Toughness and Strength; *ibid.*, pp. 218–25.

[3]C. A. Andersson and T. K. Gupta, "Phase Stability and Transformation Toughening in Zirconia," *ibid.*, pp. 184–201.

[4]R. Ruh, H. J. Garrett, R. F. Domagala, and N. M. Tallan, "The System Zirconia-Hafnia," *J. Am. Ceram. Soc.*, **51** [1] 23–27 (1968).

[5]N. Claussen and J. Jahn, "Transformation of ZrO_2 Particles in a Ceramic Matrix," *Ber. Dtsch. Keram. Ges.*, **55** [11] 487–91 (1978).

[6]A. Nagasawa, "A New Concept on the Shape Memory Effect in Metals and Alloys," *Phys. Status Solidi A*, **8**, 531–38 (1971).

[7]O. P. Maksimova and E. I. Estrin, "Change in the Kinetics of the Martensitic Transformation under the Influence of Previously Formed Martensite," *Sov. Phys.-Dokl. (Engl. Transl.)*, *7* [1] 63–65 (1962).

[8]C. H. Shih, B. L. Averbach, and Morris Cohen, "Some Characteristics of the Isothermal Martensite Transformation," *Trans. AIME*, **203** [1] 183–87 (1955).

[9]S. Jana and C. M. Wayman, "Martensite-to-Fcc Reverse Transformation in an Fe-Ni Alloy," *ibid.*, **239** [8] 1187–93 (1967).

[10]N. Claussen, R. F. Pabst, and C. P. Lahmann, "Influence of Microstructure of Al_2O_3 and ZrO_2 on K_{Ic}," *Proc. Br. Ceram. Soc.*, **25**, 139–49 (1975).

---

*A16; Aluminum Co. of America, Pittsburgh, Pa.
†ZS 2; MEL, Magnesium-Elektron Ltd., England.
‡H. C. Starck, Berlin, Germany (contains $\approx 3\%$ ZrO_2).
§Sintered at 1620°C, 1 h since A_f of the (6:4) composite is at 1600°C.
¶The attrition balls contain $\approx 9\%$ SiO_2; the wear contamination may be responsible for the creep at $T \approx 900°C$, as was also observed in K_{Ic} testing of 3% SiO_2-containing Al_2O_3 samples (Ref. 10).

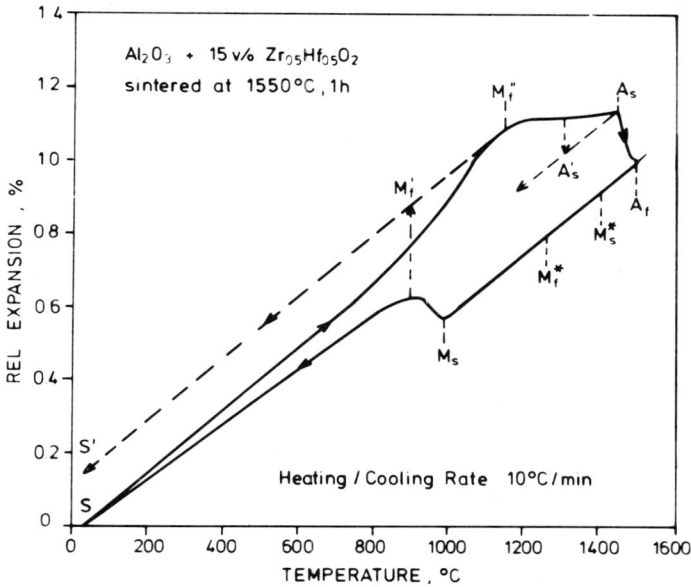

Fig. 1. Dilatometer curve of a sintered $Al_2O_3 + 15$ vol% $Zr_{0.5}Hf_{0.5}O_2$ composite. For explanation see text.

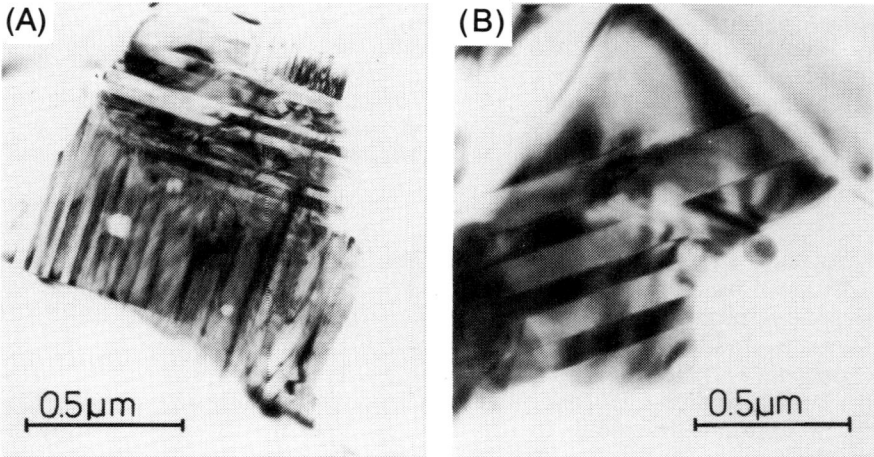

Fig. 2. TEM micrographs of $(Zr_{0.5}Hf_{0.5})O_2$ particles of same size but contained in foils which were differently heat-treated: *(A)* annealed at M_f'' for 4 h, then cooled to S' and *(B)* annealed at 1400 °C for 4 h, then cooled to room temperature. Twin coarsening occurs between M_f'' and A_s.

167

Martensite theory and twinning in composite zirconia ceramics

W. M. Kriven*

Department of Materials Science and Mineral Engineering
University of California
Berkeley, Calif. 94720

Transformation mechanisms were reviewed in thin films and bulk single crystals of zirconia and in iron particles included in dilute Fe-Cu alloys. A model was postulated for the martensitic transformation in zirconia particles included in bulk oxide ceramics, in which it may be energetically optimal for a small particle to form a single variant by lattice invariant shear (LIS) slip or twinning. Larger particles may contain pairs of self-accommodating twin-related variants produced by either LIS slip or twinning and sharing variant interfaces. By analogy with thermoelastic martensites, internal rearrangements such as twinning, interface movement, and variant coalescence simultaneously follow the transformation in response to the compressive stresses applied by the matrix. Experimental TEM microdiffraction and stereographic analyses were made of included particles in the Al_2O_3-50% pure ZrO_2 and pure ZrO_2 systems. Results were interpreted as $(100)_m$ LIS twins in ≈ 0.5 μm particles forming one martensite variant. In larger (> 1 μm) particles, four twin-related variants, each containing $(100)_m$ LIS twins and sharing a common $(110)_m$ variant interface were observed. Mechanical twins with tapered ends terminating inside the crystal were seen, as well as internal rearrangement $\{110\}_m$ twins. These and other current experimental observations were consistent with the postulated crystallography of the mechanisms.

Particles of zirconia may be metastably dispersed in a variety of solid oxide matrices. It is known that in bulk crystals of zirconia the tetragonal to monoclinic phase transformation occurs martensitically. Current principles of ceramic toughening are based on stress-induced transformations in included zirconia particles.

Bailey[1] studied the beam-induced tetragonal-to-monoclinic transformation in thin films of zirconia by transmission electron microscopy. He found that deformation twins formed in enclosed monoclinic grains, but free, unconstrained grains could be transformed into a single orientation. Monoclinic twin shear systems were determined as $(100) \langle 001 \rangle$ and $\{110\} \langle 11\bar{8} \rangle$ and their conjugates. Measured shear strains of 0.35 and 0.25, respectively, agreed closely with observed shape changes of the twins. Since the twinning shear is twice as large as the transformation shear, only half a monoclinic grain needed to be twinned on either (100) or (110) to accommodate the structure change, as was observed.[1] There was a random interface and no "macroscopic shape change" in the sense of martensite theory, but the shape change observed was directly correlated with the unit cell structure change.

Thus in thin TEM specimens, the tetragonal-to-monoclinic transformation can be achieved solely by the structure change plus deformation twins in constrained particles, without any need for a specific habit plane or average macroscopic shape change.

In bulk single crystals, however, Bansal and Heuer[2,3] proved that the transformation proceeded according to the phenomenological theory. They experimentally observed $(671)_m$ or $(761)_m$ "type A" habit planes, being 7° off $(110)_m$ and usually occurring inside a crystal. At the crystal edge "type B" lenticular-shaped $(100)_m$ or $(010)_m$ plates with a midrib formed. Coarse $(100)_m$ twins were also observed in plate-free regions and attributed to deformation twinning. Bansal and Heuer[2,3] performed martensite calculations for eight slip lattice invariant shear (LIS) systems and found that the tetragonal systems $(1\bar{1}0)[001]$ and $(1\bar{1}0)[110]$ predicted the type A and B habit planes, respectively. Martensite theory was therefore shown to apply in bulk zirconia, forming slip lattice invariant shear variants.

Recently Kriven et al.[4] reported calculations of possible martensitic transformation mechanisms in zirconia. LIS slip systems, including those previously calculated by Bansal and Heuer,[2,3] had shape strains of 11%–17%, which could be resolved essentially parallel to the habit plane. A martensitic solution was also predicted for $(100)_m$ or $(001)_m$ LIS twinning derived from the $(010)_t$ mirror plane, and had a smaller shape strain (5% per variant). Thus, on the basis of martensite calculations, a twinning LIS shear mechanism was predicted, and the shape strains of a variant were illustrated.

The martensitic transformation in dilute Fe-Cu alloys has been extensively studied.[5-7] Small, spherical iron precipitates underwent the fcc-to-bcc martensitic transformation, forming lattice invariant shear twins on $\{112\}_{bcc}$ or LIS slip variants. The orientation relation between the twinned particle and the matrix was essentially the Kurdjumov-Sachs relation. The M_s temperature in the iron precipitates was decreased by several hundred degrees from the bulk M_s temperatures.

Kubo et al.[5] elucidated the twin system within the transformed particles by taking selected area diffraction (SAD) patterns from areas <50 nm in diameter. They utilized a selected micro area diffraction technique in a high voltage electron microscope. No dislocations were observed in untransformed coherent precipitates but moiré fringes were seen in some twins. The transformation was induced by anisotropy of constraint around precipitates, which was achieved by electrolytic thinning, deformation, or extraction of precipitates.

Kinsman et al.[6] reported that the iron precipitates usually transformed to a twinned martensite of one variant, although particles larger than ≈ 200 nm could also form more than one variant. Internally twinned configurations, where twins terminated within a crystal, were present. The transformation mechanism was temperature dependent, since between ambient temperatures and 200°C, particles transformed by LIS twinning, while at 400°C they formed single orientation slipped variants. At 200°–300°C both twinned and slipped variants were observed. In the thin foils examined, the matrix accommodated the shape change by dislocation tangles about a particle.

Kato et al.[7] applied a stress to dilute Fe-Cu alloys and found that it aided the preferential formation of specific martensite variants among crystallographically equivalent ones. They concluded that the applied stress interacted with the shear deformation associated with the lattice change rather than the total shape change. Particles larger than 150 nm showed complicated internal structures and consisted of more than two variants of martensite. Only single variant particles of average diameter ≈ 95 nm were therefore analyzed, and they showed a strong dependence on the sense of loading. Applied stress thus induced the martensitic transformation in Cu-Fe precipitate alloys.

The aim of this paper is to discuss the Invariant Plane Strain (IPS) model of a martensitic transformation as it applies to an included particle in a composite ceramic. Then experimental observations of twinning in Al_2O_3-ZrO_2 ceramic are presented and discussed in the light of the theory qualitatively postulated.

Application of Martensite Theory to Included Zirconia Particles

Martensite Theory

The IPS model of a martensitic transformation[8] states that an interface can be found on which there is optimum fit of both parent and product structures. This interface or habit plane is on average unrotated and undistorted in space and is usually irrational but close to a low index plane. To satisfy this criterion, martensite transformations have the following characteristics:

(1) There is a homogeneous, shear-like *shape change* in the product.
(2) An interface or *habit plane* of optimum fit between parent and product may be found.
(3) This leads to a precise *orientation relation* between the two phases.
(4) There is often fine subtexture in the product.

A more comprehensive classification of martensitic transformations in the context of displacive transformations is given by Cohen *et al.*[9]

The mechanics of satisfying the above criteria are simply illustrated in Fig. 1.[10] A parent lattice undergoes a lattice variant deformation such as a Bain strain or Buerger deformation[11] to become the product lattice. The macroscopic shape change, however, is not the same as the unit cell shape change. Instead, it results from the coupling of the lattice variant deformation with a lattice invariant deformation of slip or twinning. When the LIS is twinning the product has a twinned subtexture, while LIS slip variants have no subtexture but dislocations form at the habit plane.[12]

Deformation Twinning

Following the work of Bailey[1] on constrained grains, one may expect that deformation twinning is the simplest transformation mechanism for an enclosed tetragonal grain. The included particle could relieve any post-transformational strains by varying amounts of deformation twins. Evans *et al.*[13] showed that most of the strain is then localized to the matrix-twinned particle interface, and there is no long range strain field associated with the particle.

One Variant by LIS Slip or Twinning

A particle enclosed in a matrix experiences a variety of forces; for example, the particle-matrix interfacial energy, tension or compression due to thermal expansion coefficient mismatch at room temperature, and applied stress or strain in the matrix. To achieve the minimum energy state, it may be

170

energetically favorable for a particle of a given size to form one variant according to the IPS model and to optimally orient to an applied stress.

One may postulate that the variant undergoes LIS slip or twinning and forms a habit plane and macroscopic shape change. The habit plane is not observed except during the transformation, since it disappears when all of the tetragonal phase is transformed. Due to being confined in the matrix, however, the particle cannot undergo its macroscopic shape change, so it exerts a strain on the matrix in the direction of its desired shape change. Thus, in comparison with the localized strain field surrounding deformation twins, there exists a long range strain field in the matrix surrounding the particle. Furthermore, if the variant transforms by LIS twinning, there will also exist a localized strain field at the matrix-particle interface, as described by Evans *et al.*[13]

If the matrix has too high a modulus of elasticity to accommodate the shape strain, the particle must then rearrange itself by mechanical twinning or other internal relaxation phenomena (see later), so as to achieve a lower energy state.

Self-Accommodating Pairs of Variants

The crystallography and phenomena associated with shape memory alloys may be relevant to larger included zirconia particles containing more than one variant. The shape memory effect[14-16] is associated with thermoelastic martensitic transformations found in Cu, Ag, and Au based alloys. In thermoelastic martensite transformations the chemical driving force is balanced by nonchemical elastic forces.[17]

Saburi and Wayman[14] analyzed the crystallography of the self-accommodating transformation in terms of four cooperative twin-related variants whose net macroscopic shape change is essentially zero. Figure 2[14] illustrates the crystallographic relations in the close-packed structures showing habit planes (A to D) and the variant interface traces (100) and (01$\bar{1}$). These habit planes, variant interface traces, and LIS twin traces were experimentally observed. Specific twin relations exist among the four variants.[18,19] On a stereogram the habit planes are clustered symmetrically about the low index (01$\bar{1}$) plane which becomes the variant interface.

An alternative configuration of variants which avoids accumulation of long range stress arises when they share a common habit plane but have opposite shape changes. The variants are then twin-related across the habit plane. The crystallographic degeneracy in higher symmetry systems leading to such a configuration was analyzed by Basinski and Christian.[20]

Under an applied stress or strain the martensite preferentially induced is the one whose shape strain is most favorably oriented to interact with the sense of the applied force.[14] Autocatalytic nucleation of variants then proceeds to achieve or to optimize self-accommodation.

In lower symmetry monoclinic zirconia, therefore, self-accommodating variants may be twin-related across a variant interface or a common habit plane. Applying the analysis of Evans *et al.,*[13] the strain field surrounding the particle is again localized to the matrix-particle interface but is larger than in the case of deformation twins due to the comparatively larger width of the variants.

In summary, the alternative transformation mechanisms postulated for included zirconia particles are illustrated in Fig. 3. A small particle may transform simply by deformation twinning with strains localized at the matrix-particle interface, and no long range strain field (Fig. 3(A)). Alternatively (Fig. 3(B)), it may be energetically optimal for the particle to transform into a single variant by LIS twinning with an internal subtexture or by LIS slip with no subtexture. There are localized strains at the matrix-particle interface, as well as a long range strain in the matrix due to the desired macroscopic shape change of the martensite variant. In larger particles (Fig. 3(C)) pairs of self-accommodating variants form by either LIS slip or twinning and are twin-related across a variant interface or common habit plane. Two variants may be mutually perpendicular when they arise from symmetry equivalent options of the parent phase, such as a_t and b_t.

Post-Transformational Rearrangement

In shape-memory alloys,[14-16] post-transformational deformation in the martensite results in internal rearrangement such as twinning, interface movement, and variant coalescence. A single transformed crystal thus grows to be most favorably oriented with respect to the applied stress and has a large net shape change in the stress direction.

Post-transformation relaxations were studied in the martensitic transformation in $RbNO_3$ by Kennedy and Kriven.[21-23] Twin blunting with deduced movement of dislocations along subgrain boundaries, twin thickening, and twin ballooning were observed by optical polarized light microscopy within minutes of the transformation. Recrystallization of a single orientation from a crystal of multiple variants was quantitatively observed by X-ray precession techniques.

In zirconia ceramics, when a transformed variant exerts a shape strain on the matrix, the particle may respond by internal deformation such as mechanical twinning, as discussed earlier, or by faulting.[24] Saburi and Wayman[14] noted internal bands near junctions of variants. Kinsman et al.[6] also reported internally twinned particles with twins terminating within the particle. Both observations are consistent with mechanical twinning.

Thus, assuming an IPS-type martensite mechanism, accommodation twins, which form after the transformation, are distinguished from transformation or LIS twins, which are part of the intrinsic mechanism.

Internal rearrangement with twinning and other relaxation phenomena may also occur in zirconia. Bailey[1] often observed twin growth behind the phase front. Claussen et al.[25] examined the thermal characteristics of a 15% $(Zr_{0.5}Hf_{0.5}O_2)$-Al_2O_3 material by dilatometry and high voltage electron microscopy. Transformed monoclinic grains contained numerous fine twin configurations, which on annealing at 1140°C coalesced into a few wide bands.

In shape memory alloys, the martensite transformation is complete before the application of an external stress which causes twinning, interface movement, and variant coalescence.[14-16] However, for particles included in a ceramic matrix, the matrix compressive stress is exerted simultaneously with the transformation. The mechanism is thus quite complex and the model here proposed separates it into two stages merely to facilitate understanding. In the process of adopting the lowest energy state, the relative kinetics of the fast martensite transformation coupled with slower internal rearrangement may be evident in in-situ observations, as will be discussed later.

172

Experimental Methods

A mixture of 50% Al_2O_3-50% ZrO_2 (42% volume fraction tetragonal) derived from aqueous salt solutions was hot-pressed at 1550 °C.[26] The relative tetragonal and monoclinic contents of these bulk specimens were determined by X-ray diffractometry. Thin TEM specimens were cut, ground, and carefully polished on both sides by successively finer grades of diamond paste of 6 μm, 1 μm, and 0.25 μm grit size, so as to avoid transformational artifacts due to grinding. Samples were ion-thinned, carbon-coated, and examined in a TEM/STEM electron microscope.† A beam size of 50 nm was obtained by setting up conditions for convergent beam microscopy but using a small C2 condenser aperture. Microdiffraction patterns from adjacent areas, ≈ 50 nm in diameter, were indexed and stereographically correlated with traces of edge-on planes.

Experimental Results

Particles were typically ≈ 0.5 μm in diameter and divided into sections of average width ≈ 50 nm. The sections were identified as $(100)_m$ twins (Fig. 4(A)) by taking a convergent beam microdiffraction pattern within a section (Fig. 4(B)) and across adjacent sections (Fig. 4(C)). Corresponding single surface trace analysis of edge-on sections confirmed the interface as the $(100)_m$ plane. This observation was repeated a total of four times in different particles. Irregular moiré fringes were usually observed in each section or in alternating sections.

In larger (>1 μm) particles, groups of sections were sometimes arranged symmetrically about a "boundary" plane. For example, the crystal shown in Fig. 5(A) was divided into four large sections. The left-most section was indexed as $[\bar{1}1\bar{4}]$ or 6° off $(\bar{1}13)_m$ projection, as illustrated stereographically in Fig. 5(B). The small section traces were consistent with $(100)_m$ twinning, while the trace of the edge-on boundary plane corresponded to the $(110)_m$ plane. These preliminary results indicated that alternative groups of sections contained $(100)_m$ twins and were twin-related about the $(110)_m$ plane. Internal rearrangement twins of the type $\{110\}_m$ (see later) are also seen in Fig. 5(A). Note the stepped nature of the "boundary plane," possibly due to internal rearrangement in the particle.

An example of deformation or mechanical twins[12] having tapered ends and stopped inside the zirconia particle is seen in Fig. 6.

Another crystal in a 50%Al_2O_3-50%ZrO_2 specimen with $(100)_m$ twins is illustrated in Fig. 7(A). The microdiffraction pattern in several adjacent sections (Fig. 7(B)) was indexed as the $(1,0,6.9)_m$ pole or the $[001]_m$ direction. Three rectangular-shaped small twins were found which widened as they approached the edge of the crystal. Their edge-on long and short interfaces were parallel to $(1\bar{1}0)_m$ and $(110)_m$ planes, respectively, as verified by stereographic analysis. Twinning on either $(1\bar{1}0)_m$ or $(110)_m$ on a stereogram gave the same composite diffraction pattern due to the symmetry of the $[001]_m$ pattern, and the reflections from the small twin were superimposed on the $(100)_m$ twin reflections. Due to the large spot size of this microdiffraction method, however, they could not be discerned. Thus the smaller twins were of the $(1\bar{1}0)_m$ or equivalent $(110)_m$ type.

Discussion

Recently, in-situ studies of ≈ 1 μm zirconia particles included in a mullite matrix were made in a transmission electron microscope.‡[27,28] The beam-induced transformation produced numerous fine twins forming a complex mosaic pattern in the particle.

In the work presented here, the hot-pressed samples were cooled to room temperature, sliced, and carefully polished with fine grit diamond paste to avoid inducing the transformation by grinding. Thus, the microstructures observed were representative of the transformation mechanism on cooling in the bulk.

According to the model postulated above, the $(100)_m$ twins in ≈ 0.5 μm particles (Fig. 4) may be interpreted as an LIS twin forming one martensite variant. This is also consistent with the calculations of a martensite solution for $(001)_m$ or $(100)_m$ LIS twins.[4]

The large particles of Fig. 5 may be interpreted as four twin-related variants, each containing $(100)_m$ LIS twins and separated by a low index $(110)_m$ variant interface. The stepped nature of the interface confirms that it is a variant interface rather than a habit plane.

The mechanical twins of Fig. 6 are similar to post-transformation deformation twins found in in-situ experiments in thick foils of $4\% ZrO_2$-Al_2O_3 in a 1 MeV high voltage electron microscope.[29]

The $\{110\}_m$ small twins occurring among $(100)_m$ LIS twins in Fig. 7 are an example of post-transformation rearrangement twins. Their magnitude of shear is 0.25 as compared to 0.35 for $(100)_m$ twins.[1] The observed widening of $(100)_m$ twins beyond $\{110\}_m$ twins may also be attributed to internal rearrangement of twin interfaces to accommodate the small twins and achieve the minimum energy state.

In conclusion, a thermoelastic martensitic transformation mechanism may apply in zirconia particles included in a ceramic matrix. In larger particles, the balance between chemical free energy and elastic strains may result in self-accommodating variants with internal rearrangement.

Acknowledgments

The author gratefully acknowledges stimulating discussions with A. G. Evans, who initiated this work, and with A. H. Heuer, M. M. Rühle, and C. M. Wayman. Nick Burlingame is acknowledged for making the specimens. This work was done with the support of a United States Office of Naval Research Grant, Contract No. 842456-25989.

References

[1]J. E. Bailey, *Proc. R. Soc. A*, **279**, 359–412 (1964).
[2]G. K. Bansal and A. H. Heuer, *Acta Metall.*, **20**, 1281–89 (1972).
[3]G. K. Bansal and A. H. Heuer, *ibid.*, **22**, 409–17 (1974).
[4]W. M. Kriven, W. L. Fraser, and S. W. Kennedy; this volume, pp. 82–97.
[5]H. Kubo, Y. Uchimoto, and K. Shimizu, *Met. Sci.*, **9**, 61–66 (1975).
[6]K. R. Kinsman, J. W. Sprys, and R. J. Asaro, *Acta Metall.*, **23**, 1431–42 (1975).
[7]M. Kato, R. Monzen, and T. Mori, *ibid.*, **26**, 605–13 (1978).
[8]J. S. Bowles and J. K. Mackenzie, I, *ibid.*, **2**, 129–37 (1954); II, *ibid.*, 138–47; III, *ibid.*, 224–34.
[9]M. Cohen, G. B. Olson, and P. C. Clapp, *Proc. Int. Conf. Martensitic Transforma-*

tions, ICOMAT, MIT, USA, **1979**; pp. 1–12.

[10]B. A. Bilby and J. W. Christian, "The Mechanism of the Phase Transformation in Metals," *Inst. Met., London,* **1961,** p. 121.

[11]M. J. Buerger; pp. 183–211 in Phase Transformation in Solids. Edited by R. Smoluchowski, J. E. Mayer, and W. A. Weyl. Wiley, New York, 1951.

[12]J. W. Christian, The Theory of Transformations in Metals and Alloys. Pergamon Press, Oxford, 1965; Chapters 20–22.

[13]A. G. Evans, N. Burlingame, M. Drory, and W. M. Kriven, *Acta Metall.,* **29,** 447-56 (1981).

[14]T. Saburi and C. M. Wayman, *Acta Metall.,* **27,** 979–95 (1979).

[15]T. Saburi, S. Nenno, and C. M. Wayman; pp. 619–38 in Ref. 9.

[16]L. Delaey, R. V. Krishnan, H. Tas, and H. Warlimont, *J. Mater. Sci.,* **9** [1-4] 1521–55 (1974).

[17]Z. Nishiyama; p. 276 in Martensitic Transformation. Edited by M. E. Fine, M. Meshii, and C. M. Wayman. Academic Press, New York, 1978.

[18]T. A. Schroeder and C. M. Wayman, *Acta Metall.,* **25,** 1375–91 (1977).

[19]K. Takezawa, T. Shindo, and S. Sato, *Scr. Metall.,* **10,** 13–18 (1976).

[20]Z. S. Basinski and J. M. Christian, *J. Inst. Met.,* **80,** 659 (1951/52).

[21]S. W. Kennedy and W. M. Kriven, *J. Mater. Sci.,* **11,** 1767–69 (1976).

[22]S. W. Kennedy, W. M. Kriven, and W. L. Fraser; pp. 208–13 in Ref. 9.

[23]W. M. Kriven, Ph. D. Thesis, University of Adelaide, 1976; W. M. Kriven and S. W. Kennedy; to be published in the Proceedings of the International Conference on Solid-Solid Phase Transformations, to be held at Carnegie-Mellon University, Pittsburgh, Pa., Aug. 10-14, 1981.

[24]J. W. Christian, The Theory of Transformations in Metals and Alloys. Pergamon Press, Oxford, 1965.

[25]N. Claussen, F. Sigulinski, and M. Rühle; this volume, pp. 164–67.

[26]N. Burlingame, M. S. thesis, University of California, Berkeley, 1980.

[27]N. Claussen and J. Jahn, *J. Am. Ceram. Soc.,* **63** [3-4] 228–29 (1980).

[28]E. Bischoff, M. Kirn, W. M. Kriven, and M. Rühle; for abstract see *Am. Ceram. Soc. Bull.,* **60** [3] 383 (1981).

[29]N. Claussen, W. Mader, and M. Rühle; for abstract see *Am. Ceram. Soc. Bull.,* **59** [3] 363 (1980).

---

*Currently on leave at the Max-Planck-Institut fur Metallforschung, Institut fur Werkstoffwissenschaften, Stuttgart, Germany.

†Model 400, Philips Electron Instruments, Mount Vernon, N.Y.

‡Model JEM-200, Japan Electron Optics Co., Ltd., Tokyo, Japan.

§AEI Scientific Instruments, Urmston, Manchester, England.

Parent

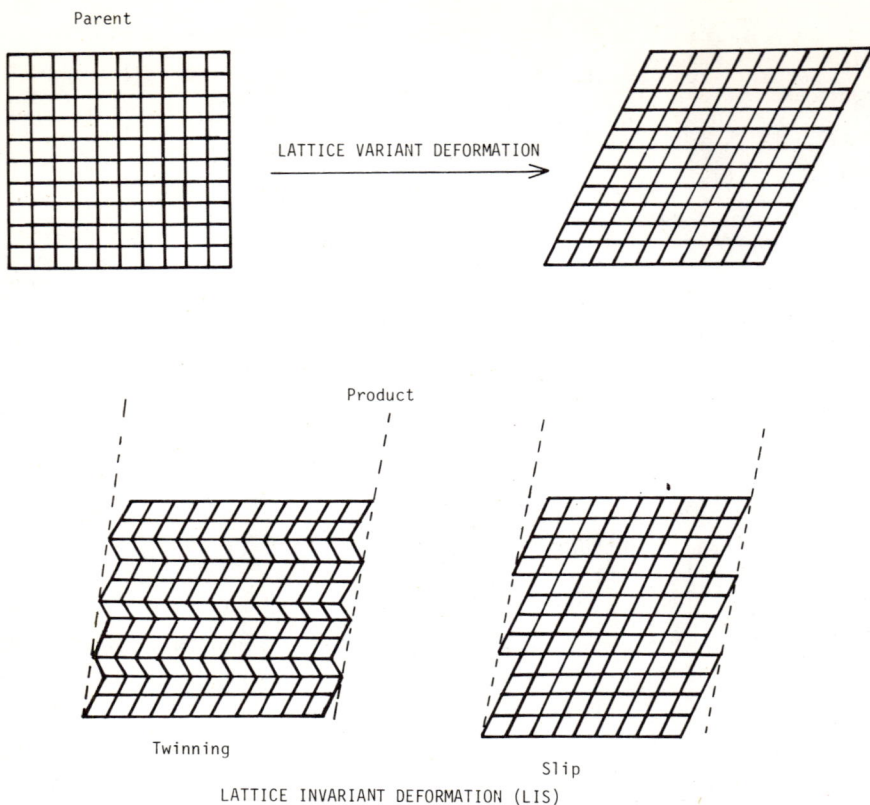

LATTICE VARIANT DEFORMATION

Product

Twinning

Slip

LATTICE INVARIANT DEFORMATION (LIS)

Fig. 1. Martensite transformation mechanism. Macroscopic shape change in the product results from a lattice variant deformation coupled with LIS slip or twinning.

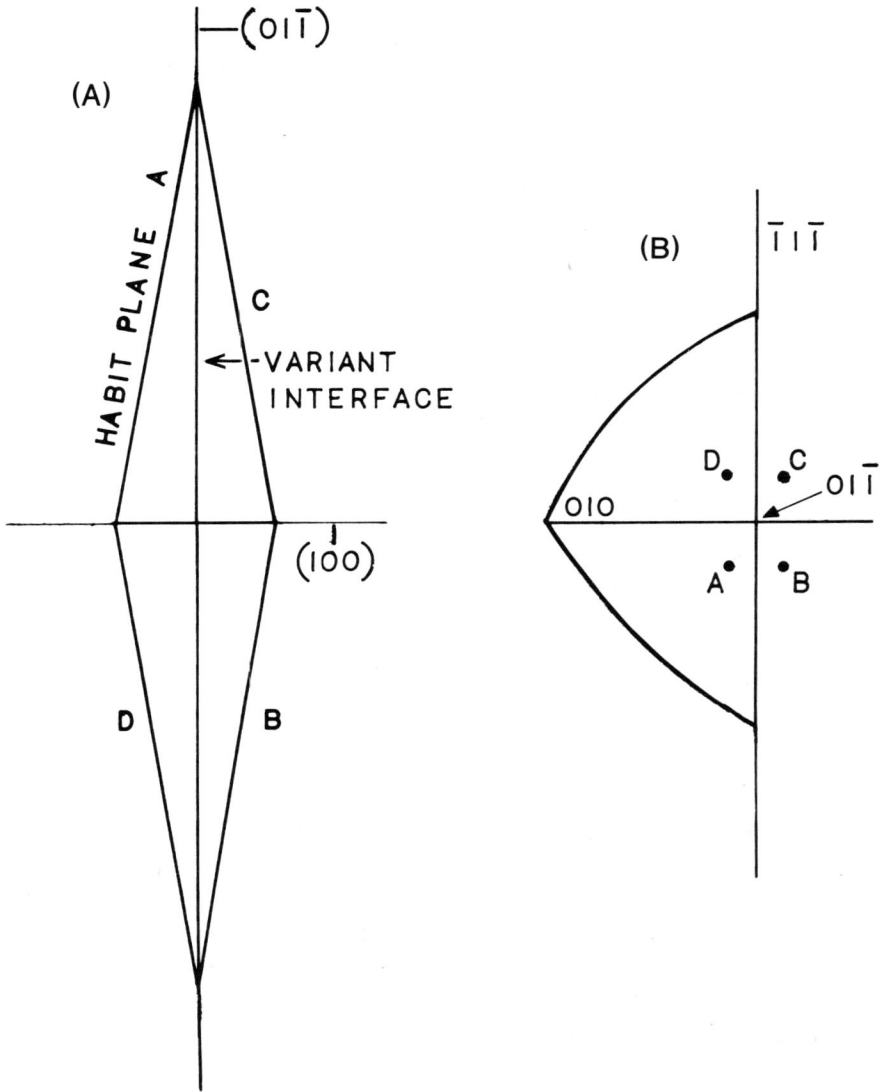

Fig. 2. (A) Crystallographic relations between four martensite variants (A to D) in a (01$\bar{1}$) plate group for close-packed β_1 structures (Ref. 14). (B) Habit plane normals of the variants clustered symmetrically about the (01$\bar{1}$) standard projection.

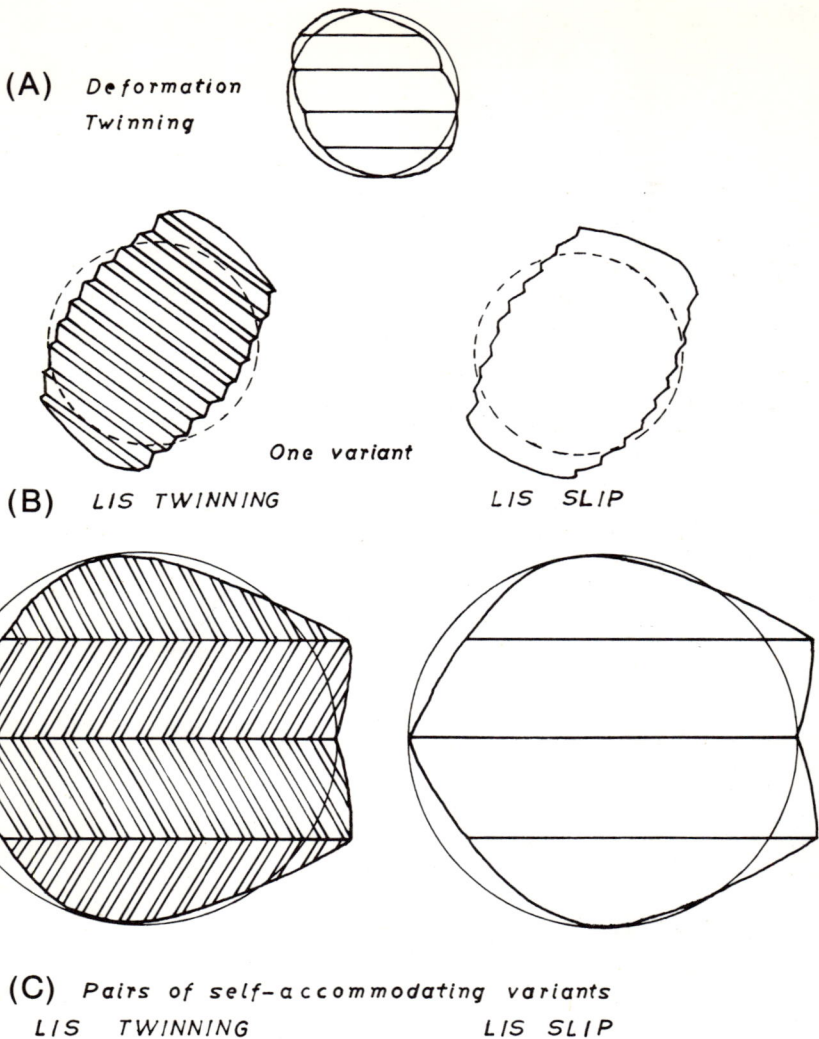

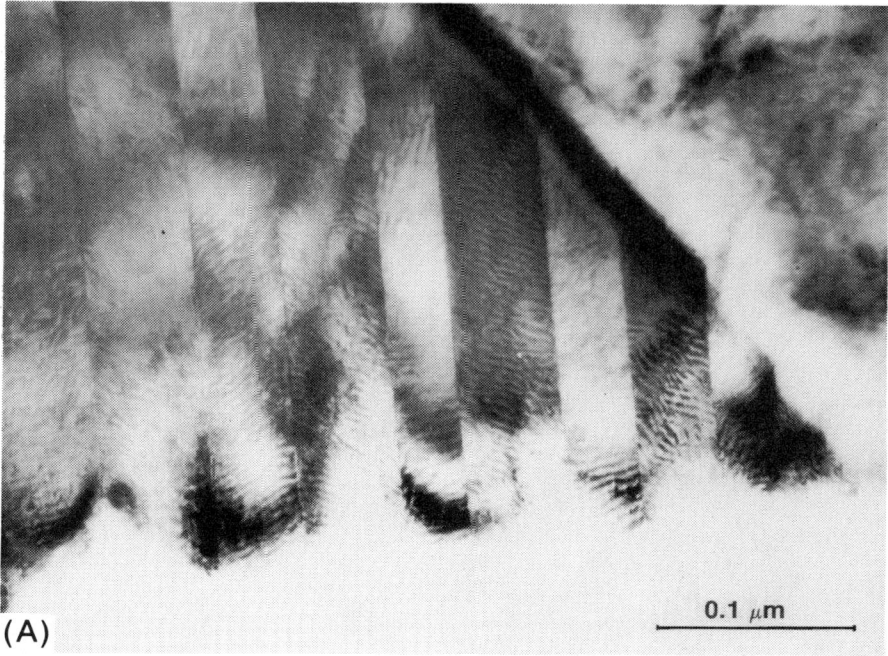

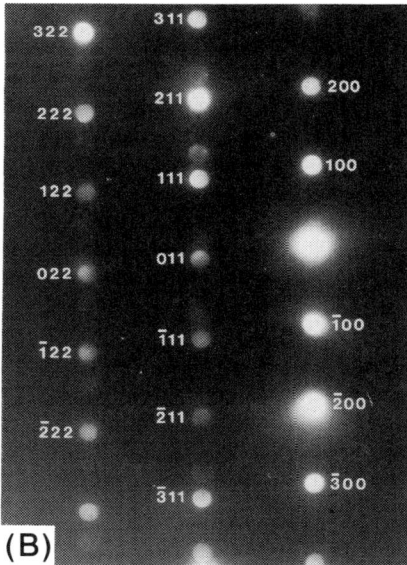

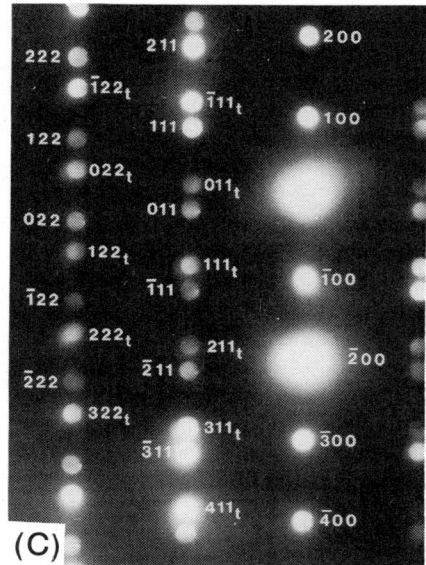

Fig. 4. (A) Sections corresponding to (100)$_m$ twins in a ZrO$_2$ particle. (B) Convergent beam microdiffraction pattern within one section or twin, as seen in a [01$\bar{1}$]$_m$ projection. (C) Composite microdiffraction pattern taken across two adjacent sections. It is twinned on (100)$_m$.

179

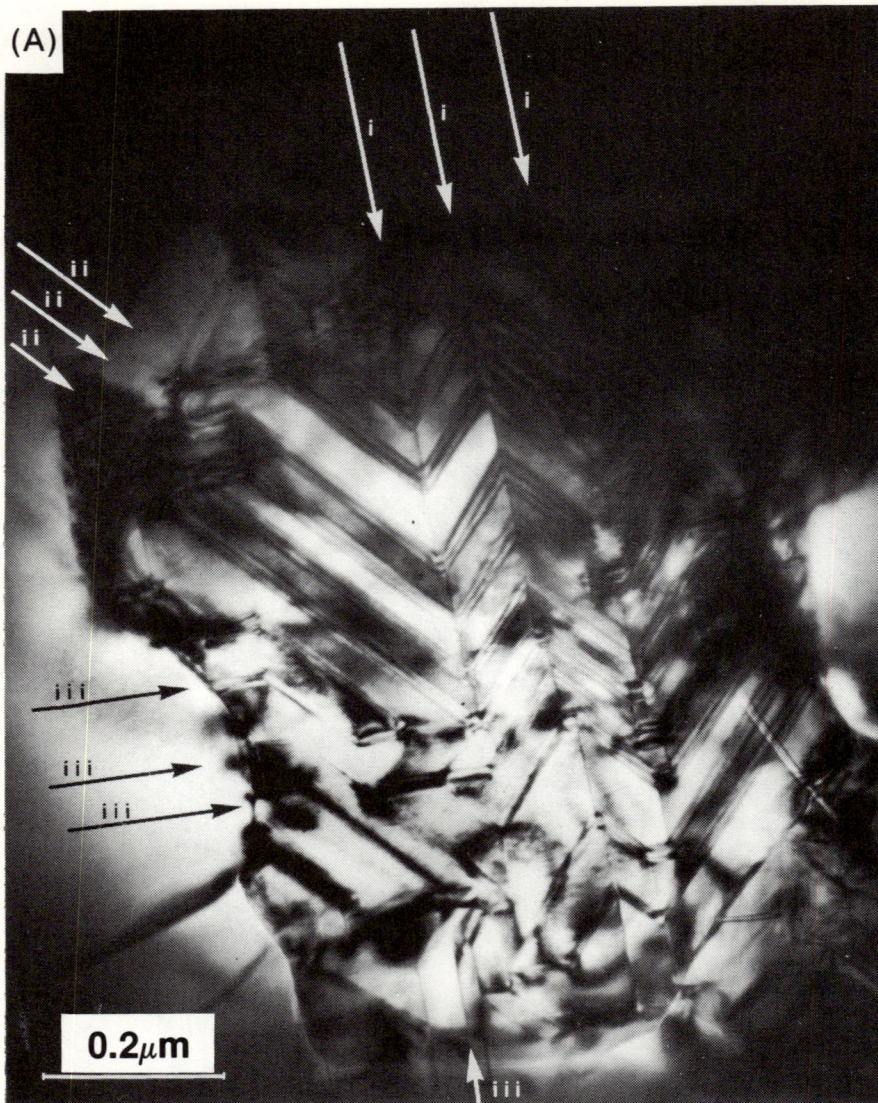

Fig. 5. (A) A large >1 μm particle of monoclinic ZrO$_2$ transformed into four large sections having (i) boundary traces // (110)$_m$, (ii) small sections with inclined traces from (100)$_m$, (iii) little internal sections with traces due to (1$\bar{1}$0)$_m$ and (110)$_m$ planes. These may be interpreted, respectively, as (i) variant interfaces, (ii) LIS (100)$_m$ twins, (iii) internal {110}$_m$-type rearrangement twins. (B) Corresponding experimental stereogram in [$\bar{1}$14]$_m$ projection illustrating the trace analysis.

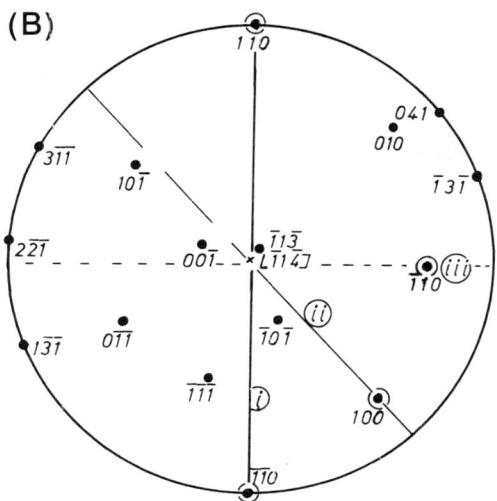

(B)

① ─ zone of large boundary traces
ⓘⓘ ── zone of small section traces
ⓘⓘⓘ ──── zone of little twin traces

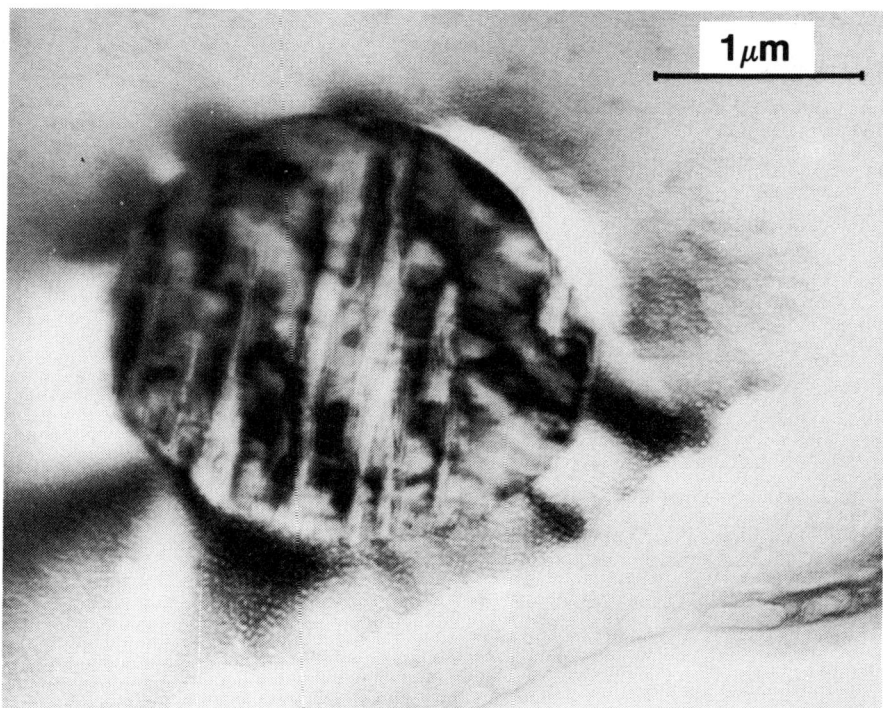

Fig. 6. An example of mechanical twins with tapered ends inside a particle.

181

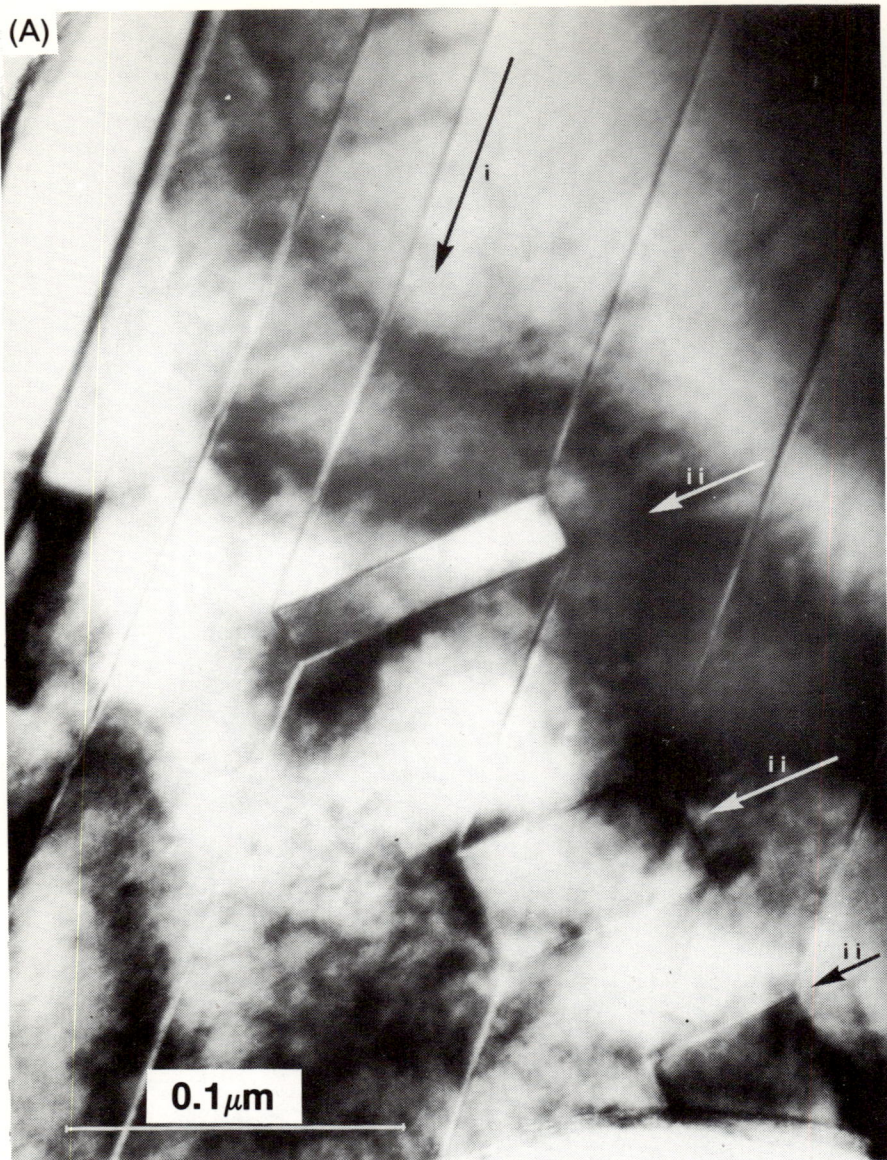

Fig. 7. *(A)* A monoclinic particle showing (i) (100)$_m$ twins, and (ii) small twins or {110}$_m$ internal rearrangement twins. *(B)*The corresponding microdiffraction pattern within and across twins, seen down a (1,0,6.9) or [001]$_m$ projection.

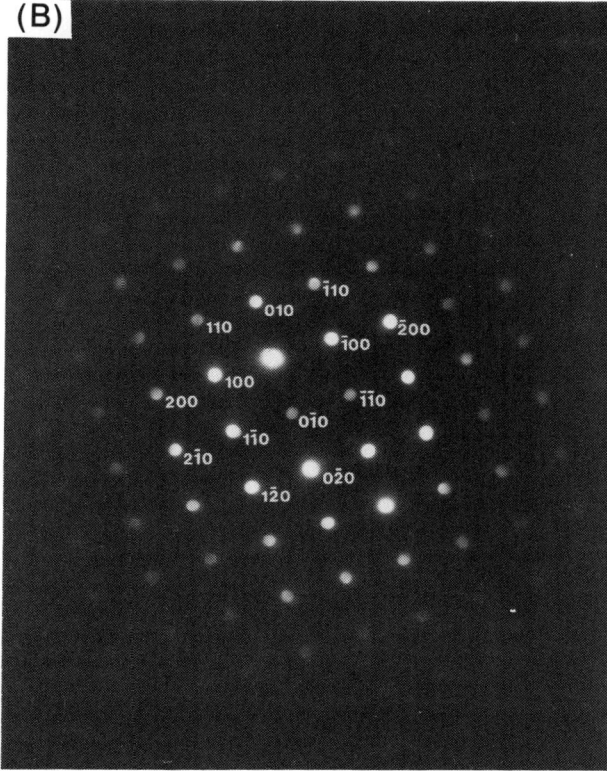

Phase stability and transformation toughening in zirconia

C. A. ANDERSSON AND T. K. GUPTA

Westinghouse Research and Development Center
Pittsburgh, Pa. 15235

An overview of martensitic transformation toughening in zirconia is presented. The thermodynamics of phase stabilization is first considered, and the distinction between diffusion-controlled and diffusionless transformations is made. The concept of a precursor martensitic embryo is used to explain the effects of particle sizes on phase stabilization. Finally, an analysis of stress-induced phase transformations in the region of crack tips and their consequences on the measured fracture toughness are presented.

Research activity in high fracture toughness materials based on or containing zirconia has been increasing. Early work on multiphased partially stabilized zirconia is being continued by a number of investigators.[1-5] There have been several recent investigations on nearly single phase, tetragonal zirconia.[6-9] Still others have reported results from mechanical mixtures of other ceramic materials with zirconia.[10,11] The improved toughness in most of these material types has been attributed to the transformation of the tetragonal phase to monoclinic symmetry in the vicinity of propagating cracks. Although some toughening can be achieved by introducing transformation-induced microcracking in the as-fabricated material, the most significant gains occur when the elevated temperature phase is retained prior to testing and the transformation is stress-induced.

It is the intent of this paper to give an overview of the subject of transformation toughening. In the first section, the effects of both chemistry and particle size on the free energy and phase stability will be discussed. In the subsequent section, stress-induced martensitic transformations and effects on fracture toughness will be considered.

Phase Stability

The tetragonal-to-monoclinic phase transformation in zirconia systems has been established as martensitic in nature,[12,13] and it has many features in common with martensitic transformation in metals. Therefore, many of the characteristics of this type of transformation reported in the metals literature are directly applicable to zirconia:

• The transformation is a military one, i.e. the structures convert by a diffusionless process through shear and dilation of the parent structure.[14]

• The transformed martensite has an invariant habit plane, with the shear occurring parallel to this plane and the dilation normal to it.[15,16] Structural accommodation between the martensite and the parent phase occurs either by twinning of the martensite and/or by the creation of dislocation arrays at the interface.

- The shear and dilation are generally accompanied by a significant increase in the strain energy of the system.[17]
- The increase in system strain energy often causes the transformation to be athermal.[17] The driving force must be further increased, e.g. by lowering the temperature, in order to allow further transformation to occur.
- Thermally activated nucleation at the transformation temperature is not considered to be a likely process considering the energetics and nuclei sizes required.[18] Instead, martensite is considered by several investigators to grow from pre-existing embryos.[19,20] Although the nature of these embryos is still in dispute even in the well studied metal systems, the precursor embryo is thought to consist of an intermediate structure of the two phases formed by dislocation arrays, stacking faults, or other faulted structures.[20]
- For an embryo to grow, it must generally overcome some strain energy barrier such as the unpinning of dislocations at its periphery.[19,20] In some cases this could be a thermally activated process leading to isothermal transformation of martensite.
- At some point in the precursor structure embryo growth, it will be energetically favorable to convert to the final martensite structure. At the start of growth, the atomic coincidence between the embryo and the parent phase would make the surface energy low but strain energy high. The latter will rapidly increase with growth. However, conversion to the less coherent stable true martensite will cause a higher surface energy but a lower strain energy, resulting in a total reduction in system energy by that conversion.[20]
- Martensite growth is stopped by barriers such as grain boundaries or other martensite platelets.[18]
- The extent of martensite transformation is often found to depend on particle or grain size.[21] Transformation can be retarded by reducing particle or grain size.
- There is generally a temperature hysteresis observed between the martensitic and reverse transformations.[21]

Although further work is required to delineate the mechanism of the martensitic transformation in zirconia, an attempt is made in this section to explain the phase stability observed in zirconia.

Embryo Growth

The basic assumption in this procedure is that the martensitic transformation in zirconia occurs by the growth of pre-existing embryos. The nature of these embryos and their origins are not known for the material. They may have been formed at higher temperatures where thermal activation is more possible and been "frozen in" as the temperature was lowered,[18,19] or they may be associated with structural defects such as stacking faults and/or dislocation arrays.[20] It is only necessary to assume that there exists some equilibrium number per unit volume as a function of their effective sizes.

The free energy difference per unit volume between the tetragonal parent phase and the embryo, Δg_v, is:

$$\Delta g_v = \Delta G/V_e = \Delta\mu_v + u_v^{str} + B\sigma^{surf}v_e^{-1/3} \tag{1}$$

where $\Delta\mu_v$ is the change in chemical free energy per unit embryo volume, u_v^{str} is the strain energy per unit embryo volume, σ^{surf} is the surface energy per unit area between the embryo and the parent phase, B is a geometric constant

185

dependent on the embryo shape, and v_e is the volume of the embryo. The values of μ_v, u_v, and σ^{surf} are dependent on the nature of the embryo, and, since the embryo is probably an intermediate structure, these will not be the same as those associated with the final martensitic phase. The $Bv_e^{-1/3}$ term is the ratio of the embryo area to its volume (for a spherical geometry B would equal $(36\pi)^{1/3}$).

Equation (1) is schematically shown in Fig. 1 where Δg_v is plotted against temperature. The line at the right is that for the chemical free energy term $\Delta\mu_v$, the middle line, which includes the strain energy effect, is for large embryo size ($v_e \to \infty$), and that at the left is for finite embryo size, and it therefore incorporates the surface energy term. The latter line will be shifted to lower temperature as the embryo size decreases. Growth of embryos of this latter size would normally be expected to occur when Δg_v is zero at temperature T_v. However, an additional strain energy barrier, u_v^{bar}, may have to be overcome before growth can proceed.[19,20] This barrier is shown as the dashed line in the figure. As a result of this additional barrier, the initiation of growth, and thus the start of the martensite transformation, occurs at a lower temperature, M_s, where the two curves intersect. In metal systems, this barrier is considered to be the strain energy required to initiate the movement of accommodating dislocations at the embryo interface. The criterion for embryo growth is thus $\Delta g_v = u_v^{bar}$.

The embryo will grow to a size beyond which the final martensitic structure becomes energetically favorable. The free energy per unit volume of the martensite phase will also be of the same form as Eq. (1). Equating the two free energies and solving for the volume, v, above which the embryo converts to the final martensite structure gives:

$$v = \left[\frac{B(\sigma_m^{surf} - \sigma_e^{surf})}{\mu_e - \mu_m + u_e^{str} - u_m^{str}} \right]^3 \tag{2}$$

where the e subscripts refer to the embryo and the m subscripts refer to the martensite phase. Thus, the contribution of the embryo to the transformation is to provide an intermediate energy state for the system for which growth is possible. The impossibly large thermal activation energy for direct conversion of martensite from the parent phase is thus bypassed.

Phase Diagrams

The relationships between thermochemistry and phase diagrams are well understood. Of particular interest to the present discussion is the distinction between diffusion-controlled transformations, in which the compositions of the phases differ from the mean composition, and military transformations, in which there are no compositional differences. These relationships are graphically depicted in Fig. 2, where the upper curves represent the free energy versus compositional diagrams at three temperatures for the schematic phase diagram at the bottom of the figure.*

In general, the phase that exists at a particular composition is determined by the phase which has the lowest free energy at a given temperature. At temperature T_3 between compositions O and C_2 (point A on the top curve), the tetragonal phase is stable. Above composition C_5 (point B on the top curve), the cubic phase is stable. Between compositions C_2 and C_5 a two-phase mixture causes the free energy of the system to be at a minimum. For diffusion-controlled transformations (the solid lines in the phase diagram), the phase boundaries are defined by the tangent points to the free energy curves such as A,B; C,D; E,F; and G,H for the three selected temperatures.

As the temperatures are lowered in materials such as zirconia, cation diffusion becomes much more difficult and diffusionless; martensitic transformations become more favorable. For the cases where diffusion is impossible, the free energies are equal at the curve intersections. Thus, at temperature T_2, the monoclinic phase would be stable over compositions from 0 to the intersection between the tetragonal and monoclinic curves at point I. Between points I and J, where the monoclinic curve intersects the cubic curve, the tetragonal phase would be stable and at higher compositions the cubic phase would be stable. Therefore, a new set of diffusionless transformation phase boundaries is defined by free energy curve intersections and shown as the dashed lines in the lower figure. The MIK curve is the boundary between the tetragonal and monoclinic phases; the JL curve is the boundary between the cubic and tetragonal phases.

The preceding discussion holds only for infinite sized quantities of materials, i.e. it ignores surface effects at phase boundaries. In real materials, surface effects contribute to the free energy and thus perturb the phase diagrams obtained experimentally. An example is shown in the free energy curve at temperature T_1. The surface energy per unit volume in Eq. (1), given as $B\sigma v^{-1/3}$, can be converted to a molar quantity by multiplying it by the molecular weight M_w, and dividing it by the density, ϱ, at the given composition. Thus,

$$u^{surf} = M_w B\sigma / \varrho v^{1/3} \tag{3}$$

in terms of energy units per mole. Adding this value to the monoclinic free energy curve results in the dot-dashed monoclinic free energy curve at T_1. The intersection of the tetragonal and monoclinic curves at point P defines the phase boundary for small volumes, v, of monoclinic in a tetragonal matrix. Thus, for finite volumes, the transformation phase boundary would be shifted from MIK to NPQ in the example given. The smaller the volume of each transformed region, the greater the shift to lower temperature. Again, it is the ratio of the surface area to volume that controls this shift. A similar argument can be made for the cubic-to-tetragonal transformation and the JL curve would be shifted to lower temperatures.

The transformations at two compositions which typify the experimental work that has been performed on transformation toughening will next be considered. The first case is that in which, by control of both particle size and composition, the tetragonal structure can be predominantly stabilized at room temperature. Consider material of composition C_2, heated to temperatures where the tetragonal phase is stable (slightly below T_3). Now, as this material is cooled below the dashed curve MIK in the phase diagram of Fig. 2, growth and martensitic transformation of the larger embryos will take place. As the temperature is further decreased, smaller and/or less effective embryos will transform. If at room temperature the predominating embryos have not reached their growth temperature, then the material will have a large fraction of retained tetragonal. The grain and particle size effects on the fraction transformed will be discussed later.

The other case to be considered is that for partially stabilized zirconia (PSZ). Consider material of composition C_3 in Fig. 2. If this material is heated into the cubic phase field and quenched to T_1 (room temperature), the tetragonal phase would be expected to be formed by diffusionless transformation. If this same composition is heated to temperature T_4 and held for equilibration, then a two-phase mixture of cubic zirconia of composition C_4 and tetragonal zirconia of composition C_1 will form. Quenching will cause diffusionless martensitic transformation of the cubic zirconia to tetragonal symmetry (composition C_4) and the tetragonal zirconia to monoclinic symmetry (composition C_1). As a third option, the material can be heated to and equilibrated at T_3 where the two-phase mixture of tetragonal zirconia of composition C_2 and cubic zirconia of composition C_5 will form. On quenching from T_3, the cubic phase will be retained since it is to the right of the JL diffusionless transformation curve. The tetragonal phase may or may not be transformed to monoclinic symmetry, depending on the embryo size and the tetragonal particle size.

It is this competition between diffusion and diffusionless transformations that has led to confusion in the phase diagram work in zirconia systems. Most of the uncertainties in these diagrams occur in temperature ranges below 1200 °C where cation diffusion becomes more difficult. The determination of free energy-composition-temperature relationships would greatly assist the interpretation of those results.

Particle/Grain Size

Experimental observations on a variety of zirconia materials have shown a dependence of tetragonal stabilization on grain size,[6-9] precipitate size,[1-4] and powder size.[5] Below some critical size, in the 0.1 to 0.3 μm range, tetragonal material can be retained at room temperature and below. Above this size, the monoclinic transformation usually occurs. It is well established in metal systems that martensite growth is stopped by obstacles such as grain boundaries or other martensite transformation regions.[18] Invocation of this phenomenon in zirconia materials, along with the concept that the martensitic transformation occurs by the growth of pre-existing embryos, when the driving force is sufficiently high, allows the particle/grain size phenomena to be assessed.

It is reasonable to presume that there exists an equilibrium number of embryos per unit volume, $n_{e/v}$. If the embryos are associated with dislocations, this would be their numbers per unit volume. If the embryos are associated with an elevated-temperature thermally activated phenomenon and "frozen in," or with stacking faults, these equilibrium numbers from kinetic theory are appropriate. The number of critical-sized embryos per particle or grain, $n^*_{e/p}$, is then:

$$n^*_{e/p} = \int_0^{v^p} n^*_{e/v} dv \qquad (4)$$

where $n^*_{e/v}$ is the number of critical-sized embryos per unit volume and v_p is the particle volume. The fraction transformed, f_{tr}, is simply the probability that a particle or grain contains a critical-sized embryo, or:

$$f_{tr} = n^*_{e/p}, \text{ for } n^*_{e/p} < f_{tr}^{max}(C,T)$$
$$f_{tr} = f_{tr}^{max}(C,T), \text{ for } n^*_{e/p} \geq f_{tr}^{max}(C,T) \qquad (5)$$

where $f_{tr}^{max}(C,T)$ is the maximum fraction of athermal transformation that can occur for a particular composition, C, and temperature, T.

For a constant or average number of embryos per unit volume, independent of the embryo size, $\bar{n}_{e/v}^*$,

$$n_{e/p}^* = \bar{n}_{e/v} v_p \tag{6}$$

For $n_{e/v}^*(v)$ which is a function of the embryo volume, Eq. (4) becomes:

$$n_{e/p}^* = \int_{v_e^{min}}^{v_p} n_{e/v}^*(v)\,dv \tag{7}$$

where $v_e^{min}(C, T)$ is the minimum embryo volume that can transform at a particular composition and temperature. If the embryos are the result of an activated growth process at a temperature T_e, then from Maxwell-Boltzmann statistics:

$$n_{e/v} = \frac{N\varrho}{M_w}\, e^{-\Delta G_e/kT_e} \tag{8}$$

where N is Avogadro's number, ϱ is the density of the material, M_w the molecular weight, and ΔG_e the free energy change accompanying the formation of an embryo:

$$\Delta G_e = (\Delta\mu_v + u_v^{str})\, v_e + B\sigma^{surf} v_e^{2/3} \tag{9}$$

The energetics are displayed schematically in Fig. 3. The critical volume, v_p^*, above which more than one critical sized embryo is contained within each particle or grain, is shown at the bottom of the figure as the intersection of the curve with f_{tr}^{max}. This lower curve can be compared with experimental results for ZrO_2 containing 1.4 mol% Y_2O_3 in Fig. 4. Here v_p^* is 11×10^{-21} m^3 (assuming spherical grains), f_{tr}^{max} is 0.85, and v_e^{min} is 5×10^{-21} m^3. This latter value appears to be fairly large since the minimum critical embryo size would contain $\approx 35 \times 10^6$ unit cells. The number of critical embryos per unit volume is 8×10^{19} m^{-3} which is a reasonable value.

Athermal Transformation

The dashed curves in Fig. 2 define the start of diffusionless transformations in the system. If the transformations are isothermal, these lines would also define the boundaries between the respective single phases and there would be no two-phase regions. Martensitic transformations such as the tetragonal-to-monoclinic transformation in zirconia usually cause the strain energy of the system to increase. Therefore, more driving force must be provided to continue the transformation, resulting in its athermal nature. If a stress-relieving mechanism is available, especially at elevated temperatures, then a component of the transformation could be isothermal. At low temperatures, the only stress-relieving mechanism would be crack growth and, barring this, the transformation would be expected to be athermal.

Transformation Toughening Mechanism

The preceding section presented concepts for fabricating materials which retain tetragonal phases at temperatures below those that would be predicted from equilibrium phase diagrams. The following section will discuss stress-induced transformations and their effects on fracture toughness.

Stress-Induced Transformations Near the Crack Tip

Since phase transformations are by definition atomic rearrangements, the associated strains in the solid state give rise to a change in the strain

189

energy of a system. For a transformed volume, V, this change has been determined by Eshelby to be[24,25]

$$U^{str} = 1/2 \int_v \sigma_{ij}^I \epsilon_{ij}^T dv \qquad (10)$$

where σ_{ij}^I are the internal stress components of the transformed particle, and ϵ_{ij}^T are the stress-free transformation strains (the repeated suffixes are summed over the values (x, y, z)). Application of negative stresses σ_{ij}^I to the precursor volume would reduce the transformation strain energy and thus promote the transformation.

Martensitic transformations in metallic systems can be induced by both uniaxial tension and compression but are usually retarded by hydrostatic compression.[26] Transformation strains have both positive dilational and shear components which increase the strain energy of the system. Uniaxial compression will assist the shear component while retarding the dilational one, whereas the more effective uniaxial tension will assist both. Hydrostatic compression has no shear component but has a negative effect on the dilation.

Martensitic transformations have been observed to occur in the regions of cracks in zirconia-containing ceramics because the stresses associated with the crack tip within an externally loaded body are extremely high. These have been given by Westergaard as[27]:

$$\sigma_{xx} = \frac{K_I}{(2\pi r)^{1/2}} \cos \frac{\theta}{2} \left(1 - \sin \frac{\theta}{2} \sin \frac{3\theta}{2}\right)$$

$$\sigma_{yy} = \frac{K_I}{(2\pi r)^{1/2}} \cos \frac{\theta}{2} \left(1 + \sin \frac{\theta}{2} \sin \frac{3\theta}{2}\right)$$

$$\sigma_{zz} = \nu(\sigma_{xx} + \sigma_{yy}) \qquad \text{(plane strain)}$$

$$\sigma_{zz} = 0 \qquad \text{(plane stress)}$$

$$\gamma_{xy} = \frac{K_I}{(2\pi r)^{1/2}} \cos \frac{\theta}{2} \sin \frac{\theta}{2} \cos \frac{3\theta}{2}$$

$$\gamma_{xz} = \gamma_{yz} = 0 \qquad (11)$$

where x is the direction in the crack plane normal to the crack front, y the direction normal to the crack plane, and z the direction parallel to the crack front; r is the radial distance from the crack tip, θ the angle in the xy plane measured from the x direction; ν is the Poisson ratio, and K_I the stress intensity factor, which is a function of applied stress σ^A and crack geometry.

Upon loading the body, the regions with strain axes most favorably oriented with respect to the stress axes will require the least amount of strain energy and will transform first,[26] and more load will be required to induce the transformation of less favorably oriented regions. The average strain energy required can be determined by averaging over all the strain orientations with respect to the stress axes. Since the strain axes, i.e. particle orientations, are generally unknown, random orientation is assumed and an average stress-free strain of transformation, $\bar{\epsilon}^T$, is determined to be the average of the principal strains, ϵ_{pii}:

$$\bar{\epsilon}^T = \epsilon_{pii}/3 \qquad (12)$$

By substituting Eqs. 11 and 12 into Eq. 10, the strain energy change of the system per unit volume of transformed material, $\bar{v}_n^{str}$, becomes:

$$\bar{u}_v^{str} = \frac{1}{2}\sigma_{ij}\bar{\epsilon}^T = \frac{(1+\nu)K_I}{(2\pi r)^{1/2}} \cos\frac{\theta}{2}\bar{\epsilon}^T \tag{13}$$

The volume of material transformed in the region of a crack tip will depend on both the volume fraction of parent material available for transformation and the temperature at which the transformation is induced. As previously described, a PSZ material can be produced with the same tetragonal phase composition and particle size as a fully tetragonal material, but the PSZ material would have a large fraction of cubic phase unavailable for transformation.

The effect of temperature can be seen by referring to the left hand curve in Fig. 1. At the martensitic start temperature, M_s, all of the applied strain energy is available for inducing transformations. At temperatures, T, above the M_s, additional strain energy, u_v^{min},

$$\Delta g_v = u_v^{min} = (T - M_s)\Delta s_v \tag{14}$$

is required before the initiation of transformation. Here Δs_v refers to the difference in entropy per unit volume between the two phases. The higher the temperature for a given amount of applied strain energy, the smaller the volume that will be transformed. At temperatures below the M_s, a certain fraction of the available volume will be autotransformed. That volume plus the stress-induced volume should be nearly the same as the volume induced at the M_s temperatures. If too large a volume is autotransformed, the material will macrocrack.

Very near the crack tip, the strain energy is extremely high and is sufficient to fully transform the available parent material. Since the strain energy falls rapidly with the distance from the crack tip, a zone of partial transformation would be expected. This is schematically depicted in Fig. 5(a). The shape of the transformation zone is a consequence of randomly oriented particles being considered. The strain energies giving rise to this zone are schematically shown in Fig. 5(b). The limits of the two zones are determined by substituting u_v^{min} and u_v^{max} into Eq. (13) and solving for r as a function of θ. The u_v^{max} is

$$u_v^{max} = u_v^{min} + \bar{u}_v^T \tag{15}$$

where, assuming random orientation, $\bar{u}_v^T$ has been given to be[24,25]

$$\bar{u}_v^T = \frac{E\bar{\epsilon}^{T^2}}{9(1-\nu)} \tag{16}$$

where E is the elastic modulus.

Transformation Toughening

Numerous derivational approaches have been used to determine the relationships between external loadings, crack sizes, and propagation, i.e. the fracture mechanics equation. Similarly, a variety of approaches can also be used to determine the effects of stress-induced martensitic transformations near crack tips on those relationships. The method used here is similar to one given by Irwin[28] which demonstrated the equivalence of the energy balance approach first proposed by Griffith[29] and the crack tip stress field approach derived by Westergaard.[27] Using the latter's equations for the stresses normal to the crack plane at $\theta = 0$,

$$\sigma_{yy} = \frac{K_I}{(2\pi x)^{1/2}} \tag{17}$$

and the crack surface displacement, v_y, normal to the plane of the crack:

$$v_y = \frac{4K_I}{(2\pi)^{1/2} E} (\alpha - x)^{1/2} \tag{18}$$

where α is a small amount of crack extension shown in Fig. 5(c). Irwin showed that the "fixed grip" loss of energy, $G_I \alpha$, from the strain energy field as the crack extends can be determined by

$$G_I \alpha = \int_0^\alpha \sigma_{yy} v_y dx = \frac{2K_I^2}{\pi E} \int_0^\alpha \left(\frac{\alpha - x}{x} \right)^{1/2} dx = \frac{K_I^2 \alpha}{E} \tag{19}$$

This amount of strain energy loss is schematically shown in the load versus deformation curve of Fig. 5(d) as the lower shaded triangle.

When the stresses of the crack tip induce phase transformations, there occurs an interaction with the strain energy field similar to that for a crack. For the "fixed grip" case, the tetragonal-to-monoclinic transformation in zirconia would cause a strain energy loss of the amount[24,25]

$$U^T = \sigma_{yy}^A \bar{\epsilon}^T V^T \tag{20}$$

where σ^A is the applied stress, remote from the crack, and V^T is the volume of transformed material. In the case under consideration, this transformation volume refers to that contained in the α-width, unit-thickness region immediately in front of the crack tips of a centrally cracked specimen. Since at $\theta = \pi/2$, $r = y$ (Figs. 5(b), (c)), the volume of the fully transformed zone is

$$V_{f.t.} = 4f_p \alpha \int_0^{Y_i} dy = 4f_p \alpha Y_i \tag{21}$$

where f_p is the fraction of matrix material available for transformation. Setting $\theta = \pi/2$ in Eq. (13), and noting that at $r = Y_i$, $\bar{u}_v^{str} = u_v^{max}$,

$$V_{f.t.} = \frac{f_p \alpha (1 + \nu)^2 K_{10}^2 \bar{\epsilon}^{T^2}}{\pi} \frac{1}{u_v^{max^2}} \tag{22}$$

The fraction transformed, f_t, in the partially transformed region will be proportional to the strain energy available,

$$f_t = \frac{\bar{u}_v^{str}(y) - u_v^{min}}{\bar{u}_v^T} \tag{23}$$

The transformation volume of the partially transformed region, $V_{f.t.}$, is then

$$V_{p.t.} = 4f_p \alpha \int_{Y_i}^{Y_0} \frac{\bar{u}_v^{str}(y) - u_v^{min}}{\bar{u}_v^T} dy \tag{24}$$

where Y_0 is the position where $\bar{u}_v^{str} = u_v^{min}$.

192

Again, by setting $\theta = \pi/2$ in Eq. (13), and integrating

$$V_{p.t.} = \frac{f_p \alpha (1+\nu)^2 K_{I0} \bar{\epsilon}^{T2}}{\pi \bar{u}_v^T} \left[2\left(\frac{1}{u_v^{min}} - \frac{1}{u_v^{max}} \right) - u_v^{min} \left(\frac{1}{u_v^{min2}} - \frac{1}{u_v^{max2}} \right) \right] \quad (25)$$

Summing the volumes and substituting into Eq. (20) gives the "fixed grip" loss of energy due to the stress-induced transformation, $T_I\alpha$:

$$T_I\alpha = \frac{f_p \alpha (1+\nu)^2 K_{I0} \sigma^A \bar{\epsilon}^{T3}}{\pi} \left[\frac{1}{u_v^{max2}} + \frac{2}{\bar{u}_v^T} \left(\frac{1}{u_v^{min}} - \frac{1}{u_v^{max}} \right) - \frac{u_v^{min}}{\bar{u}_v^T} \left(\frac{1}{u_v^{min2}} - \frac{1}{u_v^{max2}} \right) \right] \quad (26)$$

This equation is valid at temperatures above the M_s since it blows up as u_v^{min} approaches zero. As the temperatures closely approach the M_s, Eq. (24) should be solved for Y_0 equal to the outer dimensions of the test piece.

The total strain energy loss, U, due to crack propagation by an amount α is therefore

$$U = G_{I0}\alpha + T_I\alpha \quad (27)$$

In this and the previous equations, the zero subscript refers to stress intensities, K_{I0}, and the strain energy release rates, G_{I0}, that are independent of the crack tip transformations. There will also be a pair of stress intensities, K_I, and strain energy release rates, G_I, calculated from the applied stresses and crack lengths, a, that incorporate the effects of the transformation. In the present case

$$G_I = \frac{\pi \sigma^{A2} a (1-\nu^2)}{E} \quad (28)$$

and

$$K_I = \sigma^A (\pi a)^{1/2} \quad (29)$$

Noting that G_I is alternatively defined as U/α, then

$$G_I = G_{I0} + T_I \quad (30)$$

or

$$K_I^2 = K_{I0}^2 + \frac{ET_I}{(1-\nu^2)} \quad (31)$$

The failure criterion can be alternately defined as the value of G_{I0} that reaches an inherent resistance to propagation, R, or as the value of K_{I0} that reaches an inherent critical value, K_{IC}. Thus, if stress-induced transformation occurs, the "measured" values of the two parameters will be above the actual critical values. The K_I can be determined by substituting Eqs. 26 and 29 into 31, giving

$$K_I^2 = K_{I0}^2 + \beta K_{I0}^2 K_I a^{-1/2} \quad (32)$$

where

$$\beta = \frac{f_p(1+\nu)^2 E \bar{\epsilon}^{T3}}{\pi^{3/2}(1-\nu^2)} \left[\frac{1}{u_v^{max2}} + \frac{2}{\bar{u}_v^T} \left(\frac{1}{u_v^{min}} - \frac{1}{u_v^{max}} \right) - \right.$$

$$\frac{u_v^{min}}{u_v^T} \left(\frac{1}{u_v^{min^2}} - \frac{1}{u_v^{max^2}} \right) \Big] \tag{33}$$

and solving the quadratic equation:

$$K_I = \frac{K_{Io}}{2} \left[\frac{\beta K_{Io}}{a^{1/2}} + \left(\frac{\beta^2 K_{Io}^2}{a} + 4 \right)^{1/2} \right] \tag{34}$$

This then is the transformation toughening equation, i.e. the equation relating to the toughness, K_I, measured when stress-induced transformations occur to the toughness, K_{Io}, in the absence of transformations.

Experimental critical stress intensities of tetragonal zirconia as a function of temperature (Fig. 6) are in agreement with Eq. (34). As the temperature is lowered toward the M_s, the measured critical stress intensity increases. A straight line from a linear regression analysis has been drawn through the points to show the trend. Equation (34) predicts a concave upward curve whose curvature would be dependent both on the Δs_v and the M_s, neither of which are known for this material.

Summary

Two categories of phase transformations occur in zirconia materials. Above ≈ 1500 K, the diffusion of both ionic species allows diffusion-controlled transformations to occur. This results in the usual phase equilibria being functions of temperature and composition, as predicted by conventional phase diagrams. For example, phases of two compositions will form in two-phased regions. Below 1500 K, cation diffusion becomes sluggish and transformations, when they occur, are of the diffusionless (martensitic) type. In this case, the parent and the transformed phase have the same composition. In practice, both transformation types can be observed. Heating above 1500 K allows the diffusion-controlled transformations to occur, which upon cooling may under certain conditions further transform in a diffusionless manner.

The phases achieved at a given temperature and composition are not only dependent on composition and thermal history, but also on structure and the state of stress. In particular, martensitic transformations are sensitive to the particle or grain size of the parent phase and its internal strain energy. The size effect can be explained by invoking the concept that the transformation occurs by growth of a pre-existing, precursor embryo as opposed to a nucleation and growth process. Growth of an embryo of a particular size, or effectiveness, occurs when the thermodynamic driving forces are sufficient for that size. Smaller or less effective embryos will require greater driving forces. Assuming some equilibrium distribution of embryos of given effectiveness per unit volume and that the volume is subdivided into a number of particles, then, for a given amount of driving force, the probability that a particle transforms is dependent on the probability that it contains an effective embryo. That probability is inversely proportional to the particle volume. Thus, below some critical particle size, the fraction of particles transformed for given conditions also will be inversely proportional to particle size. The result is that, for particle sizes below the critical value, the stability of the elevated temperature phase is maintained to lower temperatures in a larger fraction of the particles or grains. The critical particle size is that which results in there being approximately one effective em-

194

bryo per particle.

The stress state of the parent material is important since the shear and dilational components of martensitic transformations change the strain energy state of the system. Stresses from other sources can either impede or enhance the transformations, depending on their signs and orientations with respect to the transformation stresses and strains. Tensile stresses have been shown to have the greatest effect on inducing transformation, since both the dilational and shear components are assisted.

Cracks in materials subjected to external tensile loads are known to generate high biaxial (plane stress) or triaxial (plane strain) stresses in the regions near their tips. Therefore, it is both expected and observed that transformations can be induced in these regions under certain conditions. The volume of these transformed regions increases as temperatures approach the martensite start temperatures. Therefore, the effects of stress-induced transformations will become more pronounced as the temperature decreases.

Fracture mechanics analysis of a crack within a body loaded under "fixed grip" conditions shows that there is a decrease in the strain energy of the system when the crack extends. Propagation has been demonstrated to occur when the strain energy released per unit crack extension reaches a critical value characteristic of the material. A consequence of stress-induced martensitic transformations in the "fixed grip" case is also a loss of strain energy. These transformations thus reduce the total amount of strain energy available for propagation. The larger loads required to cause crack extension in the presence of the transformations constitute an improvement in fracture toughness, i.e. the load bearing capacity of a cracked body. This then is the mechanism of transformation toughening in zirconia.

References

[1]R. C. Garvie, R. H. Hannink, and R. T. Pascoe, "Ceramic Steel?" *Nature (London)*, **258**, 703-704 (1975).
[2]D. L. Porter and A. H. Heuer, "Mechanisms of Toughening Partially Stabilized Zirconia (PSZ)," *J. Am. Ceram. Soc.,* **60** [3-4] 183-84 (1977).
[3]F. F. Lange; pp. 799-819 in Fracture Mechanics of Ceramics, Vol. 4. Edited by R. C. Bradt, D. P. H. Hasselman, and F. F. Lange. Plenum, New York, 1978.
[4]D. L. Porter and A. H. Heuer, "Microstructural Development in MgO-Partially Stabilized Zirconia (Mg-PSZ)," *J. Am. Ceram. Soc.,* **62** [5-6] 298-305 (1979).
[5]T. Mitsuhashi, M. Ichihara, and U. Tatsuke, "Characterization and Stabilization of Metastable Tetragonal ZrO_2," *ibid.,* **57** [2] 97-101 (1974).
[6]T. K. Gupta, J. H. Bechtold, R. C. Kuznicki, L. H. Cadoff, and B. R. Rossing, "Stabilization of Tetragonal Phase in Polycrystalline Zirconia," *J. Mater. Sci.,* **12**, 2421-26 (1977).
[7]T. K. Gupta, "Sintering of Tetragonal Zirconia and its Characteristics," *Sci. Sintering,* **10** [3] 205-16 (1978).
[8]T. K. Gupta, F. F. Lange, and J. H. Bechtold, "Effect of Stress-Induced Phase Transformation on the Properties of Polycrystalline Zirconia Containing Metastable Tetragonal Phase," *J. Mater. Sci.,* **13**, 1464-70 (1978).
[9]T. K. Gupta; pp. 877-89 in Ref. 3.
[10]N. Claussen, "Fracture Toughness of Al_2O_3 with an Unstabilized ZrO_2 Dispersed Phase," *J. Am. Ceram. Soc.,* **59** [1] 49-51 (1976).
[11]N. Claussen and J. Jahn, "Mechanical Properties of Sintered and Hot-Pressed Si_3N_4-ZrO_2 Composites," *ibid.,* **61** [1] 94-95 (1978).
[12]E. C. Subbarao, H. S. Maiti, and K. K. Srivastava, "Martensitic Transformation in Zirconia," *Phys. Status Solidi A,* **21**, 9-40 (1974).
[13]G. K. Bansal and A. H. Heuer, "Martensitic Phase Transformation in Zirconia (ZrO_2): I," *Acta Metall.,* **20** [11] 1281-89 (1972); "II," *ibid.,* **22** [4] 409-17 (1974).
[14]E. C. Bain, "Nature of Martensite," *Trans. AIME,* **70**, 25 (1925).
[15]M. S. Wechler, D. S. Lieberman, and T. A. Read, "On the Theory of the Formation of Martensite," *ibid.,* **197**, 1503-15 (1953).

[16]J. S. Bowles and J. K. Mackenzie, "The Crystallography of Martensite Transformations: I," *Acta Metall.*, **2**, 129-37 (1954); "II," *ibid.*, 138-47 (1954); "III," *ibid.*, 224-34 (1954).

[17]J. C. Fisher, J. H. Holloman, and D. Turnbull, "Kinetics of the Austenite-Martensite Transformation," *Trans. AIME*, **185**, 691-700 (1949).

[18]L. Kaufman and M. Cohen; pp. 165-246 in Progress in Metal Physics, Vol. 7. Edited by B. Chalmers and R. King. Pergamon, London, 1958.

[19]H. Knapp and V. Dehlinger, "Mechanics and Kinetics of Martensite Formation Without Diffusion," *Acta Metall.*, **4**, 289 (1956).

[20]G. B. Olson and M. Cohen, "A General Mechanism of Martensitic Nucleation: I," *Metall. Trans. A.*, **7**, 1897-1904 (1976); "II," *ibid.*, 1905-14; "III," *ibid.*, 1915-23.

[21]R. E. Cech and D. Turnbull, "Heterogeneous Nucleation of Martensitic Transformation," *Trans. AIME*, **206**, 124 (1956).

[22]V. Raghavan and M. Cohen; pp. 67-127 in Treatise on Solid State Chemistry, Vol. 5: Changes of State. Edited by N. B. Hannay. Plenum, New York, 1975.

[23]H. G. Scott, "Phase Relationships in the Zirconia-Yttria System," *J. Mater. Sci.*, **10**, 1527-35 (1975).

[24]J. D. Eshelby, "Determination of the Elastic Field of an Ellipsoidal Inclusion and Related Problems," *Proc. R. Soc., London,* **241**, 376-96 (1957).

[25]J. D. Eshelby; pp. 89-140 in Progress in Solid Mechanics, Vol. 2. Edited by I. N. Sneddon and R. Hill. Wiley-Interscience, New York, 1961.

[26]J. R. Patel and M. Cohen, "Criterion for the Action of Applied Stress in the Martensitic Transformation," *Acta Metall.*, **1**, 531-38 (1953).

[27]H. M. Westergaard, "Bearing Pressures and Cracks," *Trans. ASME*, **61**, A49-53 (1939).

[28]G. R. Irwin, "Analysis of Stresses and Strains Near the End of a Crack Traversing a Plate," *J. Appl. Mech.*, 361-64 (1957).

[29]A. A. Griffith, "The Phenomenon of Rupture and Flow in Solids," *Philos. Trans., R. Soc., London,* **221**, 163-98 (1920).

---

*This is Scott's diagram for the zirconia-yttria system (Ref. 23) and the present interpretation is an expansion of the one given by him.

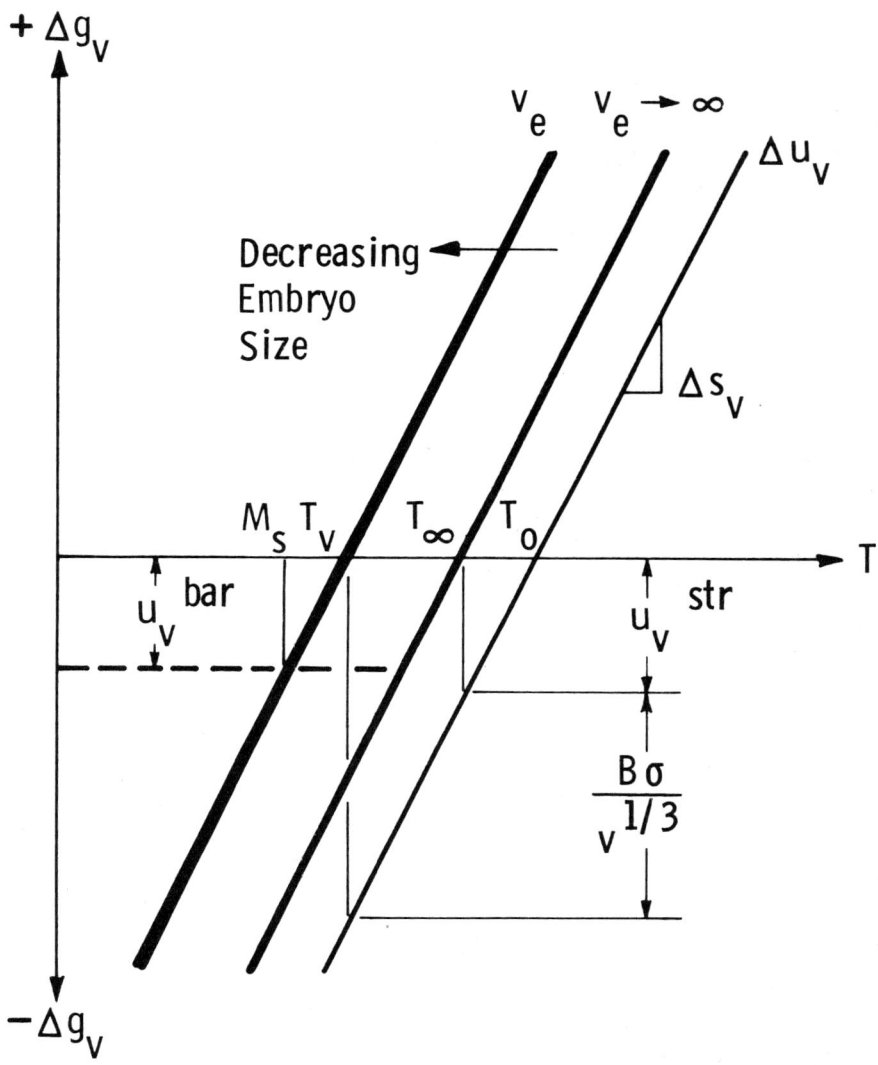

Fig. 1. Change in free energy per unit volume as a function of temperature for a hypothetical martensitic transformation.

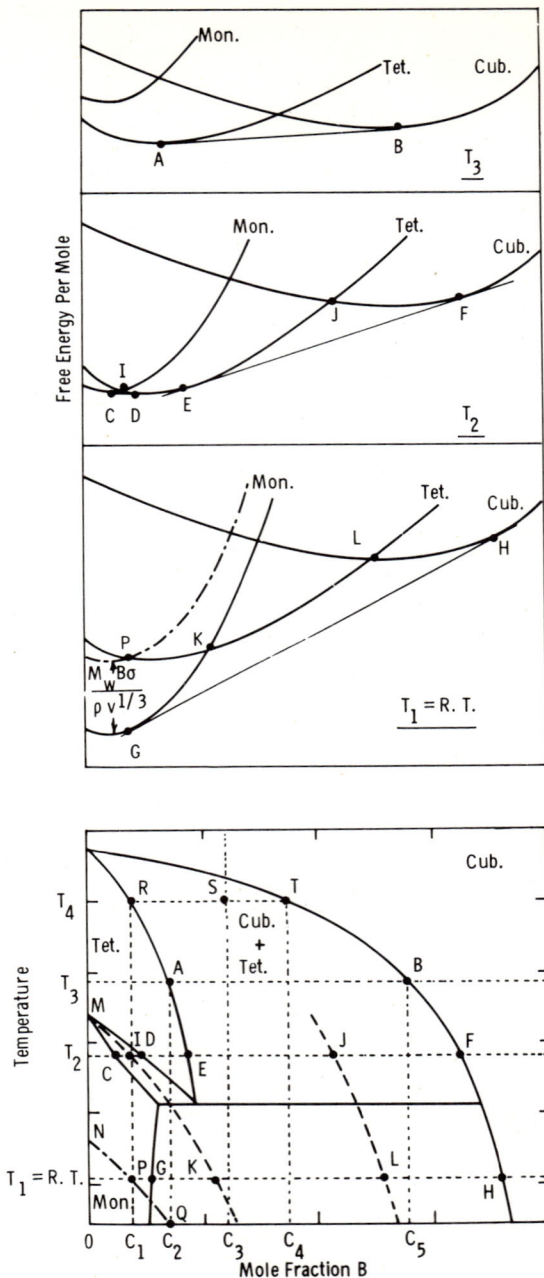

Fig. 2. Free energy vs composition curves and the related phase diagram for a typical zirconia system.

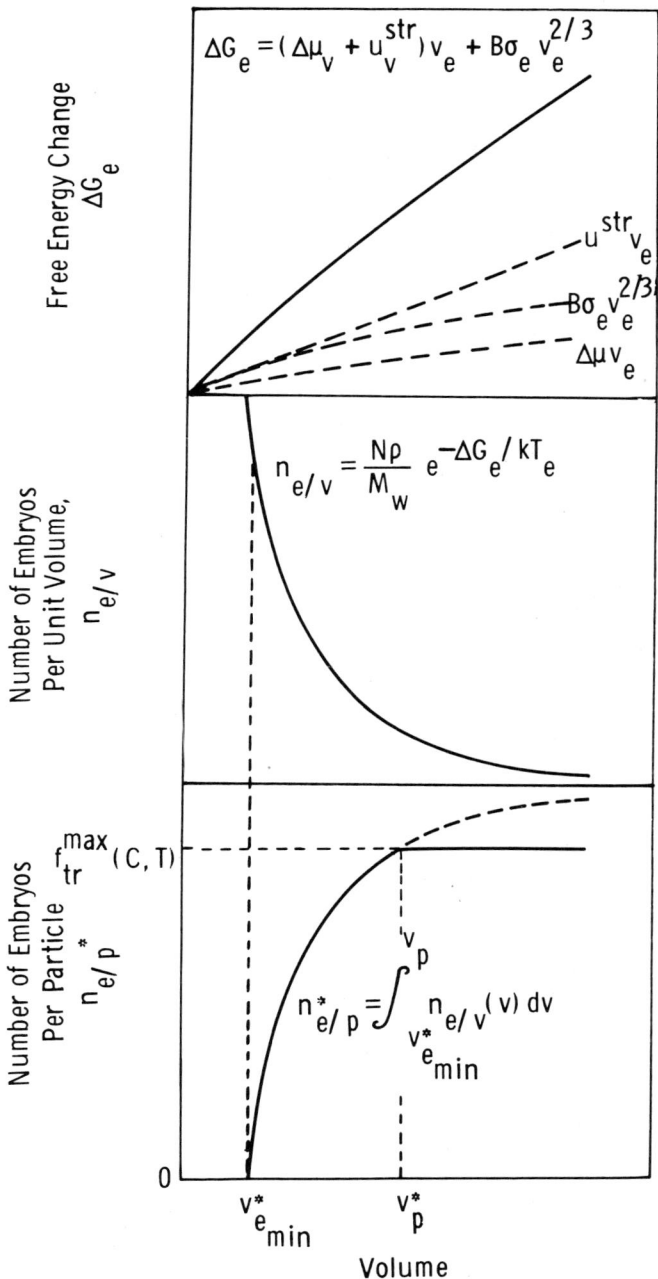

Fig. 3. Relationship between the size distribution of precursor embryos formed by an activated process and the distribution of those embryos in particles of a hypothetical material.

199

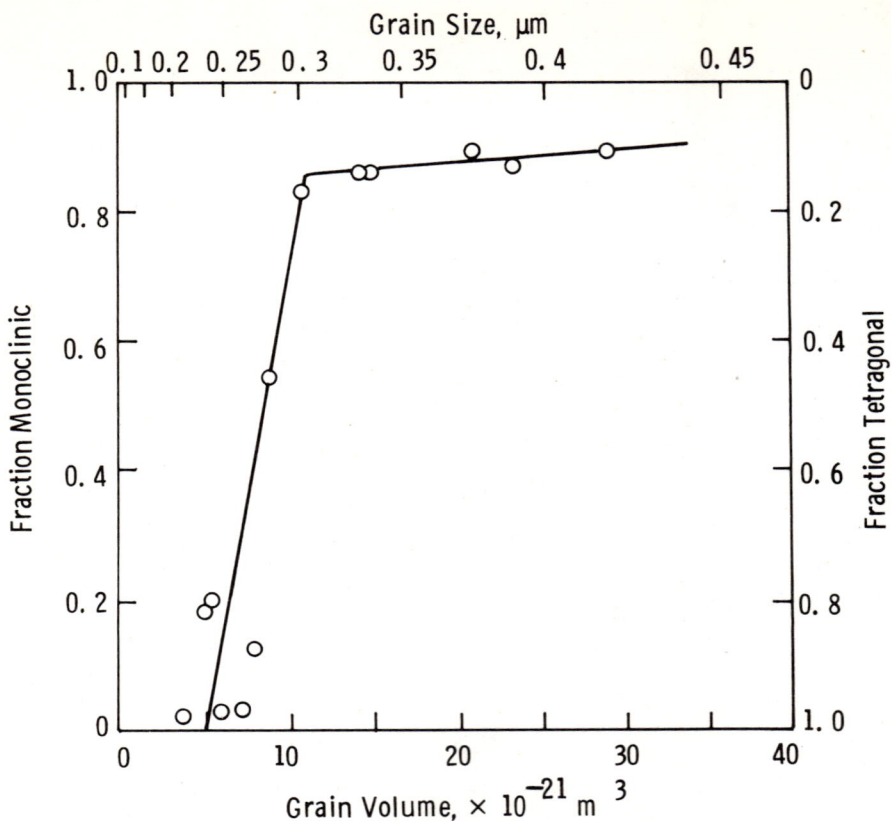

Fig. 4. Effect of particle size on the fraction of martensitic transformation for a $ZrO_2 + 1.4$ mol% Y_2O_3.

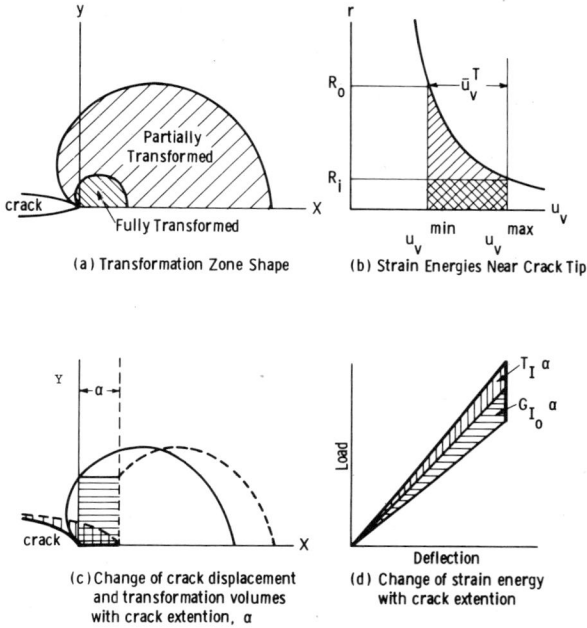

(a) Transformation Zone Shape

(b) Strain Energies Near Crack Tip

(c) Change of crack displacement and transformation volumes with crack extention, α

(d) Change of strain energy with crack extention

Fig. 5. Stress-induced phase transformations at a crack tip: (a) shape of the transformation zone, (b) strain energy as a function of distance from the crack tip, (c) crack tip and transformation zone displacements as a function of small crack extension, and (d) "fixed grip" strain energy losses due to crack tip extension and stress-induced transformation.

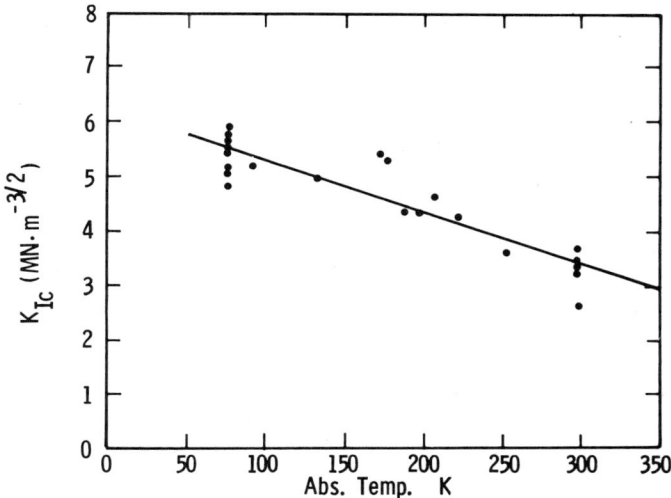

Fig. 6. Fracture toughness as a function of temperature measured on a $ZrO_2 + 10$ wt% CeO_2 material.

Transformation toughening in ceramics

A. G. EVANS, D. B. MARSHALL, AND N. H. BURLINGAME

Materials and Molecular Research Division, Lawrence Berkeley Laboratory
  and Department of Materials Science and Mineral Engineering
University of California
Berkeley, Calif. 94720

The origin of transformation toughening in ceramics is examined using two approaches: one based on the stress field ahead of the crack and the other on the changes in thermodynamic potential during a crack increment. Both approaches yield essentially similar predictions of trends in toughness with particle size, temperature, composition, etc. The stress-intensity analysis provides fully quantitative predictions of the toughness. These indicate that the shielding of the crack by the transformation zone develops only in the presence of a transformed wake, leading to R-curve behavior.

The enhancement of toughness in ceramics achieved through the controlled use of martensitic transformation mechanisms has emerged as a major research interest,[1-9] with considerable practical potential. The development of toughness requires relatively stringent control of composition and microstructure.[3,4,6,9] Ability to interpret and predict the microstructure/composition that provides the optimum toughness is thus an important aspect of the development and control of toughened materials. This capability must be based on a micromechanics model that contains the correct physical characteristics of the toughening process. The presently available models provide, in some instances, conflicting predictions of the effect of microstructure and temperature on the toughness.[6,7,9] The intent of this paper is to provide physically consistent analyses of toughening, based on the best available knowledge of the mechanisms involved in the development of toughness.

Toughening occurs through the development of a zone of transformed material ("process zone") around the crack tip (Fig. 1(a)). This process zone is a source of stress and displacement disturbance that tends to shield the crack tip from the applied stress, in a manner analogous to a plastic zone. The toughening is thus related to the dimensions of the process zone and the amplitude of the disturbance, through a crack extension criterion. The mechanics associated with the crack tip stresses and the crack surface displacements are the basis of pertinent crack extension criteria.

Alternatively, fracture can be analyzed by examining the energy changes that accompany crack extension. This approach contains inherent uncertainties associated with the modes of energy dissipation. But the energy analysis, on comparison with the mechanics analysis, may provide some useful additional insights into the mechanism of toughening.

An essential aspect of the toughening magnitude is the dimension of the transformation zone. The prediction of this zone dimension is a complex

202

problem. The analysis should be based on a critical stress amplitude $(p_{ij}^A)_c$ for transformation (analogous to a critical shear stress condition for estimating the plastic zone size and shape). However, independent experimental estimates of $(p_{ij}^A)_c$ (cf. the yield strength for plasticity) are difficult to obtain and hence are not yet available. The results obtained from a simplified analysis, based on the interaction energy for an isolated particle in a uniform stress field,[6,7] will be discussed.

Toughness/Zone Size Relations

The influence of a transformation zone on the fracture toughness resides in the modified stress field or crack opening displacement induced by the zone. The modified stress field for a stress/strain behavior typical of transformation (Fig. 1(b)) exhibits the characteristics depicted in Fig. 1(c). There is a reduction in the near tip field characterized by a local value of the stress intensity factor, K_I^{local}; the remote field (for a small transformation zone) is dictated by the applied stress intensity factor, K_I^∞; and there is an intermediate region of relatively invariant stress, with a magnitude determined by the transformation strain. Fracture is determined by either the local or the intermediate stress field, as dictated by the specific crack extension mechanism. Fracture in most ceramics (those that are not subject to microcracking) is generally considered to occur by the direct advance of the crack tip.[10] This mechanism is assumed in the present analysis, whereupon the increment in toughness can be deduced directly from the change in K_I^{local} induced by the transformation zone. However, situations in which fracture initiates in the intermediate zone should not be discounted.

A convenient procedure for calculating the stress intensity factor change induced by a process zone is illustrated in Fig. 2(a). The process zone is allowed to undergo an unconstrained transformation strain, e^{T*}. A stress σ_{ij}^W is then applied to the zone to restore it to its original dimensions. Equal and opposite tractions

$$T_j^W = n_i \sigma_{ij}^W \tag{1}$$

where n_i is the boundary normal, are then applied to the zone boundary to achieve stress continuity in the system. These tractions induce a change in stress intensity, characterized by the stress intensity factor[11]

$$\Delta K_I^T = \frac{1}{2\sqrt{2\pi}(1-\nu)} \int \underset{\sim}{T}^W \cdot \underset{\sim}{h} \, dS \tag{2}$$

where dS is an element of the zone boundary and h is a weight function. Referring to the axes X_1 and X_2 and the coordinates (r, θ) depicted in Fig. 2(b), the following relations pertain:

$$T_1^W = \sigma_{11}{}^W \cos \psi + \sigma_{12}{}^W \sin \psi$$
$$T_2^W = \sigma_{12}{}^W \cos \psi + \sigma_{22}{}^W \sin \psi$$
$$h_1 = \frac{\cos (\theta/2)}{\sqrt{r}} [2\nu - 1 + \sin (\theta/2) \sin (3\theta/2)]$$
$$h_2 = \frac{\sin (\theta/2)}{\sqrt{r}} [2 - 2\nu + \cos (\theta/2) \cos (3\theta/2)]$$
$$\left.\vphantom{\frac{\cos}{\sqrt{r}}}\right\} \quad (r \ll a) \tag{3}$$

where ψ relates the zone surface normal to the reference axis (Fig. 2(b)).

Initially, the only macroscopic particle transformation strain is considered to be the dilational component e^T. The zone restoring stress is then a uniform hydrostatic compression p, independent of zone shape, particle orientation, shape, etc. The change in stress intensity determined (in accord with the above procedure) from the transformation volume strain is found to be a function of the zone geometry selected for the analysis.[12] Specifically, zones contained ahead of the crack tip (Fig. 3(a)) do not generally result in a significant reduction in the local stress intensity; whereas, zones that extend over the crack surface (Fig. 3(b)) provide substantial crack shielding. The transformed wake thus appears to be the predominant source of toughening. Most toughness measurements pertain to cracks which, on introduction into the system, create a continuous zone over the crack surface (Fig. 3(b)), and the full toughening should thus be experienced. However, when a prior zone over the crack does not exist, crack growth should be characterized by an R curve, as schematically illustrated in Fig 3(c). This issue is discussed more extensively in the Appendix.

These toughness characteristics are demonstrated by conducting calculations of the local stress intensity for two zone profiles: one fully contained ahead of the crack (Fig. 3(a)) and the other of uniform width extending fully over the crack surface (Fig. 3(b)). First, consider a zone contour suggested by the dilational crack tip field (Fig. 3(a)):

$$r = r^* \cos^2(\theta/2) \tag{4}$$

where r^*, the zone dimension at $\theta = 0$, is a function of the applied stress intensity factor, K_I^∞,

$$r^* = \frac{2(1+\nu)^2}{9\pi} \left(\frac{K_I^\infty}{p_c^A} \right)^2 \tag{5}$$

Noting the following relations

$$dS = (r^*/\sqrt{2}) [1 + \cos \theta]^{1/2} d\theta$$
$$\cos \psi = [\cos \theta + \cos 2\theta]/\sqrt{2} (1 + \cos \theta)^{1/2}$$
$$\sin \psi = [\sin \theta + \sin 2\theta]/\sqrt{2} (1 + \cos \theta)^{1/2}$$

the stress intensity change becomes:

$$\frac{\Delta K_I^T}{p\sqrt{r^*}} = \frac{1}{8\sqrt{2\pi}(1-\nu)} \int_{-\pi}^{+\pi} [(\sin \theta + \sin 2\theta)(4 - 4\nu - \cos \theta - \cos 2\theta) \tan (\theta/2)$$
$$+ (\cos \theta + \cos 2\theta)(4\nu - 2 + \cos \theta - \cos 2\theta)] \, d\theta$$
$$= 0 \tag{6}$$

The absence of crack tip shielding is thus manifest for this zone profile.

Second, consider a transformation zone of uniform width r_0.[†] The crack is subject to a uniform compression p, due to restoration of the zone to its initial (pretransformation) shape, and to tractions, $-p$, acting at the zone boundary (Fig. 3(b)) that establish stress continuity. The change in stress intensity factor attributed to the application of these forces, obtained by employing the requisite point force functions, is given by[12]:

$$\frac{\Delta K_I^T}{p\sqrt{r_0}} = -1.13 + \frac{0.28}{(1-\nu)} + 0.094\frac{(1+4\nu)}{(1-\nu)} \tag{7}$$

where the first term derives from the uniform compression, the second term from the tractions over the zone surface (up to the crack tip), and the third term from the tractions from the zone ahead of the crack.[‡] For a typical value of the Poisson ratio, $v = \frac{1}{4}$, Eq. (7) becomes

$$\frac{\Delta K_I^T}{p\sqrt{r_0}} = -0.51 \tag{8}$$

The negative character of ΔK_I^T confirms that the zone induces crack shielding.

It now remains to determine the magnitude of the traction p in order to deduce the toughening induced by the transformation zone. For plane strain conditions, p is related to the volume strain and the concentration of particles in the zone:

$$p = \kappa \Delta V V_f \tag{9}$$

where ΔV is the unconstrained transformation volume *strain* of an individual particle, V_f is the volume concentration of particles, and κ is the average bulk modulus. The change in stress intensity factor thus becomes,

$$\frac{\Delta K_I^T}{V_f E \Delta V \sqrt{r_0}} = \frac{1}{3(1-2v)}\left[-1.13 + \frac{0.28}{(1-v)} + 0.094\frac{(1+4v)}{(1-v)}\right] \tag{10}$$

which for $v = \frac{1}{4}$ becomes:

$$\frac{\Delta K_I^T}{V_f E \Delta V \sqrt{r_0}} = -0.34 \tag{11}$$

Fracture is assumed to proceed when the local stress intensity factor attains the fracture resistance of the *pretransformed* material,[§] K_{IC}^M (a quantity that depends on the volume fraction of particles):

$$K_I^{local} \equiv K_I^\infty + \Delta K_I^T = K_{IC}^M \tag{12}$$

The level of the applied stress intensity factor at the critical condition is thus,

$$K_{IC}^\infty = K_{IC}^M + 0.34 V_f E \Delta V \sqrt{r_0} \tag{13}$$

An analogous treatment of the influence of the macroscopic transformation shear strain is substantially more complex because of the vector nature of the shear strain, i.e. the particle shape and orientation distributions become important considerations. Each analysis thus requires specific information about the particles and the transformation morphologies. Such analyses are evidently of merit only for materials in which the dilational transformation strain does not afford the principal contribution to the toughening.

The relative influence of the volume strain and the shear strain can be conveniently estimated by comparing the predictions derived from Eq. (13) with experimental results. Suitable results exist for the MgO/PSZ studied by Porter and Heuer[4] and by Schoenlein.[13] The appropriate measurements are: $r_0 \approx 0.6 \, \mu m$, $\Delta V \approx 0.058^3$, $E \sim 2 \times 10^{11}$ Pa, $V_f \sim 0.3$. The toughness ratio is thus predicted, from Eq. (13), to be $K_{IC}^\infty / K_{IC}^M = 1.4 \pm 0.1$. This compares with a measured toughening for the same material condition of $\sim 1.6 \pm 0.2$. The volume strain thus appears to be the major source of toughening in this instance. Further studies are evidently needed to establish the detailed dependence of toughness on the volume strain and the macroscopic shear strain.

The Transformation Zone Size

Prediction of the transformation zone dimension requires detailed knowledge of the nucleation of the stress-induced transformation. The specific nucleation mechanisms pertinent to ceramics, such as ZrO_2, have not been identified. An upper bound solution based on the net change in thermodynamic potential is thus presented and its use justified by comparison with zone size observations.

Zone profile observations[14] generally indicate characteristics expected from a dilation-dominated transformation, i.e. a zone extending ahead of the crack tip.[6] Hence, for present purposes, it is considered that the particles exhibit insignificant macroscopic transformation shear strain and that the transformation is activated by the hydrostatic component of the stress. The change in thermodynamic potential upon transformation has been determined for transformations in which entropy changes (other than those associated with the chemical free energy change of the particle) are neglected[7]:

$$\frac{\Delta\phi}{(4/3)\pi R^3} = -\Delta F_0 + \Gamma_T/d + 3\Delta\Gamma_i/R + 0.15E\Delta V^2 - p^A\Delta V$$
$$+ 0.13E\gamma_T^2(1.2 + R/d)^{-1} \tag{14}$$

where R is the particle radius, d the twin (variant) spacing, ΔF_0 the chemical free energy change, Γ_T the twin boundary energy, $\Delta\Gamma_i$ the change in interface energy, and γ_T the twinning (variant) shear strain. If the twin boundary energy and the change in interface energy are small, the lower bound stress needed to induce transformation ($\Delta\phi = 0$) becomes:

$$p_c^A/E\Delta V = 0.15 - \Delta F_0/E\Delta V^2 + 0.13(\gamma_T/\Delta V)^2 (1.2 + R/d)^{-1} \tag{15}$$

Substituting the lower bound critical stress into Eq. (5), the corresponding upper bound transformation zone size at the critical condition becomes:

$$r_0 \approx \frac{1}{4\pi} \left(\frac{K_{IC}^\infty}{E\Delta V}\right)^2 \frac{1}{[0.15 - \Delta F_0/E\Delta V^2 + 0.13(\gamma_T/\Delta V)^2(1.2 + R/d)^{-1}]^2} \tag{16}$$

A comparison of Eq. (16) with transformation zone size observations in PSZ indicates a predicted zone size less than twice the measured value.[7] This is similar to predictions of the plastic zone size based on the linear elastic approximation. The nucleation barrier does not appear, therefore, to dominate the crack tip transformation in PSZ. The upper bound zone size estimate should thus be an approximation suitable for anticipating toughness trends.

Toughening characteristics can be identified by inserting the zone width from Eq. (16) into Eq. (13) to give:

$$1 - \frac{K_{IC}^M}{K_{IC}^\infty} = \frac{0.7V_f}{[1 - 7\Delta F_0/E\Delta V^2 + (\gamma_T/\Delta V)^2(1.2 + R/d)^{-1}]} \tag{17}$$

This result is similar in general form to that found in previous studies.[6,7] The qualitative implications for effects of temperature, composition, etc. on the toughening are thus essentially unchanged by the present analysis.

Thermodynamic Considerations

The toughening induced by a martensite transformation can also be estimated from considerations of the changes in thermodynamic potential that accompany an increment in crack length.[6,7,9] It has already been demonstrated[6] that, since the stress state near the crack tip remains essentially unchanged during a crack increment, the net energy change associated with the transformation can be considered as the energy difference between un-transformed particles remote from the crack tip (subject to the remotely applied stress) and transformed particles placed over the crack surface (subject to zero stress), as illustrated in Fig. 4. For this problem, when the particles are stress-free prior to transformation, the toughness is given by[7]:

$$G_I^\infty = G_{IC}^M + [V_f r_0/(4/3)\pi R^3](\Delta\phi|_{p^A=0})$$
$$= G_{IC}^M + E\Delta V^2 r_0 V_f[0.15 - \Delta F_0/E\Delta V^2 + 0.13\,(\gamma_T/\Delta V)^2(1.2 + R/d)^{-1}] \quad (18)$$

Introducing the zone size from Eq. (16) the toughness becomes

$$1 - \frac{G_{IC}^M}{G_{IC}^\infty} = \frac{0.6V_f}{[1 - 7\Delta F_0/E\Delta V^2 + (\gamma_T/\Delta V)^2(1.2 + R/d)^{-1}]} \quad (19)$$

The final result is remarkably similar to that derived from the modified crack tip stress field (Eq. 17). This may be fortuitous because in both instances an upper bound thermodynamic criterion for the zone dimension has been employed to attain the final toughening expression. Nevertheless, it is of considerable interest to note that the thermodynamic result in its present form also anticipates the trends with zone profile identified from the stress intensification analysis. Notably, a zone contained ahead of the crack does not produce significant toughening; the increase in toughness is manifest only when the zone extends over the crack surface (Appendix).

The location of the energy changes during a crack increment are also of interest. The criterion selected for zone formation requires that the change in thermodynamic potential in the newly created zone (Fig. 5) be zero. The increase in potential attributed to the transformation zone (to be supplied by the enhanced G) must occur, therefore, in other segments of the zone. The principal source of the increase in potential is the increase in elastic energy that occurs at particles over the crack surface, as the stress on the particles is released by the crack advance (Fig. 5). The change in potential has a magnitude comparable to the interaction energy (with the local stress) and must be a positive quantity because the tension on the particles (which reduces the elastic energy) is being reduced in this region.

Particle Size Distribution Effect

The toughness should evidently exhibit some dependence on the particle size distribution, through the size-dependent adjacence to the critical particle size R_c for stress-free transformations. The problem can be addressed by first determining the zone tractions for each small particle size range, R to $R + dR$ (within the distribution, $f(R)dR$), deducing the resultant ΔK_I^T, and then summing over all *untransformed* particles within the distribution. This procedure yields the following relation for the toughening.

$$1 - \frac{K_{IC}^M}{K_{IC}^\infty} = \frac{V_f}{(7\Delta F_0/E\Delta V^2 - 1)<R^3>} \int_{\sim d}^{\widehat{R}} \frac{R^3(1.2d + R)}{(R_c - R)} f(R)dR \qquad (20)$$

The lower limit, d, is selected because it is improbable that small particles near the crack tip will transform into a single twin (variant).[7] The upper limit is determined by the level of the remotely applied strain relative to the transformation strain.[15] Normalization of the integral in Eq. (20) gives:

$$1 - \frac{K_{IC}^M}{K_{IC}} = \frac{V_f R_c^3}{(7\Delta F_0/E\Delta V^2 - 1)<R^3>} \int_{\sim 2/\eta_c}^{\widehat{R}/R_c} \frac{z^3(z + 2.4/\eta_c)}{(1 - z)} f(z, R_0/R_c, k)dz \qquad (21)$$

where R_0 is the scale parameter and k the shape parameter for the particle size distribution and η_c is the number of twins in a particle of critical size. Some typical results obtained by assuming an extreme value function for $f(R)dR$ are plotted in Fig. 6. Comparison with Eq. (17) indicates that the size distribution introduces an upper bound toughness. The level of the upper bound decreases as the variance increases. However, no additional variables are introduced.

The particle size distribution is also related to the ratio λ of the transformed phase content to the total (transformed plus untransformed) phase content through the critical size R_c,

$$\lambda = [1/<R^3>] \int_{R_c}^{\infty} R^3 f(R)dR \qquad (22)$$

There are thus unique relations between the predicted toughness and the transformed phase content for each value of the shape parameter k. The predictions of toughness as a function of transformed phase content should be directly comparable with experimental results.

Conclusions

Analyses of transformation toughening have been presented that permit the interpretation of toughness results, in various ceramic systems, and allow trends in the toughness with the important internal and external variables to be predicted. Preliminary comparisons between the predictions and results for partially stabilized zirconia indicate that the toughening derives primarily from the volume strain of transformation, once the transformation zone dimension has been prescribed. However, the transformation shear strain (as manifest in the twin or variant structure of the martensite) could be important in determining the transformation zone dimension. This topic requires further study.

The analyses also suggest the existence of R-curve behavior when the cracks in the material are not accompanied by a pre-existent transformation zone. This behavior resides in the observation that crack tip shielding initiates only when the transformation zone extends over the crack surface.

The trends in toughness with the important variables predicted by the present analyses are similar to those anticipated by previous studies. Specifically, important influences of temperature, chemical composition, particle size distribution, matrix toughness, and the transformation strain are predicted, in accord with experimental observations.

Appendix

An Estimate of R-Curve Behavior

The detailed characteristics of R-curves for transformation-toughened ceramics can be determined by either of the two approaches used in this paper. These details will be presented in a subsequent publication. An approximate evaluation is described here, in order to illustrate the range of crack extension over which the principal increase in crack growth resistance is established. The thermodynamic approach is used for this purpose.

Commence with an initial stress-free crack without a transformation zone. Application of an external loading causes a transformation zone to develop and the crack tip advances from A to B (Fig. 7(a)), leaving an intervening wake. The energy changes that accompany an additional increment, da, in crack length then relate to the value of G_{IC} at crack extension AB: $G_{IC} = G_{IC}^o + \Delta G_{IC}$, where $G_{IC}^o da = 2\Gamma da$ is the increase in surface energy and $\Delta G_{IC} da$ is the increase in potential associated with the transformed particles. Consider the changes in potential that occur in a strip of width dy (Fig. 7(b)). The total potential within this strip, for unit specimen thickness, is

$$\phi(y) = V_f \, dy \int_0^\infty \phi_y(x) dx \qquad (A1)$$

where x is the coordinate along the crack plane (Fig. 7(a)) and $\phi_y(x)$ is the potential per unit volume at x. The change in potential following transformation is

$$\Delta \phi = -\Delta F_0 + \Delta \phi_0 - \Delta \phi_{int} \qquad (A2)$$

where $\Delta \phi_0$ is the stress-free elastic energy and $\Delta \phi_{int}$ is the interaction energy with the local stress p. Then, if the zone boundary is defined by the thermodynamic condition, $\Delta \phi = 0$, the change in $\phi(y)$ during the crack advance, given by the shaded area in Fig. 7(b), is

$$\frac{d\phi(y)}{da} = \Delta \phi_{int}^o - \Delta \phi_{int}^A \qquad (A3)$$

where $\Delta \phi_{int}^o$ and $\Delta \phi_{int}^A$ are the magnitudes of the interaction energy at the front and rear boundaries of the transformation zone. The increase in crack growth resistance is then

$$\Delta G_{IC} = V_f \int_{-r_0}^{r_0} (\Delta \phi_{int}^o - \Delta \phi_{int}^A) dy \qquad (A4)$$

In the absence of a wake, $\Delta \phi_{int}^o = \Delta \phi_{int}^A$ and there is no increase in toughness. For a fully developed wake, $\Delta \phi_{int}^A = 0$, and the maximum toughness increase pertains:

$$\Delta \widehat{G}_{IC} = 2 V_f r_0 \Delta \phi_{int}^o \qquad (A5)$$

209

which is the result quoted in the text (with $\Delta\phi = 0$ at the zone boundary). For intermediate conditions, the trend can be estimated by computing the stress over the rear boundary and deducing the interaction energy from

$$\Delta\phi_{int}^A = p^A e^T \tag{A6}$$

The final result is:

$$\Delta G_{IC} = \Delta\widehat{G}_{IC} \left[1 - \frac{\beta\sqrt{(\alpha/r_0) + \sqrt{(\alpha/r_0)^2 + 1}}}{3^{3/4}\sqrt{(\alpha/r_0)^2 + 1}} \right]$$

where $\Delta a = r_0 \tan(\pi/6) - \alpha$ is the crack extension and β is a constant between 1 and 2. This result is plotted in Fig. 7(c) (for $\beta = 2$). It is noted that ΔG_{IC} attains an essentially invariant level when the crack advance is ≈ 3 to 4 times the zone height. Hence, for small zones, this varying crack extension resistance will have little influence on the mechanical behavior of transformation-toughened materials. Important effects will emerge only when the zone size is a significant fraction ($\gtrsim 0.1$) of the crack length.

References

[1]R. C. Garvie, R. H. Hannink, and R. T. Pascoe, *Nature (London)*, **258**, 703 (1975).
[2]N. Claussen, *J. Am. Ceram. Soc.*, **59** [1-2] 49 (1976).
[3]R. H. Hannink, *J. Mater. Sci.*, **13**, 2487 (1978).
[4]D. L. Porter and A. H. Heuer, *J. Am. Ceram. Soc.*, **60** [3-4] 183 (1977).
[5]D. L. Porter, A. G. Evans, and A. H. Heuer, *Acta Metall.*, **27**, 1649 (1979).
[6]A. G. Evans and A. H. Heuer, *J. Am. Ceram. Soc.*, **63** [5-6] 241 (1980).
[7]A. G. Evans, N. Burlingame, M. Drory, and W. M. Kriven, *Acta Metall.* **29**, 447 (1981).
[8]T. K. Gupta, F. F. Lange, and J. H. Bechtold, *J. Mater. Sci.*, **13**, 1464 (1978).
[9]F. F. Lange, Science Center Report SC 5117.6TR (Oct. 1979).
[10]A. G. Evans, D. L. Porter, and A. H. Heuer; in Fracture 1977, Vol. 1. Edited by D. M. Taplin. University of Waterloo Press, Waterloo, Ontario, 1977.
[11]P. C. Paris, R. McMeeking, and S. Tada, *ASTM STP*, 601 (1978).
[12]J. Hutchinson, R. McMeeking, and A. G. Evans; unpublished work.
[13]L. Shoenlein; M.S. thesis, Case Western Reserve University, Cleveland, Ohio, 1979.
[14]M. Rühle; unpublished work.
[15]A. G. Evans and K. T. Faber; to be published in *Journal of the American Ceramic Society*.
[16]D. B. Marshall, M. Drory, and A. G. Evans; to be published in *Fracture Mechanics of Ceramics*.

---

†The zone width r_0 is related to the zone dimension ahead of the crack r^*, by $r_0 = (\sqrt{3}/2)^2 r^*$.
‡This result exhibits some dependence on the precise shape selected for the zone ahead of the crack (Ref. 16). The shape used to obtain Eq. (7) was determined by assuming a constant zone radius, r_0, ahead of the crack.
§The toughening described in this section is relative to the toughness of the material containing transformed particles prior to crack propagation, which contains the influence of *local* interactions between the transformed particles and the crack, e.g. crack deflection effects.

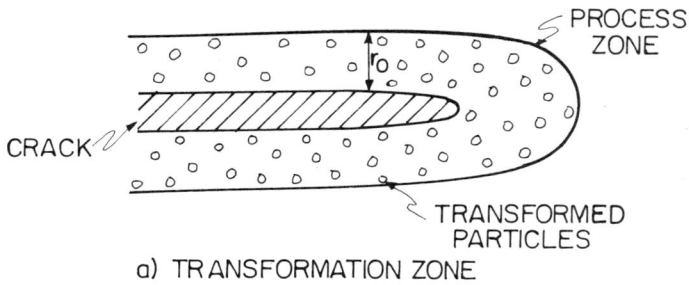

a) TRANSFORMATION ZONE

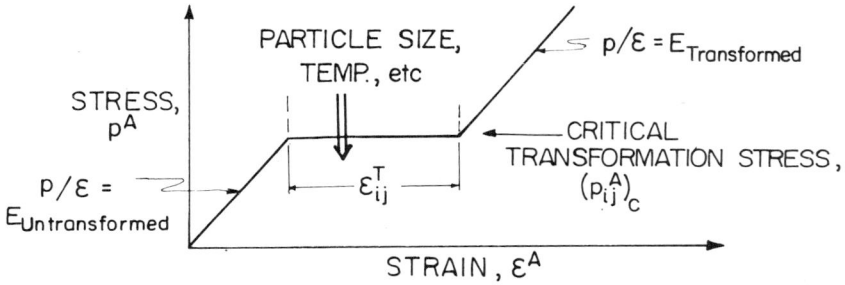

b) A TYPICAL STRESS, STRAIN BEHAVIOR
FOR A MARTENSITIC TRANSFORMATION

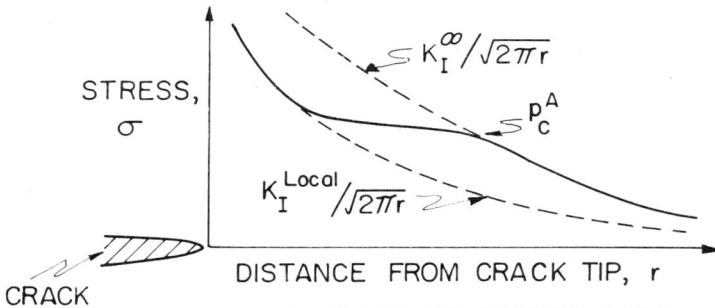

c) CRACK TIP STRESS FIELD

Fig. 1. (a) Schematic of process zone around crack induced by the martensitic transformation of particles. (b) Typical stress, strain behavior for a martensite transformation; e_{ij}^T is the transformation strain, $(p_{ij}^A)_c$ is the critical stress needed to induce transformation (which depends on particle size, etc.). Note the linearity of the stress/strain relation for stresses in excess of the transformation stress. (c) The modified stress ahead of a crack tip in the presence of a transformation zone.

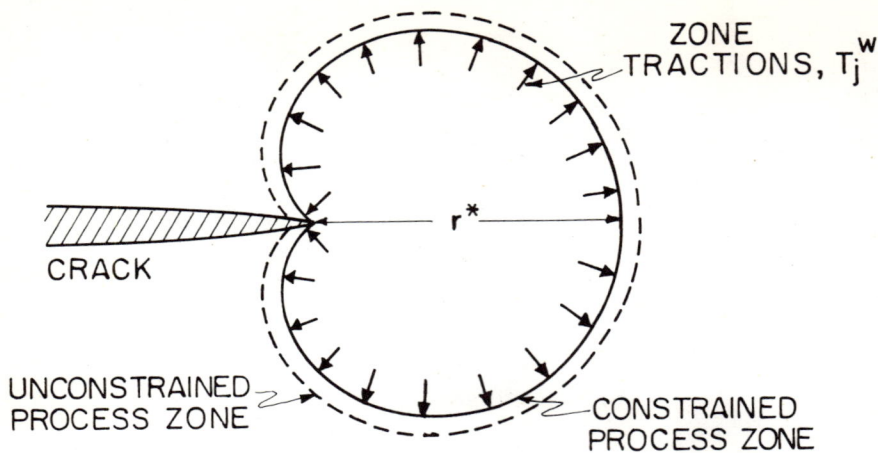

a) THE ZONE TRACTIONS T_j^w

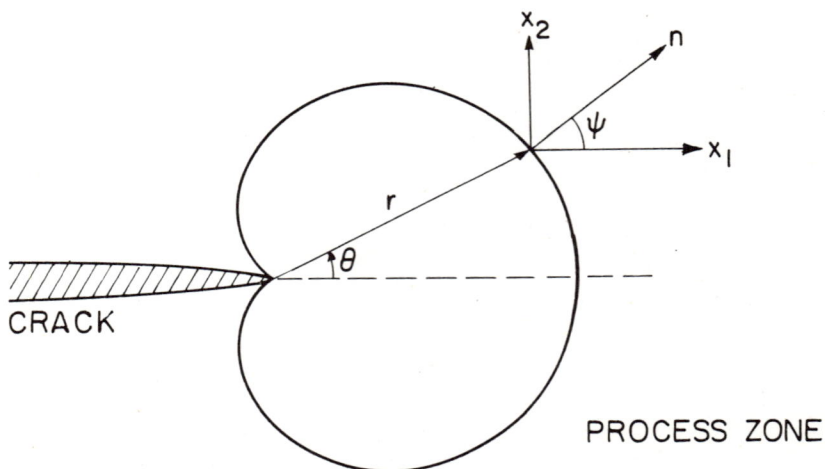

b) COORDINATES OF STRESS INTENSITY ANALYSIS

Fig. 2. (a) Unconstrained and constrained process zones, indicating the tractions needed to maintain stress continuity. (b) The coordinate system used to conduct the stress intensity analysis.

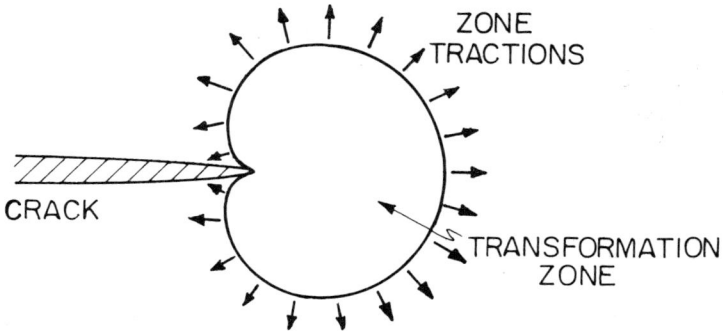

a) ZONE CONTAINED AHEAD OF CRACK TIP

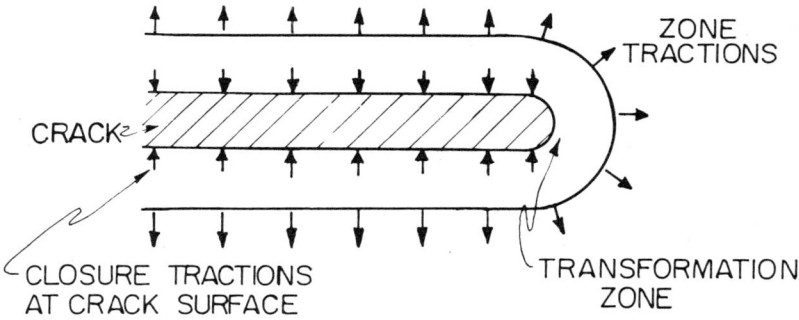

b) ZONE OF UNIFORM WIDTH OVER CRACK SURFACE

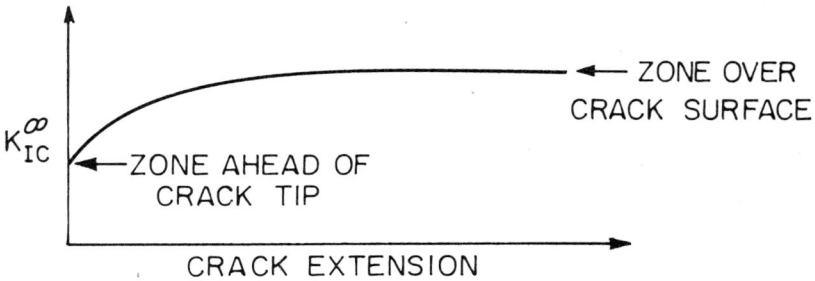

c) SCHEMATIC R-CURVE BEHAVIOR

Fig. 3. (a) Process zone contained ahead of the crack tip, with the hydrostatic contour. ΔK for this zone is zero. (b) Process zone contained over the crack surface. This zone provides crack tip shielding. (c) Schematic of R-curve behavior for transformation toughening in an annealed material.

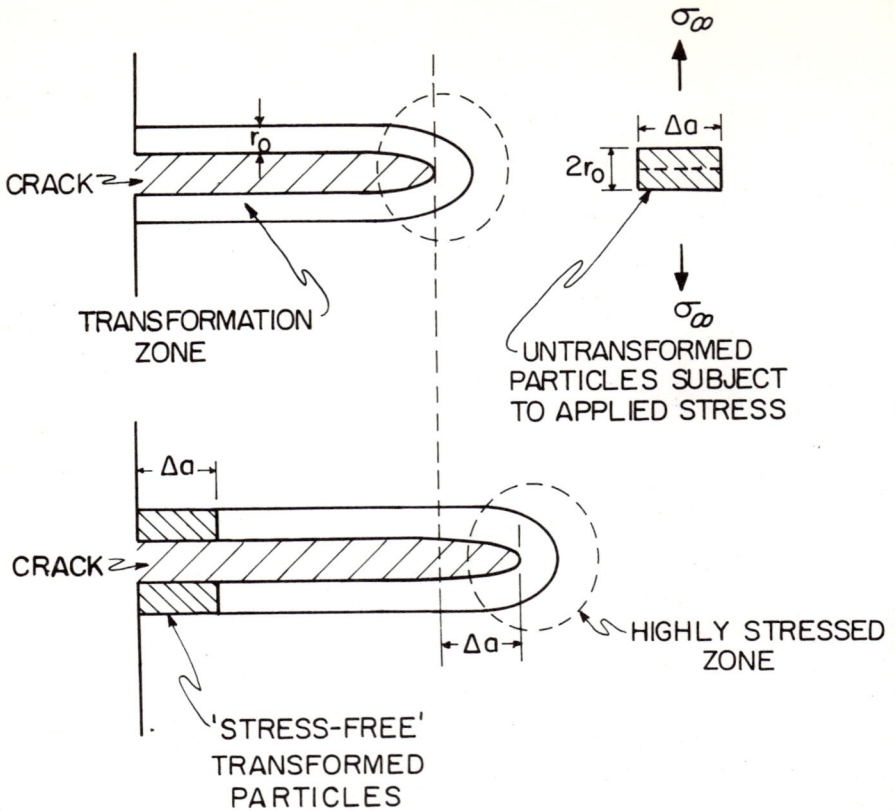

Fig. 4. Schematic indicating that the increased crack growth resistance emanates from the difference in thermodynamic potential of an untransformed region within the applied field (remote from the tip) and a transformed region over the crack surface where the stress level is essentially zero.

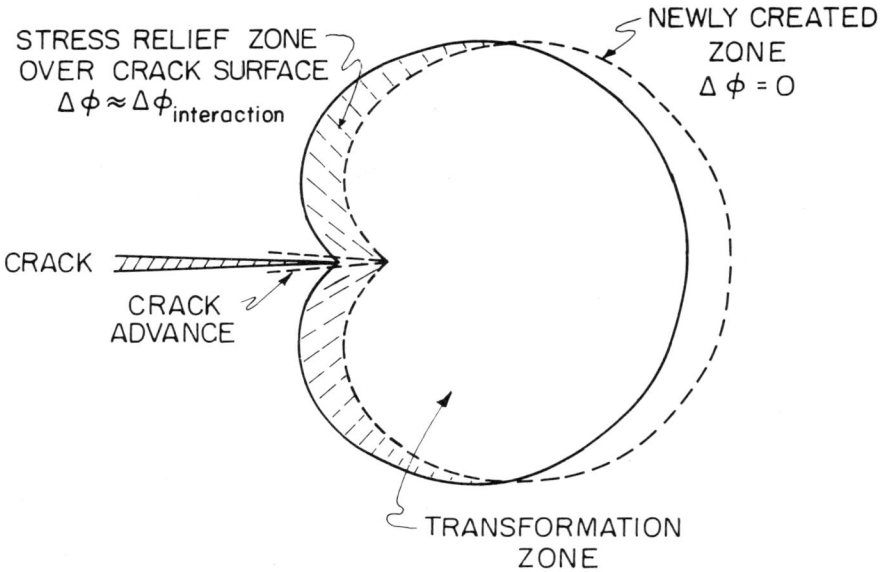

Fig. 5. Schematic indicating the regions of the process zone in which the changes in thermodynamic potential occur during a crack increment.

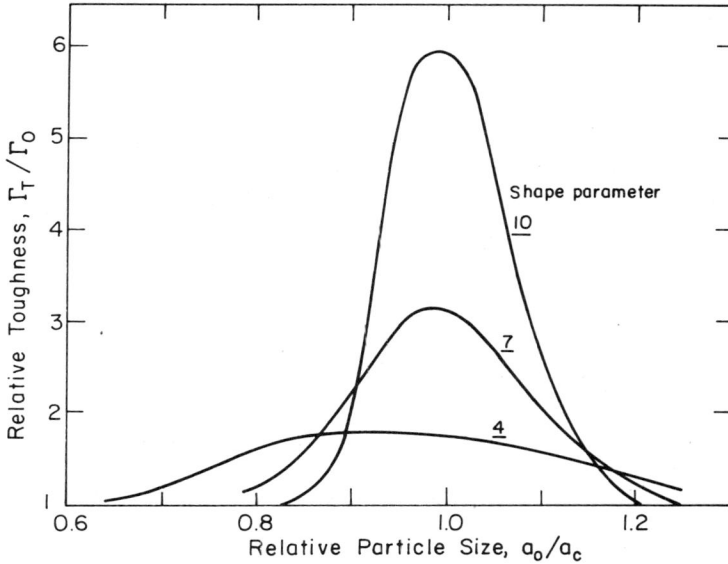

Fig. 6. Predictions of toughening as a function of particle size for several particle size distributions.

215

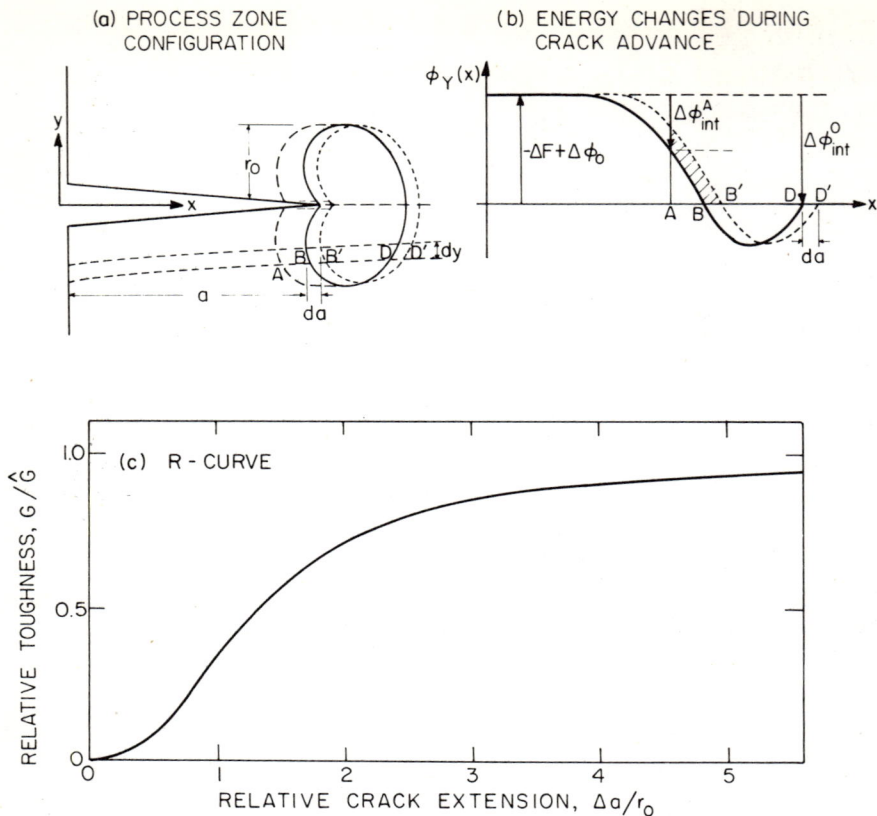

Fig. 7. The *R*-curve determined by employing a thermodynamic treatment. (*a*) The zone characteristics. (*b*) The potential change during a crack increment. (*c*) The relative toughening during a crack advance.

Effect of inclusion size on the retention of tetragonal ZrO₂: theory and experiments

F. F. LANGE AND D. J. GREEN

Structural Ceramics Group
Rockwell International Science Center
Thousand Oaks, Calif. 91360

A thermodynamic approach has been taken which shows that the transformation of an inclusion, constrained by an elastic matrix, will depend in part on its size. That is, a critical size exists below which the constrained transformation is thermodynamically improbable. This size effect is due primarily to surface phenomena (viz. twinning and/or microcracking) that can accompany the transformation to help relieve constraint and lower the strain energy associated with the transformation. The analysis results in a new type of phase diagram representing either the inclusion size-temperature or the inclusion size-composition relations required to retain the high temperature phase. The theory shows that the critical size can be increased by increasing the elastic modulus of the constraining matrix and/or by alloying to decrease the magnitude of the change in chemical free energy. Experimental results obtained for material in the systems ZrO_2/Y_2O_3 and Al_2O_3/ZrO_2 are consistent with the theoretical predictions.

Retention of tetragonal ZrO_2 by constraint and its contribution to fracture toughness have been principal subjects of study in our laboratory for the past few years. Our experimenal studies[1,2] as well as those of others[3-5] have shown that retention depends on microstructure, i.e. a critical grain size/inclusion size exists below which the martensitic ZrO_2 (monoclinic) → ZrO_2 (tetragonal) reaction can be prevented. This size effect can be explained by incorporating various surface terms into the thermodynamics of the constrained phase transformation. A thermodynamic approach can also be used to explain the contribution of the stress-induced transformation on fracture toughness.[6] The object here is to review the thermodynamic approach to explain the size effect associated with the retention of tetragonal ZrO_2 and then to present data in two systems consistent with the theoretical results.

Theory

Consider a $ZrO_2(t)$ inclusion within an elastic matrix. If the inclusion were to transform to its monoclinic structure, classical theory indicates that the free energy change per unit volume can be written as

$$\Delta G_{t \to m} = -\Delta G^c + \Delta U_{se} \pm \Delta U_s \qquad (1)$$

where ΔG^c is the chemical free energy change, ΔU_{se} is the change in strain energy due to the constrained shape and volume change associated with the transformation, and ΔU_s is the energy change associated with the inclusion/matrix interface. Below $\approx 1200\,°C$, ΔG^c is negative for the $ZrO_2(t) \rightarrow ZrO_2(m)$ reaction as indicated in Eq. (1). ΔU_{se} is always positive. ΔU_s can be either positive or negative depending on whether the interfacial energy increases or decreases after the reaction. The transformation will proceed only when $\Delta G_{t-m} < 0$.

Neglecting ΔU_s for the moment, it can be seen from Eq. (1) that the strain energy will lower the transformation temperature relative to the unconstrained case ($\Delta U_{se} = 0$), i.e. transformation will proceed only at the temperature* where $|\Delta G_{t-m}| > \Delta U_{se}$. Thus, the magnitude of ΔU_{se} will strongly affect the constrained transformation temperature. For a given inclusion shape and transformation strain, the magnitude of ΔU_{se} is controlled by the elastic properties of the constraining matrix. High modulus matrices are desirable for phase retention. Residual stresses due to thermal expansion mismatch also affect ΔU_{se}; residual stresses that oppose the shape and volume change are desirable for retention. It should be noted that the summation of ΔG_{t-m} and ΔU_{se} does not result in a size effect; i.e. both have the same dependence on inclusion volume and/or size.

The size effect is brought into the thermodynamics by including the various surface phenomena associated with the constrained transformation. The size effect considered by Garvie[7] is that due to the change in interfacial energy, which can be written as

$$\Delta U_s = \frac{A_m \gamma_m - A_t \gamma_t}{V} \tag{2}$$

where A_m and A_t are the interfacial surface areas, γ_m and γ_t are the interfacial energies per unit area for the monoclinic and tetragonal states respectively, and V is the volume of the transformed inclusion. Assuming a spherical inclusion of diameter D ($V = \pi/6D^3$), Eq. (2) can be rewritten as

$$\Delta U_s = \frac{6\,(\gamma_m - g_s \gamma_t)}{D} \tag{3}$$

where $g_s = A_t / A_m$. Since ΔU_s contains the inclusion size, on substitution of Eq. (3) into (1) and rearranging under the assumption that $\gamma_m > g_s \gamma_t$, one obtains the inclusion size required for the transformation:

$$D > D_c^s = \frac{6(\gamma_m - g_s \gamma_t)}{|\Delta G^c| - \Delta U_{se}} \tag{4}$$

Examination of Eq. (4) shows that the size effect associated with the change in interfacial energy will be observed only for temperatures where $|\Delta G^c| > \Delta U_{se}$.

Twinning and microcracking are other surface phenomena which can occur during the transformation. Unlike the effect associated with the change in interfacial energy, both twinning and microcracking relieve constraint and thus reduce the transformation strain energy. At the same time, both phenomena increase the transformation energy due to the formation of new surface, viz. twin boundaries and crack surfaces, respectively.

Let us first consider the energetics for transformation accompanied by twinning. Twinning relieves some of the shear constraint and thus reduces the strain energy by a factor $(1 - f_T)$. The surface energy per unit volume

218

associated with twinning is

$$U_T = \frac{A_T \gamma_T}{V} = \frac{6\gamma_T g_T}{D} \tag{5}$$

where A_T is the total twin boundary surface, γ_T is the twin boundary energy per unit area, and $g_T = A_T/A$ (A and V are the area and volume of the spherical inclusion). The free energy for the reaction now contains two size-dependent terms:

$$\Delta G_{t \to m} = -\Delta G^c + U_{se}f_T + \frac{6\gamma_T g_T}{D} + \frac{6(\gamma_m - g_s\gamma_t)}{D} \tag{6}$$

Rearranging Eq. (6), one obtains the inclusion size required for the transformation:

$$D > D_c^T = \frac{6[\gamma_T g_T + \gamma_m - g_s\gamma_t]}{|\Delta G^c| - \Delta U_{se}f_t} \tag{7}$$

Examination of Eq. (7) shows that the size effect associated with transformation and twinning will be observed only for temperatures where $|\Delta G^c| > \Delta U_{se}f_T$.

The conditions required for transformation and microcracking can be treated in the same manner, resulting in the inclusion size required for transformation and microcracking.

$$D > D_c^c = \frac{6[\gamma_c g_c + \gamma_m - g_s\gamma_t]}{|\Delta G^c| - \Delta U_{se}f_c} \tag{8}$$

where $(1 - f_c)$ is the strain energy relieved by microcracking, $g_c = A_c/A$ (A_c is the surface area of the crack) and γ_c is the surface energy per unit area of the crack. Again, this size effect will occur only at temperatures where $|\Delta G^c| > U_{se}f_c$.

We can now define the size requirements where all three surface phenomena are active during transformation, viz. interfacial surface change, twinning, and microcracking:

$$D > D_c^{s,T,c} = \frac{6[\gamma_T g_T + \gamma_c g_c + \gamma_m - g_s\gamma_t]}{|\Delta G^c| - \Delta U_{se}f_c f_T} \tag{9}$$

The size effect associated with these combined surface phenomena will be observed only at temperatures where $|\Delta G^c| > \Delta U_{se}f_c f_T$. Since the combined surface phenomena result in the largest reduction in strain energy, its size effect will be the first encountered during cooling.

We are now in a position to examine the relative temperatures where each size effect can be observed and the range of relative temperatures and inclusion sizes where the transformation can be accompanied by one or more of the surface phenomena. The critical size for each surface phenomenon can be normalized by dividing both the numerator and denominator by $|\Delta G^c|$. Noting that the critical size for an unconstrained particle ($\Delta U_{se} = 0$) is

$$D_{uc} = \frac{6(\gamma_m - g_s\gamma_t)}{|\Delta G^c|} \tag{10}$$

we obtain for the combined surface phenomena

$$\frac{D_c^{s,T,c}}{D_{uc}} = \frac{6[R_T + R_c + 1]}{\left[1 - \dfrac{\Delta U_{se} f_c f_T}{|\Delta G^c|}\right]} \tag{11}$$

where $R_T = g_T \gamma_T / (\gamma_T - g_s \gamma_t)$ and $R_c = g_c \gamma_c / (\gamma_m - g_s \gamma_t)$.

By choosing reasonable values for f_c, f_T, R_c, and R_T, we can use Eq. (11) and the other normalized functions to define the constrained transformation conditions. Since ΔG^c is much more sensitive to temperature and composition than ΔU_{se}, the ratio of $\Delta U_{se}/|\Delta G^c|$ is proportional to temperature and/or composition. Also, since the normalized functions are derived for $\Delta G_{t-m} = 0$, they can be used to define phase boundaries and thus phase fields in inclusion size vs temperature and/or compositional space. That is, the normalized functions result in a phase diagram to quantitatively define either the inclusion size required to retain the high temperature phase or the inclusion size where the constrained transformation will proceed with one or more of the accompanying surface phenomena.

Figure 1 illustrates the results of one plot where $f_T = 0.67$ (33% of the strain energy is relieved by twinning), $f_c = 0.90$ (10% relieved by a single radial microcrack[8]), $R_T = 1$, and $R_c = 10$ (the surface energy of the crack is 10 times that of the energy for the interfacial surface change). This figure was drawn by using the normalized results of Eqs. (4), (7), (8), and (9) and eliminating overlapping portions of each phase boundary[6] which results in the definition of the phase fields: (1) untransformed tetragonal, (2) transformed monoclinic + microcracking + twinning, (3) transformed monoclinic + twinning, and (4) transformed monoclinic without microcracking or twinning. The same phase fields would result for other values of f_T, f_c, R_T, and R_c as long as $f_T > f_c$, $R_T \geq 1$, and $R_c > R_T$.

The tetragonal phase field defines the inclusion size/temperature requirement for a fixed composition or the inclusion size/composition requirements for fixed temperature to retain the tetragonal phase in its constrained state. For example, smaller inclusion sizes are required to retain $ZrO_2(t)$ as the temperature is decreased. Likewise, at a fixed temperature (e.g. Y_2O_3 is alloyed with ZrO_2 up to the eutectoid composition of 3 mol%[9]), the inclusion size required to retain the tetragonal phase increases because $|\Delta G^c|$ is smaller. That is, the tetragonal phase can be retained for larger inclusions when a little Y_2O_3 is added than for pure ZrO_2. This information not only explains the size effect observed for phase retention but also suggests alloying routes to the fabricator who must make materials with controlled microstructures in which grain growth and/or inclusion coarsening is concurrent with the temperature/time requirements for densification.

The other phase fields define the inclusion size/temperature or size/composition conditions where one or more of the surface phenomena will occur concurrent with the transformation. For example, let us assume that we have fabricated and cooled a body with a wide distribution of inclusion sizes. Using the results shown in Fig. 1, we would predict that the very smallest inclusions would still be in their untransformed, tetragonal state. A range of slightly larger inclusions would be transformed to their monoclinic state but would be untwinned. The next largest range would be transformed and twinned, and the largest inclusions would be transformed, twinned, and microcracked. Likewise, if we were to heat-treat and cool a material which initially contained a narrow size distribution of very small, untransformed inclusions, we would expect to pass from one phase field to another (transformed, transformed + twinned to transformed + twinned + microcracked) as the inclusions coarsen.

It can be concluded that a thermodynamic reason exists to explain the critical size effect for retaining the high temperature phase in the constrained state. The most important surface phenomena that bring about the size effect are those that help relieve constraint, viz. twinning and microcracking. Qualitatively, this theory shows that the critical size can be increased by increasing the elastic modulus of the constraining matrix and/or by alloying to decrease the magnitude of $|\Delta G^c|$. The theory also predicts that, if the constrained transformation were allowed to proceed, different surface phenomena (viz. twinning and/or microcracking) may occur concurrently with the transformation, depending on the inclusion size.

Experimental Observations

Phase Retention in the ZrO₂-Y₂O₃ System (Ref. 1)

Several investigations[9,10] have shown that additions of Y_2O_3 lower the $ZrO_2(t) \rightarrow ZrO_2(m)$ transformation temperature from $\approx 1200\,°C$ for pure ZrO_2 to $\approx 565\,°C$ at the eutectoid composition of ≈ 3 mol% Y_2O_3. That is, additions of Y_2O_3 (up to 3 mol%) will decrease the magnitude of $|\Delta G^c|$.

Composite powders of submicron ZrO_2 and Y_2O_3 (added as soluble yttrium nitrate) containing 0, 1.5, 2.0, 2.5, 3.0, and 3.5 mol% Y_2O_3 were milled, calcined, cold-pressed into disks, and sintered for 2 h at temperatures ranging from 1200°† to 1600 °C. Densities and average grain sizes were measured as a function of temperature. Phase content was determined by X-ray diffraction on the sintered surfaces of each disk. The critical grain size (D_c) for retaining $\geq 90\%$ of the tetragonal phase was determined by correlating grain size measurements with phase content.

Average grain size was controlled by the sintering temperature and it was independent of the Y_2O_3 content over the range studied. Figure 2 shows that the critical grain size increased from 0.2 μm for 2 mol% Y_2O_3 to 1.0 μm for 3 mol% Y_2O_3. Retention of $>90\%$ tetragonal ZrO_2 could not be achieved for Y_2O_3 contents ≤ 1.5 mol%. It appears that grain growth at temperatures required for densification limited phase retention for compositions containing <1.5 mol% Y_2O_3.

The principal result of this study shows that the critical grain size for phase retention is controlled by the Y_2O_3 addition. Lattice parameter measurements of $ZrO_2(t)$ over the range of 2–3 mol% Y_2O_3 indicated no change ($a = 0.5090$ nm, $c = 0.5174$ nm, values in agreement with those of Scott[10]). These measurements suggest that the Y_2O_3 content does not affect the transformational strains and thus the strain energy, ΔU_{se}. It can be concluded that, with regard to the thermodynamic parameters, the Y_2O_3 has the largest effect on decreasing $|\Delta G^c|$. As discussed in the previous section, decreasing $|\Delta G^c|$ by alloying increases the critical inclusion size‡ required to retain the high temperature phase. The results shown in Fig. 2 are consistent with theoretical results.

221

Phase Retention in the Al₂O₃/ZrO₂ System

Phase equilibria studies have shown that Al_2O_3 and ZrO_2 are compatible with one another.[11] Although the solubility limit of Al_2O_3 in ZrO_2 is presently unknown, the same phase studies indicate neither extensive solubility nor any change in the ZrO_2 transformation temperature. Since Al_2O_3 has approximately twice the elastic modulus of ZrO_2 (390 GPa vs 207 GPa), theory indicates it would be a much better constraining matrix relative to ZrO_2 itself. That is, neglecting the residual strain energy due to differential thermal expansion, one would expect the strain energy (ΔU_{se}) associated with the transformation to be greater for the Al_2O_3 constraining matrix relative to the ZrO_2 matrix. Based on the above theoretical results, we would expect that the critical inclusion size would be greater in the Al_2O_3 constraining matrix than in the ZrO_2 matrix.

A series of Al_2O_3/ZrO_2 composition materials was fabricated by milling submicron powders together and hot-pressing at 1500 °C. Specimens were polished and heat-treated at 1650 °C for periods of 4, 8, and 12 h to induce growth of the ZrO_2 grains. Heat treatment also revealed grain boundaries due to thermal etching. After heat treatment each specimen was subjected to phase identification and grain size distribution measurements. The monoclinic content was determined by obtaining the area ratio of the (111) tetragonal and the $(111) + (1\bar{1}1)$ monoclinic diffraction peaks. Grain size distributions were obtained by measuring the particle diameter distribution on a plane polished section (SEM micrographs with a Zeiss particle size analyzer) and reducing the data to a three-dimensional distribution of spherical grains using the Schwarz-Saltykov method. Between 600 and 1000 ZrO_2 grains were analyzed for each heat treatment. By comparing the monoclinic content with the grain size distribution and assuming that the tetragonal grains were smaller than the monoclinic grains, the critical grain size was determined for each heat treatment. For example, if 40% of the ZrO_2 were monoclinic, then 40% of the ZrO_2 grains would be larger than the critical size; therefore, the critical size corresponds to the 40% probability level of the grain size distribution.

Table I shows the near perfect agreement obtained for the critical grain size for different heat treatments of the same composition. Further heat treatment of the 20 vol% ZrO_2 composite resulted in too few tetragonal grains to allow a reliable determination of the critical grain size in the tail of the grain size distribution curve, whereas for the 5 vol% ZrO_2 composite the monoclinic structure was not detectable by X-ray diffraction. These data indicate two important results: (1) The critical grain size in the high modulus Al_2O_3 matrix is much larger than that for the lower modulus ZrO_2 matrix. (2) The critical size decreases with increasing ZrO_2 content. These results are consistent with the theoretical prediction that the critical grain size will increase with the elastic modulus of the constraining matrix.§

Elastic modulus measurements were made to determine whether microcracking accompanied the transformation in the Al_2O_3/ZrO_2 system. As reported in Table II, the significant decrease in elastic modulus concurrent with large fractions of transformed ZrO_2 shows that microcracking does accompany the transformation. Direct observations in the SEM using backscattered electron emission of cracks decorated with $AgNO_3$ showed that the microcracks extended from the ZrO_2 grains along Al_2O_3 grain boundaries.

Transmission electron microscopy, courtesy of D. R. Clarke, also showed that the transformed grains are highly twinned. Thus, it can be concluded that both twinning and microcracking accompany the transformation of most ZrO_2 grains in the Al_2O_3/ZrO_2 composites.

Conclusion

A thermodynamic approach has shown that the conditions for a constrained transformation depend on the size of the transforming inclusion. This size effect is due primarily to accompanying surface phenomena (viz. twinning and microcracking) that help relieve constraint and lower the strain energy associated with the transformation. Thus, retention of tetragonal ZrO_2 as a toughening agent depends on fabricating a microstructure in which the ZrO_2 inclusions (or grains) are less than a critical size.

In many cases, grain growth during fabrication prevents achieving the fine grained material required for phase retention. Theory shows that the critical grain size can be increased, thus relaxing the microstructure requirements for phase retention, by alloying to decrease the magnitude of the chemical free energy change and/or by choosing a stiffer constraining matrix to increase the strain energy associated with the transformation. Experimental results presented for two systems are consistent with these theoretical results, viz. additions of Y_2O_3 to ZrO_2 and a higher modulus Al_2O_3 constraining matrix both increase the critical grain size. Thus, the theory not only explains the critical size effect, but it also directs the fabricator to explore routes which will ease the conditions to achieve phase retention.

Acknowledgment

Much of this work was supported by the Office of Naval Research under Contract N00014-77-C-0441. The critical inclusion size results on the Al_2O_3/ZrO_2 were supported by a Rockwell Independent Research and Development Program.

References

[1] F. F. Lange, "Stress Induced Martensitic Reactions: II, Experiments in the ZrO_2-Y_2O_3 System," ONR Tech. Rept. No. 3, Contract N00014-77-C-0441, July 1978.
[2] F. F. Lange, "Transformation Toughening in the Al_2O_3/ZrO_2 Composite System," ONR Tech. Rept. No. 7, Contract N00014-77-C-0441, May 1979.
[3] R. C. Garvie, R. H. Hannink, and R. T. Pascoe, "Ceramic Steels?," Nature (London), 258, 703 (1975).
[4] T. K. Gupta, F. F. Lange, and J. H. Bechtold, "Effect of Stress-Induced Phase Transformation on the Properties of Polycrystalline Zirconia Containing Metastable Tetragonal Phase," J. Mater. Sci., 13, 1464 (1978).
[5] D. L. Porter and A. H. Heuer, "Mechanisms of Toughening Partially Stabilized Zirconia (PSZ)," J. Am. Ceram. Soc., 60 [3-4] 183 (1977).
[6] F. F. Lange, "Stress Induced Transformation Toughening: Part I, Size Effects Associated with Constrained Phase Transformation," ONR Rept. No. 8, Contract N00014-77-C-0441, October, 1980.
[7] R. C. Garvie, "Stabilization of the Tetragonal Structure in Zirconia Microcrystals," J. Phys. Chem., 82, 218 (1978).
[8] Y. M. Ito; private communication.
[9] K. K. Srivastava, R. N. Patil, C. B. Chaudary, K. V. G. K. Gokhale, and E. C. Subbarao, "Revised Phase Diagram of the System ZrO_2-Y_2O_3," Trans. Br. Ceram. Soc., 73, 85 (1974).
[10] H. G. Scott, "Phase Relations in the Zirconia-Yttria System," J. Mater. Sci., 10, 1527 (1975).
[11] E. M. Levin and H. F. McMurdie, Phase Diagrams for Ceramists, 1975 Supple-

ment. Edited by Margie K. Reser. The American Ceramic Society, Columbus, Ohio, 1975; Figs. 4377 and 4378.

---

*Throughout the paper we are considering only temperatures where ΔG^c is negative. The absolute brackets are used to avoid confusion regarding the sign of ΔG^c.

†Approximately 80% of the total shrinkage during sintering was obtained after 2 h at 1200 °C.

‡Since the transformation is anisotropic, neighboring grains constrain one another as a matrix would constrain an inclusion.

§It should be noted that the addition of 2 mol% Y_2O_3 to the Al_2O_3/ZrO_2 system results in full retention of the tetragonal phase up to ≈ 80 vol% ZrO_2 (Ref. 2). The critical grain size was >2 μm, the average ZrO_2 grain size for these compositions.

Table I. Critical grain size for Al_2O_3/ZrO_2 composites

Vol fraction ZrO_2	Heat treatment Temp. (°C)	Time (h)	Critical grain size (μm)
0.05*	1650	12	>2.3
0.10	1650	4	1.35
	1650	8	1.35
	1650	12	1.41
0.15	1650	4	0.85
	1650	8	0.84
	1650	12	0.85
0.20†	1650	4	0.74

*Monoclinic ZrO_2 not detectable by X-ray diffraction. †Longer heat treatment produced too much monoclinic ZrO_2 for reliable results.

Table II. Elastic modulus of Al_2O_3/ZrO_2 composites heat-treated at 1650 °C for 12 h

Vol fraction ZrO_2	% Monoclinic	E (GPa)	E_0* (GPa)	E/E_0
0.01	0	381	390	0.98
0.05	0	375	383	0.98
0.075	13	370	378	0.98
0.10	47	282	374	0.75
0.125	92	231	369	0.63
0.150	92	229	364	0.63
0.20	95	227	355	0.64

*E_0 is the expected modulus without microcracks as measured with a series of Al_2O_3/ZrO_2 ($+2$ mol% Y_2O_3) in which the ZrO_2 is retained in its tetragonal structure (Ref. 2).

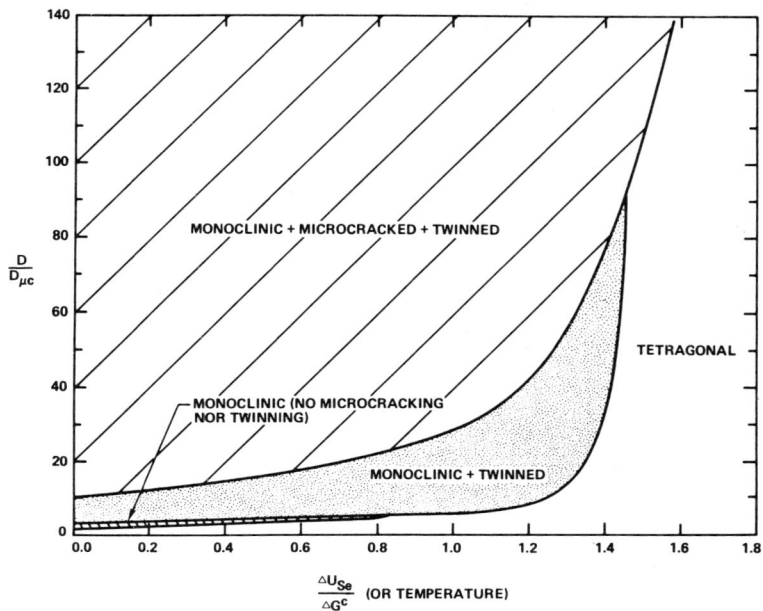

Fig. 1. Descriptive phase diagram representing inclusion size/temperature (or composition) requirements for either retaining high temperature phase or obtaining transformation with one or more of the accompanying surface phenomena.

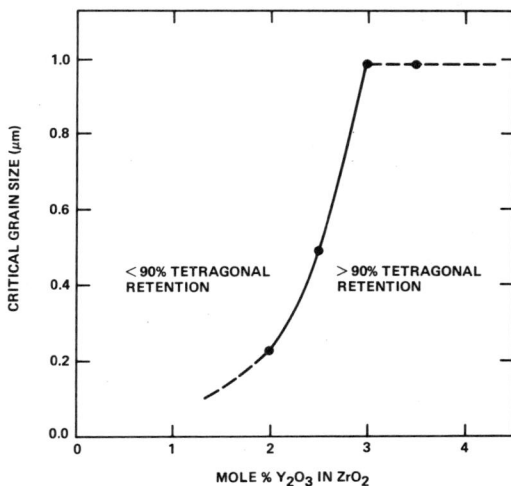

Fig. 2. Critical grain size vs Y_2O_3 content in ZrO_2.

Zirconia thermal barrier coatings

R. J. Bratton and S. K. Lau

Westinghouse Research and Development Center
Pittsburgh, Pennsylvania 15235

Ceramic thermal barrier coatings are currently under active development in the United States for both aircraft and industrial/utility gas turbine operation. These coating systems generally consist of an oxidation-corrosion resistant metal bond coat of the MCrAlY type and either a single thick layer ceramic overcoat or a graded ceramic/MCrAlY overcoat. The plasma spray deposition technique, which has been most widely used to deposit thermal barrier coatings, can result in desirable or undesirable coating characteristics, depending on the application. Recent evaluations, primarily of plasma-sprayed zirconia-based coatings, are reviewed as well as the current understanding of the interaction of these coatings with corrosion contaminants found in clean distillate and heavy petroleum fuels.

The concept of thermal barrier coatings is to provide insulation to critical air-cooled metal components. An important application is in turbine engines where there are several ways to take advantage of ceramic insulation at high temperatures,[1-4] e.g. increased engine reliability by reducing metal temperature by 50 to 220 °C, increased engine efficiency and power by maintaining current metal temperature and increasing gas temperature, and reduced fabrication cost by eliminating elaborate cooling schemes. Durability and fuel flexibility might also be achieved if the ceramic coatings were resistant to inorganic salt contaminants.[5] It should be noted that plasma-sprayed ceramic coatings have been used for over ten years to reduce metal temperatures in aircraft burners and afterburners. But because of the more stringent environmental requirements, turbine airfoils with ceramic coatings have not been successfully applied.

A contemporary two-layer (duplex) thermal barrier coating is shown schematically in Fig. 1. Typically this consists of a thick (up to 0.06 cm) plasma-sprayed ceramic layer deposited over a thin (0.01 cm) metallic bond coat. Another version not shown is to introduce a graded intermediate region of ceramic/metal that is used to minimize thermal expansion mismatch between the materials that make up the coating system.

The coating properties required for turbine engine application are achieved through the selection of material properties and coating process parameters. Desired ceramic properties include low thermal conductivity, a relatively high coefficient of thermal expansion, thermodynamic stability in the gas turbine environment, and mechanical stability toward thermal cycling. The plasma spray process[6,7] helps achieve increased thermal protection and improved thermal stress resistance due to the coating porosity that is characteristic of this process. On the other hand, gas permeability and condensed salt penetration into the porous coating structure may induce coating degradation. Plasma-sprayed zirconia compositions have been of most interest for thermal barrier coatings and have been under periodic development for about the past 30 years. Since the primary function of the metal bond coat is to provide a substrate for adherence of the ceramic coating, it is very important that the bond coat be oxidation and/or corrosion resistant at the metal[8-10] operating temperature. In this respect MCrAlY alloys developed for coating turbine airfoils are proving very useful as the bond coat for the thermal barrier coating concept and, in part, are responsible for recent successes compared to earlier work.

Most of the early thermal barrier coating research which extended from the 1950's to early 1970's[11-14] was conducted at NASA-Lewis Research Center. This work focused on surface protection of both aircraft engine and rocket engine components. By mid 1970's encouraging results were achieved with a zirconia-based two-layer system as shown in Fig. 1.[15-18] From several stabilized zirconia compositions, which included CaO, MgO, and Y_2O_3 stabilizers, the $ZrO_2 \cdot 12$ wt% Y_2O_3/NiCrAlY system was chosen on the basis of adherence to the NiCrAlY bond coat, thermal stress resistance, erosion resistance, and stability in the aircraft turbine environment. This work, as well as that of others,[19-22] stimulated further development of ZrO_2-Y_2O_3 coating compositions and compatible MCrAlY bond coats. Figure 2[23,24] shows the recent improvements made as judged by durability to cyclic testing using a very clean-burning fuel. Yttria content in the ZrO_2-Y_2O_3 coating and yttrium content in the NiCrAlY bond coat are obviously important to coating life. However the reasons for the improved durability have not been established. Other factors, such as bond coat roughness, are also important to coating adherence and durability.[20]

It should be noted that the most durable zirconia coatings are those formed from the two-phase (partially stabilized) region of the zirconia-yttria phase system.[25] This equilibrium system would predict that the stable phases are cubic ZrO_2 (high Y_2O_3) and monoclinic ZrO_2 (low Y_2O_3) at room temperature where the latter transforms from the tetragonal ZrO_2 (low Y_2O_3) that is stable above $\approx 600\,°C$. However, this equilibrium case does not exist during the rapid cooling of plasma-sprayed $ZrO_2 \cdot 8$ wt% Y_2O_3 and other compositions.[5] The room temperature phases consist primarily of a nonequilibrium high-yttria tetragonal phase and minor amounts of cubic phase and monoclinic phase similar to that obtained with rapidly quenched bulk material. A study of the stability of these phases is reported in these proceedings.[26]

Although the durability of zirconia coatings has been exceptional in tests conducted under relatively clean high temperature gases such as natural gas and jet aircraft fuel, preliminary studies conducted at Westinghouse[5] and NASA[4] suggested potential adherence problems when the coatings are subjected to combustion environments that contain impurities. Impurities are often found in turbine systems that enter the engine through ingested runway dust or sea salt aerosols.[27] This may lead to the deposition of the highly corrosive sodium sulfate salt onto turbine blades and vanes. In nonaircraft application, a similar result may occur as well as impurities introduced through the use of dirty fuels such as crude and residual oils. Major impurities found in these fuels include sodium, potassium, vanadium, iron, lead, phosphorus, and sulfur. Also, minimally processed coal-derived fuels may be burned in turbines in the future. Although trace metal impurity levels in these fuels are uncertain, the durability of thermal barrier coatings will need to be established as they become available.

The purpose of this paper is to review recent work, conducted at Westinghouse and other laboratories, that addresses the resistance of present day zirconia thermal barrier coatings to combustion gases that contain corrosive metal impurities. Atmosphere burner rig tests were used for this purpose. The recent work by Bratton et al.[28] was conducted for NASA and EPRI under contract NAS3-21377.

Experimental Procedure

Table I shows the thermal barrier coating systems that were prepared by plasma spray deposition together with the as-deposited phases present in the oxides as determined by X-ray diffraction analysis. With the exception of Ca_2SiO_4, the oxide coatings are zirconia stabilized with Y_2O_3 or MgO. For reference, the locations of the ZrO_2 (Y_2O_3) compositions are shown in the ZrO_2-rich phase regions of the ZrO_2-$YO_{1.5}$ system (see arrows in Fig. 3).

As may be seen, the as-deposited phases are nonequilibrium. Both the partially stabilized $ZrO_2 \cdot 8$ wt% Y_2O_3 and fully stabilized $ZrO_2 \cdot 20$ wt% Y_2O_3 compositions consist primarily of tetragonal ZrO_2 (Y_2O_3) with different amounts of cubic ZrO_2 (Y_2O_3). The partially stabilized $ZrO_2 \cdot 8$ wt% Y_2O_3 also contains a minor quantity of monoclinic ZrO_2 (Y_2O_3). Not shown in Table I are the substrate alloys which are Udimet 720, a nickel-based blade alloy, and ECY 768, a cobalt-base vane alloy.

Typical metallographic cross sections of the duplex (2 layer) and graded coating systems are shown in Figs. 4 and 6, respectively. The elemental characterization of the coatings was determined by electron microprobe mapping. Typical scans obtained on cross sections of the duplex and graded coatings are shown in Figs. 5 and 7, respectively. Although the elements analyzed were Zr, Y, Mg, Al, Ni, Co, Si, Fe, Na, S, Cu, only the major elements are given to illustrate the as-coated U720 substrate/NiCrAlY bond coat/oxide structure of the duplex and grade coatings.

For burner rig testing, the alloy specimen geometry was an air-cooled cylinder 7.6 cm long, 1.3 cm in diam., with a 0.6 cm in diam. cooling hole centrally located. Four thermocouple wells equally spaced at 90° were located between the central cooling hole and specimen surface to accommodate thermocouples for monitoring metal temperature. The cylindrical specimen was coated from end to end with the bond coat and the central 5 cm with the thermal barrier coating which is tapered at its terminals.

Both clean-fuel and impurity-doped fuel sensitivity tests were conducted in atmospheric burner rigs. These rigs consisted of a furnace box with a fuel-burning combustor attached to one end. The burner was operated at a predetermined fuel:air ratio, and secondary diluent air was injected in the post combustion zone to obtain the desired combustion gas temperature. Eight instrumented air-cooled specimens mounted on a platform were inserted from the bottom so that the hot gas from the burner impinged at a right angle to the cylindrical specimens. The platform was mounted on a pneumatic cylinder so that the specimens could be inserted and retracted once an hour from the furnace. The maximum metal temperature which was located nearest the cooling air exit of the specimen was used as a control throughout a test. The variation of this temperature within the eight specimen pack was ±8 °C.

Burner Rig Test Results and Discussion

Clean Fuel Tests

Clean fuel tests were conducted using GT No. 2 fuel (ASTM 2880-76). Three tests were run in which the gas temperature was maintained at 1150 °C but the metal substrate temperature was varied from 800° to 845° to 900 °C. All tests were terminated after 500 1-h cycles.

Figure 8 shows the results for coatings tested in terms of the minimum number of 1-h cycles to cause a spalling failure for each of the three metal temperature cases studied. Only one coating, a duplex (D) $ZrO_2 \cdot 15Y_2O_3$, did not survive the 1150 °C gas/800 °C metal test condition (Fig. 8(a)). The failure has been attributed to thermal stress since insignificant impurity condensate could be found in post-test analyses.

Figures 8(b) and 8(c) show that the duplex $ZrO_2 \cdot 8Y_2O_3$ coating survived the high metal temperature test conditions while the graded (G) coatings $ZrO_2 \cdot 8Y_2O_3$ and $ZrO_2 \cdot 24.65MgO$ did not. Post-test analyses revealed that the cause of the graded coating failures was internal oxidation of the NiCrAlY in the graded zone. This is depicted in Fig. 9 by the depletion of metallic particles in the graded zone. X-ray diffraction analyses of the oxidized graded region revealed NiO as the major phase, Cr_2O_3 as the minor phase, and Al_2O_3 as the trace phase, which is consistent with oxidation products of NiCrAlY.

It should be noted that, although the duplex $ZrO_2 \cdot 8Y_2O_3$/NiCrAlY coating did not fail when the metal temperature was 900 °C, EMP analyses revealed that the NiCrAlY bond coat surface had oxidized to form a thin layer of Al_2O_3. This could cause adherence problems during long exposures and therefore needs to be studied further.

Fuel Sensitivity Tests

The fuel sensitivity tests were conducted by maintaining the gas/metal temperatures at 1150°/800 °C, respectively, and varying the fuel impurities by doping GT No. 2 fuel. Two cases will be discussed: (1) GT No. 2 fuel doped to (in ppm) 1 Na, 2 V, 2 P, 0.5 Ca, 2 Fe, and 6 Mg, which represents a water-washed crude oil, and (2) GT No. 2 fuel doped to (in ppm) 1 Na, 50 V, 2 P, 0.5 Ca, 2 Fe, and 150 Mg, which represents a water-washed residual oil. In both simulated fuels the Mg/V ratio was 3 as used in commercial practice and the S content was 0.25%. Testing was terminated after 500 1-h cycles.

Figure 10(a) shows the results of one of the fuel sensitivity tests in terms of the minimum number of 1-h cycles to initiate spalling. Under these conditions (1 ppm Na, 2 ppm V), it is clear that the graded coatings outperformed the duplex coatings regardless of their compositions. An exception was the duplex Ca_2SiO_4 coating which also survived the 500 1-h cycles test. NASA[12] has also found success with this coating. In the more severe test with 1 ppm Na, 50 ppm V (Fig. 10(b)), the graded $ZrO_2 \cdot 8Y_2O_3$ coatings performed best. The duplex Ca_2SiO_4 coating was not available for this test.

A general result of the fuel sensitivity test is that the 50 ppm V fuel caused accelerated failures compared to the 2 ppm V fuel when all the other contaminants including the Mg/V ratio remained constant (compare Figs. 10(a) and 10(b)). However, post-test analyses conducted to date have revealed that Na$_2$SO$_4$-MgSO$_4$ salts likely play an important role in the failure mechanisms, whereas the role of vanadium and its salts is less clear.

Figure 11 shows an EMP scan for a failed duplex ZrO$_2$·20Y$_2$O$_3$ specimen. The deep penetration of Na, Mg, and S into the coating that remained after spalling can be seen. No V was detected because, in all specimens analyzed, it was located only at the surface; i.e. it was associated with that part of the coating which spalled away in the case of Fig. 11. Spalling of the duplex coatings generally occurred just above the oxide/bond coat interface leaving a residual oxide layer.

The deep penetration of Na, Mg, and S into the porous oxide coating has been connected to the formation and penetration of a liquid Na$_2$SO$_4$-MgSO$_4$ salt. Chemical analysis of water-washed samples showed that the soluble condensate consisted of Na$_2$SO$_4$:MgSO$_4$ in the molar ratio of 0.72, assuming that the soluble sodium and magnesium were present as sulfates. According to the NaSO$_4$-MgSO$_4$ phase system,[29] this mixed salt should be molten at 765 °C. Since the highest temperature gradient in the oxide samples is 900 °C surface and 800 °C bond coat, the salt was molten during testing where these conditions prevailed. Those parts of the specimen that were cooler than 765 °C should not be exposed to molten salt and therefore should not be attacked. This behavior has been generally consistent with experimental observations both at Westinghouse and NASA-Lewis. Using thermochemical calculations of condensate dew points and melting points combined with temperature profiles of test specimens, Miller[30] was able to qualitatively explain observed coating failures at NASA. The most severe conditions for a porous plasma-sprayed ceramic coating occur when a corrosive liquid like Na$_2$SO$_4$-MgSO$_4$, V$_2$O$_5$, Na$_2$V$_2$O$_6$, etc. can condense at the surface and completely penetrate the coating. This condition is met when the dew point (T_{dp}) of the condensate is above the surface temperature of the ceramic and the melting point (T_{mp}) of the condensate is below the bond coat temperature. Figure 12 illustrates several hypothetical cases ranging from complete liquid salt penetration to no salt condensation.

Other evidence of molten salt corrosion of plasma-sprayed ZrO$_2$-Y$_2$O$_3$ coatings has been recently described. Palko et al.[31] showed from atmospheric burner rig tests that molten sodium sulfate can penetrate the porous oxide and react with the underlying bond coat at 870 °C to cause spalling. Barkalow and Pettit[32] suggested from furnace corrosion tests that hot corrosion occurs by acidic dissolution of Y$_2$O$_3$ from stabilized ZrO$_2$·Y$_2$O$_3$ according to a reaction of the form:

$$Y_2O_3(\text{in } ZrO_2) + 3SO_3(\text{in } Na_2SO_4) \rightarrow 2Y^{3+}(\text{in } Na_2SO_4) + 3SO_4{}^{2-}(\text{in } Na_2SO_4)$$

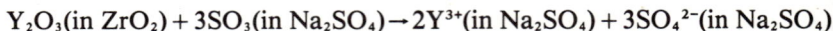

The solidified salt on cooling in this case is a hydrated sodium-yttrium sulfate [Na$_2$Y$_2$(SO$_4$)$_4$·2H$_2$O]. This reaction was significant at the relatively low temperature of 700 °C and high P_{SO_3} pressure of 7×10^{-4} atm. As shown in this work and earlier by Bratton et al.[5] the salt reactions lead to destabilization of the stabilized ZrO$_2$-Y$_2$O$_3$ coating—an effect that seems important in the failure mode.

However, in the present study, it is clear from the EMP scans shown in Fig. 11 that the Na$_2$SO$_4$-MgSO$_4$ phases are located in the pores and crevices of the oxide coating with no correspondence found between Na and Y or Zr, indicating that there is no significant chemical reaction between the molten salt and the coating. It thus seems that the failure mode is mechanical in nature; i.e. on thermal cycling the stress generated by the thermal expansion mismatch between the $ZrO_2(Y_2O_3)$ coating and the entrapped solidified salt condensate causes the coating to crack and spall. Equally possible is coating densification at least locally in the presence of molten Na_2SO_4 that would result in the reduction of coating thermal stress resistance.

While it is apparent that vanadium dopant is very detrimental to zirconia-yttria thermal barrier coatings (Fig. 10), the mechanism of vanadium attack that leads to coating failure has not been established. Molten V_2O_5 was shown by Bratton et al.,[5] Zaplatynsky,[33] Hodge et al.,[34] and Palko et al.[31] to react with the Y_2O_3 (in $ZrO_2 \cdot Y_2O_3$) to form YVO_4 according to reactions such as:

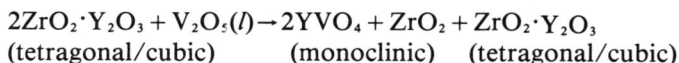

$$2ZrO_2 \cdot Y_2O_3 + V_2O_5(l) \rightarrow 2YVO_4 + ZrO_2 + ZrO_2 \cdot Y_2O_3$$
$$\text{(tetragonal/cubic)} \qquad \text{(monoclinic)} \quad \text{(tetragonal/cubic)}$$

More importantly, the result of this reaction is also destabilization of the tetragonal/cubic $ZrO_2 \cdot Y_2O_3$ to monoclinic ZrO_2. The effect is a disruptive volume change of $\approx 5\%$ when the martensitic tetragonal $\rightarrow$ monoclinic transformation occurs and, hence, eventual coating failure. However, in the present burner rig tests, the calculated dew point of V_2O_5, conservatively assuming all V in the fuel forms V_2O_5, is 690 °C for 50 ppm V and 545 °C for 2 ppm V, which are far below the 900 °C specimen surface temperature. This means that molten V_2O_5 cannot condense on the test specimens. EMP scans (Fig. 13) confirm that there is no deep penetrating liquid vanadium salt. However, destabilization of tetragonal/cubic $ZrO_2 \cdot Y_2O_3$ has still been found to occur. Obviously, an explanation other than the liquid V_2O_5 attack described earlier is needed.

The condensates found on the present burner rig test specimen surfaces are $MgSO_4$, MgO, and $Mg_3V_2O_8$ which are in qualitative agreement with those expected from the equilibrium MgO-V_2O_5-SO_3 phase diagram.[35] It is noteworthy, as demonstrated in the EMP scans of an unfailed but severely cracked graded $ZrO_2 \cdot 8Y_2O_3$ specimen (Fig. 13), that while the Na, S, and Mg have clearly penetrated both the outer oxide layer and graded layer of the coating, the major concentration of vanadium is not associated with magnesium. Instead, it is located beneath the surface $MgSO_4$ and $Mg_3V_2O_8$ deposits and is distributed rather evenly inside the bulk of the ZrO_2-Y_2O_3 top layer. A possible reaction among others that are being investigated is the formation of a monoclinic ZrO_2-Y_2O_3-V_2O_5 ternary solid solution from the attack of gaseous V_2O_5 on ZrO_2-Y_2O_3. The effect of this ternary phase formation on coating integrity is also under study.

Summary

Recent studies of evaluating zirconia-based thermal barrier coatings suggest that long lives may be achieved with clean turbine engine fuels. Thus the prospects are bright that thermal barrier coatings will be used not only in the combustion systems as they are today but also for blades and vanes in the near future. For application where fuel or air impurities are present, the level of understanding and development is less advanced. The destabilization of zirconia coatings in the presence of inorganic salt contaminants has been established and the detailed coating degradation mechanisms are currently being investigated. Programs are also in progress to develop improved zirconia coatings through composition and microstructure modifications.

References
[1] D. J. Amos, "Analytical Investigations of Thermal Barrier Coatings on Advanced Power Generation Gas Turbines," NASA CR-135146, March 1977.
[2] N. Carlson and B. L. Stone, "Thermal Barrier Coatings on High Temperature Gas Turbine Engines," NASA CR-135147, February 1977.
[3] J. S. Clark, J. J. Nainiger, and R. E. Heyland, "Potential Benefit of a Ceramic Thermal Barrier Coating on Large Power Generation Gas Turbines," NASA TM 73712, June 1977.
[4] S. R. Levine and J. S. Clark, "Thermal Barrier Coatings—A Near Term High Payoff Technology," NASA TM X-73586, 1977.
[5] R. J. Bratton, S. C. Singhal, and S. Y. Lee, "Ceramic Gas Turbine Components Research and Development: Part 2, Evaluation of ZrO_2-Y_2O_3/NiCrAlY Thermal Barrier Coatings Exposed to Simulated Gas Turbine Environments," Final Rept., EPRI RP421-1, June 1978.
[6] H. S. Ingham and G. P. Shephard, Flame Sprayed Handbook: Vol. III, Plasma Flame Process. Metco, Inc., 1965.
[7] S. J. Grisaffe, "A Simplified Guide to Thermal Spray Coatings," Mach. Des., 39 [17] 174–81 (1967).
[8] S. J. Grisaffe; pp. 341–70 in The Superalloys. Edited by C. Sims and W. C. Hagel. John Wiley & Sons, Inc., New York, 1972.
[9] D. Chatterji, R. C. DeVries, and G. Romeo; pp. 1–87 in Advances in Corrosion Science and Technology: Vol. 6. Edited by M. G. Fontana and R. W. Staehle. Plenum Press, New York, 1976.
[10] G. W. Goward; pp. 806–23 in Symposium on Properties of High Temperature Alloys. Edited by Z. A. Foroulis and F. S. Pettit. The Electrochemical Society, Inc., 1976.
[11] L. J. Schafer, Jr., "Comparison of Theoretically and Experimentally Determined Effects of Oxide Coatings Supplied by Fuel Additives on Uncooled Turbine-Blade Temperature During Transient Turbojet-Engine Operation," NASA RM E53A19, 1953.
[12] E. R. Bartoo and J. L. Clure, "Experimental Investigation of Air-Cooled Turbine Blades in Turbojet Engine XIII—Endurance Evaluation of Several Protective Coatings Applied to Turbine Blades of Nonstrategic Steels," NASA RM E53E18, 1953.
[13] L. N. Hjelm and B. R. Bornhorst, "Development of Improved Ceramic Coatings to Increase the Life of XLR 99 Thrust Chamber," Research Airplane-Committee Rept. on Conf. on Progress of the X-15 Project. NASA TM X-57-72, 1961; pp. 227–53.
[14] A. N. Curren, S. J. Grisaffe, and K. C. Wycoff, "Hydrogen Plasma Tests of Some Insulating Coating Systems for the Nuclear Rocket Thrust Chamber," NASA TM X-2461, 1972.
[15] C. H. Liebert and S. Stecura, "Ceramic Thermal Protective Coating Withstands Hostile Environment of Rotating Turbine Blades," NASA Tech Brief B75-10290, 1975.
[16] S. Stecura, "Two-Layer Thermal Barrier Coating for Turbine Airfoils—Furnace and Burner Rig Test Results," NASA TM X-3425, 1976.
[17] S. Stecura and C. H. Liebert, "Thermal Barrier Coating System," U.S. Pat. 4 055 705, Oct. 25, 1977.
[18] S. Stecura, "Two-Layer Thermal Barrier Coating for High-Temperature Components," Am. Ceram. Soc. Bull., 56 [12] 1082–85, 1089 (1977).
[19] N. M. Nijpjes; pp. 481–99 in Sixth Plansee Seminar on High Temperature Materials. Edited by F. Benesovsky. Springer-Verlag, 1969.
[20] J. R. Cavanaugh et al., "The Graded Thermal Barrier—A New Approach for Turbine Engine Cooling," AIAA Paper 72-361, April, 1972.
[21] R. C. Tucker, Jr., T. A. Taylor, and M. H. Weatherly, "Plasma Deposited MCrAlY Airfoil and Zirconia/MCrAlY Thermal Barrier Coatings," Proc. Conf. Gas Turbine Materials in Marine Environment, 3rd, Sept. 20–23, 1976.
[22] W. O. Gaffin and D. E. Webb, "JT8D and JT9D Jet Engine Performance Improvement Program—Task 1, Feasibility Analysis," (PWA-5518-38, Pratt and Whitney Aircraft Group; NASA Contract NAS3-20630). NASA CR-159449, 1979.
[23] S. Stecura, "Effects of Compositional Changes on the Performance of a Thermal Barrier Coating System," NASA TM-78976, 1978.
[24] S. Stecura, "Effects of Yttrium, Aluminum, and Chromium Concentrations in Bond Coatings on the Performance of Zirconia-Yttria Thermal Barriers," NASA TM-79206, 1979.

[25]H. G. Scott, "Phase Relationships in the Zirconia-Yttria System," *J. Mater. Sci.*, **10** [9] 1527–35 (1975).

[26]R. A. Miller, J. L. Smialek, and R. G. Garlick, "Phase Stability in Plasma-Sprayed, Partially Stabilized Zirconia-Yttria"; this volume, pp. 241–53.

[27]J. Stringer, "High Temperature Corrosion of Aerospace Alloys," AGARDograph No. 200, NATO, August 1975.

[28]R. J. Bratton, S. K. Lau, S. Y. Lee, and C. A. Andersson, "Ceramic Coating Evaluations and Developments," in *Proc. Conf. Advanced Materials for Alternative Fuel Capable Directly Fired Heat Engines, First,* Castine, Maine, July 31–August 3, 1979.

[29]A. S. Ginsberg; Fig. 1117 in Phase Diagrams for Ceramists. Edited by E. M. Levin, C. R. Robbins, and H. F. McMurdie. The American Ceramic Society, Columbus, Ohio, 1964.

[30]R. A. Miller, "Analysis of the Response of a Thermal Barrier Coating to Sodium- and-Vanadium Doped Combustion Gases," NASA TM-79205, June 1979.

[31]J. E. Palko, K. L. Luthra, and D. W. McKee, "Evaluation of Performance of Thermal Barrier Coatings Under Simulated Industrial/Utility Gas Turbine Conditions," Final Rept., General Electric Co., 1978.

[32]R. Barkalow and F. Pettit; p. 704 in Ref. 28.

[33]I. Zaplatynsky, "Reactions of Yttria-Stabilized Zirconia with Oxides and Sulfates of Various Elements," DOE/NASA/2593-78/1, NASA TM-78942, 1978.

[34]P. E. Hodge *et al.*, "Thermal Barrier Coatings: Burner Rig Hot Corrosion Test Results," DOE/NASA/2593-78/3, NASA TM-79005, 1978.

[35]K. W. Lay, "Ash in Gas Turbines Burning Magnesium-Treated Residual Fuel," Paper No. 73-WALCD-3, ASME Winter Annual Meeting, Detroit, Michigan, Nov. 11–15, 1973.

Table I. Thermal barrier coating systems

Oxide* thermal barrier	Coating† description	Oxide phase
1. $ZrO_2 \cdot 8Y_2O_3$	Duplex Two layers: 5 mil NiCrAlY bond coat 15 mil oxide overcoat	95 Tetragonal/cubic 5 monoclinic
2. $ZrO_2 \cdot 15Y_2O_3$		Tetragonal/cubic
3. $ZrO_2 \cdot 20Y_2O_3$		Tetragonal/cubic
4. $ZrO_2 \cdot 24.65MgO$		Tetragonal/cubic and free MgO
5. Ca_2SiO_4		Ca_2SiO_4
1. $ZrO_2 \cdot 8Y_2O_3$	Graded— Three layers: 4 mil NiCrAlY 8 mil graded zone 8 mil oxide overcoat	Tetragonal/cubic monoclinic
2. $ZrO_2 \cdot 15Y_2O_3$		Tetragonal/cubic
3. $ZrO_2 \cdot 20Y_2O_3$		Tetragonal/cubic
4. $ZrO_2 \cdot 24.65MgO$		Tetragonal/cubic and free MgO

*Nominal oxide compositions in wt%. †Nominal NiCrAlY compositions (wt%): Ni-20Cr-11Al-0.4Y.

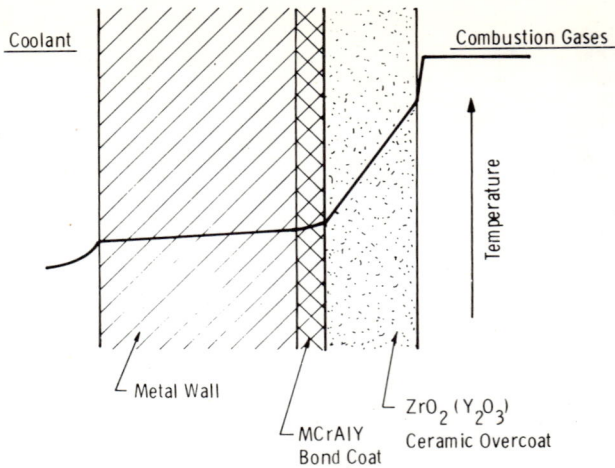

Fig. 1. Schematic representation of the duplex thermal barrier coating.

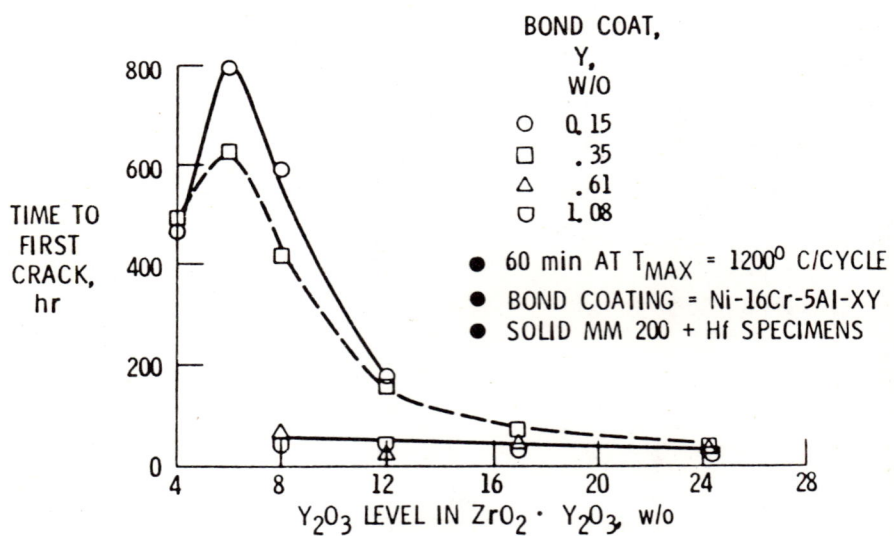

Fig. 2. TBC cracking in natural gas torch tests varied with oxide and bond coat composition (Refs. 23 and 24).

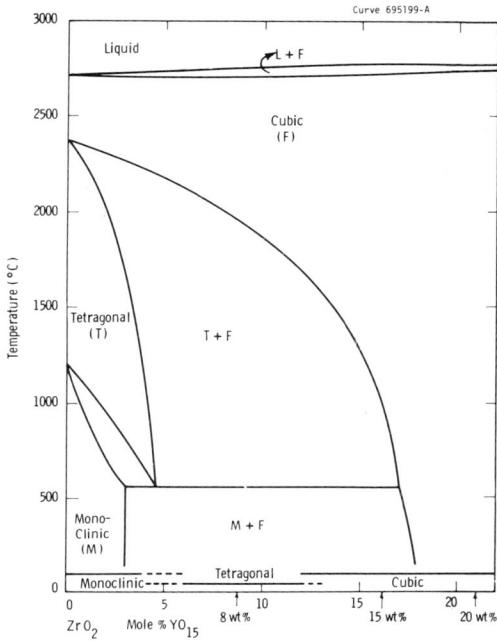

Fig. 3. Phase diagram for the zirconia-rich region of the ZrO_2-Y_2O_3 system after Ref. 25.

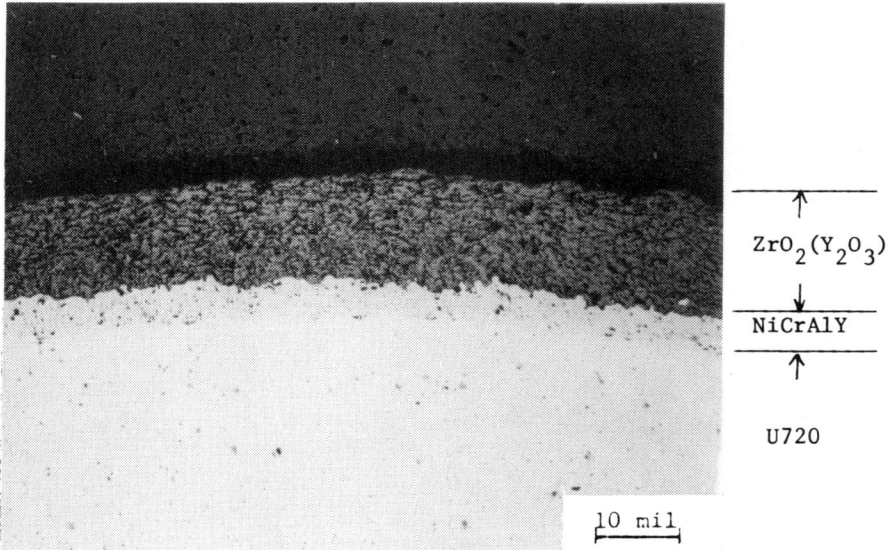

Fig. 4. Metallographic cross section of the duplex $ZrO_2 \cdot 15Y_2O_3$ coating.

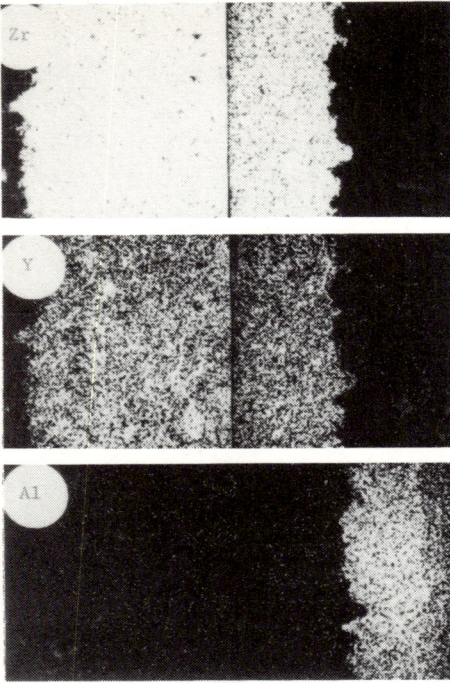

Fig. 5. Electron microprobe scans from the duplex $ZrO_2 \cdot 15Y_2O_3$ specimen.

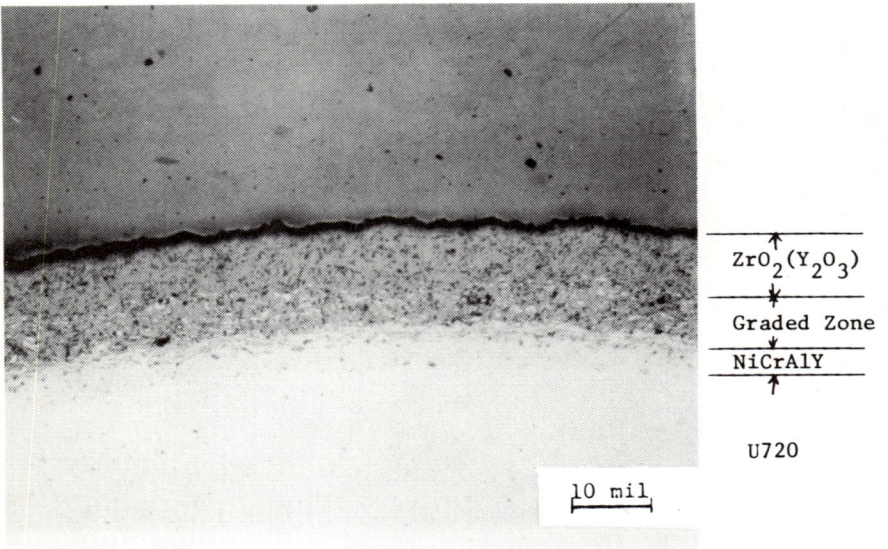

$ZrO_2(Y_2O_3)$

Graded Zone

NiCrAlY

U720

10 mil

Fig. 6. Metallographic cross section of the graded $ZrO_2 \cdot 15Y_2O_3$ coating.

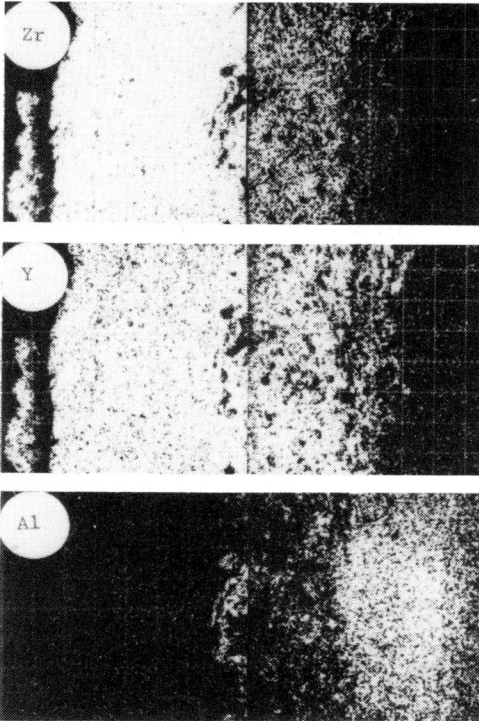

Fig. 7. Electron microprobe scans from the graded ZrO_2-$15Y_2O_3$ specimen.

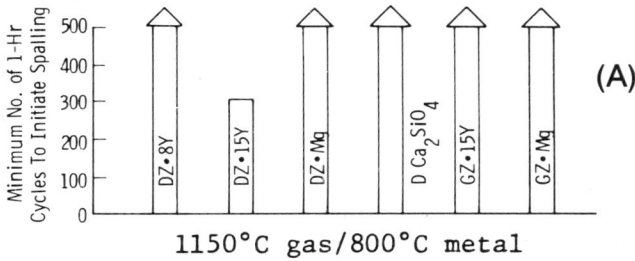

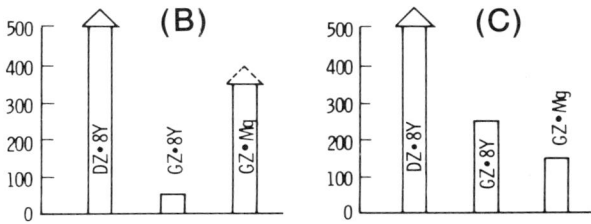

(A) 1150°C gas/800°C metal

(B) 1150°C gas/900°C metal

(C) 1150°C gas/845°C metal

Fig. 8. Cycles to failure in a 500-h cycle burner rig test using clean fuel (GT No. 2).

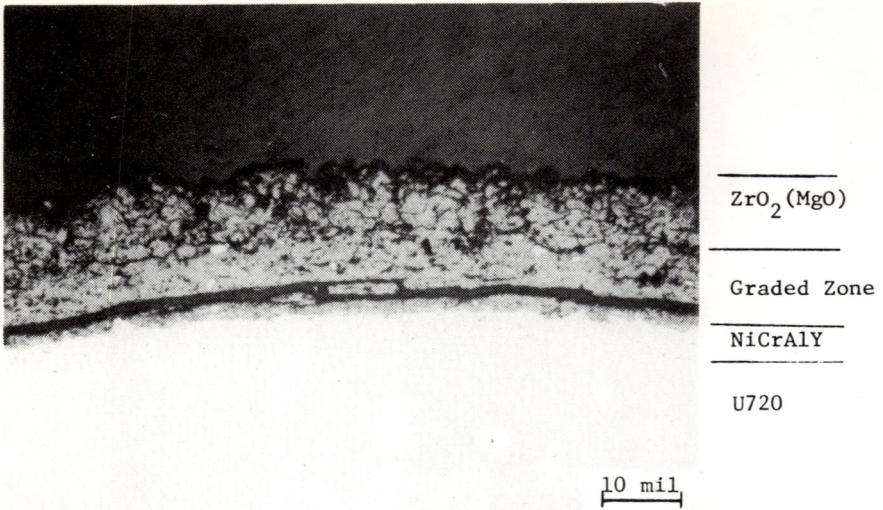

ZrO$_2$(MgO)

Graded Zone

NiCrAlY

U720

10 mil

Fig. 9. Graded ZrO$_2$·24.65 MgO coating tested at 1150 °C gas/900 °C metal using GT No. 2 fuel (350 cycles).

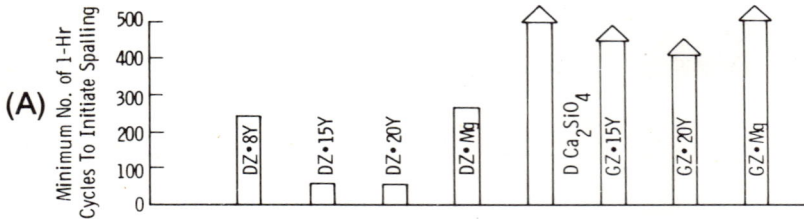

Test Temperature: 1150°C gas/800°C metal
Fuel: GT No. 2 doped to (ppm) 1-Na, 2-V, 2-P, 0.5-Ca, 2-Fe, 6-Mg

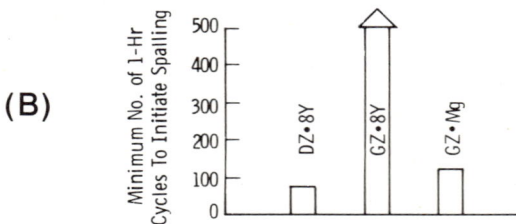

Test Temperature: 1150°C gas/800°C metal
Fuel: GT No. 2 doped to (ppm) 1-Na, 50-V, 2-P, 0.5-Ca, 2-Fe, 150-Mg

Fig. 10. Fuel sensitivity test results.

238

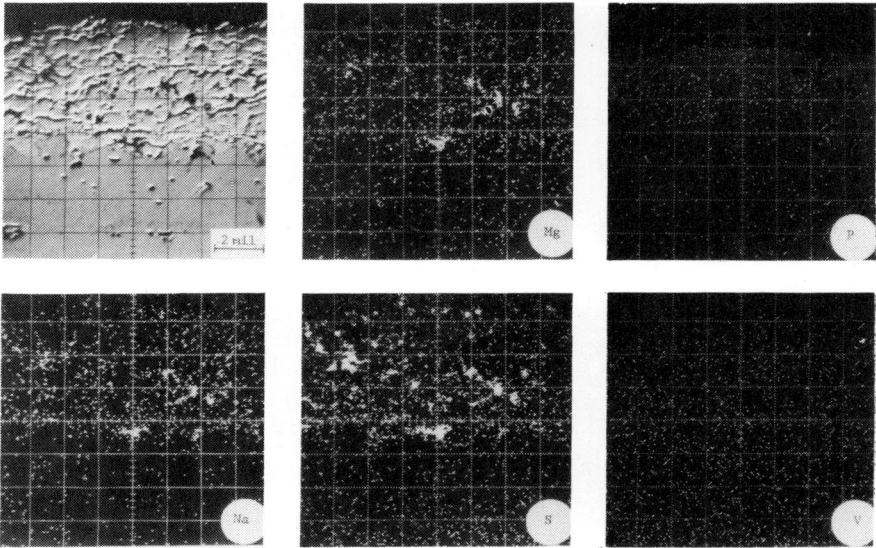

Fig. 11. Electron microprobe scans of failed duplex $ZrO_2 \cdot 20Y_2O_3$ specimen. Test temperature, 1150° gas/800°C metal; fuel, GT No. 2 doped to (ppm) 1 Na, 2 V, 2 P, 0.5 Ca, 2 Fe, and 6 Mg; failure time, 59 1-h cycles.

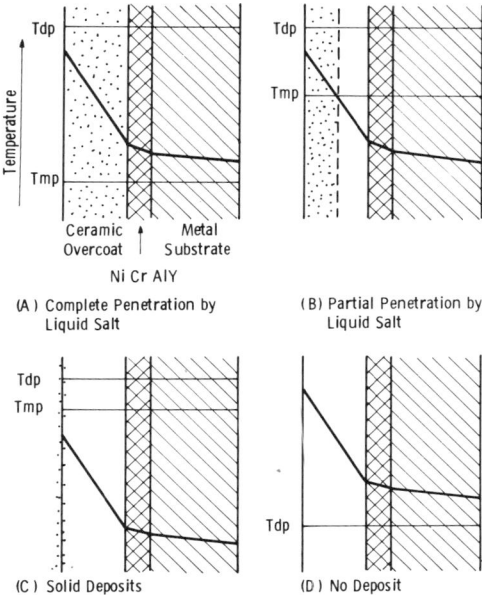

Fig. 12. Schematic representation of relationships between (T_{dp}), melting point (T_{mp}), and nature of deposit.

239

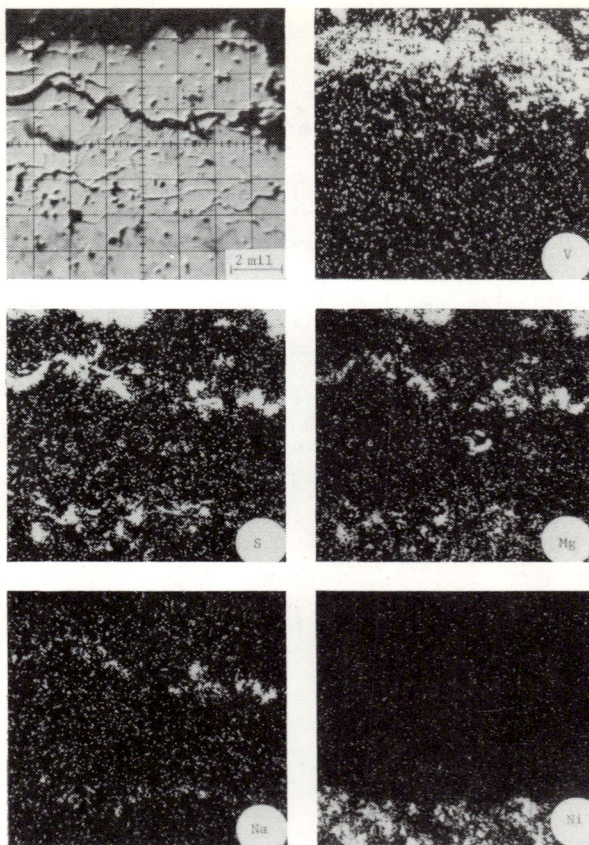

Fig. 13. Electron microprobe scans on unfailed graded $ZrO_2 \cdot 8Y_2O_3$ specimen. Test temperature, 1150°C gas/800°C metal; fuel, GT No. 2 doped to (ppm) 1 Na, 50 V, 2 P, 0.5 Ca, 2 Fe, and 150 Mg; test time, 500 1-h cycles.

Phase stability in plasma-sprayed, partially stabilized zirconia-yttria

ROBERT A. MILLER, JAMES L. SMIALEK, AND RALPH G. GARLICK

National Aeronautics and Space Administration
Lewis Research Center
Cleveland, Ohio 44135

The phase stability of plasma-sprayed ZrO_2-8.6%$YO_{1.5}$ was examined by X-ray diffraction. The amounts of tetragonal, cubic, and monoclinic phases were determined from corrected intensity ratios in the {400} and {111} regions of the diffraction patterns. The extent of {400} peak splitting was used to estimate the composition of the tetragonal phase. The as-sprayed material was primarily a nontransformable tetragonal phase having about the same composition as the starting powder. Aging at 1200° and 1400°C progressively increased the amounts of equilibrium cubic and transformable tetragonal phases at the high temperatures and this resulted in cubic and monoclinic phases at room temperature. The yttria content of the remaining nontransformable tetragonal phase was lowered to ≈ 5% after 100 h at 1400°C. Aging at 1600°C for 100 h also increased the amount of the equilibrium cubic phase at the high temperature, but fast-cooling transformed this phase into a tetragonal phase having the same yttria content, ≈ 12.5%, as the parent cubic phase. The results are discussed with reference to a published ZrO_2-$YO_{1.5}$ phase diagram which takes account of both the diffusion-controlled and diffusionless phase transformations.

Plasma-sprayed zirconia-yttria ceramics are currently being developed for use as thermal barrier coatings for gas turbine airfoils and other heat engine components.[1] Such coatings can reduce the metal temperature of the air-cooled components 50° to 200°C over current levels. Through such reductions, component durability may be improved. Alternatively, current metal temperatures may be maintained as gas temperatures are increased or cooling air requirements are relaxed, thereby improving engine efficiency.

The most durable coatings are those formed from compositions taken from the two-phase (or partially stabilized) region of the zirconia-yttria equilibrium phase diagram.[1,2] This diagram predicts a phase separation during cooling into a high-yttria cubic (fluorite) phase, F, and a low-yttria tetragonal phase, T, between $\approx 2300°$ and 600°C.[3] With further cooling to room temperature this equilibrium low-yttria tetragonal phase is expected to transform martensitically to the monoclinic structure, which has a lower free energy at lower temperatures. Under certain conditions the tetragonal phase may be retained metastably at room temperature.[4-8] In partially stabilized plasma-sprayed zirconia-yttria coatings,[2] as in quenched bulk material,[3] the main phase observed at room temperature is a nonequilibrium, high-yttria tetragonal phase, T'. This phase differs from the metastable low-yttria tetragonal phase because it is stable with respect to the monoclinic transformation.[6] This phase forms because the quenching process prevents compositional adjustments required for development of equilibrium amounts of the low-yttria tetragonal and high-yttria cubic phases. Thus only minor amounts of those phases are observed.[2] This T' phase may be denoted as "non-transformable" because, unlike the T phase, it will not transform when quenched to room temperature.

The T' phase is unstable with respect to diffusion at high temperatures. Therefore, additional phase separation into the equilibrium high-yttria cubic and low-yttria tetragonal phases can be expected with high temperature exposure. On cooling, a large volume expansion is associated with the transformation of the high temperature, low-yttria, tetragonal structure to the low temperature monoclinic structure.[9] Because of stresses associated with the volume change, the formation of increasing amounts of the "transformable," low-yttria tetragonal phase should be detrimental to the integrity of the plasma-sprayed coating.

Thus the purpose of the present investigation is to characterize the thermal stability of a plasma-sprayed coating, ZrO_2-8.6%$YO_{1.5}$ (in mol%), whose composition is in the partially stabilized two-phase field. The phases present after spraying and after aging at temperatures typical of the surface temperature of present and future coated components are quantitatively determined by X-ray diffraction.

Experimental Procedure

Commercially prepared ZrO_2-8.6%$YO_{1.5}$ powder was used as the starting material. This "standard" purity material contains 1.6% Hf by weight plus other minor impurities as reported in Ref. 1. Thin (≈ 0.04 cm) wafers of this ceramic were prepared by plasma spraying directly onto smooth stainless steel strips which were then flexed to remove the coating. Chips of these coatings were aged in air in a rapid-response electric furnace at 1200°, 1400°, and 1600°C for 1, 10, and 100 h. After aging, the specimens were allowed to cool in the furnace. The furnace cooling rate was relatively rapid; for example, cooling from 1600° to 1400°C occurred in 2.1 min. Specimens were analyzed for phase composition using X-ray diffraction techniques, primarily diffractometer scans. The X-ray analysis concentrated on the "top" side of the specimen; that is, the side that had faced the spray gun. The as-sprayed specimen and the specimen aged for 100 h at 1400° and 1600°C were also analyzed after grinding to powder.

Two regions of the diffractometer scans were analyzed extensively. In both regions, the peaks were scanned at the rate of 1/8° 2θ/min using an 8 s time constant. In the first region, the ratio of the area under the $(11\bar{1})$ plus (111) monoclinic peaks to the area under the (111) cubic plus nearly coincident (111) tetragonal peaks was determined using a planimeter. In the other region, the ratio of the area under the (400) cubic peak to that of the (400) plus (004) tetragonal peaks was determined. Because this cubic peak is centered between the two tetragonal peaks a curve resolver was used to separate the three overlapping peaks. This instrument allows the component peaks, represented by symmetric or skewed Gaussian distributions, to be systematically varied in position, width, and intensity until the summation curve coincides with the experimental curve. Relative peak areas are read from an integrating meter on the curve resolver. A preliminary transmission electron microscopy (TEM) analysis was also conducted on several specimens.

Results and Discussion
As-Sprayed ZrO_2-8.6% $YO_{1.5}$

The diffraction pattern in the (400) region for the as-sprayed material is presented at the top of Fig. 1(A). At the bottom of Fig. 1(A) the pattern is resolved into two tetragonal peaks and an intermediate cubic peak. Since the peaks are broad, skewed Gaussians were used to represent the $K\alpha_1$ and $K\alpha_2$ components of each peak. The area under the (400) tetragonal peak was found to be 3.0 times the area under the (004) peak, which is in excellent agreement with the theoretical value calculated in the Appendix. This value was used for all subsequent curve analyses. The area under the cubic peak relative to the area of all three peaks (i.e. $I_F/I_{F,T'}$) and the angular separation between the two tetragonal peaks $(\Delta(2\theta)_{T'})$ are given in Table I. The ratio of the area under the (11$\bar{1}$) plus (111) monoclinic peaks to the cubic plus tetragonal (111) peaks is also presented in the table. This ratio, when measured on the bottom side of the specimen, which is the side expected to have received the fast quench, was about one-half the ratio measured on the top side. When the specimen was ground to powder, a value which was intermediate between the top and bottom values was obtained. This indicates that grinding did not increase the monoclinic intensity of the as-sprayed specimen.

The intensity ratios in Table I were converted to approximate mole fractions by the method described in the Appendix. Also, d values, c/a ratios, and the composition of the high-yttria, nontransformable tetragonal phase were calculated from peak positions as described in the Appendix. These calculated values are presented in Table II. As given in this table, 80% of the as-sprayed material is in the tetragonal phase having an yttria content equal (within experimental error estimated to be about $\pm 1\%$ $YO_{1.5}$) to the bulk composition of the starting material. The existence of this nontransformable tetragonal phase in plasma-sprayed ZrO_2-8.6% $YO_{1.5}$ is consistent with the phase diagram of Ref. 3. This version of the phase diagram, presented in Fig. 2, was selected because it includes both the equilibrium phases and the phases observed after quenching from the liquid or the single-phase cubic field. The quenched phases are indicated by the bars at the bottom of the diagram. An "X" drawn in the liquid phase region represents ZrO_2-8.6% $YO_{1.5}$ shown ideally as a homogeneous liquid after the material has been melted by the plasma spray gun. The molten material cools very rapidly both in transit to and after striking the substrate. The cooling process is indicated by the vertical cooling line originating from the liquid phase field. This cooling line intersects the horizontal bar corresponding to the tetragonal phase, which is the phase having the lowest free energy for the diffusionless, or nearly diffusionless, transformation.[6]

Some transformable tetragonal zirconia also existed, as shown by a small amount of monoclinic (8% in Table II) at room temperature. The origin of the transformable tetragonal phase (observed as monoclinic after cooling) in the as-sprayed material is unclear. Its presence could result from the plasma spray process or from compositional inhomogeneities in the starting powders. Possibly those powder particles that are injected into the cooler edge region of the plasma arc gas do not melt. Thus some of the monoclinic phase originally present in the as-received material may be retained. The relative monoclinic intensity in the as-received material was measured to be 0.49 which corresponds to a 0.38 mole fraction. If the existence of the monoclinic phase in the as-sprayed material is due to the spray process, then the amount observed must be very sensitive to process parameters and starting material characteristics.

The as-sprayed nontransformable tetragonal phase (T') was stable toward grinding. This is because this tetragonal phase can transform only after its yttria level is lowered by a diffusion-controlled phase separation at high temperature. Grinding alone cannot cause the necessary diffusion. This is in contrast to the metastable transformable tetragonal phase which is observed as the major phase for very low-yttria compositions,[7] where the mechanical energy from grinding is sufficient to cause the diffusionless, martensitic transformation to the equilibrium monoclinic phase.

Aged Specimens

The experimentally measured quantities and derived parameters for specimens aged at 1200°, 1400°, and 1600°C are also presented in Tables I and II, respectively. The X-ray diffraction patterns and the curve-resolved tracings for specimens aged for 100 h at each of the three temperatures are shown in Figs. 1 $(B-D)$. The effect of aging on the phase constituents and the composition of the nontransformable tetragonal phase is shown in Fig. 3.

(A) 1200° and 1400°C aging: At 1200°C, 100 h was required before a significant increase (75%) in the relative amount of the cubic phase was observed (Fig. 1(B) and Table II). This was accompanied by a decrease in the yttria content of the remaining nontransformable tetragonal phase and a minor increase in the amount of transformable tetragonal phase (which is observed as the monoclinic phase at room temperature). After only 1 h of aging at 1400°C, the extent of the transformation was comparable to that observed after 100 h at 1200°C. Further phase separation was observed after 10 h, and by 100 h over half of the specimen was cubic (Fig. 1(C) and Table II). The remaining nontransformable tetragonal phase after 100 h contained only $\approx 5\%$ yttria and the fraction of the monoclinic phase had increased to 0.19. Grinding this material did not cause the monoclinic intensity to increase. This is in agreement with Ref. 10 where grinding partially stabilized zirconia-yttria which had been aged at 1300°C did not cause the monoclinic intensity to increase.

The decomposition of the nontransformable tetragonal phase into cubic and transformable tetragonal phases at 1400°C and below are consistent with the zirconia-yttria phase diagram[3] shown in Fig. 2. For example, when heated to 1400°C, the as-sprayed, predominantly tetragonal material is expected to gradually revert to the two phases given by the 1400°C tie line. Upon rapid cooling, the cubic phase will be retained, because the vertical cooling line intersects the horizontal bar corresponding to the cubic phase. However, the vertical cooling line for the low-yttria tetragonal phase intersects the horizontal bar corresponding to the monoclinic phase, thus indicating that the martensitic transformation may proceed.

The amount of the transformable tetragonal phase formed at 1400°C, even after the 100 h aging, remains well below the equilibrium value, i.e. about 0.2 vs 0.5. Note also that as the specimens are aged there is a gradual decrease in the yttria level of the nontransformable tetragonal zirconia. After

100 h at 1400 °C a signficant fraction (0.26 in Table II) of moderately low-yttria, nontransformable tetragonal zirconia remains.

(B): 1600 °C aging: The overall peak width obtained after aging for 100 h at 1600 °C, presented in Fig. 1(*D*), is narrower than those obtained at lower temperatures. In this case the tetragonal peaks are closely spaced and a minor cubic peak is located between the tetragonal peaks. For this analysis, instead of using asymmetric peaks, each tetragonal and cubic peak is represented as the sum of two symmetric Gaussian peaks representing the $K\alpha_1$ and $K\alpha_2$ contributions. Because these peaks are closely spaced and overlapping there is greater uncertainty in the determination of the intensity ratios and angular separations.

The patterns obtained after aging for 1 and 10 h were somewhat broader overall but otherwise similar to the pattern obtained after 100 h. Quantitative analysis of the 1 and 10 h specimens was hindered by a poor signal-to-noise ratio caused by small sample size.

The mole fraction of the monoclinic (nee tetragonal) phase increases on aging to ≈ 0.14 at 100 h. However, this value is still much lower than the equilibrium value of 0.4 predicted by the phase diagram. When this material was ground, the relative monoclinic intensity increased from 0.19 to 0.31 which corresponds to an increase in the mole fraction of the monoclinic phase to 0.20. This indicates that the metastable transformable tetragonal phase was present and lessens the discrepancy between the present value and that predicted by the phase diagram. However, the difference (0.2) after the 100 h aging is still significant.

According to the phase diagram in Fig. 2, the equilibrium phases at the ends of the 1600 °C tie line are a high-yttria ($\approx 12.6\%$) cubic phase and a low-yttria ($\approx 3.1\%$) tetragonal phase. As with the 1400° specimen, the vertical tetragonal cooling line intersects the bar corresponding to the monoclinic phase. The yttria content of the cubic phase is sufficiently low that it may now convert to a high-yttria tetragonal phase on rapid cooling (Fig. 2). The yttria composition of this phase, $\approx 12.5\%$ (Table II), agrees very well with the phase diagram value.

The transformation from cubic to tetragonal must be extremely rapid and diffusionless, involving primarily a shift in the positions of the oxygen ions. These slight shifts, when based in terms of a face centered unit cell, involve four oxygen positions of the type $(u, v, w) = (\pm \frac{1}{4}, \frac{1}{4}, \frac{1}{4} \pm x)$ and $(\pm \frac{1}{4}, \frac{1}{4}, \frac{3}{4} \pm x)$ where x changes from 0 to 0.065 in going from cubic to tetragonal symmetry.[11]

The temperature effects of the 100 h aging treatments are summarized graphically in Fig. 3, which illustrates that the as-sprayed material is primarily nontransformable tetragonal, having a composition equal to the overall composition of the material. When heated at 1400 °C or below, the amount and yttria composition of this phase decrease as the equilibrium cubic and monoclinic phases increase in amount. After heating at 1600 °C, most of the equilibrium cubic phase formed at temperature converts by a diffusionless transformation to a higher yttria tetragonal phase on rapid cooling.

The results of the preliminary transmission electron microscopy study are in agreement with the above X-ray results. Large (1–5 μm) areas of the tetragonal phase were found in the as-sprayed material as indicated by the existence of (112) tetragonal reflections. The cubic phase existed as a submicron fine-grained material. After aging at 1400 °C for 100 h, the fine-grained cubic

245

structure ((112) reflections absent) became more widespread. After aging at 1600 °C for 100 h, the tetragonal phase again predominated. It existed as fine 20–40 nm rectangular precipitates as revealed by (112) dark field imaging. A very fine mottled structure was associated with all of the tetragonal areas. The monoclinic phase was observed for all treatments as large 1–5 μm heavily twinned areas.

Implications for Coating Performance

Surface temperatures up to 1245 °C were calculated for thermal barrier coated turbine blades in a laboratory version of an advanced engine operated at turbine inlet temperatures of 1340 °C.[12] Future engines will run at even higher temperatures. Thus the phase transformations described in this paper will play an important role in the long term durability of zirconia-yttria thermal barrier coatings. A particularly important concern is the possible adverse effects of a large increase in the amount of the transformable tetragonal phase. That is, while a 0.1 mole fraction of this phase may be beneficial, or at least tolerable, an increased amount may cause detrimental cracking of the coating due to the martensitic transformation.

There are some factors which tend to lessen the severity of thermal degradation of the coating. First, the nontransformable tetragonal phase decomposes rather slowly, especially at lower temperatures. Second, only the outer surface of the coating is exposed to the highest temperatures and temperatures drop through the coating thickness. Thus most of the coating may be able to avoid the increased stresses brought about by the transformation.

The reason for the superior durability of partially stabilized zirconia-yttria coatings has not yet been established. Possibly, expansion of the small fraction of transformable tetragonal precipitates, as they convert to the monoclinic phase on cooling, acts to create a highly favorable microcrack network. Microcracked structures are, in general, common to all plasma-sprayed ceramics. However, the structure arising from the partially stabilized compositions may be especially favorable. Other toughening models such as precipitation "hardening" cannot yet be ruled out.

A stress-induced transformation toughening model has been used to explain the high fracture toughness of a very low-yttria 100% tetragonal zirconia-yttria material.[6,8] This model cannot apply to compositions richer in yttria where the tetragonal phase does not transform under stress. Transformation toughening in partially stabilized ZrO_2-CaO[4] and ZrO_2-MgO[5] is thought to be due to stress-induced transformation of retained metastable tetragonal precipitates. There is no evidence that this model applies for as-sprayed ZrO_2-8.6%$YO_{1.5}$ but the possibility that this model could play a role in the material aged at 1600 °C requires further investigation.

Since the major phase in the as-sprayed material is the nontransformable, quenched tetragonal material, we suggest that the intrinsic properties of this phase may play a key role in the success of partially stabilized ZrO_2-$YO_{1.5}$ coatings.

Conclusions

A nonequilibrium tetragonal phase forms when partially stabilized zirconia-yttria is plasma-sprayed. This phase forms by a diffusionless process during the rapid quench associated with the plasma spray process. As a result the yttria content of this phase is near the overall yttria content of the material, the c/a ratio is close to unity, and the phase is stable with respect to the monoclinic phase. This phase decomposes slowly at high temperatures via cation diffusion into a high-yttria cubic phase and a low-yttria tetragonal phase. This latter tetragonal phase transforms on cooling to the monoclinic phase. The cubic phase is either retained on cooling or it transforms to a new tetragonal phase, depending on the aging temperature.

Thus, the nonequilibrium makeup of plasma-sprayed thermal barrier coatings must be recognized if the behavior of these coatings is to be fully understood. Furthermore, the decomposition of the nonequilibrium tetragonal phase into the equilibrium phases can be expected to play an important role in long term coating durability.

Appendix

Calculation of Mole Fraction from Intensity Ratios and Composition of the Tetragonal Phase from Peak Separation

The intensity ratios in Table I were converted to approximate mole ratios using the following expressions. The ratio of the moles of the monoclinic phase (M) to moles of the cubic (F) plus tetragonal (T') phases is:

$$\frac{M_M}{M_{F,T}} = 0.82 \frac{I_M\,(11\bar{1}) + I_M\,(111)}{I_{F,T}\,(111)} \tag{1}$$

This expression is based on Eq. (A-5) in Ref. 5, modified using the ratio of densities to give mole ratio rather than volume ratio. The ratio of moles of the cubic phase to moles of the tetragonal phase is given by:

$$\frac{M_F}{M_{T'}} = 0.88 \frac{I_F\,(400)}{I_{T'}\,(400) + I_{T'}\,(004)} \tag{2}$$

This expression was obtained from the same type of analysis used to obtain Eq. (1), using structure factors calculated as described in Ref. 13 from ion position parameters based on those from Ref. 11. The {400} indices pertain to a unit cell which is face-centered tetragonal, after having been transformed from the body-centered tetragonal cell of Ref. 11. The ratio of the densities of the cubic and tetragonal cells was estimated from the d values given in Table II. This ratio is very close to unity. The mole fractions of the three phases were calculated from the ratios in Eqs. (1) and (2) and by assuming that the total mole fraction was unity. Metastable, transformable tetragonal phase such as was observed after aging at 1600 °C for 100 h was not considered in constructing Table II. (This metastable transformable tetragonal phase requires further attention.)

The ratio of the intensity of the (400) tetragonal peak to the (004) tetragonal peak can also be calculated as twice the ratio of the appropriate structure factors squared times a very small Lorentz polarization correction. The calculated ratio is 3.15.

The d values for the (400) and (004) tetragonal peaks were calculated from the curve-resolved peak separation. The absolute value of the tetragonal peak locations was obtained by assuming a fixed position of 73.79° ($d = 0.1283$ nm) for the cubic peak of each pattern. This value for the cubic (400) reflection was measured by Guinier-deWolff and Debye-Scherrer techniques for the specimens aged 100 h at 1400 °C and ground to powder. As shown in the top of Fig. 1 of Ref. 3, the d value for the (400) cubic reflection is relatively insensitive to small changes in yttria composition.

The c/a ratio was taken to be the ratio of the calculated tetragonal d values. These were converted to the yttria composition of the tetragonal phase using the following expression based on Fig. 1 of Ref. 3:

$$\%\,YO_{1.5} = (0.10223 - c/a)/0.001309 \tag{3}$$

The calculated yttria composition is quite sensitive to small changes in the c/a ratios. Thus the errors in this value are relatively large, probably on the order of $\pm 1\%\,YO_{1.5}$.

References

[1]S. Stecura, "Effects of Compositional Changes on the Performance of a Thermal Barrier Coating System," NASA TM-78976, 1978.
[2]R. J. Bratton and S. K. Lau, "Zirconia Thermal Barrier Coatings"; this volume,pp. 226–40.
[3]H. G. Scott, "Phase Relationships in the Zirconia-Yttria System," *J. Mater. Sci.,* **10**, 1527–35 (1975).
[4]R. C. Garvie, R. H. Hannink, and R. T. Pascoe, "Ceramic Steel?" *Nature (London)*, **258**, 703–704 (1975).
[5]D. L. Porter and A. H. Heuer, "Microstructural Development in MgO-Partially Stabilized Zirconia (Mg-PSZ)," *J. Am. Ceram. Soc.,* **62** [5-6] 298–305 (1979).
[6]C. A. Andersson and T. K. Gupta, "Martensitic Transformation Toughening in Zirconia"; this volume, pp. 184–201.
[7]T. K. Gupta, J. H. Bechtold, R. C. Kuznicki, and L. H. Cadoff, "Stabilization of Tetragonal Phase in Polycrystalline Zirconia," *J. Mater. Sci.,* **12**, 2421–26 (1977).
[8]T. K. Gupta, F. F. Lange, and J. H. Bechtold, "Effect of Stress-Induced Phase Transformation on the Properties of Polycrystalline Zirconia Containing Metastable Tetragonal Phase," *ibid.,* **13**, 1464–70 (1978).
[9]F. H. Brown, Jr., and P. Duwez, "The Zirconia-Titania System," *J. Am. Ceram. Soc.,* **37** [3] 129–32 (1954).
[10]R. H. J. Hannink, "Growth Morphology of the Tetragonal Phase in Partially Stabilized Zirconia," *J. Mater. Sci.,* **13**, 2487–96 (1978).
[11]G. Teufer, "The Crystal Structure of Tetragonal ZrO₂," *Acta Crystallogr.,* **15**, 1187 (1962).
[12]W. R. Sevcik and B. L. Stoner, "An Analytical Study of Thermal Barrier Coated First Stage Blades in a JT9D Engine (Pratt and Whitney Aircraft)," NASA CR-135360, 1978.
[13]M. J. Buerger, Crystal-Structure Analysis. John Wiley and Sons, Inc., New York, 1960; pp. 259–62.

Table I. Experimental intensity ratios for ZrO_2-8.6% $YO_{1.5}$

Aging temp. (°C)	Aging time (h)	$\dfrac{I_M(11\bar{1}) + I_M(111)}{I_{F,T}(111)}$	$\dfrac{I_F(400)}{I_{F,T}\{400\}}$	$\Delta(2\theta)_{T'}$ (degrees)
As sprayed	As sprayed	0.11	0.15	0.87
1200	1	0.14	0.15	0.92
1200	10	0.13	0.14	0.91
1200	100	0.15	0.26	1.05
1400	1	0.14	0.31	0.95
1400	10	0.17	0.50	1.24
1400	100	0.28	0.71	1.35
1600	1	0.15		
1600	10	0.21		
1600	100·	0.19	0.26	0.5

Table II. Derived phase compositions for ZrO_2-8.6% $YO_{1.5}$

Aging temp. (°C)	Aging time (h)	M_M	M_F	$M_{T'}$	$d_{T'}(400)$ (nm)	$d_{T'}(004)$ (nm)	c/a for T'	%$YO_{1.5}$ in T'
As-sprayed	As-sprayed	0.08	0.12	0.80	0.12766	0.12896	1.0102	9.2
1200	1	0.10	0.12	0.78	.12763	.12901	1.0108	8.8
1200	10	0.10	0.11	0.79	.12766	.12902	1.0107	8.9
1200	100	0.11	0.21	0.68	.12750	.12907	1.0123	7.6
1400	1	0.10	0.25	0.65	.12760	.12902	1.0111	8.5
1400	10	0.12	0.41	0.47	.12740	.12919	1.0141	6.2
1400	100	0.19	0.55	0.26	.12727	.12928	1.0158	4.9
1600	1	0.11						
1600	10	0.15						
1600	100	0.14	0.20	0.66	.12793	.12868	1.006	12.5

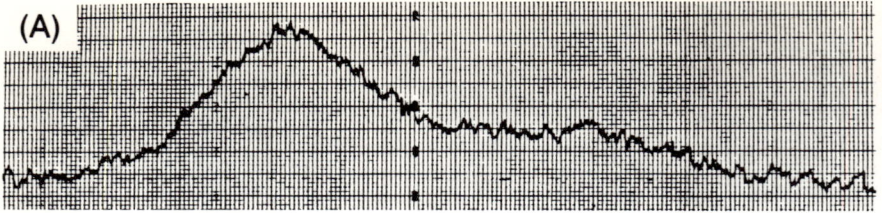

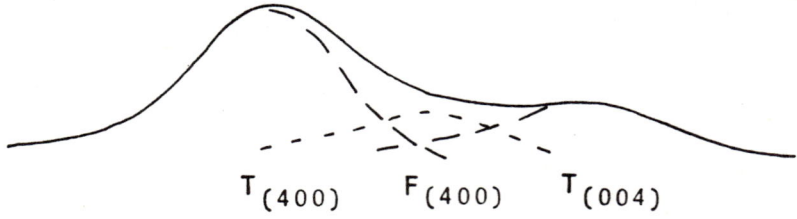

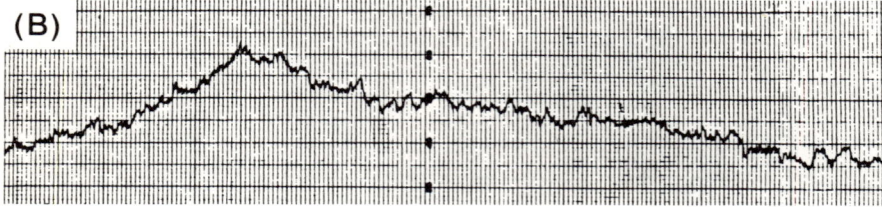

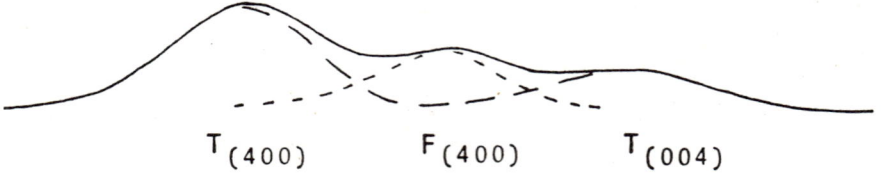

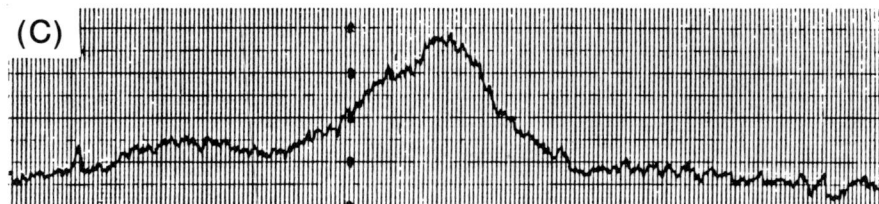

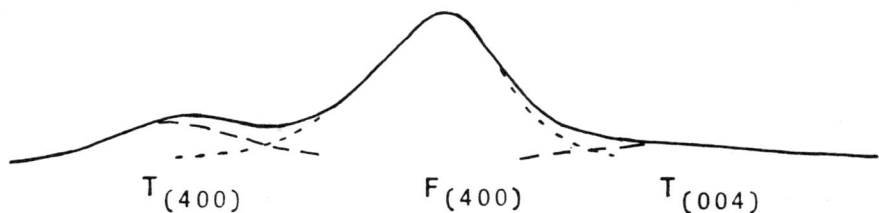

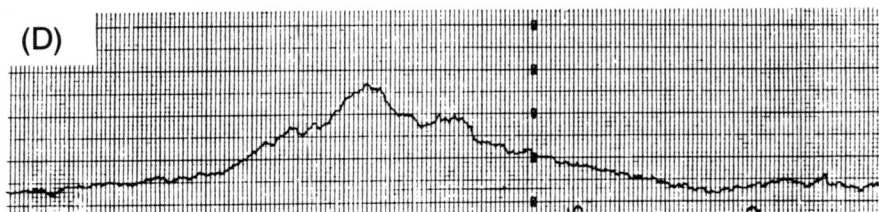

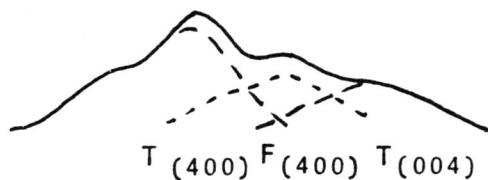

Fig. 1. Experimental X-ray diffraction patterns and resolved peaks in the {400} region for plasma-sprayed ZrO_2-8.6%$YO_{1.5}$. (A) As-sprayed. (B) Aged at 1200°C for 100 h. (C) Aged at 1400°C for 100 h. (D) Aged at 1600°C for 100 h.

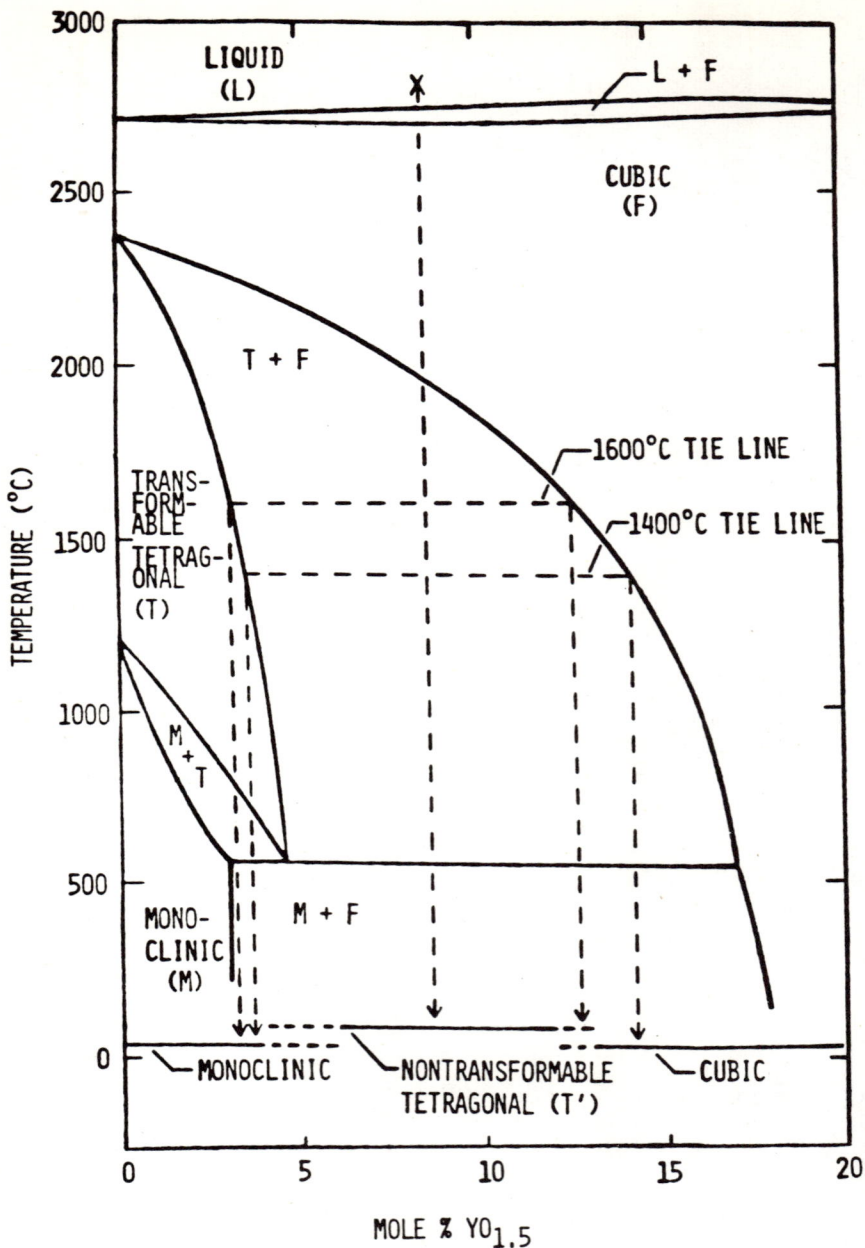

Fig. 2. Low-yttria region of ZrO_2-$YO_{1.5}$ phase diagram (Ref. 3). Horizontal bars indicate phases observed on quenching. Dashed lines indicate tie lines and composition lines for quenched phases.

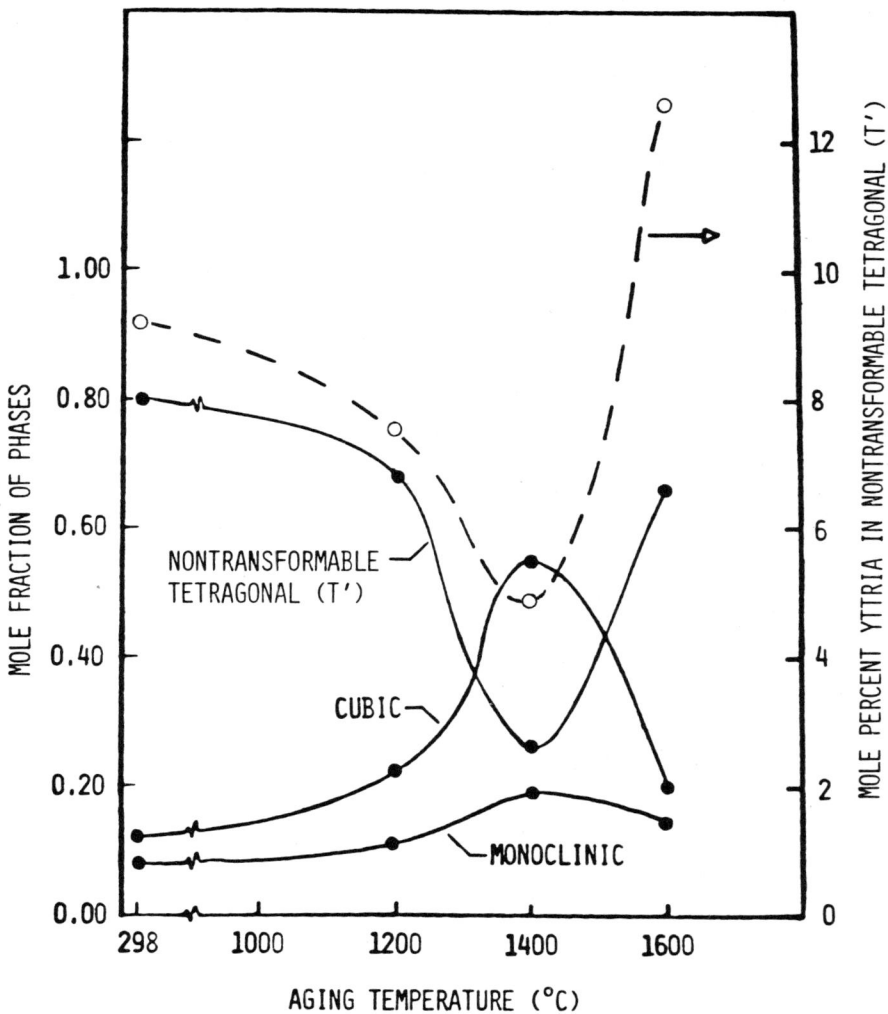

Fig. 3. Measured mole fractions and percent yttria in the non-transformable tetragonal phase for as-sprayed and 100-h aged ZrO_2-8.6%$YO_{1.5}$ specimens.

Transport theory of oxide electrolytes

H. SCHMALZRIED*

Department of Materials Science and Engineering
Cornell University
Ithaca, New York

The behavior of oxide electrolytes ($t_{el} \ll 1$) is investigated theoretically to determine if, under isothermal conditions, electrochemical potential gradients are applied. Point defect thermodynamics is discussed, and the boundary conditions of the electrolytes AO and (A,B)O are systematically varied with regard to the coexisting electrodes. The pertinent equations are evaluated with respect to open circuit and load conditions, and it is shown under which conditions meaningful results can be obtained. Corrections are presented for the reaction of electrodes with the electrolyte and for demixing of (A,B)O electrolytes.

This paper aims to give a systematic survey on electrical dc-conduction behavior of solid electrolytes in situations which occur in practice. This systematic analysis can help in understanding how solid oxide electrolytes may be used in many ways and it can help to avoid incorrect conclusions from electrochemical measurements. An oxide electrolyte is a one-phase, binary or multicomponent crystalline material, the majority anionic constituent being oxygen ions; it exhibits predominantly ionic electrical conduction, at least in a limited range of component activity. In view of this, the following relation should hold for oxide electrolytes:

$$\sigma_{el}/(\sigma_{el} + \sigma_{ion}) = t_{el} \ll 1 \tag{1}$$

σ_{el} is the partial conductivity which results from electronic point defects (i.e. electrons e' and holes h). The σ_{ion} results from all ionic constituents present. Inserting $\sigma_i = z_i F c_i u_i$ into Eq. (1), z_i being the net electrical charge, c_i the molar concentration, and u_i the mobility, Eq. (1) can be written as

$$c_{e'} u_{e'} + c_h u_h \ll \sum_n z_i c_i u_i \tag{2}$$

the summation going over all the n ionic constituents. If one excludes the small contributions from one- and two-dimensional defects, all the ions in the bulk crystal are rendered mobile through point defects. It can be shown (see later) that in terms of jump frequencies (Γ) one has

$$x_i \Gamma_i = x_p \Gamma_p \tag{3}$$

where x denotes the mole fraction of ionic constituents (i) and of those point defects (p) that render the i's mobile. In view of Eq. (3), one can replace the right-hand side of Eq. (2) and then the electrolyte condition reads:

$$c_{e'} u_{e'} + c_h u_h \ll \sum z_p c_p u_p \tag{4}$$

Generally, mobilities of electronic defects are much higher than those of ionic point defects. This gives the necessary conditions for (oxide) electrolytes:

$$c_{el} \ll c_p \tag{5}$$

In that case, ionic point defects are the majority defects, which constitute the disorder type. Two disorder types can occur: (1) thermal (intrinsic) ionic disorder and (2) disorder due to heterovalent doping. Any chemical reaction of the electrolyte component O_2 with structural elements of the crystal results in electronic defects. If, for example, AO is an electrolyte with a Frenkel type of thermal disorder, one can formulate the defect reaction with O_2 either as

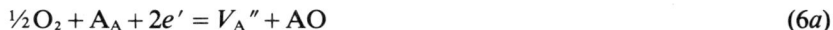

$$\tfrac{1}{2}O_2 + A_A + 2e' = V_A'' + AO \tag{6a}$$

or as

$$\tfrac{1}{2}O_2 + A_i^{..} = V_i + 2\dot{h} + AO \tag{6b}$$

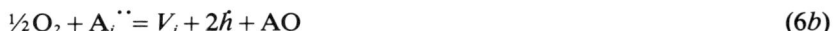

Since $c_{el} \ll c_p$, the chemical potential of the majority ionic defects is not affected by reactions (6a) and (6b) and is therefore independent of $\mu_{O2}(p_{O2})$. One can then read from these equations that, in equilibrium:

$$d\mu_{e'} = -\tfrac{1}{4}d\mu_{O2}; \quad d\mu_{\dot{h}} = \tfrac{1}{4}d\mu_{O2} \tag{7}$$

This is equivalent to

$$x_{e'} x_{\dot{h}} = K_{el} = \exp - \frac{\Delta G_i^{\circ}}{RT} \tag{8}$$

provided that the electronic defects are ideally diluted. As long as all the defects are fully ionized, the difference in the mole fractions of electrons and holes does result in a nonstoichiometric oxide $A_{1-\delta}O$:

$$\delta = x_{e'} - x_{\dot{h}} \tag{9}$$

From Eqs. (6)–(9) it is concluded that

$$\delta = 2\sqrt{K_{el}} \sinh \left[\frac{\mu_{O_2} - \mu_{O_2}^*}{4RT} \right] \tag{10}$$

$\mu_{O_2}^* = \mu_{O_2} (\delta = 0)$. From Eq. (10) it follows

$$x_{e'} (\delta = 0) = x_{\dot{h}} (\delta = 0) = x^* = \sqrt{K_{el}} = 2RT \left(\frac{d\delta}{d\mu_{O_2}} \right)_{\delta = 0} \tag{11}$$

δ, which can be measured by coulometric (or gravimetric) titration, is then given in terms of the oxygen chemical potential:

$$\delta = 4RT \left(\frac{d\delta}{d\mu_{O_2}} \right)_{\delta = 0} \sinh \left[\frac{\mu_{O_2} - \mu_{O_2}^*}{4RT} \right] \tag{12}$$

In the present context it is not necessary to extend these relations in terms of degenerated electronic defects. At concentrations c_{el} where degeneracy occurs, t_{el} is no longer smaller than t_{ion}, and the oxide material cannot be used as an electrolyte. Rearranging Eqs. (7), one finds

$$x_e' = x^* \left(\frac{p_{O_2}}{p_{O_2}^*} \right)^{-1/4} = x^* \exp - \left[\frac{\mu_{O_2} - \mu_{O_2}^*}{4RT} \right] \tag{13a}$$

$$x_{\dot{h}} = x^* \left(\frac{p_{O_2}}{p_{O_2}^*} \right)^{1/4} = x^* \exp - \left[\frac{\mu_{O_2} - \mu_{O_2}^*}{4RT} \right] \tag{13b}$$

$p^*_{O_2}$ being an arbitrarily chosen reference pressure, usually $p^*_{O_2} = p_{O_2}$ ($\delta = 0$). Reinserting Eqs. (13) into Eq. (1), t_{el} is found to be a function of the oxygen potential.

$$\frac{t_{el}}{1 - t_{el}} = \frac{\sigma_{el}}{\sigma_{ion}} = \left(\frac{p_{O_2}}{p_\oplus}\right)^{1/4} + \left(\frac{p_{O_2}}{p_\ominus}\right)^{-1/4} \tag{14a}$$

$$p_\oplus = p^*_{O_2}\left[\frac{\sum_n z_i x_i u_i}{u_{h^.} x^*}\right]^4; \quad p_\ominus = p^*_{O_2}\left[\frac{x^* u_{e'}}{\sum_n z_i x_i u_i}\right]^4 \tag{14b}$$

As can be seen from Eq. (14a), $p_\oplus = p^*_{O_2}$ ($t_{h^.} = \frac{1}{2}$) or $p_\ominus = p_{O_2}$ ($t_{e'} = \frac{1}{2}$), as long as $p_\oplus$ and $p_\ominus$ are sufficiently different. Rearrangement gives

$$\frac{\sigma_{el}}{\sigma_{ion}} = 2\left(\frac{p_\ominus}{p_\oplus}\right)^{1/8} \cosh\left[\frac{\mu_{O_2} - \overline{\mu}_{O_2}}{4RT}\right] \tag{15}$$

with

$$\overline{\mu}_{O_2} = \mu^\circ_{O_2} + \frac{RT}{2}[\ln p_\oplus + \ln p_\ominus] \tag{16}$$

if $u_{h^.} = u_{e'}$, Eq. (14b) shows that $p_\oplus \cdot p_\ominus = p^*_{O_2}{}^2$. In this special case,

$$\frac{\sigma_{el}}{\sigma_{ion}} = \left(\frac{p_\ominus}{p_\oplus}\right)^{1/8}\left[\left(\frac{p_{O_2}}{p^*_{O_2}}\right)^{1/4} + \left(\frac{p^*_{O_2}}{p_{O_2}}\right)^{1/4}\right] = 2\left(\frac{p_\ominus}{p_\oplus}\right)^{1/8}\cosh\left[\frac{\mu_{O_2} - \mu^*_{O_2}}{4RT}\right] \tag{17}$$

Defining $\overline{\mu}_{O_2} = \mu^*_{O_2}$ ($\delta = 0$) $- \Delta\overline{\mu}_{O_2}$, Eq. (15) is rewritten as

$$\frac{\sigma_{el}}{\sigma_{ion}} = \left(\frac{p_\ominus}{p_\oplus}\right)^{1/8}\cosh\left[\frac{\mu_{O_2} - \mu^*_{O_2}}{4RT}\right]\left\{A + B \tanh\left[\frac{\mu_{O_2} - \mu^*_{O_2}}{4RT}\right]\right\} \tag{18a}$$

where

$$A = \cosh\left[\frac{\Delta\overline{\mu}_{O_2}}{4RT}\right]; \quad B = \sinh\left[\frac{\Delta\overline{\mu}_{O_2}}{4RT}\right] \tag{18b}$$

A and B can be written in terms of $u_{e'}$ and $u_{h^.}$. Since $\Delta\overline{\mu}_{O_2} = \mu^*_{O_2} - \overline{\mu}_{O_2} = \frac{1}{2}RT\ln(p_\oplus p_\ominus/p^*_{O_2}{}^2)$, one obtains

$$\Delta\overline{\mu}_{O_2} = 2RT\ln\left(\frac{u_{h^.}}{u_{e'}}\right) \tag{19}$$

and inserting in Eqs. (18b):

$$A = \cosh\left[\frac{1}{2}\ln\left(\frac{u_{h^.}}{u_{e'}}\right)\right]; \quad B = \sinh\left[\frac{1}{2}\ln\left(\frac{u_{h^.}}{u_{e'}}\right)\right] \tag{20}$$

Equations (10) and (18) are the fundamental equations for the treatment of electronic conduction in solid oxide electrolytes. In Fig. 1, σ_{el} is plotted as a function of the oxygen potential. Both coordinates are normalized. It is seen that Eq. (18a) is a somewhat distorted $\cosh$ curve. The minimum is shifted to smaller or larger values of μ_{O_2}, depending on the values of the mobilities of electrons and holes. The shift of the minimum in σ_{el} relative to $\mu^*_{O_2}$ (where the minimum is found if $u_{h^.} = u_{e'}$) namely $\Delta\mu^{\dagger\dagger}_{O_2} = \mu^{\dagger\dagger}_{O_2}$ (min) $- \mu^*_{O_2}$, can be calculated from $d\sigma_{el}/d\mu_{O_2} = 0$. One finds

$$\frac{A}{B}\sinh\left[\frac{\Delta\mu^{\dagger\dagger}_{O_2}}{2RT}\right] + \cosh\left[\frac{\Delta\mu^{\dagger\dagger}_{O_2}}{2RT}\right] + 1 = 0 \tag{21}$$

256

or, in a linearized form as a first order approach:

$$\Delta\mu_{O_2}^{++} = -4RT \; \tanh\left[\tfrac{1}{2}\ln\left(\frac{u_{h'}}{u_{e'}}\right)\right] \tag{22}$$

If $(u_{h'}/u_{e'})<1$, which is normally the case, one expects this minimum to occur at positive values of $\mu_{O2} - \mu_{O2}^{*}$.

Up to this point, partial conductivities and transference numbers have been worked out as a function of the oxygen chemical potential. Application of relations due to Einstein, Nernst, and Planck leads to the various diffusivities. In what follows we have to investigate the behavior of ions and electronic defects under the action of driving forces X_i. In the context of oxide electrolytes, the most important driving force is the gradient of the electrochemical potential $\eta_i = \mu_i + z_iF\phi$, z_i being the electrical charge number and F the Faraday constant. Conceptionally it is convenient to formulate the fluxes in terms of irregular structural elements (point defects p, e', and h) since they are the primary mobile species. Ions are then rendered mobile through the diffusional jumps of ionic point defects p. In a linear approach ($\Delta\eta \ll kT$ per jump) one has

$$_\omega j_p = -\Sigma_{k\omega}L_{pk}X_k \qquad (P,T) \tag{23}$$

ω denotes the reference frame. The most natural frame of reference for structural elements is the crystal lattice. In this frame, the ω index is omitted. If the mobility of electronic defects is large compared to ionic defects, their motion is decoupled ($L_{pe}=0$). In addition, at sufficiently low concentrations of point defects, the individual elementary jumps are independent. Thus

$$j_p = -L_{pp}X_p \tag{24}$$

In one-dimensional experiments, $X_p = (RT)/x_p\;(dx_p/d\xi) + z_p\,F\,d\phi/d\xi$, ξ being the space coordinate. By comparison with Fick's first law, $j_p = -D_p\,(d\phi/d\xi)$, it is seen that

$$L_{pp} = \frac{c_pD_P}{RT} = \frac{\sigma_p}{(z_pF)^2} \tag{25}$$

The right-hand side of Eq. (25) is found if one compares the partial electrical current density $\sigma_p \cdot d\phi/d\xi$ with z_pFj_p. It has already been stated that the ions (on a particular sublattice of an oxide electrolyte) are rendered mobile through the motion of point defects. If there are several sorts of ions A, B. . . on this sublattice, one has (by site conservation):

$$j_p + j_Aj_B + \cdots\cdots = 0 \tag{26}$$

Introducing $j_i = -\sigma_i/(z_iF)^2\,z_iF(d\phi/d\xi)$ in the case that an electrical field acts on the homongeneous crystal, it is seen that $-\sigma_p/z_p = \sigma_A/z_A + \sigma_B/z_B + \cdots\cdots$ For the sake of simplicity we continue with only A ions on the sublattice. It then follows from Eq. (4) that

$$c_pD_p = c_AD_A; \qquad x_pD_p \cong D_A \tag{27}$$

and also that $|d\eta_p| = d\eta_A$. In view of the foregoing, the flux equations of ions and of electrons (holes) in the oxide crystal are formulated as:

$$j_{ion} = -\frac{D_{ion}c_{ion}}{RT}\frac{d\eta_{ion}}{d\xi} = -\frac{\sigma_{ion}}{(z_iF)^2}\frac{d\eta_{ion}}{d\xi} \tag{28}$$

$$j_{el} = -\frac{D_{el}c_{el}}{RT}\frac{d\eta_{el}}{d\xi} = -\frac{\sigma_{el}}{F^2}\frac{d\eta_{el}}{d\xi} \tag{29}$$

In what follows, these flux equations are applied to different situations in which different oxide electrolytes may be used. One notes that L_{el} and σ_{el} in Eqs. (28) and (29) are dependent on the component chemical potentials as deduced in Eq. (15) ff.

It should be mentioned that, in 1975, Wagner[19] most adequately discussed those situations in which the cross coefficients in Eq. (23) ought to be taken into account when treating transport problems in oxide systems. To the author's knowledge, no unambiguous example for $L_{ei} \neq 0$ has yet been found, and for solid electrolytes one would not expect to find them.

Oxide Electrolytes AO in Thermodynamic Potential Gradients

Table I gives a systematic survey of the transport situations that occur with oxide electrolytes. They can all be treated with the transport equations of the previous section provided that the boundary conditions are properly taken into account.

The basic experiment is shown in Fig. 2. The simplest electrolyte AO is brought into an oxygen potential gradient (as is depicted $a/4$, $a/5$, and $a/6$ in Table I). A transport of (neutral) components results. If one disregards the extremely small charge separation needed for the build-up of diffusion potentials, local electrical neutrality must be maintained:

$$\Sigma z_i j_i = 0 \tag{30}$$

The summation goes over all ions and electronic defects. Explicitly one has from Eq. (30):

$$-Fd\phi = t_A\frac{d\mu_{A^{++}}}{z_A} + t_O\frac{d\mu_{O^{--}}}{z_O} + t_{el}\frac{d\mu_{el}}{z_{el}} = t_A\frac{d\mu_A}{z_A} + t_O\frac{d\mu_O}{z_O} + \frac{d\mu_{el}}{z_{el}} \tag{31}$$

since $t_{el} = 1 - (t_A + t_O)$. Inserting Eq. (31) into the flux equations, Eq. (28), it is found that the ionic fluxes are

$$j_{A^{++}} = \frac{\sigma_A t_{el}}{(z_A F)^2}\frac{d\mu_{O_2}}{2d\xi} = -\frac{\sigma_A t_{el}}{(z_A F)^2}\frac{d\mu_A}{d\xi} \tag{32}$$

$$j_{O^{--}} = -\frac{\sigma_O t_{el}}{(z_O F)^2}\frac{d\mu_{O_2}}{2d\xi} = \frac{\sigma_O t_{el}}{(z_O F)^2}\frac{d\mu_A}{d\xi} \tag{33}$$

$$j_{O^{--}} = -\left[\frac{\sigma_O t_{el}}{(z_O F)^2}\frac{d\mu_O}{dc_O}\right]\frac{dc_O}{d\xi} \tag{33a}$$

The bracketed term in Eq. (33a) has the dimension of a diffusion coefficient. It is the chemical diffusion coefficient for all chemical relaxation processes in the case where $D_A \ll D_O$, and it includes the so-called thermodynamic factor ($d\ln a_i/d\ln x_i$).

$$j_{ion} = \frac{\sigma_A + \sigma_O}{8F^2}t_{el}\frac{d\mu_{O_2}}{d\xi} \tag{34}$$

In a simplified form, Eq. (34) may be written as follows:

$$j_{ion} = \frac{\sigma_{ion}\bar{t}_{el}}{8F^2}\frac{\Delta\mu_{O_2}}{\Delta\xi} \tag{34a}$$

$\bar{t}_{el}$ is the average over the oxide electrolyte thickness $\Delta\xi$. However, an exact integration has to make use of the $t_{el}(\mu_{O2})$ equations which have been derived in the first section. Equations (31)–(34) cover the transport situations $a/4$, $a/5$, and $a/6$ of Table I. Since AO is binary and $d\mu_A = -d\mu_O$ (Gibbs-Duhem equation), $a/4$ and $a/6$ give identical results. Note that $j_{O^{--}}$ does not lead to a shift of the AO crystal, whereas $j_{A^{++}}$ induces a shift of the crystal toward the high oxygen pressure side with this velocity:

$$v_{AO} = j_{A^{++}} V_m = \frac{V_m \sigma_A \bar{t}_{el}}{8F^2} \frac{\Delta\mu_{O_2}}{\Delta\xi} \tag{35}$$

Equation (35) may be applied to obtain data on $\sigma_A \cdot \bar{t}_{el}$. The situation designated $a/5$ in Table I is of course the basic metal oxidation experiment. Therefore, Eq. (34) is the kinetic equation for metal oxidation.[1] In terms of the thickness growth one has

$$\frac{d\Delta\xi}{dt} = j_{ion} V_m = \frac{V_m \sigma_{ion} \bar{t}_{el}}{8F^2} \frac{\Delta G_{AO}}{\Delta\xi} \tag{36}$$

or, after integration, one obtains $\Delta\xi^2 = 2kt$. The parabolic rate constant k is found from Eq. (36) to be

$$k = \frac{V_m \sigma_{ion} \bar{t}_{el} \Delta G_{AO}}{8F^2} \tag{37}$$

ΔG_{AO} is the Gibbs energy of formation of AO from A and O_2, O_2 having the chemical potential $\mu_{O2} = \mu_{O_2}^0 + RT \ln (p_{O2}/p_{O_2}^0)$.

In the first part of this section we discussed the ionic fluxes under open circuit conditions. The next step is a discussion of the electrical open circuit potentials under the same conditions. Figure 3 explains the situation. If one adds to the so-called diffusion potential of Eq. (31) the electrical potential jumps across the electrode/electrolyte interfaces, one arrives at the emf E of the galvanic cells. (Obviously, cells $a/1$–$a/3$ do not give a definite emf.) The emf of cells $a/4$–$a/6$ can be obtained in this way: Let us assume that we deal with reversible electrodes. Then one derives from

$$\eta_{el} \text{ (electrode)} = \eta_{el} \text{ (electrolyte)} \tag{38}$$

$$\mu_{el} \text{ (electrode 1)} = \mu_{el} \text{ (electrode 2)} \tag{39}$$

$$\mu_{O2} = 2\mu_{O^{--}} - 4\mu_{e'} \tag{40}$$

the electrical potential jumps $\Delta\phi(1) + \Delta\phi(2)$ across the interfaces (1) and (2). When added to the diffusion potential (Eq. (31)), the well known relation for the emf is found as

$$4FE \cong \int t_{ion} d\mu_{O2} = 2 \int t_{ion} d\mu_A \tag{41}$$

By Eq. (14a), $t_{ion} = 1 - t_{el}$ is a function of μ_{O2}. The correct integration of the integrals for oxide electrolytes has been performed by the author previously.[2] There is, however, a slight uncertainty involved in the derivation of Eq. (41), since the "inert" electrode, by necessity, dissolves some A in different amounts at the two electrodes. As a consequence, the chemical potential of the electrons in the two electrodes is not exactly the same, as it was assumed in the derivation of Eq. (41). Theoretically, one must know the Fermi level as a function of x_A in the electrode (Me,A)-alloy in order to make quantitative corrections. Figure 4 shows the E vs μ_{O_2} curve as calculated from Eq. (41), starting with an oxygen potential $p_{O_2} \gg p_\oplus$ as reference electrode.[3]

The next step in this discussion is a treatment of the cells $a/1$ through $a/6$ when an outer voltage is applied. The experimental situation is depicted in Fig. 5. In the case of cell $a/1$, $j_{A^{2+}}$ and $j_O{}^{2-}$ must vanish. From Eq. (28) it follows (electrons are minority defects!) that $(d\phi/d\xi) = 0$. Thus

$$j_{el} = - \frac{\sigma_{el}}{F^2} \frac{d\mu_{el}}{d\xi} \tag{42}$$

The applied voltage U thus amounts to $\Delta\phi(1) + \Delta\phi(2)$, and since $\mu_e(\text{Me } 1) = \mu_e(\text{Me } 2)$, one finds

$$FU = F[\Delta\phi(1) + \Delta\phi(2)] = - [\mu_e(1) - \mu_e(2)] \tag{43}$$

$\mu_e(1)$ and $\mu_e(2)$ being the chemical potentials of electrons in the electrolyte. Hence:

$$F j_{el} = - \frac{\bar{\sigma}_{el}}{F} \frac{\Delta\mu_e}{\Delta\xi} = \bar{\sigma}_{el} \frac{U}{\Delta\xi} = i_{el} \tag{44}$$

i_{el} is the electronic current density. Therefore, cell $a/1$ measures the average of the electronic (partial) conductivity under load, and the definition of $\bar{\sigma}_e$ is

$$\bar{\sigma}_{el} = \frac{1}{\Delta\mu_{el}} \int \sigma_{el} d\mu_{el} \tag{45}$$

However, cell $a/1$ defines AO thermodynamically very poorly. Therefore, cells $a/2$ and $a/3$ are to be preferred for a correct determination of the partial electronic conductivity of AO. $\mu_e(2)$ in the electrolyte is now fixed through the coexisting A (O_2), at interface (2). Integration of Eq. (42) gives

$$j_{e'} = \frac{RT}{F} \frac{u_{e'}}{\Delta\xi} c_{e'}(2) \left\{ 1 - e^{-\frac{UF}{RT}} \right\}; \quad j_{h\cdot} = \frac{RT}{F} \frac{u_{h\cdot}}{\Delta\xi} c_h(2) \left\{ e^{\frac{UF}{RT}} - 1 \right\} \tag{46}$$

These equations have been used on several occasions to determine small partial electronic conductivities.[4,5] From the i_{el} vs U curve one can distinguish between electron and hole conduction. The discussion of cells $a/4$ and $a/6$ under load is straightforward. The $(j_{ion} + j_{el})$ is measured, and through the shift of the electrolyte it is possible to evaluate the partial conductivity of the A^{2+} ions. (Let $\mu_{O_2}(1) = \mu_{O_2}(2)$. Then $\mu_O{}^{--}(1) = \mu_O{}^{--}(2)$, and $d\mu_O{}^{--}(\text{AO})$, $d\mu_{A^{++}}(\text{AO})$ and $d\mu_{e'}(\text{AO})$ are zero. The driving force for the ion and electron fluxes is $\Delta\phi/\Delta\xi = U/\Delta\xi$ in the electrolyte AO.)

The discussion of cell $a/5$ under load is somewhat more complicated. Again Eqs. (28) and (29) have to be applied. One notes that the condition of reversible electrodes fixes the $\Delta\phi$ values across the interfaces as well as the $\mu_{e'}(\text{AO})$ values at these interfaces. Since $\mu_O{}^{2-}$ and $\mu_A{}^{2+}$ are constant throughout the electrolyte, $\Delta\phi(\text{AO})$ is given as $U - \Delta\mu_{O_2}/4F$. The partial ionic conductivity is given by two terms which stem from the chemical and electrical potential gradient, respectively

$$j_{A^{++}} = - \frac{\sigma_A}{2F} \left\{ \frac{U}{\Delta\xi} - \frac{\Delta\mu_{O_2}}{4F \Delta\xi} \right\} \tag{47}$$

The definition of $\bar{\sigma}_e$ is given in Eq. (45) and can be calculated from the first section, since $d\mu_e = -\frac{1}{4}d\mu_{O_2}$,

$$j_{e'} = -\frac{\bar{\sigma}_e}{F}\frac{U}{\Delta\xi} \tag{47a}$$

Oxide Electrolytes $(A,B)_{1-y}O$ in Thermodynamic Potential Gradients

Normally, oxide electrolytes are heavily doped and have to be treated as solid solutions, which means that the cationic sublattice is the host of A and B ions. This adds one more flux equation (for the B ions) to Eqs. (28), and one more composition-dependent variable to the unknowns. To solve the transport problem, one has to formulate, in addition to the initial and boundary conditions as given, one more equation, which is the steady state condition

$$v = \frac{j_{A^{++}}}{c_A} = \frac{j_{B^{++}}}{c_B} = \cdots \tag{48}$$

(For simplicity, the equations are given for $(A,B)O$ solid solutions only. Other stoichiometries do not change the basic structure of the following equations.) In the context of this presentation, the most important phase sequence is that which has oxygen of different chemical potential on the two electrolyte/electrode interfaces.

In analogy to Fig. 2, Fig. 6 shows again the basic experiment. $(A,B)O$ is brought into an oxygen potential gradient under open circuit conditions as depicted in Table I ($b/6$). A demixing of the initially homogeneous $(A,B)O$ takes place, due to the different values of σ_A and σ_B (i.e., $u_A \neq u_B$ and $D_A \neq D_B$). If the thermodynamics of the $(A,B)O$ solid solution is known and the defect thermodynamics as well, the problem can be solved completely. This has recently been performed.[6,7] In analogy to Eqs. (32) one obtains:

$$j_{A^{++}} = -\frac{\sigma_A}{(z_A F^2)}\left[(1 - t_A)\frac{d\mu_{AO}}{d\xi} - t_B\frac{d\mu_{BO}}{d\xi} - t_{el}\frac{d\mu_O}{d\xi}\right] \tag{49}$$

and the analogous equation for $j_{B^{2+}}$. In the case that $t_{el} = 0$ there is no demixing, since $j_{A^{++}}$ (and $j_{B^{++}}$) are then functions only of μ_{AO}. The full demixing occurs if $t_{el} = 1$. Introducing Eq. (49) into Eq. (48), one finds for this case:

$$\frac{\sigma_A}{c_A}\left[\frac{d\mu_{AO}}{d\xi} - \frac{d\mu_O}{d\xi}\right] = \frac{\sigma_B}{c_B}\left[\frac{d\mu_{BO}}{d\xi} - \frac{d\mu_O}{d\xi}\right] \tag{50}$$

Along with the Gibbs-Duhem equation, this is the differential equation for steady state demixing in an oxygen potential gradient. Of course, the reverse is also true: Inhomogeneous $(A,B)O$-crystals build up an oxygen potential gradient at opposite sides, which can be detected experimentally and from which conclusions can be made as to the diffusivities of the cations. For the explicit solution, σ_A and σ_B must be known as functions of composition and oxygen potential. For extended oxide solid solutions, the following expression is often applicable:

$$\sigma_{ion} = \sigma_{ion}^0(p_{O_2}/p_{O_2}^0)^{1/n}\exp(A_1 + A_2 x_{BO}) \tag{51}$$

Explicit calculations are given in Ref. 6. Note that the demixing process is not due to the flux of oxygen ions. The phase sequences $b/4$ and $b/5$ from Table I are not treated here further for open circuit conditions. The first one seems to be rather unimportant from a practical viewpoint, the second one refers to alloy oxidation and has been treated to some extent elsewhere.[8,9] The next step of this discussion deals with the diffusion potential and the open circuit potential of cell $b/6$. In analogy to Eq. (31) one has

$$-F \cdot d\phi = t_A \frac{d\mu_A}{z_A} + t_B \frac{d\mu_B}{z_B} + t_O \frac{d\mu_O}{z_O} + \frac{d\mu_{el}}{z_{el}} \tag{52}$$

Intregration and the addition of $F(\Delta\phi(1) + \Delta\phi(2))$ yields

$$z_A FE = \tfrac{1}{2}\int t_{ion} d\mu_{O_2} - \int (x_{BO}t_A - x_{AO}t_B) \frac{d\mu_{AO}}{1 - x_{AO}} \tag{53}$$

From Eq. (53) one concludes that the addition of BO to the electrolyte AO gives the same emf only as long as either $d\mu_{AO} = 0$ or $t_{ion} = t_0$. This latter condition is approximately true for stabilized zirconia. If, however, demixing has occurred since $\sigma_A \neq \sigma_B$, $d\mu_{AO}$ from Eq. (50) can be inserted into Eq. (53) to give

$$z_A FE = \tfrac{1}{2}\int t_{ion} d\mu_{O_2} - \tfrac{1}{2} \int \frac{(x_A t_B - x_B t_A)^2}{x_A^2 t_B + x_B^2 t_A} \, d\mu_{O_2} \tag{54}$$

The last part of this discussion refers to cells $b/1$–$b/6$ if a voltage U is applied. One can show by proper application of the flux equations that cases $b/1$–$b/3$ can be discussed in the same way that was done in the case of $a/1$–$a/3$. It is essentially the partial electronic conductivity that can be determined in this way. Also, the cell $b/4$ poses no problems; the total conductivity can be determined in this way. Cells $b/5$ and $b/6$ are more complicated to treat under load. One obtains a total conductivity, but the electrolyte after attaining steady state is inhomogeneous. The inhomogeneity can be evaluated according to the lines given by (A,B)O electrolytes under open circuit conditions, if the proper boundary conditions are introduced. In consequence, the $\bar{\sigma}_{el}$ and $\bar{\sigma}_0$ values have to be averaged over the inhomogeneous samples.

Oxide Electrolytes with Polarizing Electrodes, Open Circuit

The basic situation is shown in Fig.7. Although in principle all possible sorts of polarization might occur, the formulas are given only for diffusion—plus exchange current polarization. All polarization phenomena finally stem from the ionic "leakage" current through the electrolyte due to the nonvanishing electronic flux (if $t_{el} \neq 0$). This has, for open circuit conditions, been evaluated in Eq. (34a).

The ionic leakage current under open circuit conditions of a galvanic cell leads to an oxygen diffusion in the electrolyte that consumes a part of the available driving force ($\Delta\mu_{O2}/\Delta\xi$) as does the polarization stemming from the migration across the electrode/electrolyte interface. Since the current is position-independent, we have, in a linear approach:

$$j_{ion} = -\frac{c_O D_O}{RT} \frac{\Delta\mu_{O_2}^D}{2\Delta l} = j_{ex}^\circ \frac{d\mu_{O_2}^P}{RT} = \frac{\sigma_{ion} t_{el}}{8F^2} \frac{\Delta\mu_{O_2}^{El}}{\Delta\xi} \tag{55}$$

where $\Delta\mu_{O_2}^D$, $\Delta\mu_{O_2}^P$, and $\Delta\mu_{O_2}^{El}$, designate the respective driving forces for diffusion in the electrode, interface polarization, and diffusion in the electrolyte. In view of

$$\mu_{O_2}(2) - \mu_{O_2}(1) = \Delta\mu_{O_2} = \Delta\mu_{O_2}^D + \Delta\mu_{O_2}^P + \Delta\mu_{O_2}^{El} \qquad (56)$$

one can readily calculate the reduced emf when polarization occurs if $t_{el} \neq 0$. One obtains instead of Eq. (41):

$$4FE = \frac{\bar{t}_{ion}}{1 + \dfrac{RT\sigma_{ion}\bar{t}_{el}}{8F^2\Delta\xi}\left\{\dfrac{1}{j_{ex}^o} + \dfrac{2\Delta l}{c_o D_o}\right\}}\,\Delta\mu_{O_2} \qquad (57)$$

Depending on the numerical values of the various terms in the denominator, the emf can be reduced considerably by polarization.

Polarization phenomena for identical electrodes, if an outer voltage is applied to the cell, will not be treated here. They are treated adequately in textbooks.

Oxide Electrolytes with Reacting Electrodes

Often, in practical applications, oxide electrolytes are contacted with electrodes which are not inert; they may either dissolve in, or react with, the electrolyte. A typical situation is depicted in Fig. 8. On the left-hand side, the electrolyte BO is contacted with the oxygen electrode $AO + O_2$. AO and BO react to form ABO_2.

Since we assume local equilibrium everywhere, under open circuit conditions one has for every phase $\Sigma z_i j_i = 0$, which leads to Eqs. (31). The emf E is obtained through integration across ABO_2 and BO plus the addition of the various interface potentials $\Delta\phi$, which in turn are determined by $\eta_e^I = \eta_e^{II}$ etc. at these interfaces. The final result reads:

$$4FE = \bar{t}_{ion}^{III}\mu_{O_2}(6) - \bar{t}_{ion}^{II}\mu_{O_2}(1) + (\bar{t}_B^{II} - \bar{t}_A^{II})\Delta G_{ABO_2}^o + (\bar{t}_{ion}^{II} - \bar{t}_{ion}^{III})\mu_{O_2}(4)$$

$$4FE = \bar{t}_{ion}^{III}[\mu_{O_2}(6) - \mu_{O_2}(4)] + \bar{t}_{ion}^{II}[\mu_{O_2}(4) - \mu_{O_2}(1)] + (\bar{t}_B^{II} - \bar{t}_A^{II})\Delta G_{ABO_2}^o \qquad (58)$$

In case that $t_{el} \cong 0$, Eq. (58) is simplified and yields

$$4FE = \Delta\mu_{O_2}(1/6) + (\bar{t}_B^{II} - \bar{t}_A^{II})\,\Delta G_{ABO_2}^o \qquad (58a)$$

If, however, t_{el} cannot be neglected, one has to determine $\mu_{O_2}(4)$, which is the oxygen potential at the interface ABO_2/BO. To this end it is necessary to formulate the electronic current in the various phases and equate them: $j_{el}^{.II} = j_{el}^{.III}$. Using Eqs. (29), (31), and (52), the result is:

$$\sigma_{ion}^{III}\Delta\xi^{II}\bar{t}_{el}^{III}\Delta\mu_{O_2}^{III} = \sigma_{ion}^{II}\Delta\xi^{III}t_{el}^{II}\left\{\Delta\mu_{O_2}^{II} + 2\Delta G_{ABO_2}^o\,\frac{\bar{t}_A^{II} - \bar{t}_B^{II}}{1 - \bar{t}_{el}^{II}}\right\} \qquad (59)$$

After some rearrangement:

$$4FE = \Delta\mu_{O_2}(1/6)\left\{1 - \frac{\Delta\xi^{III}\bar{\sigma}_{ion}^{II} + \Delta\xi^{II}\bar{\sigma}_{ion}^{III}}{\Delta\xi^{III}\bar{\sigma}_{ion}^{II}/\bar{t}_{el}^{II} + \Delta\xi^{II}\bar{\sigma}_{ion}^{III}/\bar{t}_{el}^{III}}\right\} +$$

$$\Delta G_{ABO_2}^o(\bar{t}_B^{II} - \bar{t}_A^{II})\left\{1 - \frac{2}{1 - \bar{t}_{el}^{II}}\,\frac{\bar{t}_{el}^{III} - \bar{t}_{el}^{II}}{\Delta\xi^{II}\bar{\sigma}_{ion}^{III}\bar{t}_{el}^{II} + \Delta\xi^{III}\bar{\sigma}_{ion}^{II}\bar{t}_{el}^{II}}\right\} \qquad (60)$$

It is concluded from Eqs. (58) and (60) that it is, in principle, possible to correct the measured emf in galvanic cells with reacting electrodes (as long as local equilibrium is established throughout) in order to obtain the oxygen potential difference between these electrodes.

Concluding Remarks

The concepts and equations outlined in order to describe the behavior of oxide electrolytes in chemical and electrochemical potential gradients were phenomenological in nature. There is little basic knowledge on the atomistic background of the transport coefficients which were introduced. This of course does not affect the conclusions which have been drawn from the phenomenological theory. However, it may be appropriate to add some remarks on the models for transport in oxide electrolytes.

Most oxide electrolytes are heavily doped in order to achieve sufficient ionic conductivity and low electronic transference. It is found in many cases that the addition of a heterovalent dopant leads to a maximum in the σ_{ion} vs x_{dopant} curves. This holds not only for anionic conductors (see Fig. 9(A)), but has also been ascertained for semiconducting oxides with predominantly cationic diffusion[10,11] (see Fig. 9(B)).

Obviously, any simple modeling with point defects must be ruled out, if the heterovalent dopant is of the order of 0.1. In some way it is astonishing that models are proposed at all in view of the fact that these solutions are neither diluted not ideal; they are partly ordered and exhibit long range forces. (Remember that the Ising model has not yet been solved for three-dimensional crystals.) Any kinetic treatment must, by necessity, be still more difficult to do than the thermodynamics. The motion of ions in these solid solutions can neither be unassociated nor uncorrelated in space and time.

The decrease of ionic (oxygen) conductivity with increasing heterovalent dopant has been ascribed to clusters and associates,[12,13] to repulsive interaction,[14] to ordering or Debye-Hueckel-type effects.[15,16] Recently the author[10] has shown that Debye-Hueckel-type interactions, as well as ordering, lead generally to a marked decrease in the correlation factor and an increase in activation energy. A still more complicated mode of motion was proposed[17] as a multimode mechanism through and between different degrees of dopant-defect associates. All these models help, of course, in curve fitting.

To my knowledge, no one has yet explained the partial electronic conductivities on an atomistic basis in heavily doped oxide electrolytes. To this end it would first of all be necessary to determine σ_e values with the help of polarization methods at high T as functions of dopant concentrations, μ_{O_2} and T. This is experimentally quite a difficult task. Theoretically, these data then will have to be interpreted in terms of electronic processes in quasi non-crystalline materials as treated by Mott and Davis.[18] An oxide like ZrO_2 which is doped with a two-valent metal oxide up to 0.1 or 0.2 in mole fraction can (in view of the ionic point defects, associates, clusters, and other partial ordering) no longer be treated as a periodic potential crystal.

Acknowledgment

This work was done while the author was visiting professor at the Department of Materials Science and Engineering, Cornell University. The opportunity of doing research there is greatly appreciated.

Appendix
Equivalent Circuits

The idea of an equivalent circuit in order to describe the behavior of a galvanic cell was presented many years ago (Jost). Its construction and notation are given in Fig. 10. The following relations hold:

$$R_{ion} = \Delta\xi/\sigma_{ion} \qquad (=R_i) \tag{A-1}$$

$$R_{el} = \Delta\xi/\sigma_{el} \qquad (=R_e) \tag{A-2}$$

$$i = i_1 + i_2 \tag{A-3}$$

$$0 = iR_a + U + i_2 R_e \tag{A-4}$$

$$-E = iR_a + U + i_1 R_i \tag{A-5}$$

The following quantities are of interest: i, i_1, i_2, E, $i^2 R_a$. From (A-3)–(A-5) one has

$$i = -E^0 \frac{R_e}{R_a(R_i + R_e) + R_i R_e} - U \frac{R_i + R_e}{R_a(R_i + R_e) + R_i R_e} \tag{A-6}$$

$$i_1 = -E^0 \frac{R_a + R_e}{R_a(R_i + R_e) + R_i R_e} - U \frac{R_e}{R_a(R_i + R_e) + R_i R_e} \tag{A-7}$$

$$i_2 = E^0 \frac{R_a}{R_a(R_i + R_e) + R_i R_e} + U \frac{R_i}{R_a(R_i + R_e) + R_i R_e} \tag{A-8}$$

$$E = \frac{R_e}{R_a(R_i + R_e) + R_i R_e} (E^0 R_a + UR_i) \tag{A-9}$$

$$i^2 R_a = R_a \frac{[E^0 R_e + U(R_i + R_e)]^2}{[R_a(R_i + R_e) + R_i R_e]^2} \tag{A-10}$$

From Eqs. (A-6)–(A-10), the characteristic cell quantities can immediately be derived for limiting cases. For example, (1) $U = 0$, $R_a = \infty$. Open circuit galvanic cell: $i = 0$, $i_1 = -i_2$.

$$E = E^0 \frac{R_e}{R_i + R_e} ; \quad i^2 R_a = 0$$

Introducing Eqs. (A-1) and (A-2), one obtains:

$$E = E^0 (1 - \bar{t}_{el})$$

$$i_1 = \frac{j_{ion}}{z_{ion}F} = -\frac{\sigma_{ion}\bar{t}_{el}}{z_{ion}F^2} \frac{E^0 F}{\Delta\xi} = -\frac{\sigma_{ion}\bar{t}_{el}}{z_{ion}F^2} \frac{\Delta\mu_{O_2}}{4\Delta\xi} \tag{A-11}$$

These equations ought to be compared with Eqs. (34) and (41).
(2) $U = 0$, $R_a = 0$. Short circuited galvanic cell: $E = 0$.

$$i = i_1 = -\frac{E^0 \sigma_{ion}}{\Delta\xi}$$

$$j_{ion} = -\frac{\sigma_{ion}}{z_{ion}F^2} \frac{E^0 F}{\Delta\xi} = -\frac{\sigma_{ion}}{z_{ion}F^2} \frac{\Delta\mu_{O_2}}{4\Delta\xi} \tag{A-12}$$

(3) $U = 0$, $R_a \neq 0$.

$$i = -\frac{1 - \bar{t}_{el}}{R_a + (\Delta\xi/\sigma)} E^0$$

265

$$j_{ion} = \frac{1 - \bar{t}_{el}}{R_a + (\Delta\xi/\sigma)} \frac{(\Delta\mu_{O_2}/4)}{z_{ion}F^2}$$

$$E = E^0(1 - \bar{t}_{el}) \frac{R_a}{R_a + (\Delta\xi/\sigma)} \tag{A-13}$$

Finally, the maximum work efficiency of a cell is attained if $R_e = \infty$. In the case that $R_e \neq \infty$, let us define the efficiency η_w as (i^2R_a/i_1E^0). Introducing Eqs. (A-6) and (A-7), one finds the general form of η_w in terms of $\bar{\sigma}_e$ and thus in terms of $\Delta\mu_{O_2}$. For a more thorough discussion of this last point, see Ref. 20.

References

[1]C. Wagner, *Z. Phys. Chem. B,* **21**, 25 (1933).
[2]H. Schmalzried, *Z. Elektrochem.,* **66**, 572 (1962).
[3]H. Schmalzried, *Z. Phys. Chem. (Leipzig),* **38**, 87 (1963).
[4]B. Illschner, *J. Chem. Phys.,* **28**, 1109 (1958).
[5]R. A. Rapp and D. A. Shores, Techniques of Metals Research IV/2, Chapt. VIC. Interscience, New York, 1970.
[6]H. Schmalzried, *et al., Z. Naturforsch. A,* **34**, 192 (1979).
[7]H. Schmalzried and W. Laqua, *Oxid. Met.* **15**, 339 (1981).
[8]B. D. Bastow, *et al., Proc. R. Soc. London,* **356**, 177 (1977).
[9]C. Wagner, *Corrosion Sci.,* **9**, 91 (1969).
[10]H. Schmalzried, *Z. Phys. Chem. (Leipzig),* **105**, 47 (1977).
[11]W. Laqua; unpublished results.
[12]F. A. Kroeger, *J. Am. Ceram. Soc.,* **49** [4] 215 (1966).
[13]D. W. Strickler and W. G. Carlson, *ibid.,* **48**, 286 (1965).
[14]W. W. Barker and O. Knop, *Proc. Br. Ceram. Soc.,* **19**, 15 (1971).
[15]T. Y. Tien and E. C. Subbarao, *J. Chem. Phys.,* **39**, 1041 (1963).
[16]R. E. W. Casselton, *Phys. Status Solidi A,* **2**, 571 (1970).
[17]A. Nakamura and J. B. Wagner, *J. Electrochem. Soc.,* **127**, 2325 (1980).
[18]N. F. Nott and E. A. Davis, Electronic Processes in Non-Crystalline Materials. Clarendon Press, Oxford, 1971.
[19]C. Wagner, *Prog. Solid State Chem.,* **10**, 3 (1975).
[20]D. S. Tannhauser, *J. Electrochem. Soc.,* **125**, 1277 (1978).

---

*Permanent address: Institut fuer Physik. Chemie, Universitaet Hannover, West Germany.

Table I. Systematic survey of transport situations with oxide electrolytes

Electrolytes: (a) AO; (b) (A,B)O
Electrodes: Me (inert); Me(O$_2$); A or (A,B)

(I) Open circuit

(a/1)	Me |AO| Me	No information
(a/2)	Me |AO| A	'' ''
(a/3)	Me |AO| O$_2$, Me	'' ''
(a/4)	A′ |AO| A″	Ion + electron flux, shift AO
(a/5)	A |AO|O$_2$, Me	Ion + electron flux, metal oxidation
(a/6)	Me, O$_2'$ |AO| O$_2''$, Me	Ion + electron flux, shift AO

(b/1)–(b/6): Replace AO by (A,B)O and A by (A,B).

(II) Under load, reversible electrodes

(a/1)	Me |AO| Me	σ_{el}, thermodyn. not defined
(a/2)	Me |AO| A	σ_{el}^o (A)
(a/3)	Me |AO| O$_2$(Me)	$\sigma_{el}(p_{O_2})$
(a/4)	A′ |AO| A″	σ_{total}, transference, shift
(a/5)	A |AO| O$_2$,Me	σ_{total}, transference, shift
(a/6)	Me, O$_2'$ |AO| O$_2''$, Me	σ_{total}, transference, shift

(A)

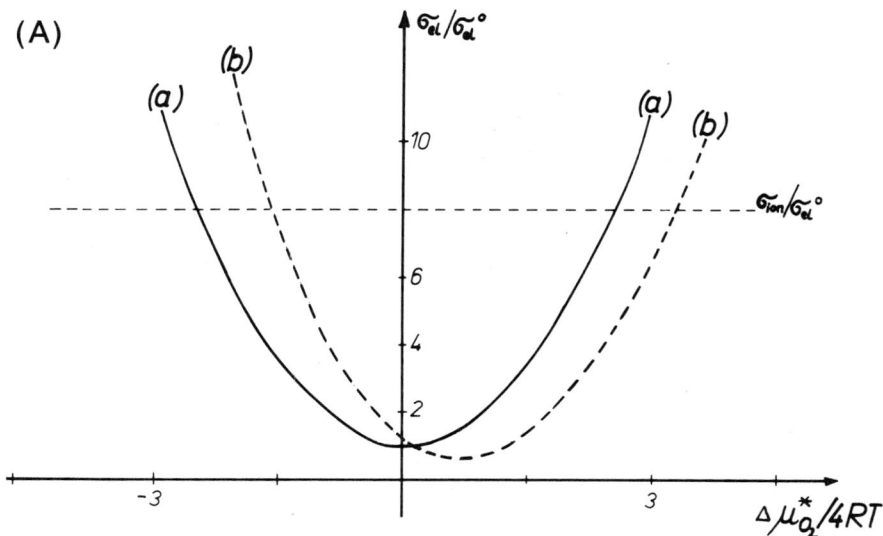

Fig. 1. See page 268 for caption.

(B)

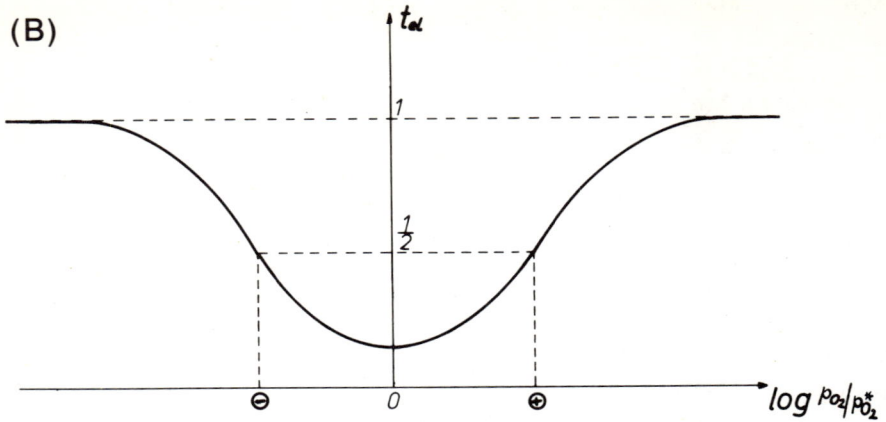

Fig. 1. (A)(a) Normalized electronic conductivity ($\sigma^0_{el} = \sigma_{el}$ ($\delta = 0$)) vs normalized oxygen potential ($\Delta\mu^0_{O_2} = \mu_{O_2} - \mu_{O_2}$ ($\delta = 0$)): $u_{e'} = u_h$; (b) is calculated for $u_{e'} = 5u_h$. (B) Electronic transference number as a function of the normalized oxygen potential.

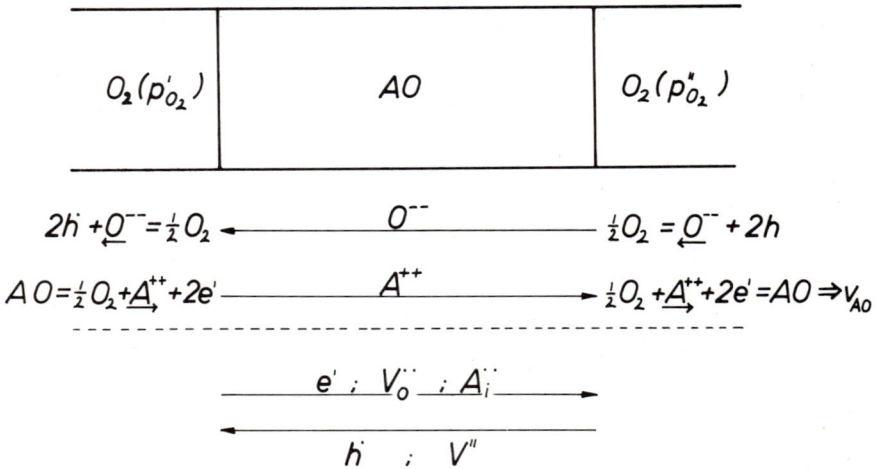

$$2h^{\cdot} + \underline{O}^{--} = \tfrac{1}{2}O_2 \longleftarrow \qquad O^{--} \qquad \longrightarrow \tfrac{1}{2}O_2 = \underline{O}^{--} + 2h^{\cdot}$$

$$AO = \tfrac{1}{2}O_2 + \underline{A}^{++} + 2e' \longrightarrow \qquad A^{++} \qquad \longrightarrow \tfrac{1}{2}O_2 + \underline{A}^{++} + 2e' = AO \Rightarrow V_{AO}$$

$$\underline{\qquad e' \; ; \; V^{\cdot\cdot}_O \; ; \; A^{\cdot\cdot}_i \qquad} \longrightarrow$$

$$\longleftarrow$$
$$h^{\cdot} \quad ; \quad V''$$

Fig. 2. AO electrolyte between different component potentials. Migrating species and phase boundary reactions are indicated.

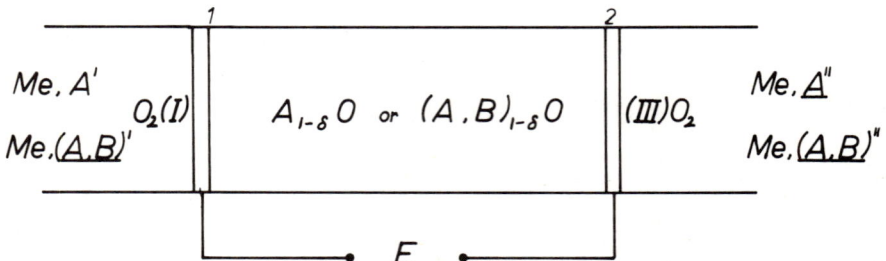

Fig. 3. Schematic setup for the determination of open circuit potentials.

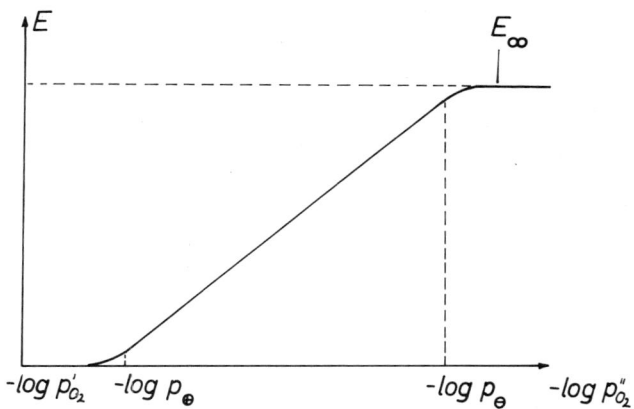

Fig. 4. E vs ln $p_{O_2}^{\delta}$ with fixed $p_{O_2}^{\delta}$ ($p_{O_2}^{\delta} \gg p_{\oplus}$; $p_{\oplus} \gg p_{\ominus}$).

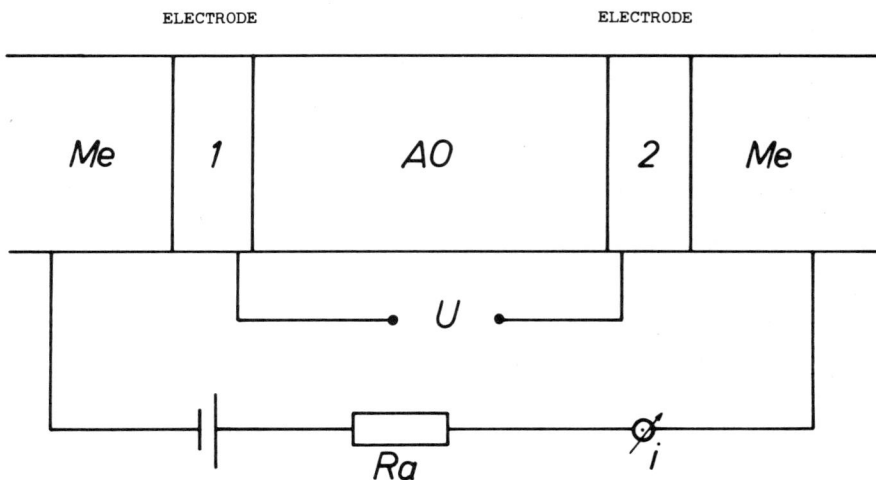

Fig. 5. Schematic setup for the investigation of cells $a/1 - a/6$ under load.

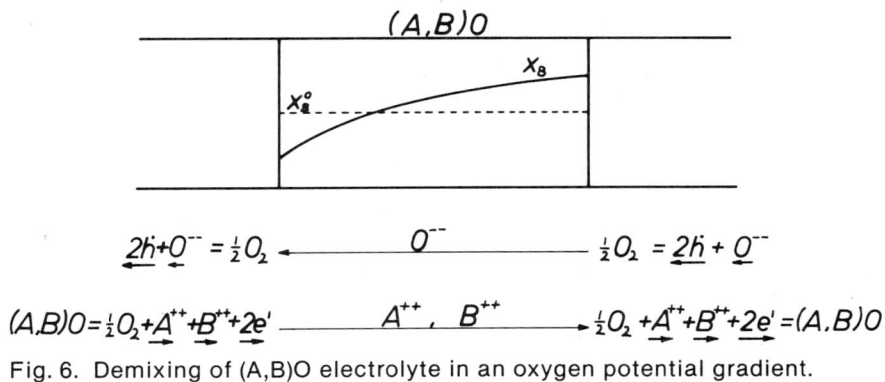

Fig. 6. Demixing of (A,B)O electrolyte in an oxygen potential gradient.

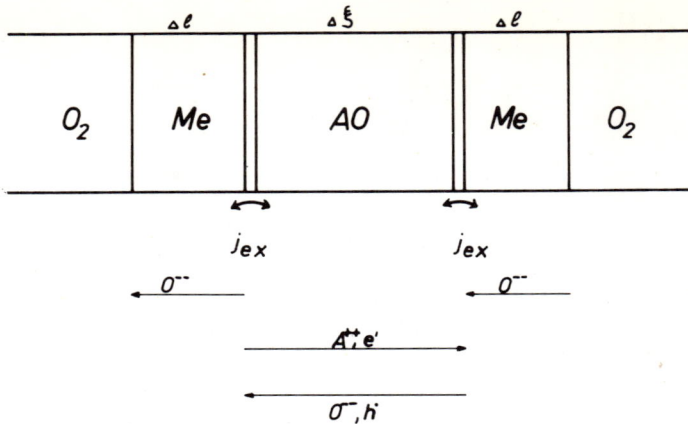

Fig. 7. Polarizing cell: fluxes and exchange current j^0_{ex}.

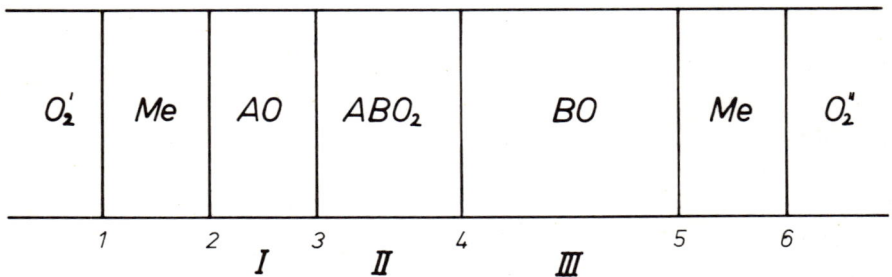

Fig. 8. Galvanic cell, if electrolyte BO can react with electrode AO. Assume $\mu_{O_2}(1) = \mu_{O_2}(3)$ and $\mu_{O_2}(5) = \mu_{O_2}(6)$.

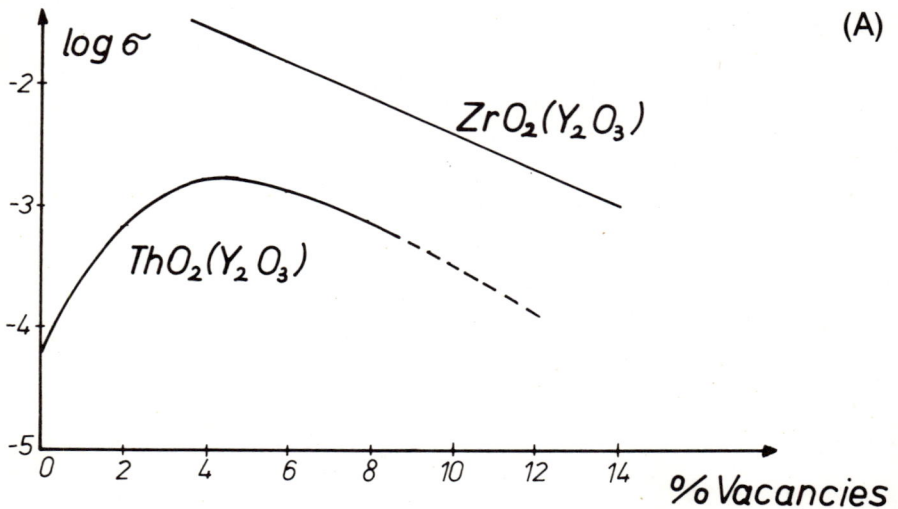

(A)

270

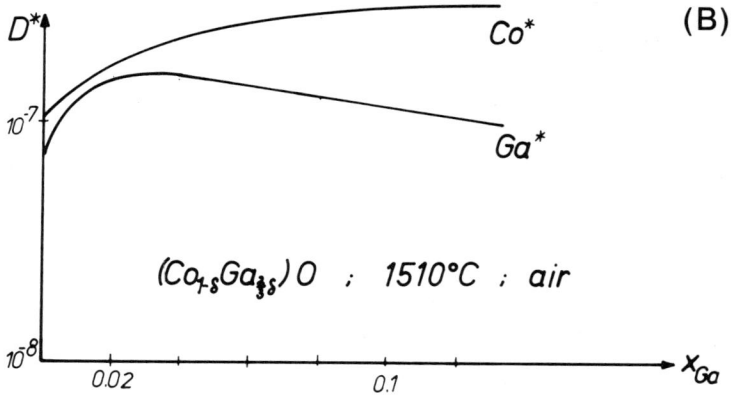

Fig. 9. (A) Dc conductivity of Y_2O_3-doped ZrO_2 and ThO_2. (B) Tracer diffusivities (D^*_{Co}, D^*_{Ga}) of Ga_2O_3-doped CoO.

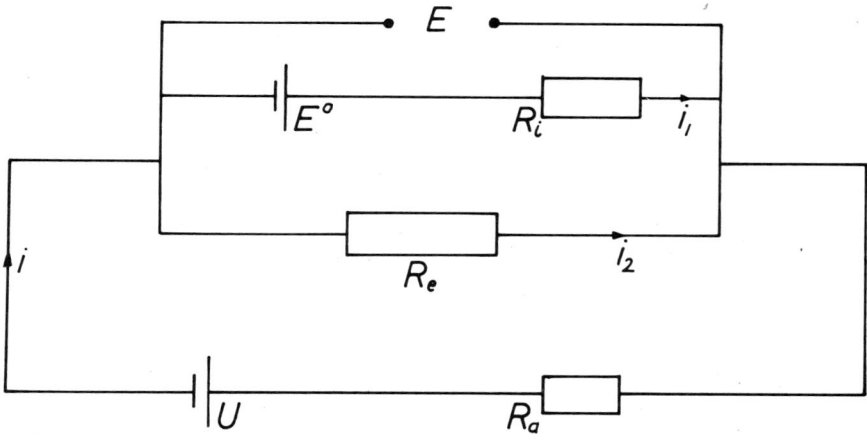

Fig. 10. Equivalent circuit for cells under load.

Preparation and electrical behavior of zirconia ceramics

R. J. BROOK

Department of Ceramics
The University
Leeds LS2 9JT
United Kingdom

Traditional procedures for ceramic fabrication (the firing of mixed oxides) require the use of high temperature when applied to zirconia; temperatures above 1800 °C are typical for the attainment of high density. To avoid this requirement, three avenues are possible, (1) the use of carefully prepared, uniform, and active powders sinterable at lower temperatures, (2) the use of liquid phase additives, and (3) the use of modified firing procedures such as hot-pressing. The respective merits of these approaches are reviewed in terms of the electrical performance and properties of the oxide.

The electrical applications that stabilized zirconia has found derive from its properties as a high performance solid electrolyte.[1] The oxides, such as CaO, Y_2O_3, or Sc_2O_3, that are added to stabilize the zirconia into the cubic structure, introduce oxygen ion vacancies as compensating defects, and the resulting high oxygen ion diffusion coefficient (Fig. 1) imparts good ionic conductivity (Fig. 2) to the stabilized oxide; since zirconia has low electronic conductivity, it acts as a predominantly ionic conductor over a wide span of oxygen pressures (Fig. 3). These features make it suitable for a variety of applications based on the use of solid electrolyte membranes such as oxygen pumps for partial pressure regulation,[2] fuel/oxygen cells for electricity generation,[3] and oxygen monitors for metals processing[4] or for combustion control.[5]

For the efficient use of zirconia in such devices, it is important that the fabricated membranes have good ionic conductivity ($> 10^{-2}$ ohm^{-1}cm^{-1} at the operating temperature for high current devices such as fuel cells), that they be ionic conductors (transport number $t_i > 0.99$) over the range of oxygen pressures encountered in the device, and that they be impermeable to prevent direct molecular transport of oxygen through the membrane (density $> 92\%$). In addition, for suitable lifetimes, the membranes should have adequate strength, thermal shock resistance, and chemical stability in the operating environment.

These requirements have resulted in the demand for an effective, reliable, and low cost fabrication route for sintered zirconia ceramics. Since these requirements apply to the membrane as a whole, that is, both to the zirconia phase itself and to any other, usually intergranular, phases that have been introduced during processing, the fabrication problem has not found an easy solution.[6]

In the following, attention is first given to the few studies in which attempts have been made to identify the controlling mechanisms during the microstructural development of zirconia ceramics. The several avenues that have been explored in the attempt to make high quality electrolytes will then be considered.

Sintering Mechanisms in Zirconia

Although the microstructural requirements of electrolyte membranes suggest that solid state sintering would be an attractive fabrication route, the sintering of zirconia has received relatively little attention in terms of fundamental kinetic studies. This stems in part from the fact that the large difference in zirconium and oxygen diffusion coefficients (Fig. 1) already helps to suggest the rate-controlling species and in part from the fact that no effective sintering additive (such as MgO has proved to be for the sintering of Al_2O_3)[7] has been identified for which explanations are then sought.

In an early study[8] of sintering kinetics (13 mol% CaO, 1550–1766 °C, 0.91 μm, H_2 atm), Jorgensen used isothermal shrinkage rates and the now questioned[9] procedure of the measurement of the time exponent, m, of the shrinkage, ($\Delta L/L_0$), to identify the controlling mechanism. The measured exponent (0.42–0.50) was interpreted as indicating volume diffusion, and the excellent agreement of the derived diffusion coefficient with those of the zirconium and calcium ions suggested lattice diffusion of one or other of these cations as the controlling mechanism.

The role of the cation as controlling species has not been questioned (simple models and the known oxygen diffusion coefficient suggest that high densities would be reached in 1 s at 820 °C and were oxygen controlling). Support for lattice diffusion processes has been given by compression creep measurements[10] (6 mol% Sc_2O_3 and 6 mol% Y_2O_3, 1378 °C and 1485 °C, 1 μm and 17 μm, respectively) in which the low stress creep rate was found to be inversely proportional to the square of the grain size and, for the Y_2O_3-stabilized material, directly proportional to the stress. Seltzer and Talty[11] (17.5–28 mol% Y_2O_3; 1480–2000°C) interpreted high temperature, low stress creep data in terms of lattice diffusion of zirconium ($V_{Zr}^{\cdots}$). A low temperature study[12] of sintering kinetics for fine-grained powders (4–33 mol% Er_2O_3, 800–1100° C, 0.2 μm), again using measurements of the time exponent of the shrinkage, suggested ($m = 0.24$–0.32) that grain boundary diffusion processes were dominant under these conditions. Although comparison of these results is made difficult by the many differences in experimental conditions (additive concentration and type, grain size, density, temperature, purity, applied stress), the available data point to the importance of the lattice diffusion of cations at high temperature for conventional (> 1 μm) grain sizes.

The effect of the concentration of the stabilizing additive has been variously interpreted. Diffusion data[13] for zirconium (12 and 16 mol% CaO) and high temperature creep data[10] (17.5–28 mol% Y_2O_3) have shown no effect from variations in the concentration of additive; against this, measurements[14] of the as-fired density of sintered samples (8 to 16 mol% CaO) have decreased from 95 to 85% of the full density value as the CaO content is raised. Compressive creep data[15] (13–20 mol% CaO, 1350 °C) have shown a factor of three variation of creep rate with CaO content, the values rising to and falling from a maximum at 15 mol% CaO close to the concentration at which complete stabilization of the single-phase cubic structure is attained. Although these data are too few on which to base any firm interpretation, the evidence for a maximum is interesting in the light of the well-recognized maximum in the conductivity occurring at this concentration (Fig. 4). Since the conductivity maximum corresponds to a maximum in the free oxygen vacancy concentration, a maximum in a process controlled by zirconium at the same point would be more consistent with a role for zirconium interstitials than with control by zirconium vacancies. Such interstitials have been recognized to exist in this material[16] but only under different (high) temperature conditions. This would seem a rewarding area for further study.

To summarize the information from the studies of sintering kinetics, it is believed that process control is by diffusion of the cation, probably of zirconium through the lattice for conventional powders and sintering temperatures. Though low, the zirconium diffusion coefficient value (approximately a factor of 10 less than that of Al in Al_2O_3) is not itself sufficient to explain the difficulties in preparing high density materials.

There have been very few studies of grain growth in stabilized zirconia, perhaps because the effect of the grain boundaries on the conductivity has been found, for single-phase materials, to be small. In two measurements of the activation enthalpy for conduction in polycrystalline samples, Tien[17] found that it increased as the grains grew (16 mol% CaO, $T < 800\,°C$, 11–100 μm), while Joffe et al.[18] found that it decreased (5.7 mol% Y_2O_3, 0.2–18 μm, 450–1000 °C). Tien and Subbarao[19] studied normal grain growth (16 mol% CaO, 1600–2000 °C, 12–100 μm) finding an activation enthalpy for the process (329 kJ mol^{-1}) somewhat less than the value found by Jorgensen[8] for sintering (396 kJ mol^{-1}).

Approaches to the Fabrication Problem

The results of the few kinetic studies made suggest that zirconia follows closely the classical qualitative picture of microstructural development during heat treatment, namely that densification is controlled by a mechanism involving lattice diffusion of the slower species. From study of micrographs of sintered products, it is also apparent that grain growth occurs during densification and, further, that trapping of porosity within the growing grains readily develops.[19,11]

If conventional ceramic processing methods are employed in which ball-milled powders of the pure component oxides are mixed, dry-pressed, and heated, then zirconia is not easy to fabricate; the first stages of densification occur but even high temperatures result in disappointing final, fired densities (Table I). Since the diffusion coefficient of zirconium cannot itself account for this and since there is no evidence for the pronounced coarsening that would occur if either surface or gas phase transport processes were strongly interfering with densification, the likely explanation for the poor sintering behavior is the one most commonly quoted for such problems in oxides, namely that the onset of abnormal grain growth and the trapping of porosity within grains bring pore size reduction and densification to an end well before the attainment of high densities.[20] The microstructures commonly encountered for single-phase zirconias clearly show the development of abnormal growth (Fig. 5).

The features that characterize abnormal grain growth are: (1) the existence in the microstructure of a bimodal grain size distribution consisting of isolated large grains in a matrix of small grains and (2) the separation of pores from the boundaries of the large, growing grains with the result that these grains contain isolated, entrapped porosity. To prevent this process: (1) the initial condition of the powder must be such that there are no large variations of particle size, of particle packing density (agglomerates), or of composition, all of which may encourage the nucleation of abnormal growth; (2) the development of the microstructure during processing must be such that the pores remain attached to the boundaries notwithstanding the boundary movements that occur while normal grain growth takes place.

The factors that determine the interactions between pores and boundaries include microstructural features such as grain size, pore size, and pore separation and kinetic factors such as the rates of pore and boundary migration.[21] While the many uncertainties surrounding these items make a quantitative interpretation inappropriate, the main pattern of the reaction can be qualitatively presented in diagrammatic form. As an example, Fig. 6 shows the behavior for an idealized system in which the pore separation is proportional to the grain size and in which the pores (isolated and spherical) move by surface diffusion. The diagram is divided into two parts by the line V_f^*; this links the values of grain size, G, and pore size, r, which correspond to the critical value for the volume fraction of porosity described by Zener.[22] To the lower right of the line, the boundaries are thermodynamically unable to leave the pores; to the upper left, they satisfy the energy criteria to leave the pores but will only do so under kinetically favorable conditions, i.e. where the pores are large and hence slow-moving.[23] The objective, therefore, is so to control the grain size and pore size during processing that the microstructure never develops r and G values corresponding to points within the shaded region.

Procedures to avoid abnormal grain growth can be selected on the basis of the diagram. These are: (1) to use fine particle size powders (point A rather than point B) so that the separation region is avoided as densification proceeds, i.e. as the r value is reduced; (2) to enhance densification mechanisms (trajectory C) at the expense of grain growth mechanisms; and (3) to increase the mobility of the pores relative to that of the boundaries (I rather than II) so that larger pores are able to migrate with the boundaries.

The steps that have been taken to allow successful fabrication of zirconia ceramics have broadly followed the above lines. They are: (1) the use of fine powders; (2) the use of additives, most commonly to provide liquid phase at the firing temperature in order to accelerate densification; and (3) the use of processes which emphasize densification such as hot-pressing. The use of additives to alter the relative boundary and pore mobilities (the role often attributed[24] to MgO in the sintering of Al_2O_3) has not been successfully developed for zirconia.

Preparation and Use of Fine Powders

The objective of this approach is to prepare powders that are fine (< 500 nm), uniform in size, free from hard agglomerates, and homogeneous in respect to composition. Two linked stages in the use of such powders have been recognized as important, namely the preparation of the fine, homogeneous powders and the treatment or beneficiation of these powders so that forming can be achieved without the development or retention of hard agglomerates.

275

Many procedures have been considered[6] for the preparation of fine powders of stabilized zirconias (Table II). Two that have proved particularly successful in terms of the densities of the finished products are illustrative of the careful processing that is required. The hydrolysis of alkoxide precursors as described by Mazdiyasni *et al.,*[25] Fig. 7, results in high purity submicrometer aggregates of the mixed oxides in suspension in benzene; if these are then dried (100–110° C, air, 24 h), ground to form 5 nm particles after breakup of the aggregates, calcined (850°C, 30 min) to form 40 nm particles of the stabilized zirconia, cold-pressed, and sintered (1450°C, 16 h), polycrystalline materials of 2–5 μm grain size and density $\approx 95\%$ are obtained. Vasilos and Rhodes[26] demonstrated the importance of residual agglomerates in such powders by grinding the powders, separating out the agglomerates by sedimentation, centrifugally casting the remaining suspension to form a cast of 72% green density, and firing to give high density (99.5%), low grain size (0.2 μm) ceramics at temperatures as low as 1100°C.

The somewhat simpler technique of coprecipitation has also proved capable of development to the point of yielding high density products. Starting from mixed aqueous solutions of the chlorides, e.g. $ZrCl_4$ and YCl_3, coprecipitates are formed[27] by adding $6M$ NH_4OH; in the subsequent processing, the removal of the residual chloride ions and the prevention of formation of hard agglomerates are critical. The chlorine can be removed by aqueous washing, followed by washing with ethyl alcohol and drying at 120°C. The alcohol wash when followed by calcination (500–1000°C, 30 min) gives[27] loose agglomerates which break down on cold-pressing (50–200 MPa) to give powders which sinter to 99% density after some 3 h at 1150°C. Scott and Reed,[28] working with chloride-derived powders, have convincingly shown the importance of such beneficiation (Fig. 8) in the avoidance of hard agglomerates.

From such studies as these, it is clear that the attainment of sinterable powders is possible if suitable attention is given both to the precise preparation condtions and to the subsequent handling.[29] The large active surface areas of such powders mean that sensitivity of powder behavior to atmosphere,[30] to washing media,[27] and to residual chemical species[28] is to be expected, and the need for careful process control is evident. Given attention to these difficulties, however, these procedures have offered the surest avenue to the preparation of high-density, single-phase ceramics of stabilized zirconia.

Liquid Phase Sintering

A common procedure for lowering the firing temperature for systems where special attention cannot be given to powders is the use of liquid phase sintering aids, i.e. additives which at the firing temperature provide a liquid around the grains; this liquid then acts as a high diffusivity medium so that rapid densification can occur.[31]

For zirconia the additive used must meet requirements additional to those normally met (wetting liquid, limited solubility of the grains in the liquid) in that the boundary phase remaining after densification must have satisfactory electrical properties; in particular the phase should, if present as a continuous boundary film, have a reasonable ability to transport oxygen ions and a low value for the electronic conductivity. The latter feature is significant in that, unless t_i the ionic transport number for the boundary

276

phase is close to unity, the response time and the accuracy of signal from such devices as oxygen monitors are adversely affected.[32] Accordingly the properties of the boundary phase must either be carefully matched to the application or the boundary phase must be removed following densification (by vaporization, as in the case of LiF additions to MgO[33] or, as in sialon fabrication,[34] by reaction between the boundary phase and the grains).

A number of additives have been used, such as SiO_2[35] (7 mol% CaO, 1.6 μm and 16 μm, 1 wt% SiO_2) or kaolin[36] (2 wt% additions), which result in the formation of silicate phases at the grain boundaries. Kaolin is a successful densification aid for milled, fine-grained powders (15 mol% CaO, 12.5 m^2/g), allowing full density to be reached after sintering at 1600 °C, but is illustrative of the complex processes that can arise in liquid phase sintering. After firing, the boundaries contain either an insulating glassy or crystalline phase ($2CaO \cdot Al_2O_3 \cdot SiO_2$) in which, on annealing at 1250 °C, the stabilizing CaO from the grains dissolves, leaving porosity in the rims of the zirconia grains. Alumina has been found[37] to be an effective sintering additive (12 mol% CaO, 3 μm, 1700 °C, 1 wt% Al_2O_3) and, in a survey of a wide range of additive oxides made by Radford and Bratton,[38] it has proved, together with TiO_2, to be one of the less damaging in terms of electrical properties. Additions of 2 mol% Al_2O_3 or 5 mol% TiO_2 (10 mol% CaO or 20 mol% Y_2O_3, 2 μm) allowed densification to 93% at 1500 °C and gave materials whose conductivity in the 600–1000 °C range was about a factor of ten less than that of the pure, stabilized oxide.

An interesting additive system[39] is that of Bi_2O_3. When mixed with zirconia and either lime or yttria, this additive allows, after prefiring (16 h, 800–1050 °C), milling in isopropanol, isostatic pressing (400 MPa), air sintering (16–70 h, 1080 °C), and annealing (930 °C), the fabrication of 95% dense materials with small (3–4 μm) grain size. The conductivity of the best material (12 mol% Y_2O_3, 3 mol% Bi_2O_3, 85 mol% ZrO_2) is about a factor of three less than that of materials prepared without the additive. This work is a promising development in that the volatility of Bi_2O_3 and the good ionic conduction of a number of Bi_2O_3-containing phases suggest that the electrical behavior of the boundary phase can be brought to an acceptable level. The sintering conditions rival those of the best fine powder systems described earlier.

The many benefits that stem from the use of liquid phase additives (firing at low temperature, use of standard milled powders) make it an attractive process; careful attention to the nature and properties of the boundary phase together with exploration of the conditions required for its elimination subsequent to densification could bring about refinement of what is basically a low-cost and practicable approach to the sintering of zirconia.

Special Processing Techniques

A number of methods are known or have been proposed to promote densification in comparison with grain growth; these have been used in some instances with good effect in the preparation of zirconia ceramics.

Hot-Pressing

The pressing of powders during heat treatment has long been known[40] as one of the surest ways to prepare dense samples of oxides that are otherwise recognized as difficult to fabricate. The effectiveness arises because the applied pressure directly assists the densification while leaving grain growth unaffected. Although expensive and limited in terms of sample geometry, the technique is valuable in providing test samples and in allowing study of the densification mechanisms. Many instances have been reported of the hot-pressing of stabilized zirconia (Table I); commonly, temperatures above 1500 °C, when used with low applied pressures (≈ 10 MPa, graphite dies) have yielded densities above 90%. St. Jacques and Angers,[41] working with submicrometer powders (13–20 mol% CaO, 1600 °C, 5 MPa), achieved values (99.5) equivalent to the best fine powder results.

Reaction Sintering

The processing of a powder under conditions where a reaction and phase change are simultaneously occurring can lead to enhanced densification owing to the increased diffusion rates encountered at the transition. For zirconia, the clearest example is the enhancement found in the pure, unstabilized oxide at the monoclinic-tetragonal transition; Chaklader and Baker[42] achieved 99% dense samples by cycling zirconia powder at 800–1200 °C under 6 MPa pressure. A similar effect perhaps accounts for the densities achieved by Gupta[43] on sintering fine-grained, tetragonal powders, the maximum density values being found in those samples which lie at the point of instability between the tetragonal and monoclinic phases.

A more extreme instance[44] is the preparation of 98% dense monoclinic zirconia by hydrothermal (98 MPa, 1000 °C) oxidation of zirconium powder contained in a platinum capsule. Generally, the reactivitiy of such systems is not in itself sufficient to achieve high density and the simultaneous use of pressure or of fine-grained powders is required.

Fast Firing

The use of short, high temperature firing cycles[45] can enhance sintering under conditions where the activation enthalpy for the densification mechanism is greater than that for grain growth. For zirconia this condition, on the basis of the very sparse data available, is met.

Results obtained[46] on fine (20 nm) but agglomerated powder (5 μm) show that densities $> 80\%$ can be achieved for 14% Y_2O_3-stabilized material in 15 min at 1500 °C. While unsatisfactory, such density values are comparable to those of samples hot-pressed from the same powder and further work is justified to see if the process offers the possibility of achieving results equivalent to those of hot-pressing without the associated limitations and cost.

Conclusions

In the presence of a need to provide zirconia suitable for solid electrolyte applications, many different approaches have emerged. While a number of these have proved most effective, they are often associated with high cost or inconvenient limitations on sample shape and size, and no one method has won universal approval.

The clear feature is that attention to powder quality is of the utmost importance in this system. If sufficient attention is given to the achievement of fine, nonagglomerated powders, then conventional sintering is satisfactory. Even for other processes, which have been specifically adopted to counter the most stringent requirements for powder quality, these requirements should be regarded as reduced rather than avoided. Hot-pressing and liquid phase

278

sintering are both made that much more effective if submicrometer powders are used.

Future progress will probably be linked to a fuller exploration of the potential of additives to aid the sintering process. The careful choice of additive composition and of heat treatment should be a valuable line of research in respect to liquid phase sintering of zirconia. The question too of solid state sintering aids, which have after all provided the classical processing route for high quality alumina ceramics, should also remain open. The search for a suitable additive for electrolyte-grade zirconia would appear to be a most rewarding topic for research study.

References

[1](a) P. Hagenmuller and W. van Gool, Solid Electrolytes. Academic Press, New York, 1978.
  (b) J. Hladik, Physics of Electrolytes. Academic Press, New York, 1972.
  (c) S. Geller, Solid Electrolytes. Springer, Berlin, 1977.
[2](a) C. B. Alcock, Mater. Sci. Res., 10, 419 (1975).
  (b) W. A. Fischer and D. Janke, Metallurgische Elektrochemie. Springer, Berlin, 1975.
[3]L. Heyne; p. 65 in Measurement of Oxygen. Edited by H. Degn, I. Balslev, and R. J. Brook. Elsevier, Amsterdam, 1976.
[4]D. Yuan and F. A. Kroeger, J. Electrochem. Soc., 116 [5] 594 (1969).
[5]A. McDougall, Fuel Cells. MacMillan, London, 1976.
[6]M. J. Bannister and W. G. Garrett, Ceramurgia Int., 1 [3] 127 (1975).
[7]R. L. Coble, J. Appl. Phys., 32, 793 (1961).
[8]P. Jorgensen; p. 401 in Sintering and Related Phenomena. Edited by G. C. Kuczynski, N. A. Hooton, and C. F. Gibbon. Gordon and Breach, New York, 1967.
[9]R. L. Coble, Mater. Sci. Res., 6, 177 (1973).
[10]P. E. Evans, J. Am. Ceram. Soc., 53 [7] 365 (1970).
[11]M. S. Seltzer and P. K. Talty, ibid., 58 [3-4] 124 (1975).
[12]M. Heughebaert-Therasse, Ann. Chim., 2, 229 (1977).
[13]W. H. Rhodes and R. E. Carter, J. Am. Ceram. Soc., 49 [5] 244 (1966).
[14]J. R. Hague, quoted by R. C. Garvie; p. 117 in High Temperature Oxides 5-II. Edited by A. M. Alper. Academic Press, New York, 1970.
[15]R. G. St. Jacques and R. Angers, Trans. Br. Ceram. Soc., 72 [6] 285 (1973).
[16]A. Diness and R. Roy, Solid State Commun., 3, 125 (1965).
[17]T. Y. Tien, J. Appl. Phys., 35, 122 (1964).
[18]A. I. Joffe, M. V. Inozemtsev, A. S. Lipilin, M. V. Perfilev, and S. V. Kavoparkov, Phys. Status Solidi A, 30, 87 (1975).
[19]T. Y. Tien and E. C. Subbarao, J. Am. Ceram. Soc., 46 [10] 489 (1963).
[20]R. L. Coble and J. E. Burke, Prog. Ceram. Sci., 3, 197 (1963).
[21](a) R. J. Brook, J. Am. Ceram. Soc., 52 [1] 56 (1969).
  (b) F. M. A. Carpay; p. 261 in Ceramic Microstructures '76. Edited by R. M. Fulrath and J. A. Pask. Westview Press, Boulder, Colo., 1977.
  (c) M. F. Yan, R. M. Cannon, and H. K. Bowen, ibid., p. 276.
[22]C. Zener quoted by C. S. Smith, Trans. AIME, 175, 15 (1948).
[23]P. G. Shewmon, ibid., 230, 1134 (1964).
[24]A. H. Heuer, J. Am. Ceram. Soc., 62 [5-6] 317 (1979).
[25](a) K. S. Mazdiyasni, C. T. Lynch, and J. S. Smith, ibid., 48 [7] 372 (1965).
  (b) K. S. Mazdiyansni, C. T. Lynch, and J. S. Smith, ibid., 50 [10] 532 (1967).
[26]T. Vasilos and W. H. Rhodes; p. 137 in Ultrafine Grain Ceramics. Edited by J. J. Burke, N. L. Reed, and V. Weiss. Syracuse University Press, 1970.
[27]K. Haberko, Ceramurgia Int., 5 [4] 148 (1979).
[28]C. E. Scott and J. S. Reed, Bull. Am. Ceram. Soc., 58 [6] 587 (1979).
[29](a) G. Y. Onoda and L. L. Hench, Ceramic Processing Before Firing. Wiley, New York, 1978.
  (b) H. Palmour III, R. F. Davis, and T. M. Hare, Processing of Crystalline Ceramics. Plenum Press, New York, 1978.
[30]M. A. Thompson, D. R. Young, and E. R. McCartney, J. Am. Ceram. Soc., 56 [12] 648 (1973).
[31]W. D. Kingery, J. Appl. Phys., 30, 301 (1959).
[32]L. Heyne, Electrochim. Acta, 21, 303 (1976).
[33]R. W. Rice, Proc. Br. Ceram. Soc., 12, 99 (1969).
[34]K. H. Jack, J. Mater. Sci., 11, 1135 (1976).

[35]J. F. Shackelford, P. S. Nicholson, and W. W. Smeltzer, *Bull. Am. Ceram. Soc.,* **53** [12] 865 (1974).
[36]F. J. Esper and K. H. Friese, *Ber. Dtsch. Keram. Ges.,* **55**, 314 (1978).
[37]H. Takagi, S. Kuwabara, and H. Matsumoto, *Sprechsaal,* **107** [13] 584 (1974).
[38]K. C. Radford and R. J. Bratton, *J. Mater. Sci.,* **14**, 59 (1979).
[39]K. Keizer, M. J. Verkerk, and A. J. Burggraaf, *Ceramurgia Int.,* **5**, 143 (1979).
[40]T. Vasilos and R. M. Spriggs, *Prog. Ceram. Sci.,* **4**, 95 (1966).
[41]R. G. St. Jacques and R. Angers, *J. Am. Ceram. Soc.,* **55** [11] 571 (1972).
[42]A. C. D. Chaklader and V. T. Baker, *Bull. Am. Ceram. Soc.,* **44** [3] 258 (1965).
[43]T. K. Gupta, *Sci. Sintering,* **10**, 205 (1978).
[44]M. Yoshimura and S. Somiya, *Bull. Am. Ceram. Soc.,* **59** [2] 246 (1980).
[45]M. P. Harmer, E. W. Roberts, and R. J. Brook, *Trans. Br. Ceram. Soc.,* **78** [1] 22 (1979).
[46]G. A. Wood; B. Sc. Thesis, University of Leeds, 1980.
[47]H. H. Mobius, H. Witzmann, and D. Gerlack, *Z. Chem.,* **4**, 154 (1964).
[48]L. A. Simpson and R. E. Carter, *J. Am. Ceram. Soc.,* **49** [3] 139 (1966).
[49]W. D. Kingery, J. Pappis, M. E. Doty, and D. C. Hill, *ibid.,* **42** [8] 393 (1959).
[50]A. E. Paladino and W. D. Kingery, *J. Chem. Phys.,* **37**, 957 (1962).
[51]J. Patterson, *J. Electrochem. Soc.,* **118**, 1033 (1971).
[52]R. E. Carter and W. L. Roth; p. 125 in EMF Measurements in High Temperature Systems. Edited by C. B. Alcock. IMM, London, 1968.

Table I. Fabrication of stabilized zirconias

Ref.	Material	Grain size (μm)	Temp. (°C)	Pressure (MPa)	Density achieved (%)[†]
*	CSZ	1–5	1600–1800		75
19	CSZ		1800		91
28	YSZ	0.02	1350		99.5
27	YSZ	0.02	1150–1300		99
40	YSZ	0.04	1100		99.5
38	YSZ;CSZ	2	1500		93 (L)
39	YSZ;CSZ		1100		95 (L)
35	CSZ	1.6; 16	1800		83 (L)
36	CSZ		1600		99 (L)
41	CSZ	<1	1600	5	99.5
10	YSZ		1500	8	92
46	CSZ	10	1700	20	91

*S-I. Pyun, *Ceramurgia Int.,* **5** [2] 61 (1979). †L denotes use of liquid phase sintering aid.

Table II. Powder preparation of stabilized zirconia (Ref. 6)

Coprecipitation of mixed hydroxides
Evaporation and decomposition of mixed salt solutions
Hydrolysis of mixed alkoxide solutions
Freeze drying of mixed salts
Dehydration of mixed oxide sols or gels
Decomposition of dry-mixed salts
Decomposition of salt-impregnated polymer fibers
Hydrothermal recrystallization
Flame spraying of salt solutions
Organic dehydration of mixed salt solutions

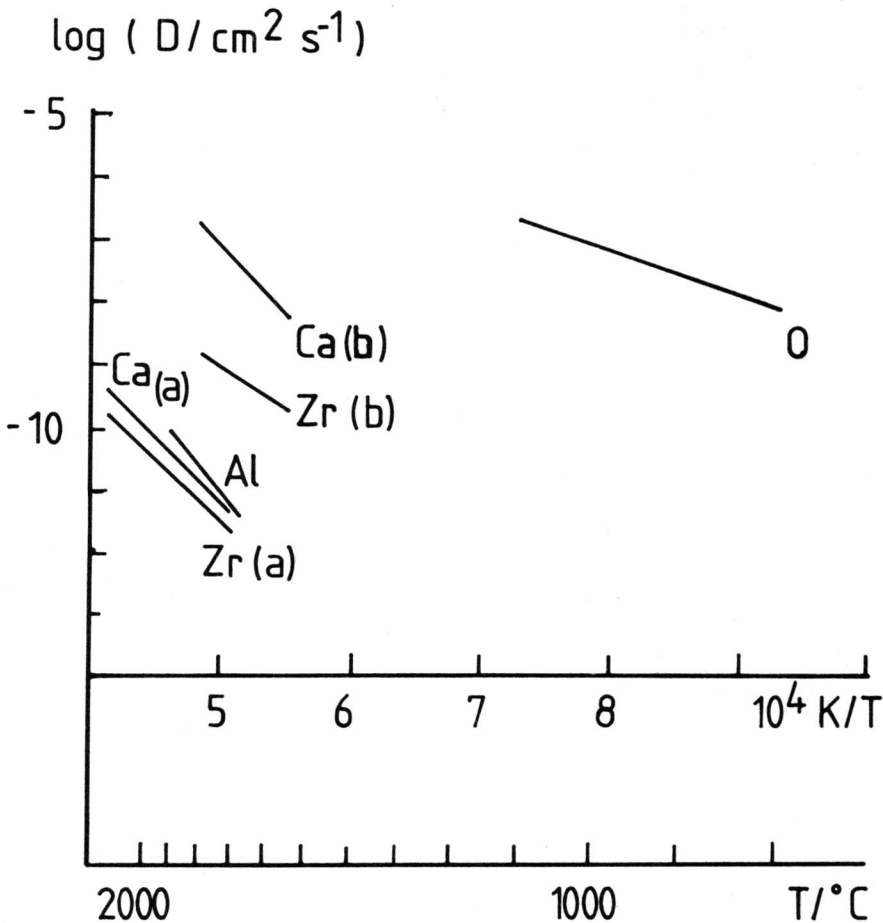

Fig. 1. Diffusion coefficients for CaO-stabilized ZrO_2. Data for Al in Al_2O_3 are included for comparison. Sources were, for Ca_a, Zr_a: Ref. 13; for Ca_b, Zr_b: Ref. 47; for O: Refs. 48 and 49; for Al: Ref. 50. The differences between the two sets of Ca and Zr data are possibly explained by sample differences or by the interferences of the decay products in the interpretation of the tracer results.

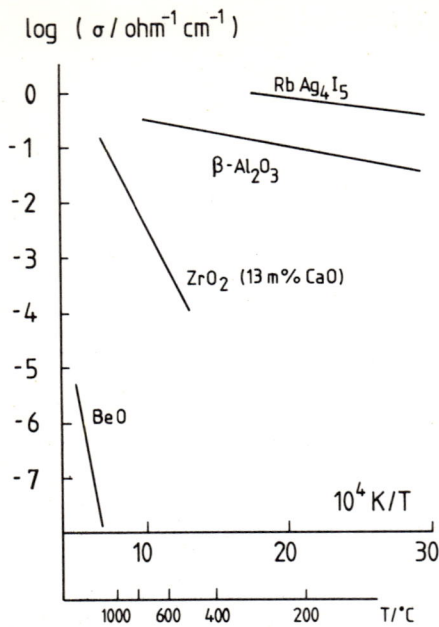

Fig. 2. Conductivity of CaO-stabilized ZrO_2 in comparison with one poor and two good ionic conductors.

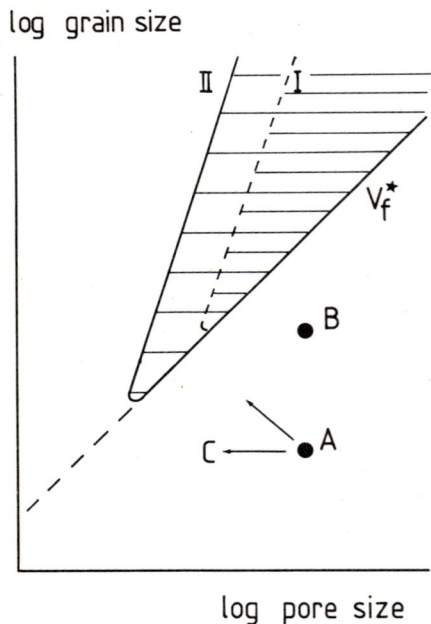

Fig. 3. The ionic range of CaO-stabilized ZrO_2 (Ref. 51). Electrolytic devices based on the material can be operated under the conditions of p_{O_2} and T within the shaded region.

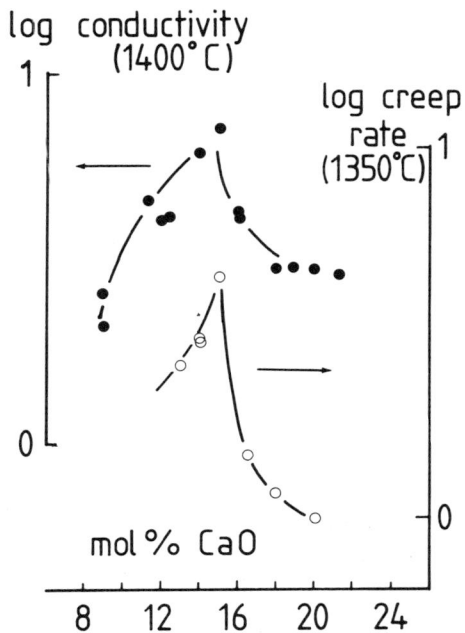

Fig. 4. The dependence of creep rate and ionic conductivity on the CaO content of zirconia ceramics. Data from Refs. 15 and 52.

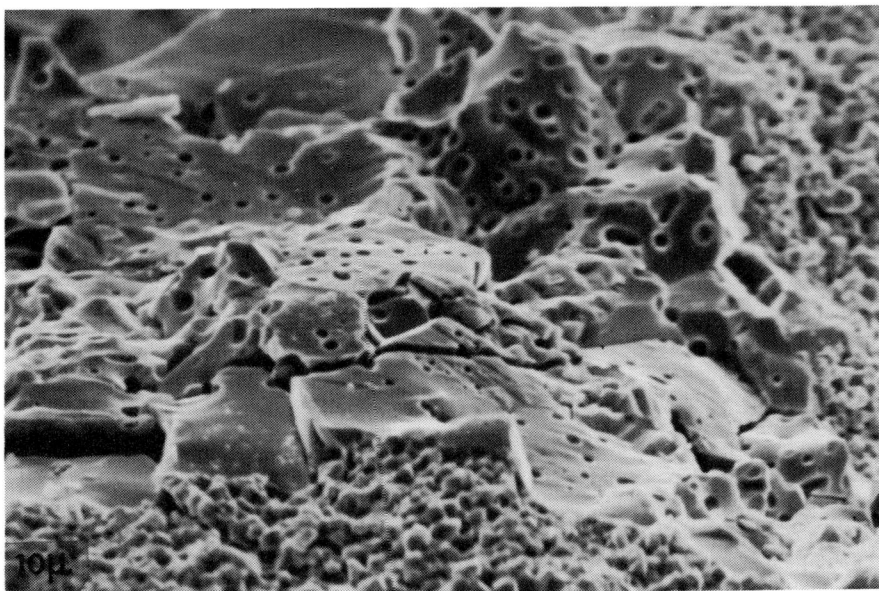

Fig. 5. Abnormal grain growth in CaO-stabilized ZrO_2.

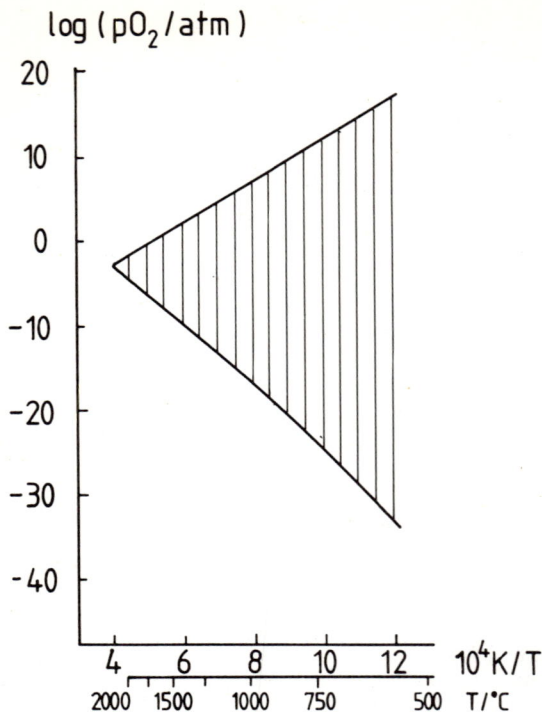

Fig. 6. The influence of pore and grain size on the interactions between pores and grain boundaries.

Preparation of alkoxides (YSZ)

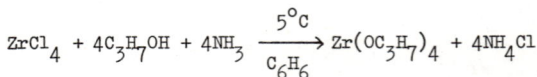

$$ZrCl_4 + 4C_3H_7OH + 4NH_3 \xrightarrow[C_6H_6]{5^{\circ}C} Zr(OC_3H_7)_4 + 4NH_4Cl$$

react, filter, distil off C_6H_6, recrystallise

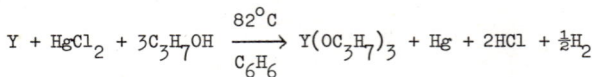

$$Y + HgCl_2 + 3C_3H_7OH \xrightarrow[C_6H_6]{82^{\circ}C} Y(OC_3H_7)_3 + Hg + 2HCl + \tfrac{1}{2}H_2$$

react, filter, distil, recrystallise

Hydrolysis of alkoxides;   add water to benzene solutions of alkoxides

$$2Y(OC_3H_7)_3 + 3H_2O \longrightarrow Y_2O_3 + 6C_3H_7OH$$

$$Zr(OC_3H_7)_4 + 2H_2O \longrightarrow ZrO_2 + 4C_3H_7OH$$

Dry (100 – 110, air, 24 hr)$\longrightarrow$ Mill $\longrightarrow$ Calcine (850°C, 30 min)

Fig. 7. The preparation of stabilized zirconia from alkoxides (Ref. 25).

```
YSZ powder 19 nm                          1 μm solid agglomerates
|                                         100 μm loose agglomerates
|                                            |
|                                            |
wash as aqueous suspension                   |
6 times to remove Cl⁻                        |
|                                            |
|                                            |
centrifuge                                   |
|                                            |
|                                            |
dry                                       1 - 6 μm agglomerates
|                                            |
|                                            |
mill in 2-propanol with                      |
ZrO₂ beads                                   |
|                                            |
|                                            |
centrifuge                                   |
|                                            |
|                                            |
dry                                       weak 1 μm agglomerates
|                                            |
|                                            |
disaggregate by shaking                      |
with fluorocarbon media                      |
|                                         uniform < 1 μm agglomerates
calcine (500°C, 5 min, O₂)
```

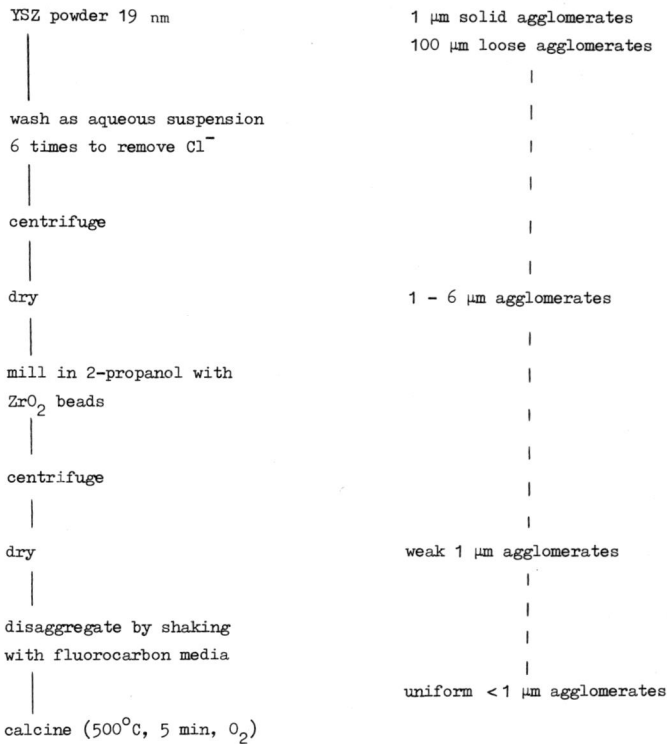

Fig. 8. Beneficiation of zirconia powders prepared from chlorides (Ref. 28).

Factors influencing the performance of zirconia-based oxygen monitors

B. C. H. STEELE, J. DRENNAN, R. K. SLOTWINSKI,
N. BONANOS, AND E. P. BUTLER

Wolfson Unit for Solid State Ionics
Imperial College
London SW7
England

At moderate temperatures (>600 °C) zirconia-based oxygen monitors incorporating porous Pt electrodes should respond to changes in oxygen partial pressure within a few milliseconds. However, the performance of such devices can be degraded by the microstructure of the zirconia ceramic, the type and morphology of the electrode system, and specific design features of commercial monitors. The influence of microstructural parameters is discussed with particular reference to silicate-containing grain boundary phases detected by electron microscopy. Data are also presented showing how the presence of precipitated tetragonal and monoclinic zirconia can modify the ac response of partially stabilized zirconia-based electrolytes. The different behavior of catalytic and noncatalytic electrodes is summarized and specific design features of oxygen monitors for flue gas analysis and for heat treatment furnaces are discussed. Finally some alternative concepts and materials which might challenge the current supremacy of zirconia-based electrolytes are presented.

Many factors can influence the performance of zirconia-based oxygen monitors and the relative importance of these factors is often determined by the particular application proposed for the monitor. The interaction between the structure-property relationships of the ceramic zirconia electrolyte device and operating parameters such as temperature, type of gaseous environment, and specific design features can often be complex. Furthermore, analysis of this complex situation is complicated by the fact that very little quantitative information has been published about the behavior of oxygen monitors in industrial environments. Although perhaps understandable because of proprietary interests, this situation has undoubtedly retarded the development of zirconia oxygen monitors optimized for specific applications.

It should be recalled that the incorporation of zirconia-based electrolytes in oxygen monitors was first demonstrated more than 20 years ago.[1,2] By 1967, after 10 years of investigations in many laboratories, the principal features of the transport properties of zirconia-based electrolytes were established.[3-6] These features included the variation in conductivity with increasing concentration of dopant, the decrease in conductivity with time due to defect ordering processes, and the limits of the ionic domain due to the occurrence of hole and electron conductivities (Fig. 1). Data on these properties were collated in a useful review[7] in 1970.

Commercially available oxygen monitors incorporating zirconia-based electrolytes began to be produced about 1965. It should be emphasized, however, that until the past two or three years there has been little cooperation between the instrument firms and ceramic manufacturers to optimize the properties of the ceramic zirconia electrolyte for specific applications. In an effort to improve and standardize the properties of zirconia electrolytes the Commission on High Temperatures and Refractory Materials of the International Union of Pure and Applied Chemistry (IUPAC) initiated a collaborative program in 1975 to examine the properties of commercially available zirconia electrolytes. Further details of this program are summarized in the Appendix and some of the information relating to the microstructural parameters of selected samples will be discussed in a later section.

The technological applications of zirconia-based oxygen monitors are now widespread. This paper is principally concerned with the performance of oxygen monitors in flue gases for combustion control and their incorporation in heat treatment furnaces for atmosphere control.

Performance of Zirconia-Based Oxygen Monitors

General

In order to consider the factors influencing the performance of zirconia-based oxygen monitors it is useful to consider the ideal situation and then introduce those complications which can arise in technological devices. The cell

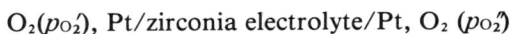

$$O_2(p_{O_2}'), Pt/zirconia \ electrolyte/Pt, O_2 \ (p_{O_2}'')$$

generates an emf according to the relationship

$$E = \frac{RT}{4F} \int_{p_{O_2}'}^{p_{O_2}''} t_{ion} d \ln p_{O_2} \tag{1}$$

If we assume the electrolyte is impermeable and is operating well within the ionic domain ($t_{ion} \approx 1$), then the emf is simply given by:

$$E = \frac{RT}{4F} \ln \frac{p_{O_2}''}{p_{O_2}'}$$

Implicit in the preceding derivations is the assumption of local thermodynamic equilibrium at the electrolyte-electrode-gas three-phase interphase, that is the electrodes are behaving reversibly. The rate at which the electrodes attain equilibrium will depend on the gaseous species in contact with the electrode and the equilibration kinetics of the interfacial electrolyte space charge region. The simplest gaseous atmosphere consists of oxygen-inert gas mixtures, and the easiest way to examine the electrode kinetics is to measure the impedance (or admittance) over a wide range of frequency and temperature of a ceramic zirconia electrolyte sample incorporating identical electrodes of sputtered platinum on opposite faces.

Following Bauerle,[8] the simplest equivalent circuit for this arrangement is given in Fig. 2 and the corresponding impedance behavior in the complex plane representation is shown in Fig. 3. Although all three semicircles are rarely visible at a particular temperature and are often distorted and depressed below the horizontal axis, it is often possible to obtain the relevant values for the appropriate R-C combination and the corresponding characteristic frequency. Interpretation of these impedance measurements is discussed in detail by Kleitz et al. (this issue, pp. 310–36). Note, however, at this stage, that providing the oxygen partial pressure is greater than $\approx 10^{-3}$ atm, the characteristic frequency ($\omega_0{}'{}'$) of the electrode system at 700 °C, incorporating well prepared sputtered electrodes, is in the range 10^2–10^3 Hz. As the frequency response can be transformed into the time domain, this implies that under these conditions the electrode system will respond to changes in oxygen partial pressure within a few milliseconds. This time scale is more than adequate for most technological applications and so attention will be focused in the following sections on those factors which can degrade the performance of zirconia-based oxygen monitors.

Influence of Grain Boundary Phases

It is well known[9] that at present most commercially available zirconia-based ceramic electrolytes contain impurities, particularly SiO_2 and Al_2O_3. These impurities may be deliberately added to control grain growth and aid densification during the sintering process or may be inadvertently introduced via the original raw material or during the fabrication process. These impurities segregate at the grain boundaries and give rise to a grain boundary impedance which is associated with a characteristic semicircle in the complex plane representation (Fig. 3). If there is only a small concentration of these impurities present then a high temperature anneal may cause dissolution of the grain boundary phase and the associated semicircle vanishes.[10] However, in the IUPAC samples examined so far, the relatively high concentration of SiO_2 and Al_2O_3 present ensures that the grain boundary phases are still present at elevated temperatures (≈ 1200 °C). The thickness of these grain boundary phases is such that they can be examined with modern electron microscopes which allow microdiffraction and microanalysis to be carried out on selected areas only 20 nm in diameter. The microanalysis scan in Fig. 4, for example, shows the relative concentrations of Zr, Ca, Si, and Al in the CaO-stabilized ZrO_2 matrix and the grain boundary phase of sample 4. It is immediately apparent that the matrix phase has been depleted of CaO which has dissolved in the grain boundary silicate phase. During the high temperature densification process, the distribution of CaO will approach the equilibrium situation given by equal thermodynamic activity values for CaO in the matrix and grain boundary phase. According to available data,[11] the activity of CaO in stabilized zirconia, $Zr_{0.85}Ca_{0.15}O_{1.85}$, has a value of 0.015 at 1000 °C and so this composition would be in equilibrium[12] with a binary calcium silicate of composition $Ca_{0.4}Si_{0.6}O_{1.6}$ or a ternary calcium alumina silicate of composition $Ca_{0.5}Al_{0.3}Si_{0.7}O_{2.35}$. In fact, if sufficient silica is present, then the extent of CaO depletion from the zirconia matrix can produce local destabilization and Kirkendall pores[13] near the grain boundaries.

The grain boundary phases may be crystalline or glassy, depending on the particular composition and heat treatment as illustrated in Figs. 5 and 6, which reproduce electron micrographs of grain boundary phases in IUPAC samples 2 and 3, respectively. Obviously, the precise nature of the grain boundary phase will influence the electrical and mechanical properties of the ceramic electrolyte, and it is essential therefore that if these phases are present in commercial material they should be fully characterized by modern physical instrumental techniques.

The electrical behavior of grain boundary phases and regions is of particular importance to the performance of zirconia-based oxygen monitors. Most discussions of the grain boundary impedance follow the idealized model first proposed by Bauerle[8] and depicted in Fig. 7(a). Two zirconia electrolyte grains A and B are assumed to be in direct contact in the region x-y. Elsewhere, the grains are in contact via an impurity second phase which is a poor oxygen ion conductor. Possible electrical equivalent circuits are given in Figs. 7(b) and (c) in which R_{GB}, C_{GB} refer to the grain boundary phase and R_C is the constriction resistance produced by the limited direct control between the stabilized zirconia grains. It is assumed that R_C is purely geometrical and exhibits the same activation energy as the bulk stabilized zirconia material. The equivalent circuit shown in Fig. 7(c) was proposed by Schouler et al.[11] to obtain a more consistent interpretation of the various values obtained for R_C, R_{GB}, C_{GB} in their complex plane analysis of the impedance of single and polycrystalline stabilized zirconia samples. The structure of actual grain boundaries is obviously more complex than that envisaged in these simple models, but it is difficult at present to incorporate the effects of space charge, grain boundary segregation, and mobility differences between grain boundary and bulk,[15] although it should be noted that attempts have been made[16,17] to model complicated conduction pathways in beta alumina ceramics. The simple models, however, do have the merit of being able to assess fairly easily by complex plane analysis the relative importance of the grain boundary impedance.

The results of such an analysis are presented in Figs. 8 and 9 for two commercial samples of zirconia-yttria: samples 5 and 6. From the details given in the Appendix and the optical micrograph reproduced in Fig. 8, it is evident that sample 5 is a relatively coarse-grained dense ceramic (30–50 μm) with a significant concentration of impurity phase uniformly distributed around the grain boundaries. In contrast, sample 6 appears to be free of impurities, and the optical micrograph in Fig. 9 reveals a fine-grained material (≈ 1 μm) incorporating a significant proportion of small closed porosity which effectively reduces the intergranular contact area producing a constriction resistance. Both samples were subjected to complex plane impedance analysis to separate the bulk and grain boundary contributions to the total conductivity. Inspection of the log conductivity vs $1/T$ plots in Fig. 10 shows that the bulk conductivity of the purer material (sample 6) is slightly higher than that of the more impure sample 5. However, by the time the respective grain boundary contributions are included, the total conductivities for the two samples are similar. Thus, although the purity and microstructure of these two samples are very different, their dc resistances do not differ greatly.

The question arises whether the performance of oxygen monitors incorporating samples 5 and 6 might be expected to be different. If we confine our considerations to potentiometric (zero current) devices operating at $\approx 700\,°C$, then for similar electrode structures the response of both oxygen monitors to changes in oxygen partial pressure should also be very similar (i.e. a few milliseconds). However the oxygen monitor incorporating the less pure zirconia electrolyte (sample 5) could be subjected to relatively long term emf changes due to the presence of the impurity second phases. This phenomenon has been discussed by Beekmans and Heyne[17] and Heyne and den Engelsen[18] and may arise because the different transport properties of the impurity phase can generate an "internal" emf in response to a concentration gradient established within the ceramic electrolyte. If we consider a homogeneous ceramic electrolyte (Fig. 11) with sputtered porous platinum electrodes on opposite faces initially in contact with the same value of oxygen partial pressure, then the oxygen concentration will be constant throughout the bulk (Fig. 11(b)) except at the very narrow space charge region (Fig. 11(a)). If the oxygen partial pressure is now rapidly increased at the right-hand electrode the space charge region (Fig. 11(c)) rapidly equilibrates to the new value of imposed oxygen partial pressure. As the space charge region* is only one or two atomic layers thick, the rate of equilibration is usually determined by the overall kinetics for the electrode reaction

$$\tfrac{1}{2}O_2 + 2e \rightleftharpoons O^{2-}$$

which is of the order of a few milliseconds at 700 °C. Although the space charge regions rapidly equilibrate, the interior of the electrolyte also adjusts its composition as shown in Fig. 11(d) and this chemical diffusion process is many orders of magnitude slower than the surface equilibration. The reason that the correct emf is observed after a few milliseconds even though a diffusion process is still occurring arises from the fact that t_{ion} in Eq. (1) is effectively a constant (≈ 1) and can be placed in front of the integral. Consequently the emf becomes independent of the changing oxygen composition within the bulk of the electrolyte and is determined by the boundary values only.

For an inhomogeneous electrolyte sample containing impurity second phases the situation is more complicated. An analysis of the model inhomogeneous electrolyte depicted in Fig. 12 shows that the monitor initially develops the correct emf value (Fig. 13(a)) corresponding to a change in oxygen partial pressure at the right-hand electrode. However the subsequent concentration changes within the electrolyte phase (Fig. 13(b)) can produce an emf across the impurity phase which contributes to the overall emf and deviations from the correct emf occur (Fig. 13(a)). Depending on the temperature, these deviations can persist for minutes or hours and erroneous results will be recorded for this period. Restoring the oxygen partial pressure at the right-hand electrode to its original value also produces deviations (Figs. 13(c), (d)) which can persist for an extended period.

The magnitude of these deviations depends very much on the transport properties of the second phase and for phases exhibiting an ionic transference number of essentially unity erroneous emf values should not be observed. In this context the use of Bi_2O_3 as a sintering aid is interesting as this material is an excellent oxygen ion conductor at high oxygen partial pressures and moderate temperatures. A preliminary study[19] of stabilized zirconia incorporating Bi_2O_3, however, revealed a complicated microstructure with additional phases produced by the reaction of ZrO_2 and Bi_2O_3. The transport properties of these new phases are not known and it is impossible at present to predict whether their presence will produce significant deviations from the expected emf.

Influence of Precipitated Phases

The discussion so far has been concerned with pure fully stabilized zirconia ceramic electrolytes and with zirconia-based electrolytes incorporating an impurity second phase around the grain boundaries. Partially stabilized zirconia (PSZ) ceramics, however, have often been incorporated into commercially available oxygen monitors because it is well known[9] that PSZ ceramics are much more resistant to thermal and mechanical stresses than the fully stabilized material. Many contributions to the First International Conference on Zirconia are concerned with the development of strong toughened zirconia ceramics. These ceramics are toughened by utilizing the tetragonal-to-monoclinic phase transition of ZrO_2 precipitates dispersed in a cubic zirconia matrix. The size of tetragonal precipitate which can be retained at room temperature, and which will transform martensitically to monoclinic symmetry in a stress field of a crack tip, is an important feature in developing optimum strength and toughness. It is appropriate therefore to discuss the influence of these precipitates on the electrical properties of PSZ.

The tetragonal precipitates occur both at grain boundaries and within the cubic zirconia grain. As discrete particles, their principal effect will be to increase the resistivity of the ceramic as the volume fraction of the cubic phase decreases, but their presence is not expected to influence the steady state emf generated according to Eq. (1). It was expected, however, that the size, distribution, and type of precipitate would influence the ac response of PSZ and possibly thus provide a nondestructive technique for monitoring the development of the optimum microstructure.

To examine this possibility, samples of 8 mol% (3.8 wt%) calcia-stabilized zirconia have been solution-treated in the single-phase (cubic) region at 1820 °C in air for 3 h. These specimens were fast-cooled (50 °C/min) to 1300 °C and then allowed to cool further to room temperature. To encourage growth of the tetragonal precipitates the samples were next aged at 1320 °C for a variety of annealing times: 10, 20, 30, 40, 50, and 100 h. Examination by X-ray diffraction (Figs. 14 and 15) indicated that the optimum concentration of tetragonal zirconia precipitates was attained after $\approx$ 40 h annealing. Further annealing at 1320 °C coarsened the particles too much and these overaged precipitates transformed to the monoclinic symmetry when cooled to room temperature. A dark field electron micrograph of a sample annealed for 20 h reveals the distribution of the tetragonal particles (Fig. 16(a)) and a bright field electron micrograph of the same sample (Fig. 16(b)) shows the characteristic "tweed-like" contrast of a high second-phase content in the cubic matrix. A bright field image (Fig. 17) of an overaged sample shows the appearance of the transformed monoclinic precipitate.

The impedance of a series of specimens aged for different periods of time were subjected to complex plane analysis and the complex resistivity results are depicted in Fig. 18, for a temperature of 275 °C. The shape of the spectra is clearly influenced by the growth of the precipitates, and a more detailed examination and interpretation is now in progress.

Influence of Electrode Parameters

Oxygen in Inert Atmosphere: As already mentioned in the section Influence of Grain Boundary Phases, for a simple oxygen-inert gas mixture the speed of response of an oxygen monitor is determined by the relevant electrode kinetics which for sputtered platinum electrodes > 700 °C produce a response time of a few milliseconds when the oxygen content of the gas exceeds 0.1%. The response of technological devices is often slower because of the necessity of fabricating a robust electrode configuration which often incorporates a porous ceramic coating[20] to protect the platinum electrode. This morphology can extend the diffusion pathway between the gaseous environment and the electrolyte-electrode-gas three-phase contact region with a consequent decrease in the response performance. It should also be noted that surface preparation of the zirconia-based ceramic can be an important parameter[21] in determining the long-term stability and adhesion of the platinum electrode exposed to industrial environments.

Oxygen in Reactive Gases: When oxygen is present in contact with reactive gaseous species such as H_2, CO, CH_4, etc. then above $\approx 600\,°C$ platinum electrodes effectively catalyze the relevant reactions such as

$$CO + \tfrac{1}{2}O_2 \rightleftharpoons CO_2$$

so that the monitor responds to the appropriate equilibrium oxygen partial pressure. At lower temperatures, the observed emf reflects the relative rates of adsorption and desorption of the reacting species on the platinum electrode and a nonequilibrium situation often prevails. At low concentrations adsorbed carbon monoxide, for example, can often produce spurious emf values which may even oscillate[22] under certain temperature and gas composition situations. In fact a device to monitor carbon monoxide levels in gases incorporates a zirconia-based electrolyte fitted with two types of platinum electrode configurations.[23] At low temperatures, therefore, care is needed in the interpretation of observed emf values which may reflect a nonequilibrium situation. Moreover it should be recognized that platinum electrodes can lose their catalytic activity when poisoned by Pb, S, and other species.

Other metals may not exhibit the high catalytic behavior characteristic of platinum and it is possible to develop noncatalytic electrodes for measuring the free oxygen content in nonequilibrium gas mixtures containing oxygen and species such as CH_4, C_3H_6, CO, H_2, etc. Silver electrodes, for example, have been found[24] to be noncatalytic to CH_4 oxidation, and Pb-poisoned Pt electrodes allow quantitative measurement of nonequilibrium oxygen concentrations in many combustible gases.[25]

Design Features

Oxygen Monitors for Flue Gas Analysis

The present survey is restricted to in-situ monitors incorporating zirconia-based electrolyte and designed to monitor equilibrium oxygen contents in boiler flue gases to improve combustion efficiency. The principal features of these devices have often been discussed,[26-29] and the factor influencing their design can best be discussed with reference to Fig. 19 which depicts the calculated equilibrium oxygen partial pressures and cell emf as a function of methane/air ratio at 700° and 500°C. The diagram is representative of all fossil fuels and the large change in oxygen partial pressure and cell emf at the stoichiometric ratio ($\lambda = 1.0$) is utilized in the car exhaust λ sensor to provide an emission control system. As it can be dangerous to have unburnt fuel in the flue gas of a boiler, it is usually desirable to operate with 0.1 to 1.0% excess oxygen in the flue gases. Without control many boilers currently operate at 5–10% excess oxygen in the flue gas which corresponds to 20–40% excess air and an associated loss of combustion efficiency. Assuming that the control is set at 0.5% excess oxygen and that the device incorporates an air reference electrode, the calculated emf at 400°C should then be 53 mV. However, flue gas temperatures often vary, for example, between 200 and 600°C and if the zirconia-based oxygen monitors were simply inserted in the flue gas then interpretation of the observed emf would be difficult for the following reasons.

(i) The slow electrode kinetics at low temperatures would produce spurious emf data.

(ii) The variation of emf with temperature $(\delta E/\delta T)$ for the cell under discussion is given by

$$\Delta S/4F = (R \ln p_{O_2'}/p_{O_2''})/4F = 0.08 \text{ mV/K}$$

Thus increasing the cell temperature to 500 °C (773 K) produces an emf change of 8 mV, which is 16% of the emf output at 400 °C. Temperature compensation of this magnitude would obviously increase the complexity of the electronic control circuitry.

(iii) In an environment where the temperature is changing rapidly it is probable that temperature gradients will exist across the zirconia electrolyte and that thermoelectric voltages will be superimposed on the emf output. The magnitude of this effect depends on the temperature difference between the two electrodes but can be calculated from empirical relationships.[30,31]

(iv) Rapid changes in temperature will subject the zirconia ceramic electrolyte to thermal stresses which in turn will demand more stringent specifications for the ceramics and certain of the other design parameters.

To overcome these problems, oxygen monitors for flue gas analysis incorporate an oven to ensure that the zirconia electrolyte operates at a constant elevated temperature (700–800 °C). This arrangement is depicted schematically in Fig. 20(a), in which it can be seen that a relatively short (5–10 cm) piece of zirconia ceramic electrolyte tube is joined to an appropriate heat and corrosion resistant alloy. The seal may be fabricated from glass, glass-ceramic, fused ceramic, or gold and it is important to ensure that the zirconia ceramic is strong enough to withstand differential thermal expansion stresses transmitted by the metal-ceramic seal. If gold is used as the seal material then mixed potentials may arise if the seal is close to the electrode region as shown in Fig. 20(b) and this configuration should be avoided; otherwise the performance and long term stability of the monitor will be impaired.

The air reference electrode may be replaced by a metal-metal oxide electrode such as Pd/PdO[32,33] which enables a miniaturized device to be fabricated (Fig. 20(c)), or alternatively both an air reference and metal-metal oxide may be incorporated in the oxygen monitor to provide an internal standard to check that the device is functioning correctly (Fig. 20(d)).

Oxygen Monitors for Heat Treatment Furnaces

The range of problems encountered when using in-situ oxygen monitors to control the furnace atmosphere in heat treatment furnaces may be illustrated by reference to carburizing furnaces.[34] In this situation the oxygen partial pressures are invariably low ($\approx 10^{-20}$ atm) and the temperatures usually in the range 800–1100 °C. Under these conditions significant ($\approx 1\%$) electronic conduction may develop due to the reaction

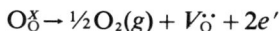

$$O_O^x \rightarrow \frac{1}{2}O_2(g) + V_O^{\cdot\cdot} + 2e'$$

and it is important therefore to ensure that the electronic conductivity is not increased due to the presence of impurity transition metal cations such as Fe^{3+} in the ceramic electrolyte. The reducing conditions prevailing in the furnace can also produce volatilization of grain boundary phases due to reactions such as the following:

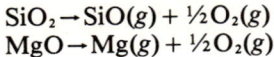

$$SiO_2 \rightarrow SiO(g) + \tfrac{1}{2}O_2(g)$$
$$MgO \rightarrow Mg(g) + \tfrac{1}{2}O_2(g)$$

Removal of grain boundary phases in this manner can make the tube permeable[35,36] allowing gaseous molecules to penetrate through the ceramic electrolyte which can produce spurious emf values. It is essential therefore that zirconia-based ceramic electrolytes should be composed of as pure constituents as possible for this application. An additional requirement is that a tough strong ceramic should be selected to withstand the thermal stresses which are often imposed when the furnace operators burn away any deposited carbon by introducing air into the furnace. This practice can produce very rapid local heating and some additional protection may be given to the zirconia ceramic by surrounding it with another ceramic such as silicon carbide which exhibits a relatively high thermal conductivity.

A further design constraint is introduced by the fact that platinum electrodes and leads become brittle after prolonged exposure in a carburizing atmosphere. Fortunately satisfactory performance can be obtained by replacing the platinum by other alloys such as Ni-Cr and Au-Pd.

Diffusion and Current Limiting Devices

So far the present survey has been restricted to potentiometric (zero current) devices. It is possible, however, to design oxygen monitors which are based on current measurement and in fact all ambient temperature electrochemical oxygen measurements are based on some type of current or diffusion limiting device to overcome the problems inherent in the lack of reversibility exhibited by the O_2/Pt electrode under these conditions. The principles of such devices have often been reviewed[26] and can obviously be extended to high temperature oxygen monitors incorporating zirconia. Although the kinetics of the $O_2/Pt/ZrO_2$ electrode are sufficiently rapid above 600 °C (section on Oxygen in Inert Atmospheres) to ensure satisfactory performance of potentiometric-based devices, there may be situations at lower temperatures in which a diffusion-limited device may demonstrate an advantage although it must be recognized that the available current flux through a zirconia-based ceramic at lower temperature will impose severe design restrictions. A typical illustration of a zirconia-based cell[37] operating in a diffusion-limited current mode of operation is shown in Fig. 21(a), and results obtained at 1000 °C (Fig. 21(b)) indicate a linear relationship between the measured current and oxygen content of oxygen-argon mixtures for different orifice areas controlling the diffusion flux.

An interesting application utilizing an oxygen ion current mode of operation is depicted in Fig. 22 and is designed to be incorporated into a zirconia-based oxygen monitor operating in a gaseous environment containing significant concentrations of gaseous sulfur species such as SO_2, H_2S, etc. Under reducing conditions deleterious reactions such as

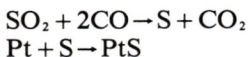

$$SO_2 + 2CO \rightarrow S + CO_2$$
$$Pt + S \rightarrow PtS$$

can occur which may cause rapid degradation of the platinum electrode assembly. To protect the electrode from such deleterious reactions, an electronic circuit[38] is designed to pump oxygen from the inner air reference electrode into the external electrode compartment whenever the oxygen monitor measures a reducing environment in this external compartment. In this manner, the platinum electrode will not suffer prolonged exposure to sulfur-containing gaseous species under reducing conditions.

Alternative Concepts and Materials

The present survey has been principally concerned with existing technological devices incorporating zirconia-based ceramics, and so it is appropriate to conclude by mentioning some alternative concepts and materials which might form the basis in the future for alternative oxygen monitors which could compete with existing zirconia-based devices.

It has been reported[39] that, for example, thin sputtered films (150 nm) of stabilized zirconia have been incorporated into experimental oxygen monitors that respond to changes in oxygen partial pressure at temperatures as low as 400 °C.

Other workers have suggested[40] utilizing the corrosion equilibrium of Pd/PdO as a function of temperature to monitor oxygen partial pressures. The formation of PdO on Pd electrodes applied to zirconia-based ceramic electrolytes produces a significant increase in the Faradaic impedance at the relevant electrode/electrolyte interface which can be registered by an appropriate electronic circuit. A separate reference electrode system is not required for such a device as the mode of operation is solely dependent on the thermodynamics and kinetics of the Pd/O_2 reaction.

It has been proposed[31] that the thermoelectric properties of stabilized zirconia and other oxygen ion conductors could be used to detect changes of oxygen partial pressure according to the relationship

$$\epsilon = \epsilon^0 + R/4F \ln p_{O_2} \ (\mu VK^{-1})$$

Taking a value for ϵ^0 of 100 μVK^{-1} for stabilized zirconia,[30] a tenfold change in oxygen partial pressure corresponds to approximately 50 μVK^{-1} which suggests that the output of a device operating on this principle would be low to overcome the electrical noise invariably generated in an industrial environment even if a temperature difference of 100 K could be established across the device.

Turning our attention to alternative materials, it should be noted that other oxygen ion electrolytes, such as Bi_2O_3-Y_2O_3[41] and CeO_2-Gd_2O_3,[42] exhibit higher ionic conductivities at moderate temperatures than stabilized zirconia which may confer some advantage for devices utilizing the diffusion or current limiting mode of operation (section on Diffusion and Current Limiting Devices).

It is also known that solid halide electrolytes saturated with oxygen, such as CaF_2/CaO,[43] $SrCl_2/SrO$,[44] and PbF_2/PbO,[45] can be used to detect oxygen, and in the form of thin film devices these materials might possess some advantages over zirconia-based oxygen monitors at low temperatures.

Considerable interest is also being shown in the development of thin film semiconductor devices based on SnO_2,[46] of thick film semiconductor devices incorporating CoO and TiO_2,[47] and of a variety of IGFET and CHEMFET configurations[48] for detecting gaseous species including oxygen and carbon monoxide.

However, the extensive experience now available relating to the performance of zirconia-based monitors in industrial environments will ensure that this material will continue to dominate the market for applications requiring operation at moderate and elevated temperatures ($> 500\,°C$).

Acknowledgment

The authors thank Stuart M. Clark for contributing some of the experimental data reported in this paper.

Appendix
Zirconia-Based Ceramic Electrolytes

The Commission on High Temperature and Refractory Materials (International Union of Pure and Applied Chemistry) has initiated a collaborative project to examine the properties of commercially available zirconia-based ceramic electrolytes and the behavior of oxygen monitors incorporating these materials. The project is coordinated by A. M. Anthony.[†]

Microstructure-property relationships of the ceramic electrolytes will be principally characterized at the Wolfson Unit for Solid State Ionics.[‡] Five samples have been received by the Unit from IUPAC, and a sixth sample has been provided directly from the manufacturer. A report of the microstructural investigations on the five IUPAC samples has been submitted to the Commission on High Temperatures and Refractory Materials and some of the information contained in this report has been used in the preceding text. The sources of the six samples and their principal features are given in Table I.

References

[1]K. Kiukkola and C. Wagner, *J. Electrochem. Soc.*, **104**, 379–87 (1957).
[2]H. Peters and H. H. Mobius, *Z. Phys. Chem.*, **209**, 298–309 (1958).
[3]H. Schmalzreid, *Z. Elektrochem.*, **66**, 572–76 (1962).
[4]B. C. H. Steele and C. B. Alcock, *Trans. A.I.M.E.*, **233**, 1359–67 (1965).
[5]B. C. H. Steele; pp. 3–27 in Electromotive Force Measurements in High Temperature Systems. Edited by C. B. Alcock. Institute of Mining and Metallurgy, London, 1968.
[6]R. E. Carter and W. L. Roth; pp. 125–44 in Ref. 5.
[7]T. H. Etsell and S. N. Flengas, *Chem. Rev.*, **70**, 339–76 (1970).
[8]J. E. Bauerle, *J. Chem. Phys. Solids*, **30**, 2657–70 (1969).
[9]R. C. Garvie; p. 160 in High Temperature Oxides, Part II. Edited by A. M. Alper. Academic Press, New York, 1970.
[10]E. Schouler, G. Giraud, and M. Kleitz, *J. Chim. Phys.*, **70**, 1309–16 (1973).
[11]S. Pizzini and R. Morlotti, *J. Chem. Soc. (Farad. Trans. I)*, **68**, 1601–10 (1972).
[12]J. H. E. Jeffes, Dept. of Metallurgy & Materials Science, Imperial College, University of London; private communication.
[13]F. J. Esper and K. H. Friese, *Ber. Dtsch. Keram. Ges.*, **55**, 314–16 (1978).
[14]S. K. Tiku and F. A. Kroeger, *J. Am. Ceram. Soc.*, **63** [3-4] 183–89 (1980).
[15]R. W. Powers and S. P. Mitoff; pp. 123–44 in Solid Electrolytes. Edited by P. Hagenmuller and W. Van Gool. Academic Press, New York, 1978.
[16]E. Lilley and J. E. Strutt, *Phys. Status Solidi A*, **54**, 639–50 (1979).
[17]N. M. Beekmans and L. Heyne, *Electrochim. Acta.*, **21**, 303–10 (1976).
[18]L. Heyne and D. den Engelsen, *J. Electrochem. Soc.*, **124**, 727–35 (1977).
[19]K. Keizer, M. J. Verkerk, and A. J. Burggraaf, *Ceramurgia Int.*, **5**, 143–47 (1979).
[20]George Kent Ltd., Brit. Pat., 1 201 806.
[21]Corning Glass Corp., Ger. Pat., OLF 216 917 (2/11/78).
[22]R. E. Hetrick and E. M. Logothetis, *Appl. Phys. Lett.*, **34**, 117–19 (1979).
[23]H. Okamoto, H. Obayashi, and T. Kudo; to be published in *Solid State Ionics*.
[24]Y. L. Sandler, *J. Electrochem. Soc.*, **118**, 1378–81 (1971).
[25]D. M. Haaland, *ibid.*, **127**, 796–804 (1980).
[26]H. Dietz, W. Haecker, and H. Jahnke; pp. 1–90 in Advances in Electrochemistry, Vol. 10. Edited by H. Gerischer and C. W. Tobias. Wiley, New York, 1977.

[27]J. Fouletier, H. Seinera, and M. Kleitz, *J. Appl. Electrochem.*, **4**, 305–15 (1974).
[28]B. C. H. Steele and R. W. Shaw; pp. 483–95 in Ref. 15.
[29]D. L. R. May, *Chem. Eng.*, **82**, 53–57 (1975).
[30]S. Pizzini, C. Riccardi, and V. Wagner, *Z. Naturforsch. A*, **25**, 559–63 (1970).
[31]G. H. J. Broers, H. T. Cahen, A. Honders, and J. H. W. de Wit, *J. Appl. Electrochem.*, **10**, 229–31 (1980).
[32]B. C. H. Steele (Electronic Instruments Ltd.), Brit. Pat. 42 546/69.
[33]C. Desportes, M. Henault, F. Tasset, and G. Vitter, Fr. Pat. 73 32671.
[34]L. H. Fairbank, *Heat Treat. Met.*, **4**, 95–106 (1977).
[35]B. C. H. Steele and M. A. Williams, *J. Mater. Sci.*, **8**, 427–38 (1973).
[36]D. Hayes, D. W. Budworth, and J. P. Roberts, *Trans. Br. Ceram. Soc.*, **62**, 507–14 (1963).
[37]Westinghouse Electric Corp., U.S. Pat. 1 523 550 (6/9/78).
[38]A. Pebler, Proc. Int. Mtg. Solid Electrolytes, 2nd, St. Andrews, Scotland, Sept. 20–22, 1978.
[39]M. Croset, J. P. Schnell, and G. Velasco, *J. Appl. Phys.*, **48**, 775–80 (1977).
[40]H. J. de Bruin and S. P. S. Badwal, *J. Solid State Chem.*, **34**, 133–35 (1980).
[41]H. T. Cahen, "Bi$_2$O$_3$-M$_2$O$_3$ Systems, Structural, Electrical and Electrochemical Aspects"; Ph.D. Thesis, University of Utrecht, 1980.
[42]T. Kudo and H. Obayashi, *J. Electrochem. Soc.*, **122**, 142–47 (1975).
[43]T. A. Ramanarayanan, M. L. Narula, and W. L. Worrell, *ibid.*, **126**, 1360–63 (1979).
[44]A. Pelloux, J. P. Quessada, J. Fouletier, P. Fabray, and M. Kleitz, *Solid State Ionics*, **1** 343–54 (1980).
[45]J. M. Reau and J. Portier; p. 330 in Ref. 15.
[46]H. Windischmann and P. Mark, *J. Electrochem. Soc.*, **126**, 627–33 (1979).
[47]E. M. Logothetis, K. Park, A. H. Meizler, and K. R. Land, *Appl. Phys. Lett.*, **26**, 209–11 (1975).
[48]J. Janata and R. J. Huber, *Ion-Sel. Electrode Rev.*, **1**, 31–79 (1979).

---

*The Debye length is given by $l_D = (\epsilon\epsilon_0 kT/q^2 n)^{1/2}$.

†Centre de Recherches sur la Physiques des Hautes Temperatures, University of Orleans, Orleans 45045, France.

‡Wolfson Unit for Solid State Ionics, Imperial College, London, SW 7.

Table I. Characterization of samples used

Sample No.	Dopant content		Color	Measured grain size (μm)	Measured specific gravity	Impurities
	(wt%)	(mol%)				
1[a]	8.0 Y$_2$O$_3$	4.7 Y$_2$O$_3$	Yellow	10–30	5.76	Not determined
2[b]	5.5 CaO	12.8 CaO	Cream	30–60	5.00	Si,Al
3[c]	7.4 CaO	15.0 CaO	White	15–30	4.99	Si,Al
4[d]	7.4 CaO	15.0 CaO	Gray	10–20	5.36	Si,Al,Mg
5[d]	13.7 Y$_2$O$_3$	8.0 Y$_2$O$_3$	White	10–30	5.49	Si,Al
6[e]	12.9 Y$_2$O$_3$	7.5 Y$_2$O$_3$	White	1–10	Not determined	Si,Al, Fe,Ca,Ti

[a]Composition 1372; Corning Glass Works, Ceramic Products Div., Solon, Ohio 44139. [b]Degussit, Postfach 7, Steinzeugstrasse, D-6800 Mannheim-71, West Germany. [c]Nuclear Research Institute, 25068 Rez, Czechoslovakia. [d]Nippon Kagaku Togyo Co., Ltd., Wako Shoken Bldg. No. 3, Kitahama 3-chome, Higashi-ku, Osaka 541, Japan. [e]Viking Chemicals, 4591 Follensley, Denmark.

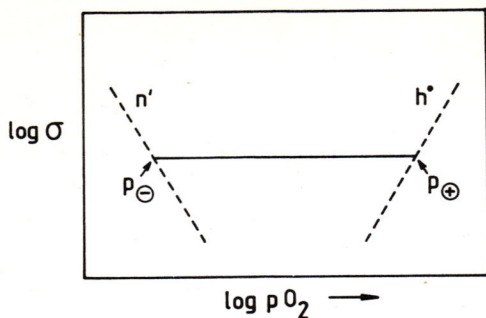

Fig. 1. Dependence of conductivity on oxygen partial pressure for a typical fully stabilized zirconia electrolyte, showing ionic conduction domain.

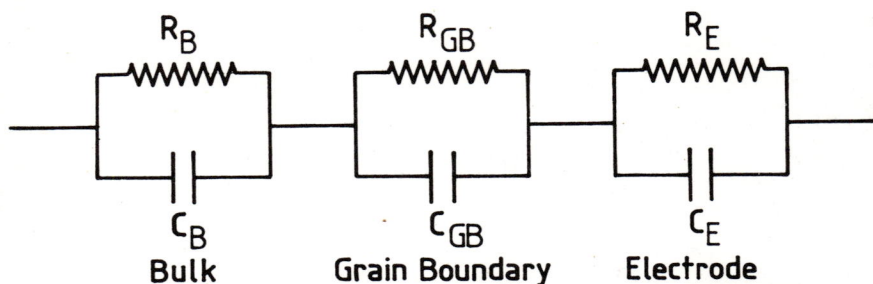

Fig. 2. Simple equivalent circuit for ceramic zirconia electrolyte (after Bauerle).

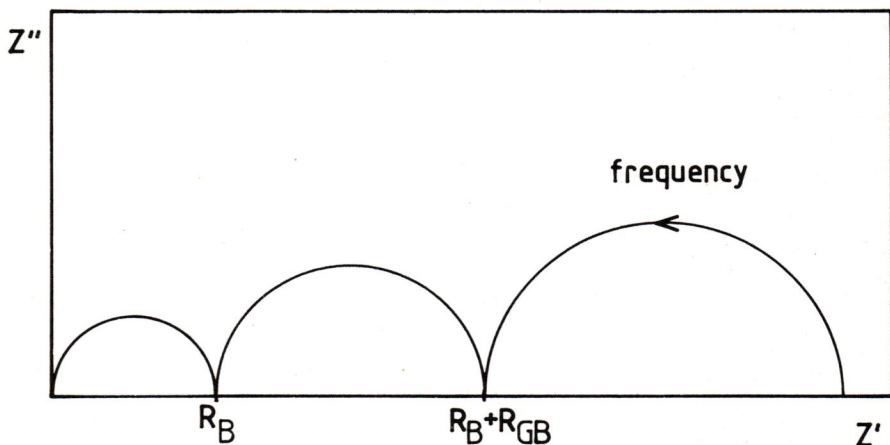

Fig. 3. Complex plane representation of the impedance of the circuit shown in Fig. 2.

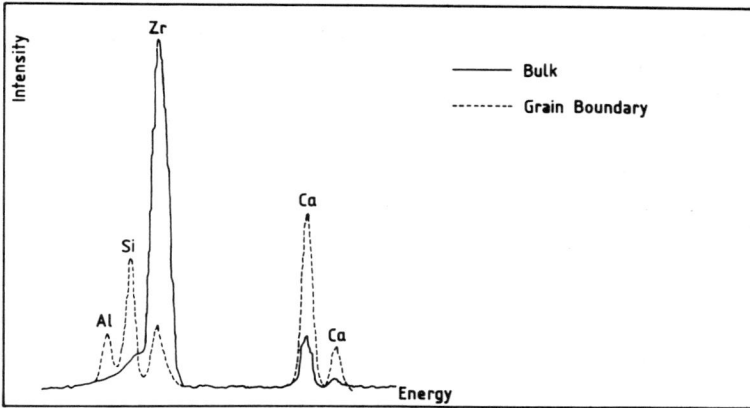

Fig. 4. Microanalysis scan on grain and grain boundary of a sample of calcia-stabilized zirconia (sample 4).

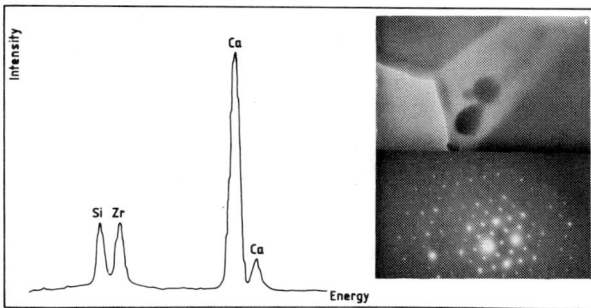

Fig. 5. Microanalysis scan, transmission electron micrograph, and electron diffraction pattern of a crystalline grain boundary phase in calcia-stabilized zirconia (sample 2).

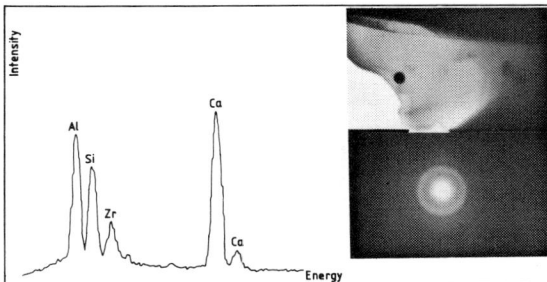

Fig. 6. Microanalysis scan, transmission electron micrograph, and electron diffraction pattern of a glassy grain boundary phase in calcia-stabilized zirconia (sample 3).

(a)

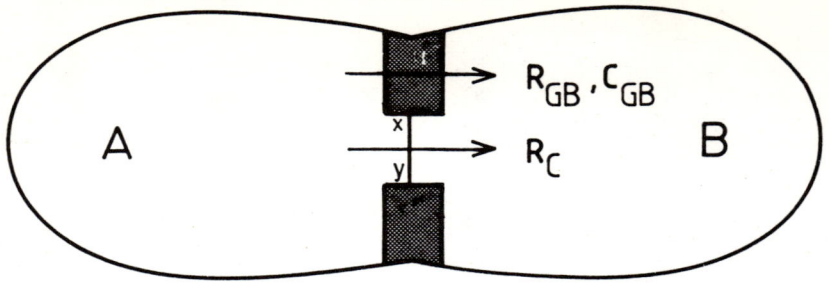

R_{GB} Grain boundary resistance
C_{GB} Grain boundary capacitance
R_C Constriction resistance

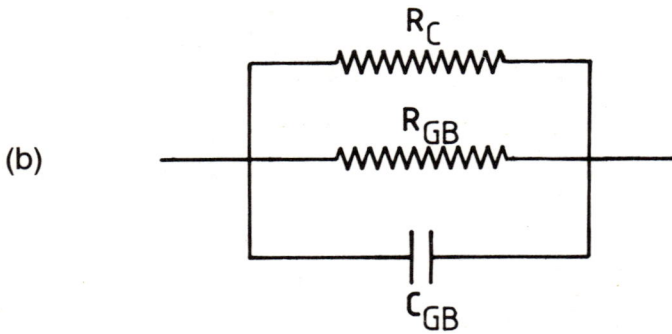

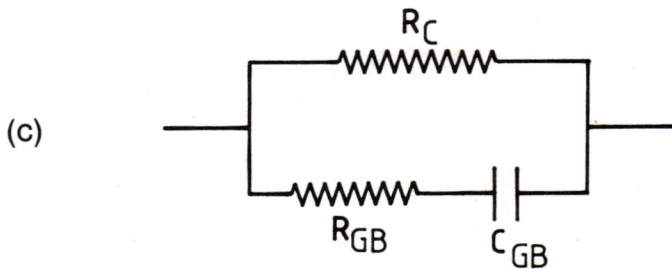

(b)

(c)

Fig. 7. Models for oxygen ion conducting electrolytes with grain boundary impedances: (a) Idealized model proposed by Bauerle. (b) and (c) Proposed equivalent circuits.

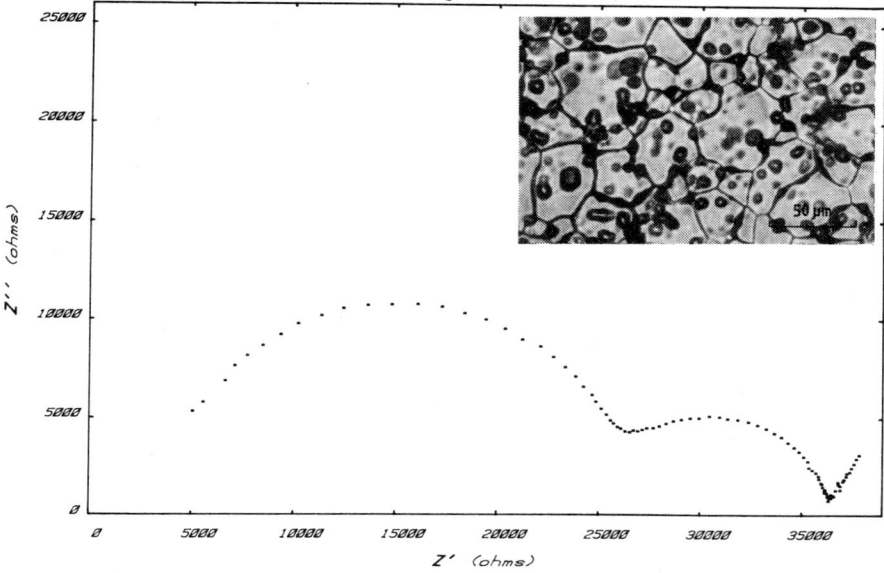

Fig. 8. Complex impedance plot (at 425 °C) and optical micrograph of commercial yttria-stabilized zirconia (sample 5).

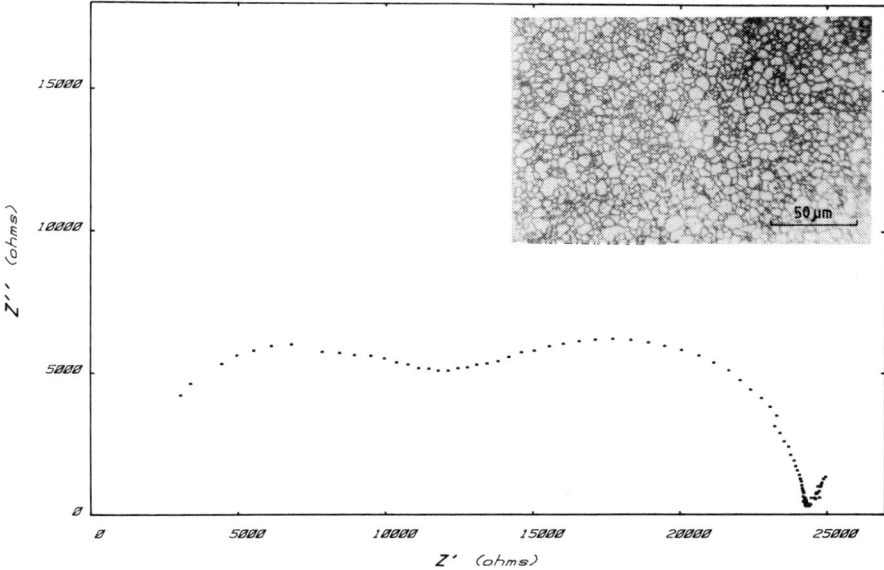

Fig. 9. Complex impedance plot (at 425 °C) and optical micrograph of commercial yttria-stabilized zirconia (sample 6).

301

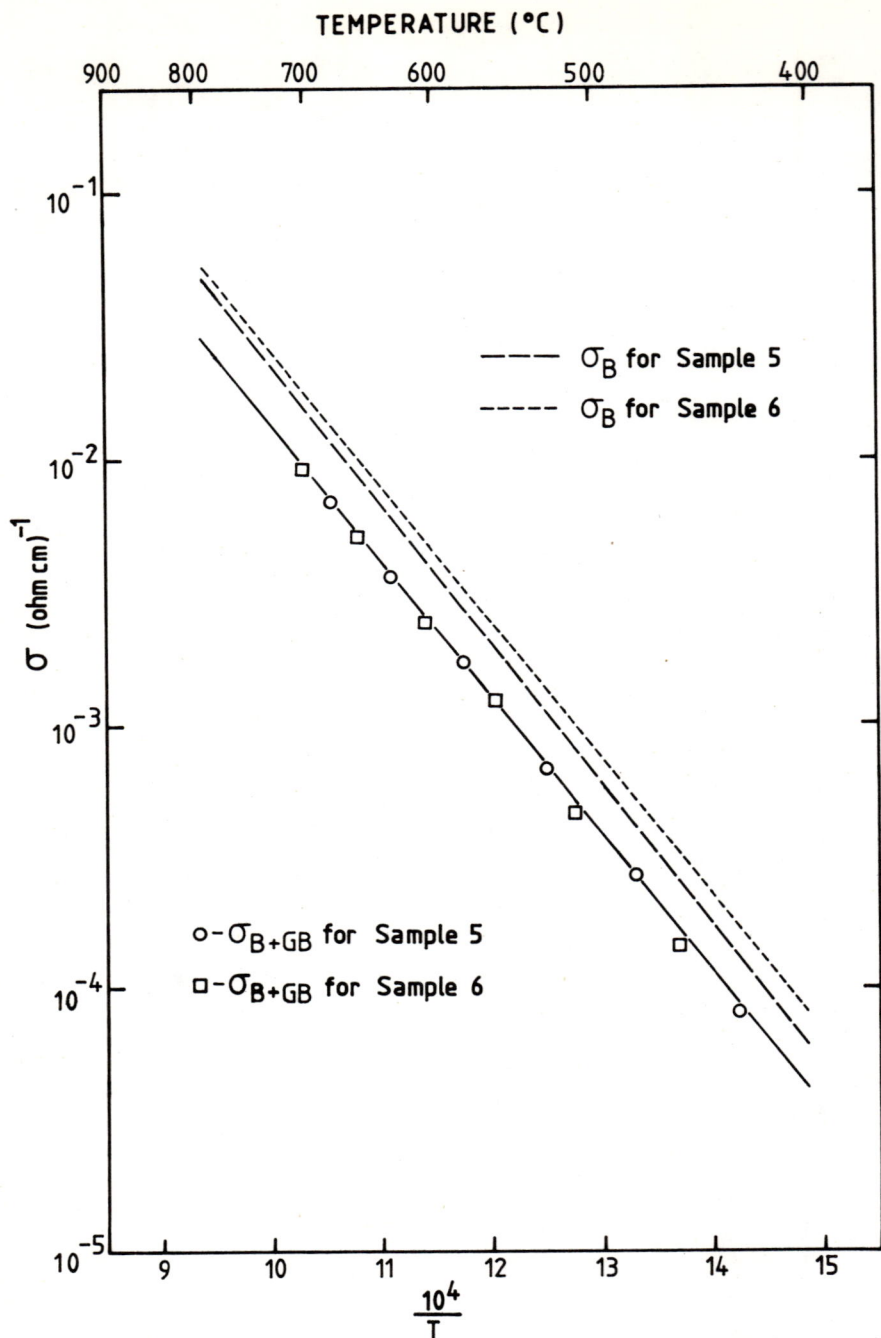

Fig. 10. Ionic conductivity plots for two samples of commercial yttria-stabilized zirconia (samples 5 and 6).

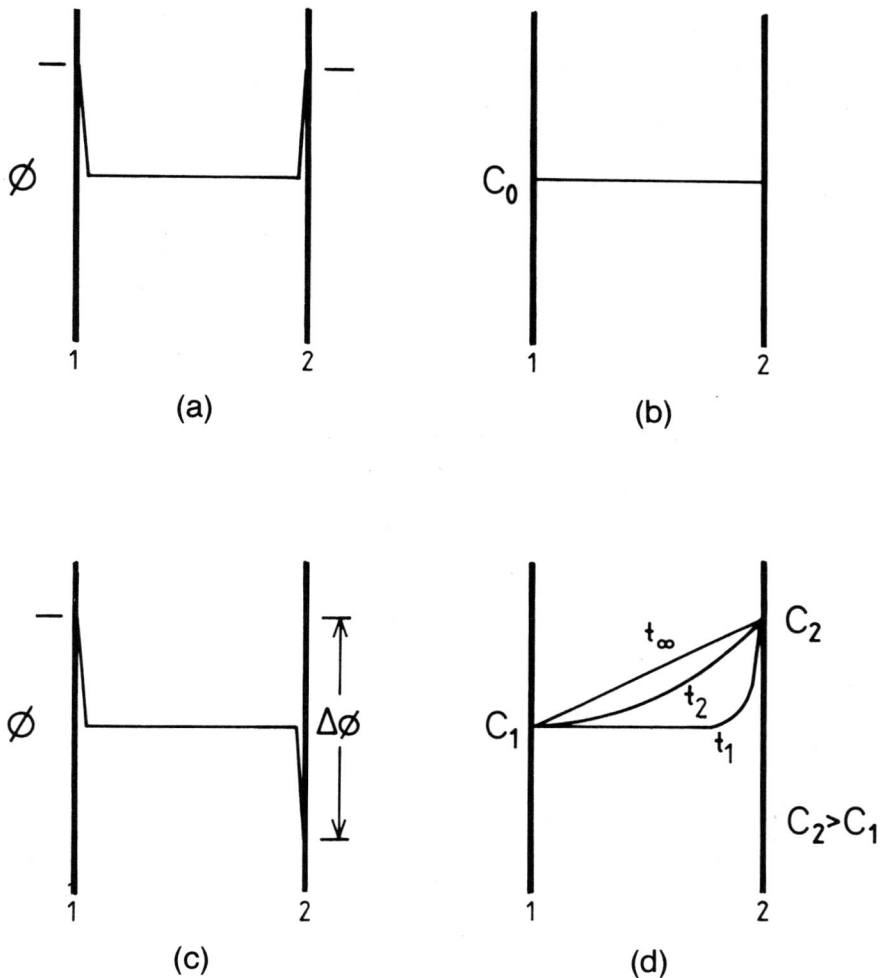

Fig. 11. Electrical potential and ion concentration profiles within a homogeneous ceramic electrolyte: (a) Potential diagram showing space charge regions; faces in contact with the same partial pressures of oxygen. (b) Oxygen concentration profile; faces in contact with the same partial pressures of oxygen. (c) Potential diagram showing space charge regions; faces in contact with different partial pressures of oxygen. (d) Oxygen concentration profile; faces in contact with different partial pressures of oxygen.

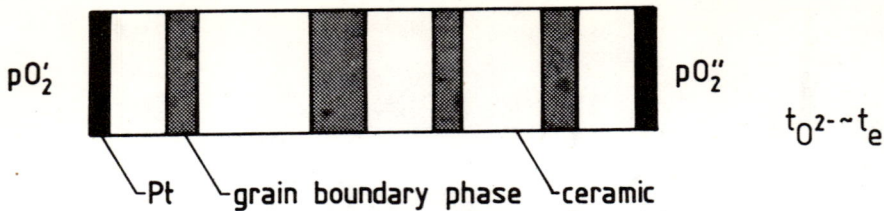

Fig. 12. Idealized model of an inhomogeneous electrolyte.

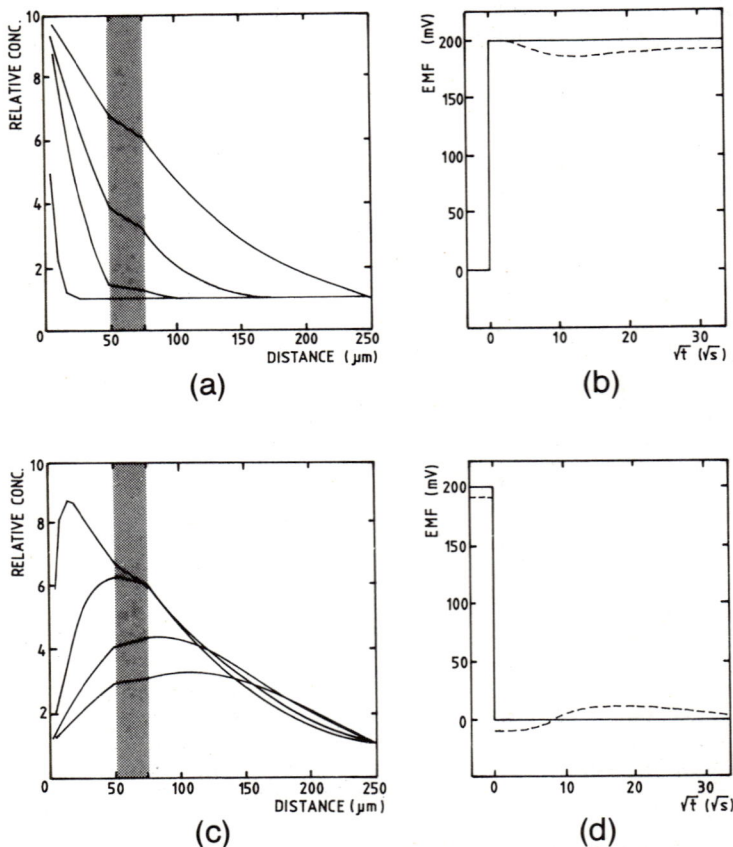

(a)

(b)

(c)

(d)

Fig. 13. Concentration profile and potential-time response of an inhomogeneous ceramic electrolyte, containing a second phase (depicted as shaded area) (after Heyne). (a) Oxygen concentration profile after different times, following change in partial pressure of oxygen. (b) Corresponding potential-time response. (c) Oxygen concentration profile after different times, following return to the original partial pressures of oxygen. (d) Corresponding partial-time response.

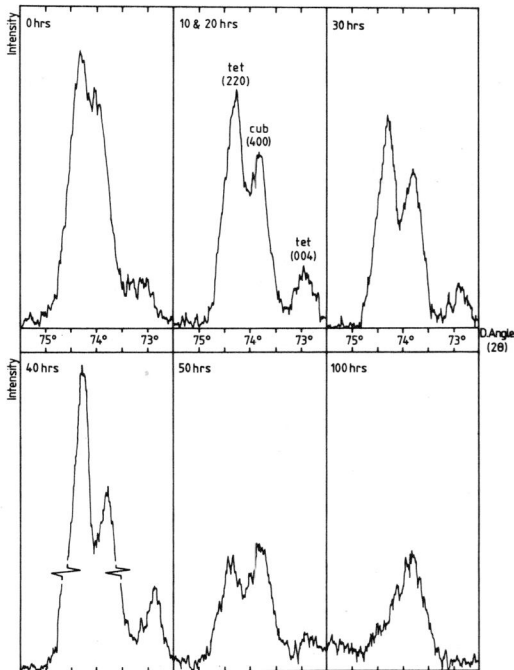

Fig. 14. X-ray diffractometer traces showing decreasing peaks for tetragonal (004) and (220) reflections, with increasing aging time; taken from 8 mol% calcia partially stabilized zirconia, CuKα radiation.

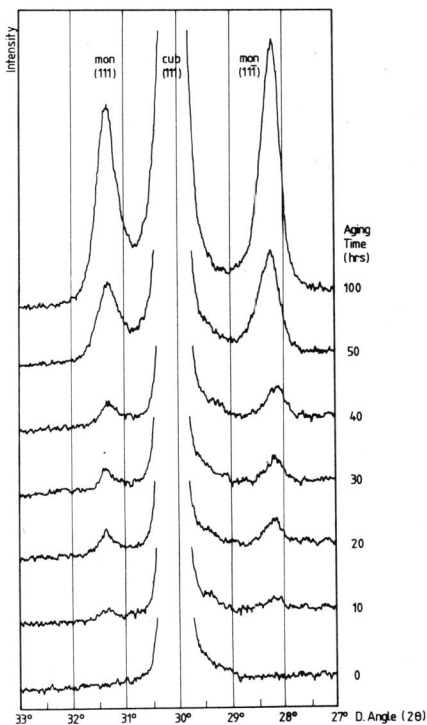

Fig. 15. X-ray diffractometer traces showing growth of peaks for monoclinic (111) and (11$\bar{1}$) reflections, with increasing aging time; taken from 8 mol% calcia partially stabilized zirconia, CuKα radiation.

Fig. 16. (a) Dark field transmission electron micrograph from 8 mol% calcia partially stabilized zirconia, solution-treated and aged for 20 h, showing the tetragonal phase distribution (1000 kV). (b) Bright field transmission electron micrograph from 8 mol% calcia partially stabilized zirconia, solution-treated and aged for 20 h, showing the "tweed-like" contrast (1000 kV).

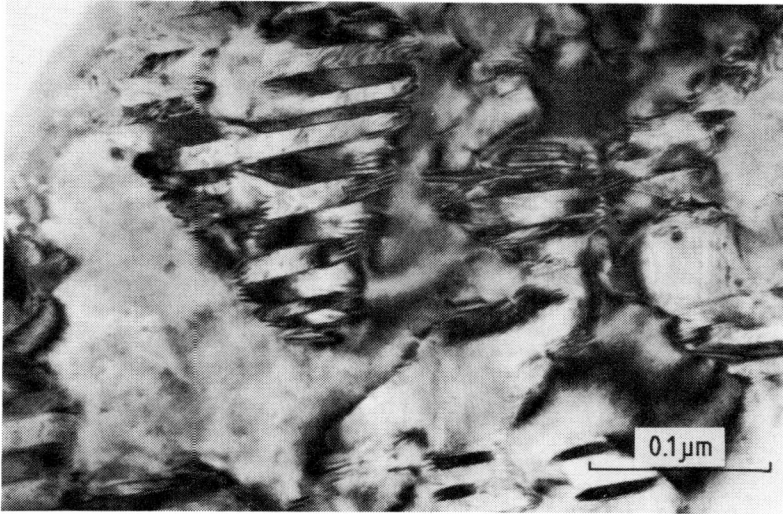

Fig. 17. Bright field transmission electron micrograph from 8 mol% calcia partially stabilized zirconia, solution-treated and aged for 50 h (overaged), showing fine twinning and strain fringes of a monoclinic precipitate (100 kV).

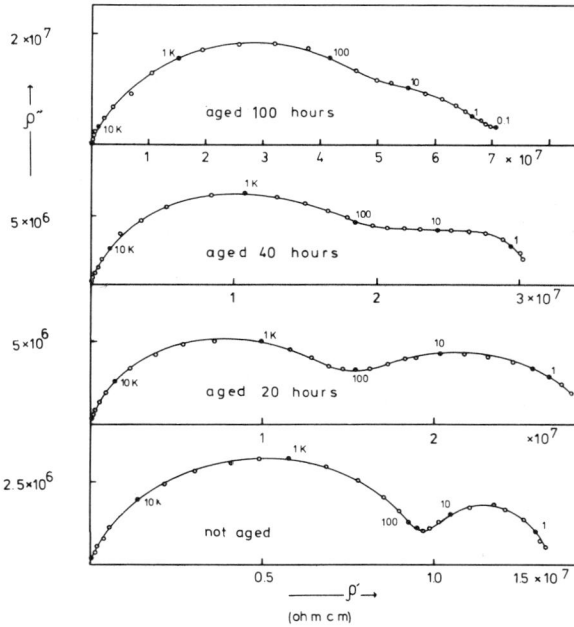

Fig. 18. Complex resistivity plots for a series of samples of 8 mol% calcia partially stabilized zirconia, solution-treated and aged for a range of times.

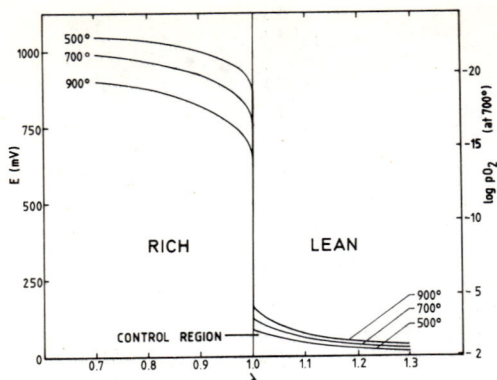

Fig. 19. Dependence of equilibrium voltage on λ at three temperatures, for a methane-air mixture, where: λ = (actual air/fuel ratio)/(stoichiometric air/fuel ratio) (after Dietz).

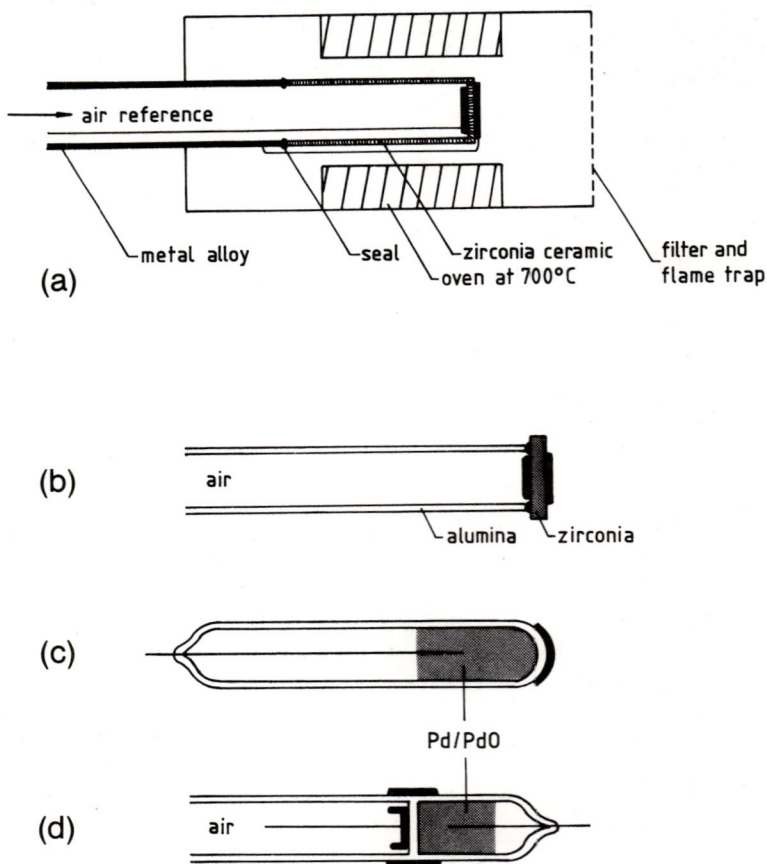

Fig. 20. Typical zirconia-based oxygen monitors: (a), (b) Devices incorporating seals. (c), (d) Devices incorporating metal oxide electrodes.

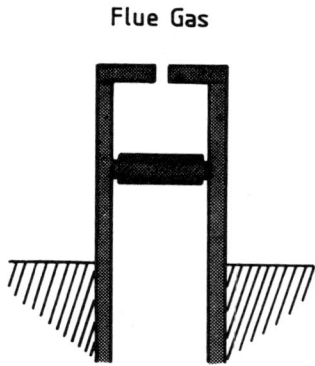

Flue Gas

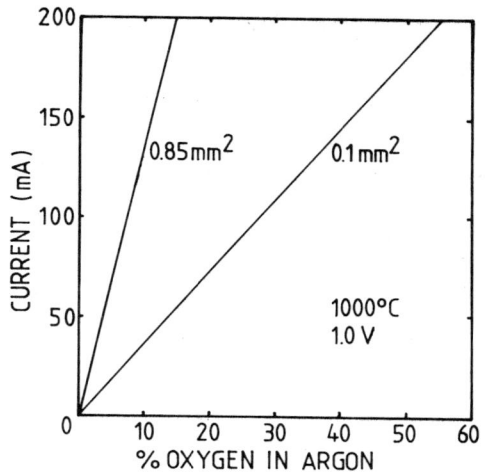

(a)   (b)

Fig. 21. Typical zirconia cell operating in diffusion mode: (a) Device. (b) Results obtained at 1000 °C.

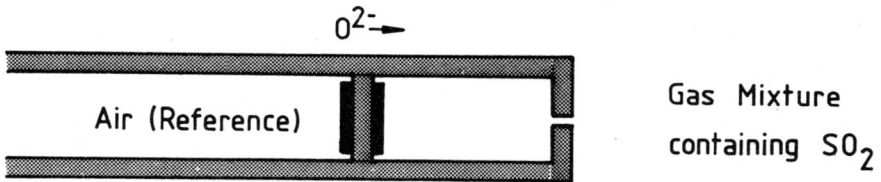

Air (Reference)   Gas Mixture containing SO_2

Fig. 22. Oxygen monitor incorporating oxygen pump for protection of platinum electrode from reducing sulfurous atmospheres.

Impedance spectroscopy and electrical resistance measurements on stabilized zirconia

M. KLEITZ, H. BERNARD,* E. FERNANDEZ, AND E. SCHOULER

Laboratoire d'Energétique Electrochimique
Institut National Polytechnique de Grenoble
E.N.S. d'Electrochimie et d'Electrométallurgie
38401 Saint Martin d'Heres
France

Since the work by Bauerle, impedance spectroscopy has been extensively used to characterize stabilized zirconia cells. A measurement gives a set of semicircles (and straight lines) and the first step of the analysis is to ascribe each of them to either bulk properties or interface phenomena. Reliable techniques to separate them are reviewed; the utilization of reference probes is a very convenient technique. Possible sources of error are also reviewed. The interface phenomena diagram consists of two semicircles. The higher frequency one is usually small with zirconia but its relative magnitude increases for electrodes with reduced coverage. It is also important with low dopant-concentration ceria or thoria electrolytes. Medium and low temperature bulk property diagrams typically exhibit a grain boundary semicircle. The relevant phenomena can be interpreted either in terms of a series connection (the grain boundary equivalent circuit is inserted between the grain circuits) or in terms of a parallel connection (here the grain boundaries are viewed as blocking a fraction of the mobile ions). The results of a thorough quantitative analysis favor the second model. The blocking effect is either due to a segregated impurity phase or to a specific property of the contact between the grains. Intragrain conductivity was found to be simply proportional to the density of the material. Using this property, accurate and reproducible specific conductivities of stabilized zirconia can be determined.

The impedance spectroscopy technique was introduced for accurate conductivity determination of solid electrolytes by Bauerle[1] in 1969, in conjunction with work on stabilized zirconia. This technique was further developed in Sverdlovsk[2-6] and in Grenoble.[7-10] Suzuki and Takahashi[11] were also among the first to use it on zirconia electrolytes. In 1976, Beekmans and Heyne[12] reported a detailed investigation of the relationship between the microstructure of the material and the impedance diagram characteristics. They confirmed the influence of segregated impurities suggested by Bauerle. Currently, most studies[13,14] on ionic conductivity of stabilized zirconia and electrode reactions use this technique. Due to its success, its use has been extended to other solid-oxide electrolytes such as ceria[15] and thoria.[16]

A typical measurement is made by forming a simple cell on a pellet of the investigated material with two appropriate metal layers coated on two opposite faces (Fig. 1). An electrical signal

$$U = U_0 e^{iwt}$$

is applied between these electrodes, and the resulting current passing through the cell

$$I = I_0 e^{i(wt + \phi)}$$

is measured, or vice versa. The impedance of the cell is calculated according to the usual formula:

$$Z = \frac{U}{I} = \frac{U_0 e^{-i\phi}}{I_0} = Z_0 e^{-i\phi}$$

and plotted on a Nyquist diagram, with the opposite of the imaginary component

$$Z'' = - Z_0 \sin\phi$$

on the y axis.

It is sometimes preferred to plot the admittance diagram ($Y = 1/Z$).

Results show that both the modulus Z_0 and the phase angle ϕ depend on the signal pulsation w. By varying the latter, the Z point is found to describe a locus which is characteristic of the cell. Figure 2 shows a typical locus obtained with single stabilized-zirconia crystals or with sintered pellets at high temperature. It consists of two semicircles with their centers below the real axis.

The distributions of the experimental points on each semicircle obey the following equation:

$$Z = R_{l-1} + \frac{R_l - R_{l-1}}{1 + (iw/w_l)^{(1 - \alpha_l)}}$$

with $l = 1$ or 2 and where α_l is the depression parameter.

As a first approximation, such a diagram can be considered as the response of the RC circuit sketched in Fig. 2(b) (assuming that $\alpha_1 = \alpha_2 = 0$). Published data clearly show that the $R_1 C_1$ circuit simulates the behavior of the electrolyte, with R_1 equal to the electrolyte resistance and C_1 the capacitance resulting from its dielectric properties. The other $R_2 C_2$ circuit characterizes the contacts at the interfaces.

In principle, the cell should be described by the circuit sketched in Fig. 2(c). However, when the electrodes are identical, their equivalent electrical circuits can be summed up in the single $R_2 C_2$ circuit shown in Fig. 2(b) with the relationships

$$R_2 = 2R_2'$$
$$C_2 = \tfrac{1}{2} C_2'$$

From the correlations between the semicircle parameters and the R and C components (cf. Fig. 2) it can be seen that the plotting of such a diagram provides a straightforward determimation of the electrolyte resistance, whatever the degree of electrode polarization (the values of the components R_2 and C_2). This is, in essence, the advantage of the impedance-spectroscopy technique. This method does not necessarily require the use of reversible electrodes nor a more complicated cell which eliminates the masking effect of electrode polarization, such as the 4-probe cell.

311

The additional determinations of the dielectric properties of the material and of the electrode polarization parameters are obvious further advantages of impedance spectroscopy.

With sintered samples, at relatively low temperatures, the situation becomes somewhat more complicated, with three semicircles frequently observed (Fig. 3). It has been well established that the third semicircle, appearing between those previously described, is characteristic of a partial blocking of the conductivity at the grain boundaries. Under such circumstances, several terms must be clearly defined to avoid any ambiguity.

The resistance measured by the first real intercept (cf. Fig. 3) and the conductivity calculated from it are usually called intragrain resistance and conductivity.

The resistance measured by the second intercept and the corresponding conductivity, which includes the blocking effect, will be called the dc resistance and conductivity. These are the values which would have been measured if the dc 4-probe technique had been used and must be taken into account in any evaluation of the ohmic drop in a dc device such as a fuel cell or an electrolyzer. To minimize this ohmic drop, it is as important to reduce the blocking effect (through the use of the appropriate purity of material, grain size, etc.) as it is to select the proper stabilizer to induce the highest intragrain conductivity.

A typical Arrhenius diagram, obtained by plotting the intragrain and dc conductivities as functions of temperature, is shown in Fig. 4. At high temperatures the blocking phenomenon vanishes and the intragrain and dc curves merge.

A careful examination of the results has recently shown[17] that, in fact, the electrode characteristic also consists of two semicircles. At present, the real situation therefore appears to be as schematized in Fig. 5. These two semicircles will be designated the high frequency and the low frequency electrode semicircles.

It is only recently that the high frequency semicircle has been recognized for zirconia, as it is relatively small. However, with the other oxide electrolytes, such as ceria[18] and thoria,[16] it is much larger, especially with low dopant concentrations. It is almost negligible with yttria concentrations corresponding to fully stabilized zirconia. It becomes easily detectable with concentrations of 2 or 3% and below 0.1% this semicircle is predominant and overlaps the grain boundary semicircle. This can be a source of misinterpretation.

Experimentation

Equipment

Different types of equipment can be used for measuring zirconia cell impedance (or admittance) and plotting the relevant impedance diagram.

Regular and semiautomatic bridges provide the most accurate determination over the broader frequency and resistance intervals; however, systematic measurements with such units are tedious. This equipment is especially suited to calibration.

Correlation function analyzers are at present the most convenient machines for such measurements. They use either an ac signal, based on the conventional theory, or an erratic noise. This type of stimulus provides a faster response; however, the measured impedance is less accurate on the low frequency side of the diagram. Lock-in amplifiers are also rather successful because of their relatively low cost.

Interfacing these units with computers significantly improves their capabilities.[19,20] Among other advantages, this is the only simple way to correct the errors introduced at high frequency by the spurious capacitances of the sample holder, the circuitry, and the input circuit of the measuring unit.

Overlapping Semicircles

Once a diagram has been plotted, the first problem which may be presented is the separation of partially overlapping semicircles.

For conductivity determinations for stabilized zirconia this never presents a major problem as there is only slight overlapping of the pertinent semicircles. When the real intercepts that measure the resistance (cf. Fig. 3) are not directly obtained by plotting, they can be determined by extrapolation using a compass.

On the other hand, for electrode characteristics and for other solid oxide electrolytes, such as ceria and thoria with low dopant concentrations, overlapping is frequently a major problem.

As a first example, Fig. 6 shows typical cases obtained with various electrodes on zirconia. Depending on the nature of the electronic conductor and the coverage ratio of the electrode, the relative magnitude of the component semicircles varies and makes the determination of the intermediate intercept more or less difficult.

As another example, Fig. 7 shows typical overlappings between the grain boundary semicircle and the high-frequency electrode semicircle. These diagrams were recorded with yttria-doped thoria samples. Diagrams (A) and (B) differ only in the length of the samples and diagrams (B) and (C) differ in the dopant concentrations.

Under such circumstances, extrapolation with a compass leads to significant errors and mathematical resolution must be resorted to.[19]

In extreme cases, the overlapping may be so important that it is even difficult to determine whether there is one semicircle or two. One solution is to check the frequency distribution using either the Cole-Cole approach or by reciprocating the semicircle.[17] The conventional Cole-Cole technique consists in measuring the distances u and v for all experimental points as shown in Fig. 8. Log u/v is then plotted as a function of the logarithm of the signal frequency and a straight line will be obtained if there is only one semicircle.

Semicircle Identification

The second type of experimental problem is the identification of the recorded semicircles. Because of the limits of the measurement equipment, the diagram shown in Fig. 3 is rarely obtained in full. For instance, in some cases only one semicircle may be seen within the "observation range" of the unit. It may then be difficult to tell which semicircle is which.

As a rule, the "observable range" shifts from the left to the right of the complete impedance diagram as the temperature of the sample increases (Fig. 9). Bulk properties are more easily observed at low temperatures and electrode characteristics at high temperatures. One sure way to separate bulk property semicircles from electrode semicircles is to vary the length of the sample. Real intercepts corresponding to bulk resistances vary proportionally while electrode resistances are independent of this parameter.

Reciprocally, by varying the oxygen pressure in equilibrium with the sample only the electrode semicircle varies (Fig. 10).

Another method is to measure the relaxation frequency at the top of the analyzed semicircle and to refer to a diagram such as the one shown in Fig. 11. The relaxation frequencies are characteristic of the relevant phenomena and are independent of the sample dimensions.

Similarly, it is also possible to calculate the capacitance of the approximately equivalent circuit or the related dielectric constant. These values provide other accurate identification criteria.

A simple and convenient technique consists in varying the amplitude of the signal applied to the cell. Bulk properties are found to be perfectly linear and the relevant part of the diagram is independent of the signal amplitude. Conversely, the electrode responses are far from linear and are highly dependent on this amplitude, especially for applied voltages greater than 100 mV.

A more sophisticated technique has been introduced recently and consists in using reference probes, as in certain dc measurements. The three-electrode cell sketched in Fig. 12 was investigated.[22] Impedances were determined according to Eq. (1) from the voltage (U_1) between the extreme electrodes and the voltages (U_2, U_3) between each of these and the intermediate probe. An example of the results is shown in Fig. 13. A check was made to ensure that at each frequency the total impedance is exactly equal to the sum of the half-cell impedances. This demonstrates that the reference probe does not suffer any interface polarization and that it only detects the bulk effects.

Tuller and his coworkers[23] developed a four-electrode technique based on the same principle. Two probes are wrapped around the samples between the extreme electrodes. The impedance is determined from the voltage measured between these probes. The diagram plotted from this method shows only the bulk property semicircles. The electrode semicircles are reduced to a point. Comparing diagrams obtained in this way to the conventional diagrams from measurements between the extreme electrodes is a tangible means of separating the bulk property semicircle and the electrode characteristics.

Accuracy and Sources of Error

Conventionally, conductivity measurements have been performed at a fixed frequency, typically at 10 kHz. Examination of the position of the 10 kHz point on an impedance diagram shows that it stays very near the dc resistance intercept[17,21] over a certain high temperature interval. The measurement therefore gives the correct value of the sample dc resistance (as defined in Fig. 3). On the other hand, at low temperatures, the 10 kHz point shifts over the grain boundary semicircle and even partly onto the intragrain semicircle (Fig. 9). In this case, the real component (or modulus) of the corresponding impedance is likely to be different from the sample resistance and the measurement at this fixed frequency is erroneous. Figure 14 compares the resistances correctly measured by impedance spectroscopy to the values obtained at 10 kHz. There is, in fact, an important deviation in the low temperature range.

In addition to the systematic error, it must be stressed that the 10 kHz measurement at medium temperature on sintered samples includes the blocking effect at the grain boundaries. As previously mentioned, this measurement provides the dc resistance. The blocking effect is highly dependent on the conditions of preparation (cf. the section Blocking Effect). It is therefore not surprising that under such conditions the reproducibility of the results on conductivity of stabilized zirconia has been rather poor.[21]

Reproducibility tests were recently performed by impedance spectroscopy on 6 samples[21] prepared from coprecipitated powders. The conductivities were found reproducible to within 2% at a temperature below T_2 (cf. Fig. 4). The deviations of the activation energies were less than 1% below T_2 and less than 2% above this temperature. The results were also compared to former results for a single crystal and sintered pellets prepared in different ways, but also measured by impedance spectroscopy. Agreement was to within 10%.

Concerning irreproducibility, it must be stressed that the conductivities of ytterbia and gadolinia-stabilized zirconia were measured by impedance spectroscopy[21] and, in contrast to published data,[24] ytterbia was not found to give a significantly higher conductivity than yttria, and gadolinia gave a lower conductivity. The proposed equations for the high-temperature ($T > T_2$) conductivities are, respectively:

$$9\% \text{ ytterbia: } \sigma(\Omega \cdot \text{cm})^{-1} = 3.90 \times 10^2 \exp - \frac{0.84}{kT} \text{ (eV)}$$

$$9\% \text{ gadolinia: } \sigma(\Omega \cdot \text{cm})^{-1} = 2.83 \times 10^3 \exp - \frac{1.08}{kT} \text{ (eV)}$$

The fact that impedance spectroscopy permits a perfect separation between bulk and interface responses eases the requirement of reversible electrodes. In principle, any electrode should be acceptable for conductivity measurement. Uncontrolled use of this possibility may, however, lead to errors. For example, Fig. 15 shows conductivity data[17] obtained with several gauze simply pressed against the sample surfaces. As can be seen, use of electrodes with insufficient coverage results in marked deviations. Furthermore, Fig. 16 shows that the deviation depends on the oxygen pressure. Note that this does not, however, mean that zirconia is a mixed conductor in contact with a loose electrode. The correct explanation is illustrated in Fig. 17. With insufficient coverage the current lines are not perpendicular to the surfaces and show a marked constriction resulting in an enhancement of the measured resistance. With increasing oxygen pressure, the spreading of the electrode reaction zones[25] broadens and reduces the constriction effect.

Concerning the risk of errors, the following point must also be emphasized. When the diagram is found to depend on the amplitude of the applied signal, as is frequently the case with electrode semicircles, it may also depend on the value of a reference resistance connected in the measuring circuit. In this case the measured impedance is not characteristic of the investigated cell alone and the equation corresponding to its variations may be an artifact.[20]

As previously mentioned, small capacitances associated with the sample holder and the other elements of the circuitry are possible sources of error. This is especially true at frequencies higher than 100 kHz, and their main effect is to alter the distribution of the experimental points on the intragrain semicircle.[20] In principle, they have no significant effect on the intragrain resistance. On the other hand, they significantly alter the measured dielectric

315

constant of the material. Their effect is easily detected by anomalously high measured value.

In this frequency range, it is recommended to perform an initial measurement on a dummy sample made of a pure resistor and a capacitor with values close to those expected for the sample. These components must be connected exactly at the sample location.

Semicircle Depression

One feature which has attracted attention is the depression of the semicircles. As a rule, circular arcs with centers below the real axis are obtained instead of true semicircles. The depression angle α (cf. Fig. 2) is usually very small for the intragrain semicircles. It can, however, be very great (up to 45°) for electrode semicircles.

Several interpretations have been proposed. A general explanation can be offered in terms of heterogeneity of the relevant properties and of a statistical distribution of the corresponding relaxation frequency. Along these lines, mathematical analysis of the bulk property semicircles of glassy solid electrolytes[26] has provided a convincing demonstration. A qualitative explanation of this type was also proposed to explain the depression of electrode semicircles on another solid electrolyte.[27]

Jonscher[28] considers the semicircle depression to be a specific property of dielectrics and of ionic conductors. It is viewed as an essential property of the elementary jump of an ion and is due to a local energy-storage effect associated with a slow relaxation of the polarization induced in its surroundings by the mobile ion. Cales and Abelard[29] are attempting to exploit the depression of the intragrain semicircle of stabilized zirconia within the framework of this theory to characterize the elementary jump of the ions.

For stabilized zirconia electrode semicircles, a model of the capacitance effect was developed based on an electrochemical pseudocapacitance associated with concentration variations in the adsorbed layer.[30] It was suggested[10] that semicircle depression results from a small additional contribution of a diffusion process in an adjoining phase.

Exploratory measurements[9] indicated that the variation of the depression angle with temperature might provide interesting complementary information.

At the data-acquisition level, an important consequence of the depression must be pointed out: the calculation of capacitance according to the usual formula

$$RCw_0 = 1$$

may be misleading. It is generally recognized that due to the depression the relevant phenomena cannot be described exactly by an RC circuit as shown in Fig. 2. However, experimenters still calculate a relevant capacitance according to the above formula as a first-order approximation. Because of the depression, the calculated capacitance does not vary in the same way when deduced from the impedance as when deduced from the equivalent admittance diagram.[9] Therefore, depending on the form of the data, different results may be obtained. A typical comparison of such results is shown in Fig. 18.

Electrode Reactions

This subject is still controversial even at the level of elementary interpretations. For this reason only the salient features obtained by impedance spectroscopy will be mentioned here. It should be noted that Schouler[17] has recently presented an important experimental study devoted to the oxygen electrode reactions on stabilized zirconia.

As mentioned in the introduction, with standard electrodes, the relevant characteristic is composed of two semicircles. Usually the low-frequency semicircle resistance $R\eta_A$ (cf. Fig. 5) is found to vary as $P_{O_2}^{1/2}$ and the high-frequency semicircle resistance $R\eta_B$ as $p_{O_2}^{1/4}$. An example is given in Fig. 19 for a thoria electrolyte for which the high-frequency semicircle is plotted more accurately.

The pre-exponential factor and the activation energy of $R\eta_A$ depend greatly on the texture of the electrode and on the nature of the material of which it is made. The activation energy reaches a minimum with porous solid silver electrodes. In this case the activation energy is very close to that of the conductivity of the electrolyte. The Arrhenius diagram of $R\eta_A$ even exhibits the same break as the conductivity diagram near 700 °C (Fig. 20). This correlation between the electrode polarization resistance and the electrolyte resistance is even more clearly demonstrated in Table I. This table compares the values of the ratio of the electrode resistance to the electrolyte resistance obtained for various electrolytes. Despite a variation of the electrolyte resistance over approximately two orders of magnitude, the calculated ratio remains significantly constant. This confirms an earlier study[31] which concluded that both the electrode material and the electrolyte usually contribute to the kinetics of the limiting step of the electrode reaction. This conclusion applies especially to porous metallic electrodes.

When diffusion of oxygen is the limiting step, as is the case with molten silver, the electrode characteristic is a well defined Warburg impedance (Fig. 21):

$$Z = \frac{RT}{n^2 F^2 \sqrt{2}} \ \frac{1}{CD^{1/2}} \ (1+i)w^{-1/2}$$

with C and D equal to the concentration and the diffusion coefficient of oxygen. This law is verified with no ambiguity for frequencies over several orders of magnitude.

The capacitance of the usual low frequency semicircle is far too high (typically of the order of several mF) to be interpreted in terms of a double layer or a space charge capacitance. The only reasonable explanation is in terms of an electrochemical pseudocapacitance.[30]

The three-electrode setup previously described can be used to investigate the influence of a dc bias on the electrode characteristic. In general the results obtained confirm the conclusion reached by varying the oxygen pressure in equilibrium with the electrodes. With water vapor electrolysis, a new feature was observed, which is illustrated in Fig. 22. As the dc bias potential is made more cathodic, the electrode resistance initially increases, as is usually observed with oxygen electrodes; with a further increase in the bias, it decreases to zero. When this point is reached, the interface electrolyte-electrode is perfectly ohmic. One possible interpretation is the onset of a predominant electronic conductivity of the electrolyte. However, the resistance of the electrolyte as measured by the impedance diagram does not show any variation consistent with a large increase in the electronic conductivity. It must therefore be concluded that the electronic conductivity is limited to the electrode surface. This phenomenon is observed at approximately -1500 mV/air where the bulk electronic conductivity is actually still small and where the reduction of a dissolved redox couple was detected.[32] This redox couple could be responsible for the displacement of the electrons along the surface where they react directly with, and thereby reduce, adsorbed water molecules. This type of electrode reaction may be viewed as an electrocatalyzed reaction.

Electrolyte Resistance

The following is a summary of a systematic investigation[21] of the influence of impurities, porosity, and grain size on the conductivity of sintered stabilized zirconia. The samples were prepared from coprecipitated ultrafine powders. Following the standard procedure, the samples were pressed under 200 MPa and sintered at 1300 °C to a density of 93% of the theoretical value. The nominal composition was 9 mol% Y_2O_3. The porosity was varied from 25 to 5% by varying the forming pressure and the grain size (from an average submicrometer diameter to 16 μm) by annealing at a higher temperature. Samples with Al_2O_3 and Fe_2O_3 added as impurities and samples prepared from technical-quality powders containing SiO_2 were also compared to the standard samples.

Intragrain Conductivity

The intragrain conductivity was found to be proportional to the density of the sample (Fig. 23). The proportionality factor is 1.2. This can be used to easily correct any results for the effect of porosity. The extrapolation to 100% theoretical density was equal, within less than 10%, to that of a single crystal used as a reference. As expected, intragrain conductivity is independent of grain size. These two results confirm that intragrain conductivity is a specific property of the stabilized zirconia phase.

This conductivity depends on the impurity content of the sample. Silica, reputed to be a severe blocking agent, also has a marked negative influence on intragrain conductivity. This clearly indicates SiO_2 solubility in stabilized zirconia and merits further investigation, as the magnitudes of the resulting conductivity variations are not easily interpreted within the framework of the usual theories. It is now well established that the Arrhenius diagram of intragrain conductivity exhibits a break near 500 °C. This indicates a change in the conductivity mechanism. This point also merits further investigation.

Below this temperature, the proposed equation for the intragrain conductivity is

$$\sigma(\Omega \cdot cm)^{-1} = 8.67 \times 10^3 \, \exp\, -\frac{1.06}{kT} \quad (eV)$$

and above this temperature

$$\sigma(\Omega \cdot cm)^{-1} = 5.68 \times 10^2 \, \exp\, -\frac{0.89}{kT} \quad (eV)$$

Blocking Effect

In terms of electric circuits, the blocking effects at the grain boundaries can be described in two ways. A fraction of a grain boundary is sketched in Fig. 24. Following the usual assumptions, it is characterized by a zone where the O^{2-} ions are free to cross the boundary and a zone where they are blocked.

First, within the framework of the series model proposed by Bauerle,[1] which is generally referred to, the contacting grains are decomposed in three "layers": the intragrain layers (1 in Fig. 24) and a boundary layer (2 in Fig. 24). They are physically in series. A parallel R_1C_1 circuit is associated with the intragrain layers as described in the introduction. Another parallel circuit R_2C_2 is connected in series to simulate the electrical behavior of the boundary layer. Usually the blocking effect is considered to be caused by an impurity phase segregated at the boundary. The resistance R_2 is accordingly considered as the resistance of this segregated phase. Sometimes it is also described as a constriction resistance, expressing that the area of the conducting electrolyte is smaller than the theoretical value in the boundary layer. The capacitance C_2 is also either associated with dielectric properties of the segregated phase or considered to be a space charge effect resulting from the blocking of some ions.

Most published experimental results have pointed out that the activation energy of R_2 is always very close to that of the intragrain resistance. This rules out any general interpretation in terms of a segregated phase resistance. Similarly, the systematic study of Bernard[21] showed that C_2 does not significantly depend on the nature of the segregated phase. This also rules out any correlation between C_2 and a specific property of the impurity phase.

The connection between all the circuits representing the grains in a sintered sample is clearly a difficult problem and was introduced by Beekmans and Heyne.[12]

The blocking effect can be described in another way. The mobile oxide ions in the grains can be separated into two groups: those which will cross the boundary and those which will be blocked. Naturally, they should be represented by two parallel resistances r_1 and r_2. The blocking effect, whatever it is, is simulated by a capacitance c_2 in series with the resistance associated with the blocked ions (Fig. 24(b)). In this model, an essential parameter is the ratio β of the quantity of blocked ions to the total number of mobile ions.

This ratio β is related to the impedance diagram parameters (Fig. 3) by the equation:

$$\beta = (B - A)/B$$

The comparison by Bernard[21] of the plots and equations expressing his results in terms of R_2 or β (and also in terms of C_2 or c_2), showed conclusively that the parallel model gives a more consistent description of the electrical behavior of sintered stabilized zirconia.

The main conclusions of Bernard's systematic study are:

•The temperature T_2 (Fig. 4) at which the blocking effect vanishes markedly depends on the level of sample purity. With nominally pure materials it is $\approx 700\,°C$; with technical-quality materials it can be as high as $1000\,°C$.

•In most technical-quality products, impurities play a predominant role

319

in the blocking effect. Up to 80% of the mobile ions are typically blocked below T_2.

•The roles of the segregated phases may be drastically different. In the absence of silica, alumina segregated at rather low temperatures does not significantly block conduction up to a concentration of 4 mol%. Silica already has a tremendous blocking effect at a level of 1%. The key point may be the formation of a molten phase during sintering.

•In pure materials where no segregated phases are detectable (even by high resolution microscopy) a marked blocking effect remains which appears to be a specific property of the contact between the grains. It could be associated with mismatching between grains. This result is in part confirmed by an earlier investigation of a macroscopic contact between two sintered samples.[33] Such a contact seems to behave simply as a grain boundary.

•This specific blocking effect is proportional to the area of the boundaries and therefore increases significantly as the grain size of the sintered material decreases. A similar result was obtained by Ioffe et al.[6] In technical applications, the loss in conductivity becomes important with grain sizes smaller than 5 µm (20% at 850°C between 5 µm and 0.7 µm).

•The pores contribute an additional blocking effect which is proportional to their density (Fig. 25).

Aging

Aging of yttria-stabilized zirconia, which results in a detrimental decrease in conductivity (especially between 800 and 1000°C), was reexamined using impedance spectroscopy[17,21] to separate the intragrain and grain boundary contributions.

In technical-quality products, aging is important and involves mainly gradual enhancement of the blocking effect. A true square-root law was found (Fig. 26), indicating a diffusion-controlled process which could simply be the segregation at the grain boundaries of oversaturated impurities. This is likely caused by higher solubilities of the impurities at the sintering temperature than at the utilization temperature.

To avoid any interference by the blocking effect, the intragrain resistance aging was investigated on a single crystal by Schouler.[17] Usually it is slight and obeys an exponential law (Fig. 27). This would correspond to a reorganization of the crystal, probably associated with a partial ordering of the cations.

Bernard[21] found that the intragrain aging rate depends on the surrounding atmosphere (Fig. 28). Aging is slower in hydrogen than in air. A similar observation was made by Baukal.[34] It was also found that the aging rate increases as the grain size decreases (Fig. 28).

References

[1] J. E. Bauerle, *J. Phys. Chem. Solids*, **30**, 2657 (1969).
[2] M. V. Perfil'ev, *Elektrokhimiya*, **7**, 792 (1971).
[3] A. E. Zupnik, M. V. Perfil'ev, and S. V. Karpachov, *ibid.*, **7**, 1188 (1971).
[4] M. V. Inozemtsev and M. V. Perfil'ev, *ibid.*, **11**, 1031 (1975).
[5] A. V. Smirnov, M. V. Siminova, and E. G. Shubanova, *ibid.*, p. 1836.
[6] A. I. Ioffe, M. V. Inozemtsev, A. S. Lipilin, M. V. Perfil'ev, and S. V. Karpachov, *Phys. Status Solidi A*, **30**, 87 (1975).
[7] E. Schouler, M. Kleitz, and C. Déportes, *J. Chim. Phys.*, **70**, 923 (1973).
[8] E. Schouler, G. Giroud, and M. Kleitz, *ibid.*, p. 1309.
[9] E. Schouler and M. Kleitz, *J. Electroanal. Chem.*, **64**, 135 (1975).
[10] M. Kleitz, P. Fabry, and E. Schouler; p. 1 in Electrode Processes in Solid State Ionics. Edited by M. Kleitz and J. Dupuy. Reidel Publ. Co., Dordrecht, Holland, 1976.
[11] Y. Suzuki and T. Takahashi, *Denki Kagaku oyobi Kogyo Butsuri Kagaku*, **39**, 406 (1971).
[12] N. M. Beekmans and L. Heyne, *Electrochim. Acta*, **21**, 303 (1976).
[13] S. H. Chu and M. A. Seitz, *J. Solid State Chem.*, **23**, 297 (1978).
[14] T. M. Guer, I. D. Raistrick, and R. A. Huggins; p. 113 in Fast Ion Transport in Solids. Edited by P. Vashishta, J. N. Mundy, and G. K. Shenoy. North Holland, Amsterdam, 1979.
[15] A. S. Nowick, Da Yu Wang, D. S. Park, and J. Griffith; *ibid.*, p. 673.
[16] E. Schouler, A. Hammou, and M. Kleitz, *Mater. Res. Bull.*, **11**, 1137 (1976).
[17] E. Schouler, Thesis, Grenoble, 1979.
[18] K. el Adham, Thesis, Grenoble, 1978.
[19] M. Kleitz and J. H. Kennedy; p. 185 in Ref. 14.
[20] M. Kleitz, J. R. Akridge, and J. H. Kennedy; to be published in *Solid State Ionics*.
[21] H. Bernard, Thesis, Grenoble, 1980.
[22] E. Fernandez, Thesis, Grenoble, 1980.
[23] T. G. Stratton, D. Reed, and H. L. Tuller; pp. 114–23 in Advances in Ceramics 1. American Ceramic Society, Columbus, Ohio, 1981.
[24] H. Obayashi and T. Kudo; p. 327 in Solid State Chemistry of Energy Conversion and Storage. Advances in Chemistry Series No. 163. Edited by A.S. Whittingham and J.B. Goodenough. American Chemical Society, Washington, D.C., 1977.
[25] M. Kleitz, Thesis, Grenoble, 1968.
[26] D. Ravaine, Thesis, Grenoble, 1976.
[27] R. D. Armstrong, T. Dickinson, and P. M. Willis, *J. Electroanal, Chem.*, **53**, 389 (1974).
[28] A. K. Jonscher, *Nature (London)*, **267**, 673 (1977).
[29] B. Cales and P. Abelard; "Lattice Defects in Ionic Crystals," Conf. Canterbury (1979).
[30] M. Kleitz, P. Fabry, and E. Schouler; p. 439 in Fast Ion Transport in Solids. Edited by W. Van Gool. North Holland, Amsterdam, 1973.
[31] P. Fabry and M. Kleitz, *J. Electroanal. Chem.*, **57**, 165 (1974).
[32] P. Fabry, M. Kleitz, and C. Déportes, *J. Solid State Chem.*, **6**, 230 (1973).
[33] P. Fabry, E. Schouler, and M. Kleitz, *Electrochim. Acta*, **23**, 539 (1978).
[34] W. Baukal, *ibid.*, **14**, 1071 (1969).

*Centre d'Etudes Nucleaires, B.P. 85X - 38041 Grenoble, France.

Table I. Variations of the ratio $r = R\eta_A/\varrho_{dc}$ (780 °C, $P_{O_2} = 10^{-4}$ atm)

	ThO$_2$-Y$_2$O$_3$						ZrO$_2$-Y$_2$O$_3$
YO$_{1.5}$ mol%	0.2	0.5	1	2	10	15	17
$r \times 10$	7	7	6	2.5	2	8	7

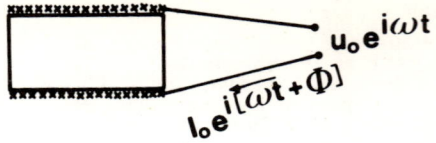

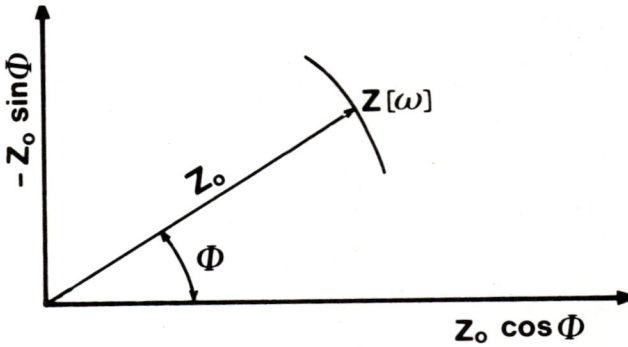

Fig. 1. Principle of the impedance spectroscopy.

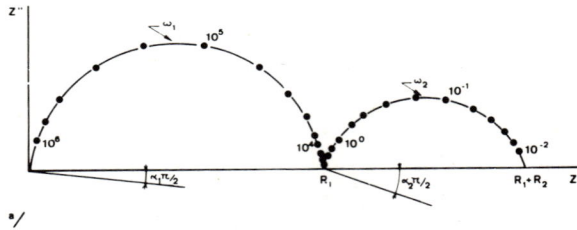

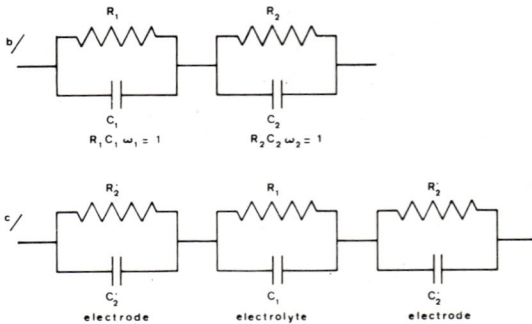

Fig. 2. Typical impedance diagram obtained with a single crystal and equivalent circuits.

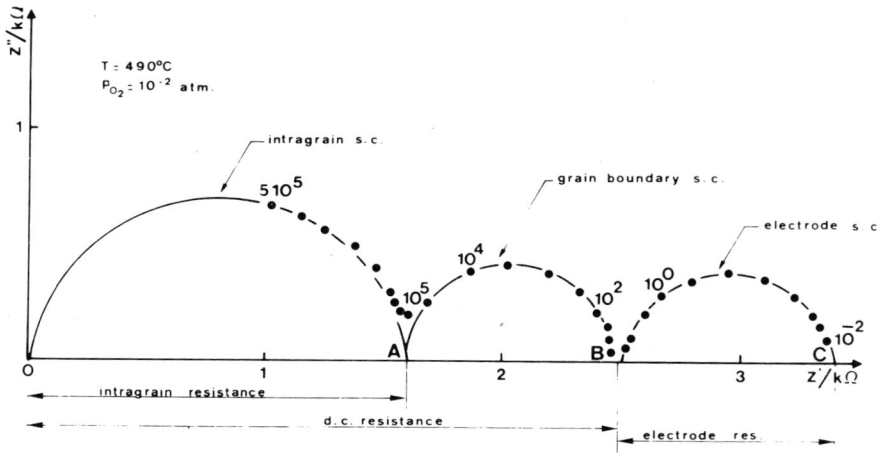

Fig. 3. Impedance diagram obtained with a sintered sample at low temperature.

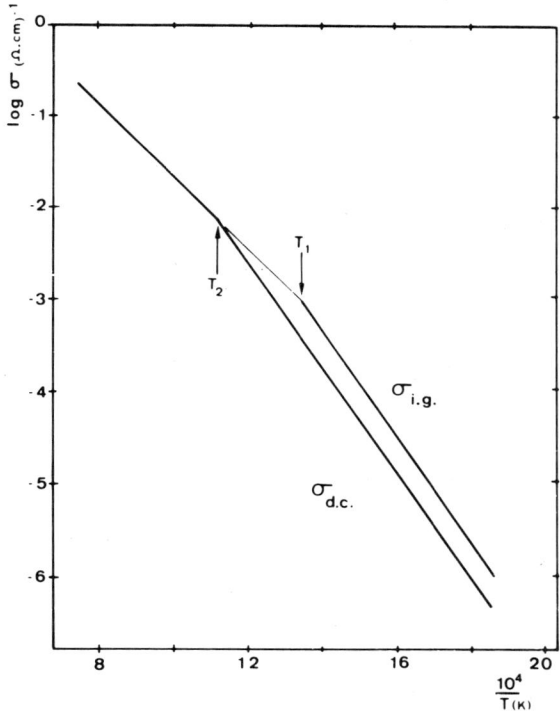

Fig. 4. Arrhenius diagram of the intragrain (i.g.) and dc conductivities.

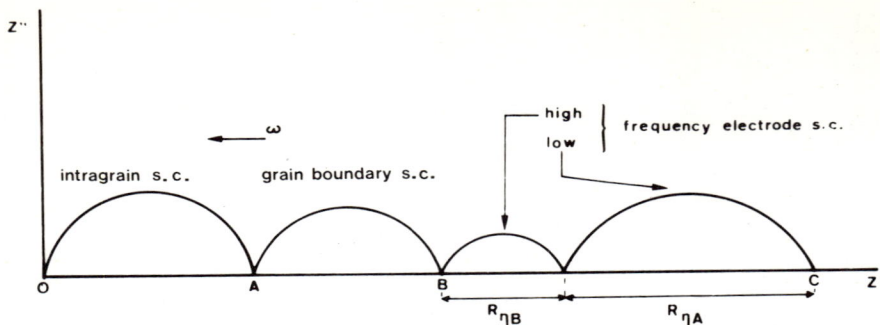

Fig. 5. Complete impedance diagram of a symmetrical cell formed on a sintered pellet. R_{η_A}, low frequency semicircle resistance; R_{η_B}, high frequency semicircle resistance.

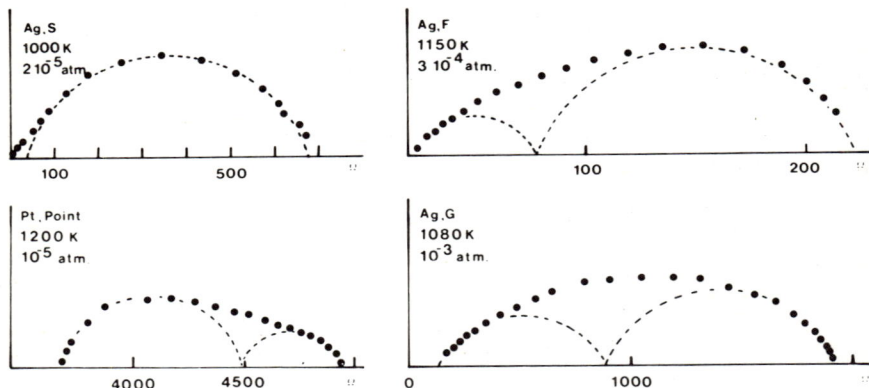

Fig. 6. Shapes of the electrode characteristics with different types of electrode (Ref. 17). Ag,S: rf sputtered silver; Ag,F: silver foil pressed against the sample; Ag,G: silver screen pressed against the sample; Pt, Point: platinum point electrode.

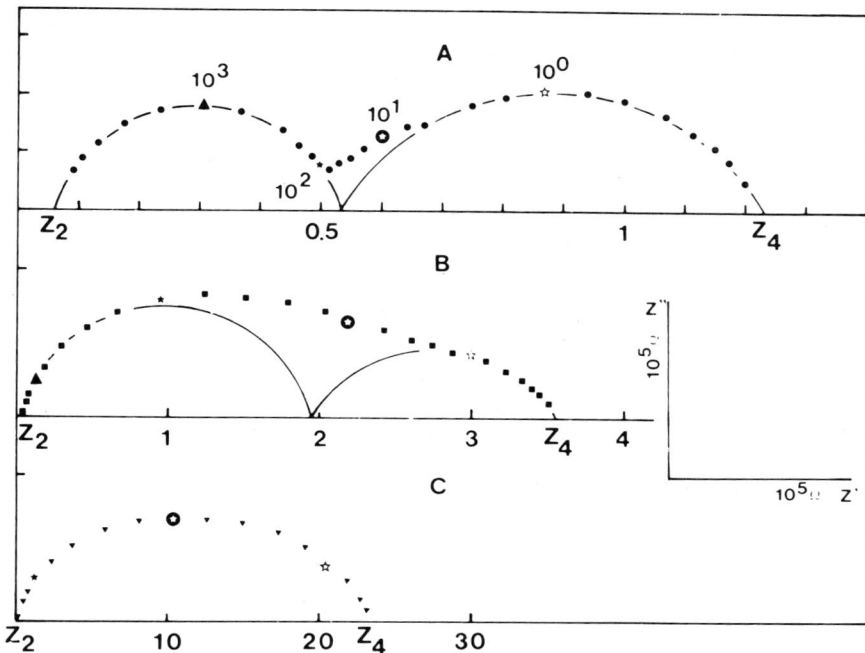

Fig. 7. Impedance diagram of the cell (Ref. 17). CO-CO_2, Pt/ThO_2 $(1-x)$-$Y_2O_3(x)/Pt$,CO-CO_2. $T = 700\,°C$, $CO/CO_2 = 1/99$; (A) $x = 2 \times 10^{-2}$, $l/s = 1.69$ cm^{-1}; (B) $x = 2 \times 10^{-2}$, $l/s = 0.87$ cm^{-1}; (C) $x = 5 \times 10^{-3}$, $l/s = 0.87$ cm^{-1}.

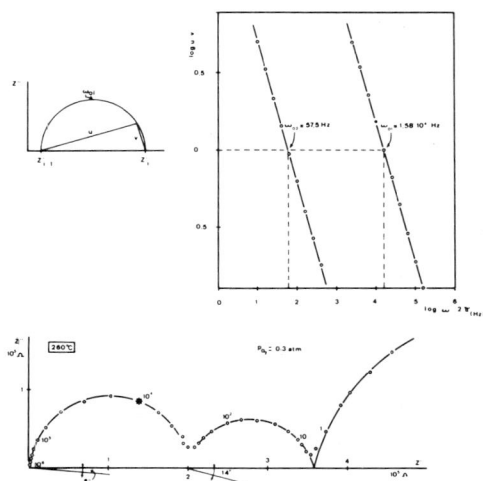

Fig. 8. Verification of the frequency distribution of the experimental points on the intragrain and grain boundary semicircles according to the Cole-Cole method (Ref. 21).

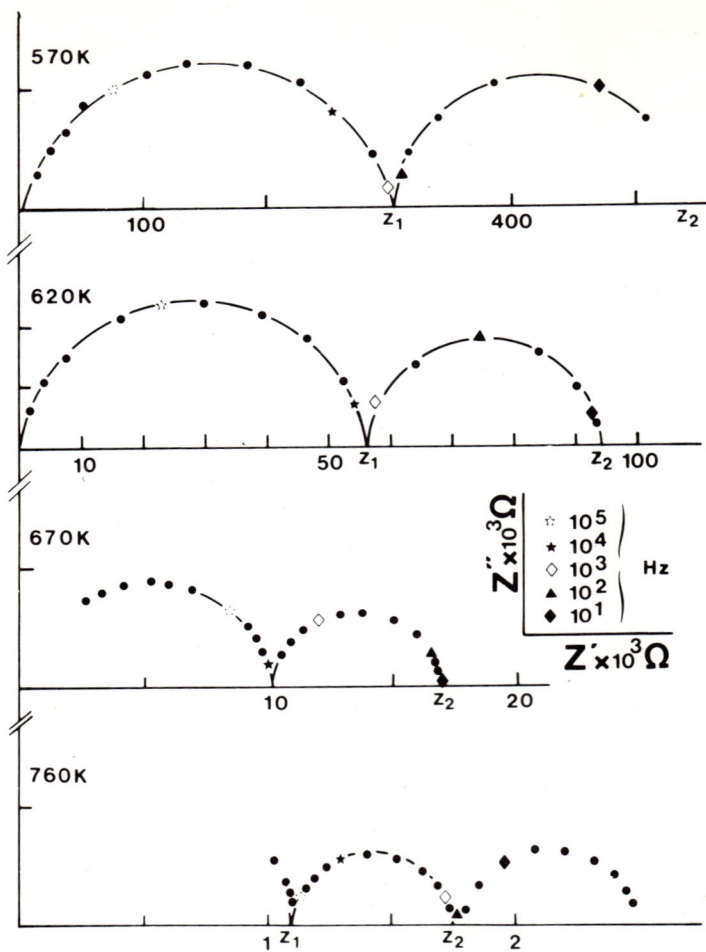

Fig. 9. Variation of the "observable range" with the temperature of the sample (Ref. 17).

326

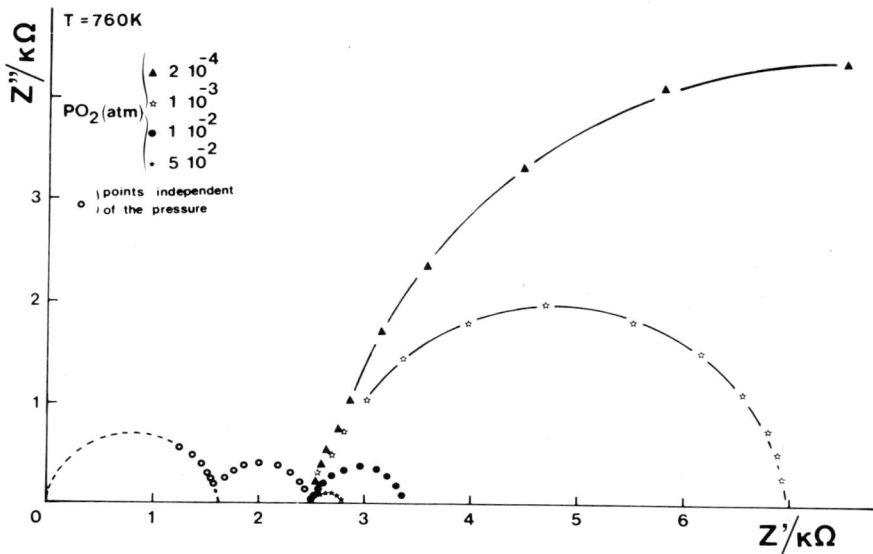

Fig. 10. Influence of the oxygen pressure in equilibrium with the cell on the impedance diagram.

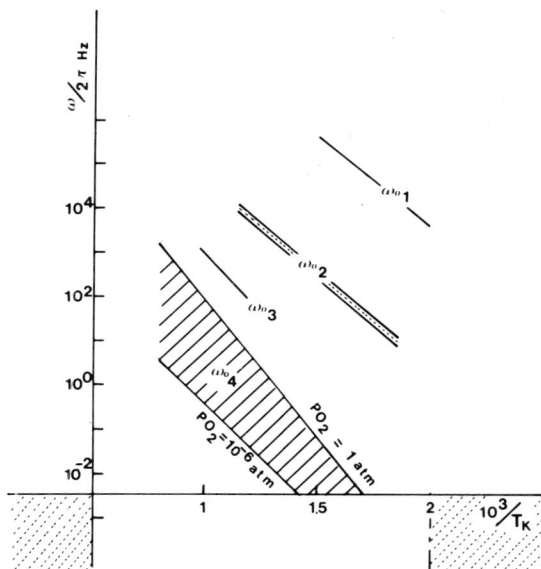

Fig. 11. Variation of the relaxation frequencies with temperature (Ref. 17). ω_1^0, intragrain semicircle; ω_2^0, grain boundary semicircle; ω_3^0, high-frequency electrode semicircle; ω_4^0, low-frequency electrode semicircle.

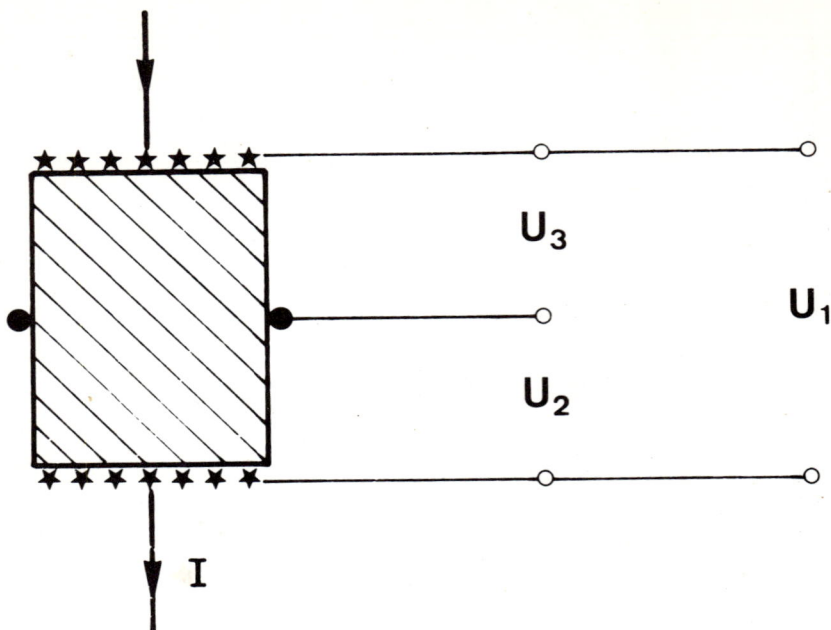

Fig. 12. Three-electrode cell.

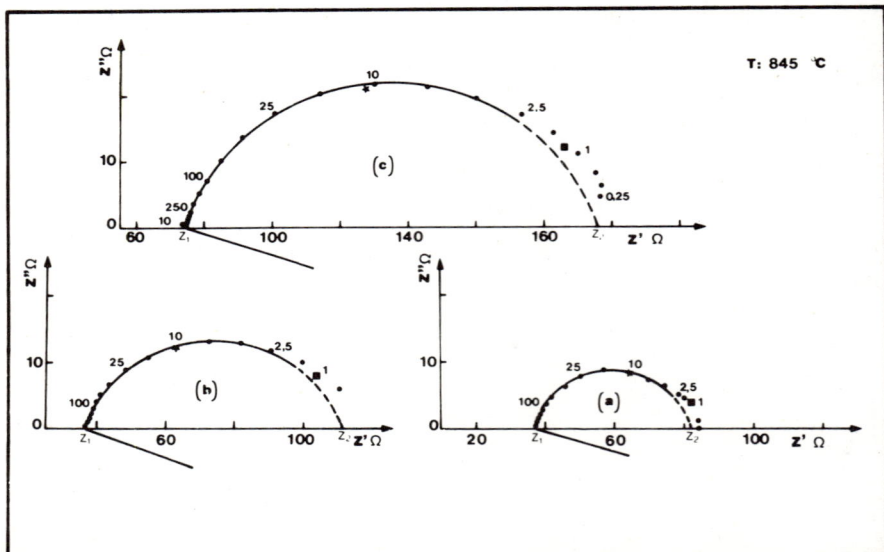

Fig. 13. Impedance diagrams determined with the three electrode cell (Ref. 22). (a) and (b) Diagram deduced from U_2 and U_3 (cf. Fig. 12); (c) global diagram deduced from U_1.

328

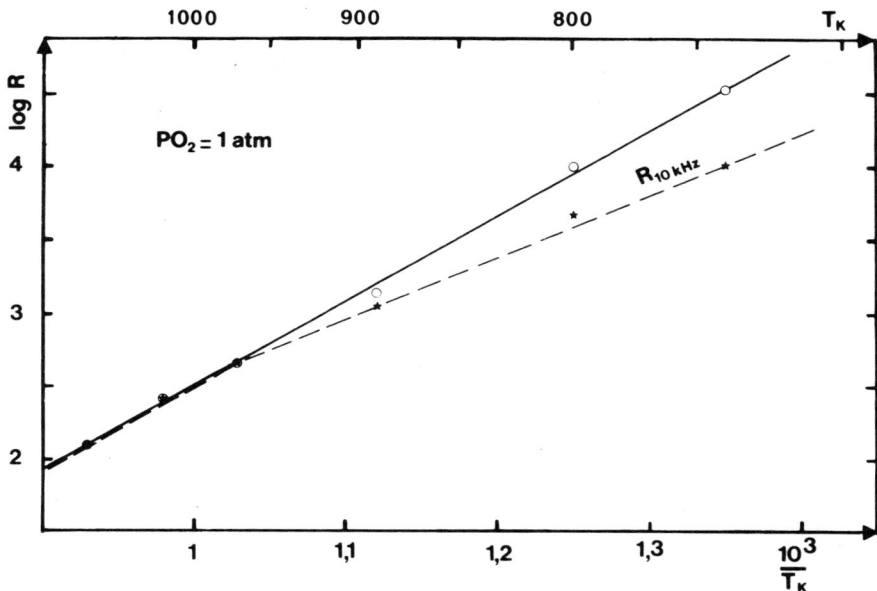

Fig. 14. Dc resistance of a ZrO_2-Y_2O_3 (9 mol%) sample measured by impedance spectroscopy and at a fixed frequency of 10 kHz (Ref. 8).

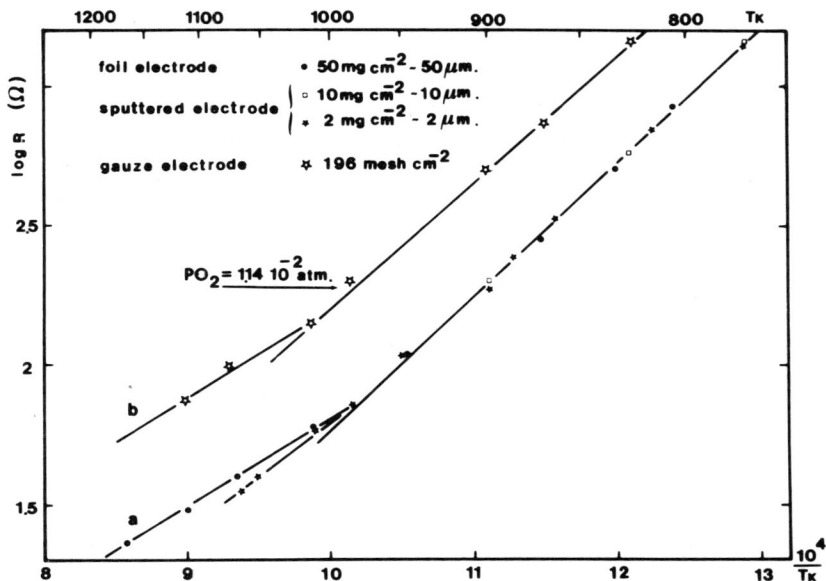

Fig. 15. Conductivity of a sample measured with electrodes of different coverages (Ref. 17).

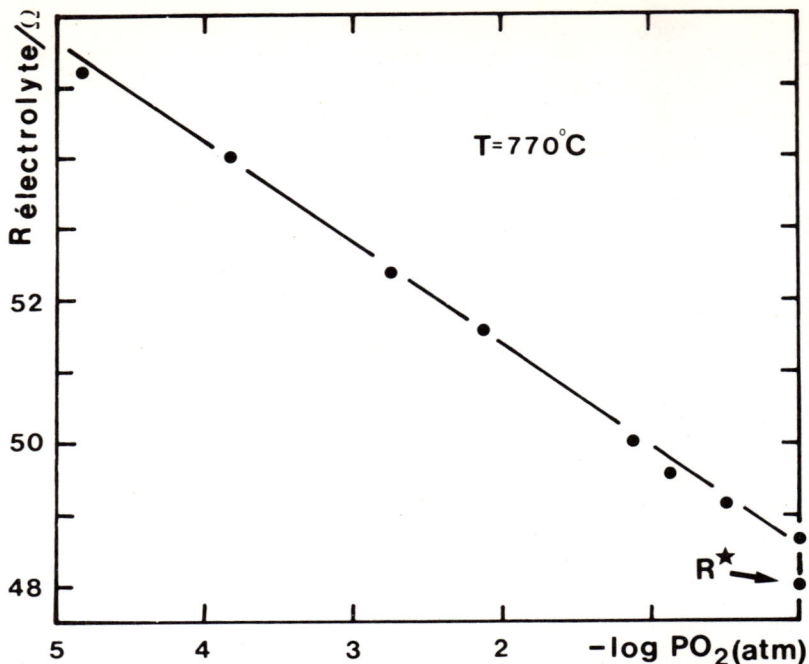

Fig. 16. Variation of the conductivity of a sample measured with silver-gauze electrodes (cf. Fig. 15) as a function of the oxygen pressure in equilibrium. $R^{\star}$, resistance measured with electrodes ensuring sufficient coverage (Ref. 17).

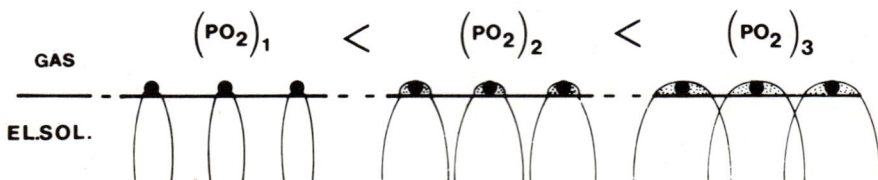

Fig. 17. Variation of the spreading of the electrode reaction zone and the resulting constriction of the current lines as a function of the oxygen pressure (Ref. 17).

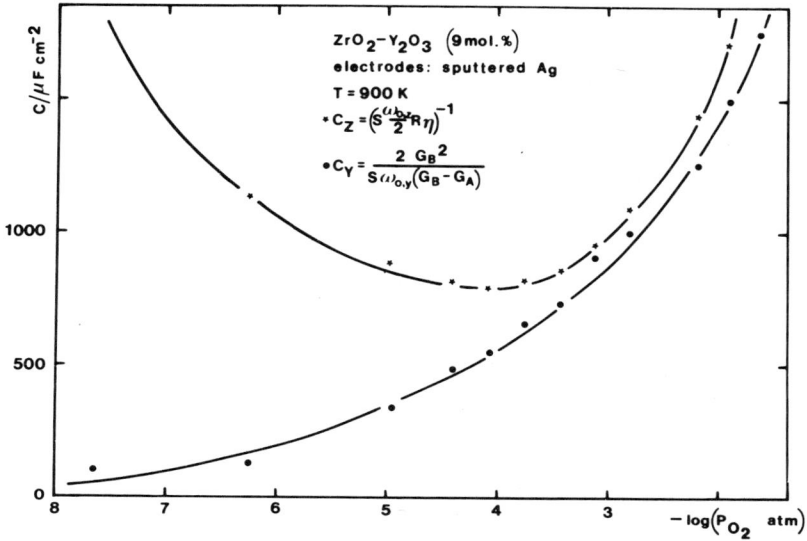

Fig. 18. Typical variation of the electrode capacitance (Ref. 17). C_Z, capacitance calculated from the impedance diagram; C_Y, capacitance calculated from the equivalent admittance diagram.

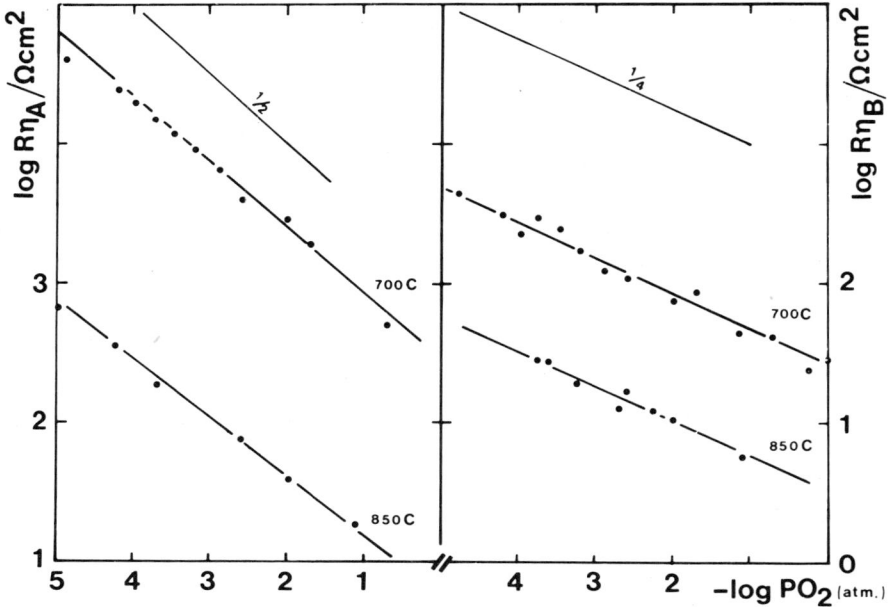

Fig. 19. Variation of the platinum electrode semicircle resistances (Ref. 17) ThO_2-10% $YO_{1.5}$.

331

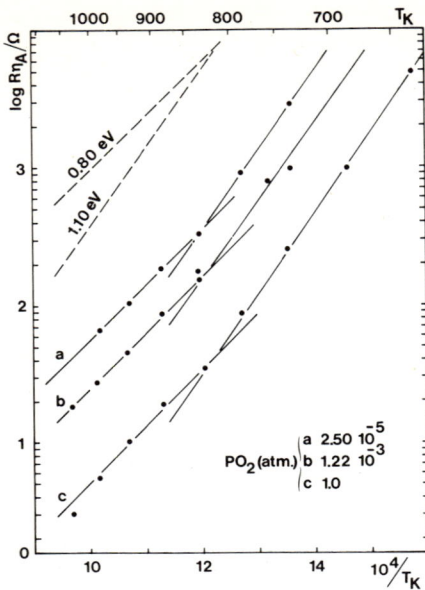

Fig. 20. Arrhenius diagram of a low frequency electrode resistance (Ref. 17). Sputtered silver electrode on ZrO_2-Y_2O_3 (9 mol%).

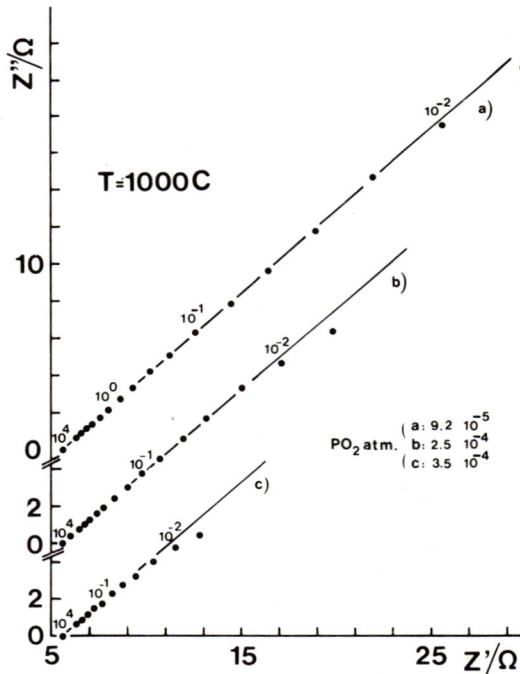

Fig. 21. Electrode characteristics of molten silver electrodes in contact with ZrO_2-Y_2O_3 (9 mol%).

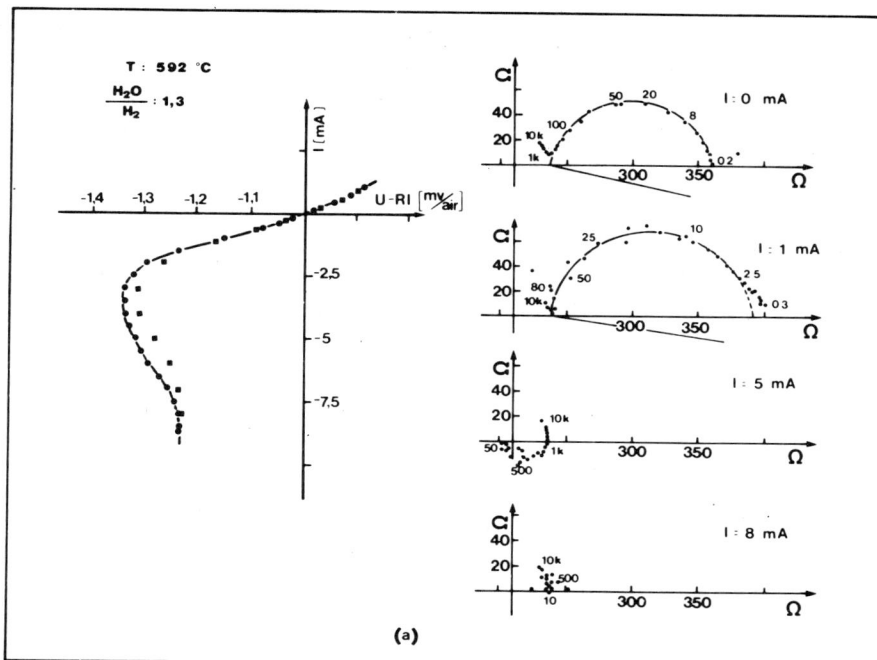

Fig. 22. Variation of the electrode semicircle with the dc bias Pt, H_2-H_2O/ZrO_2-Y_2O_3.

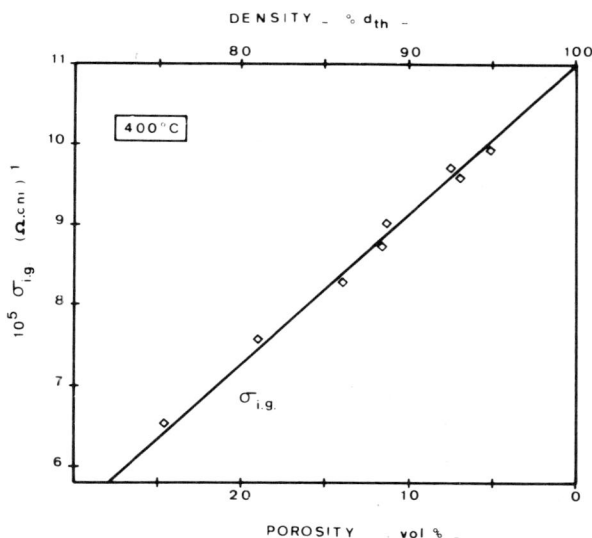

Fig. 23. Variation of the intragrain conductivity as a function of the porosity of the sample (Ref. 21).

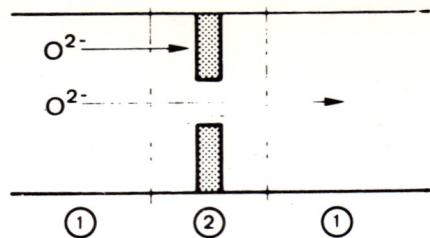

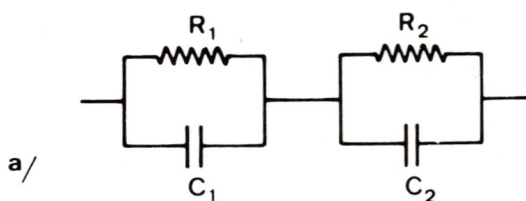

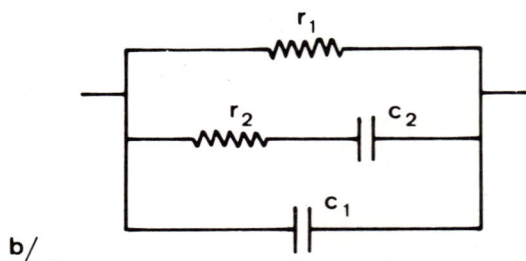

Fig. 24. Diagram of a grain boundary; (a) equivalent series circuit, (b) equivalent parallel circuit.

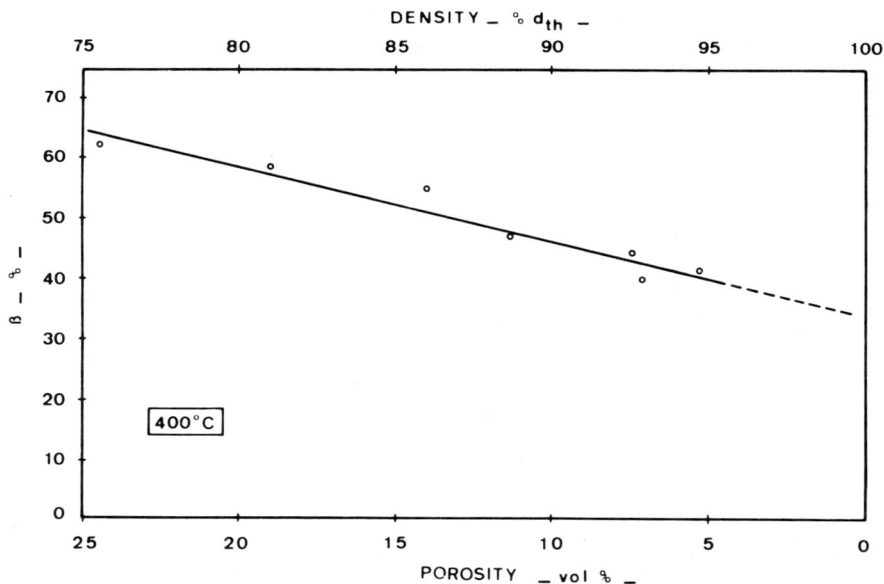

Fig. 25. Variation of the blocking ratio with the porosity (Ref. 21).

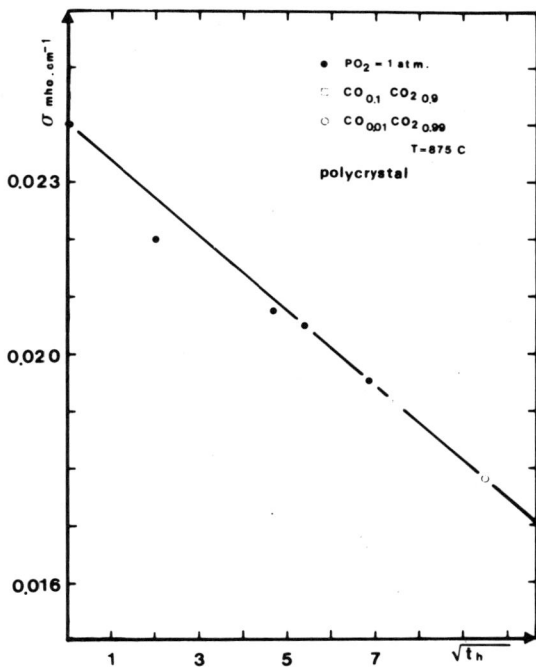

Fig. 26. Aging of the dc conductivity of a sample of technical purity.

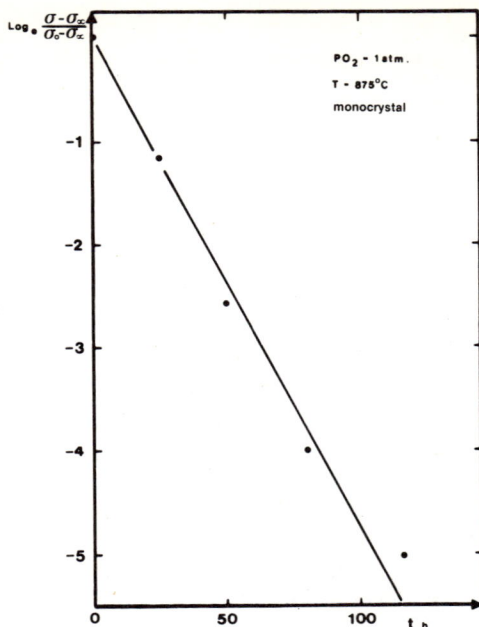

Fig. 27. Aging of the single crystal conductivity.

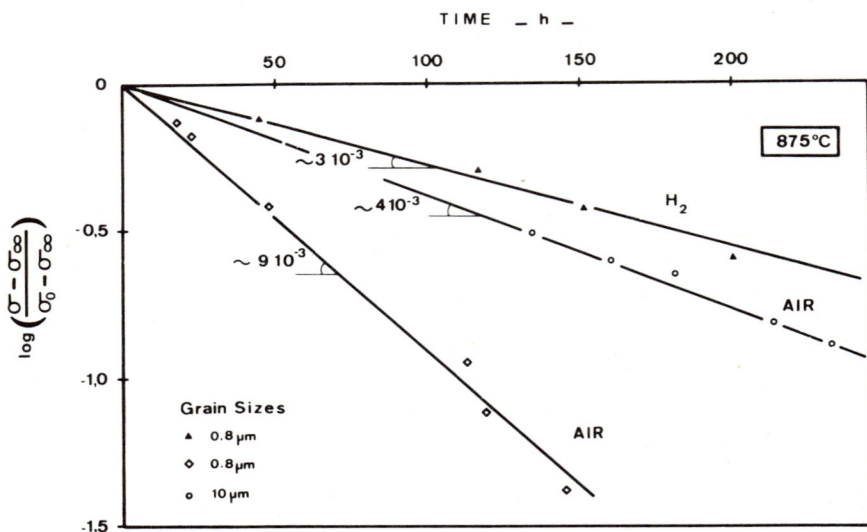

Fig. 28. Aging of the intragrain conductivity under different atmospheres and with different grain sizes (Ref. 21).

336

Determination of electronic conductivity in ZrO_2-Y_2O_3 by electrochemical reduction

M. Kleitz, M. Levy, J. Foulétier, and P. Fabry

Laboratoire d'Energétique Electrochimique
Institut National Polytechnique de Grenoble
E.N.S. d'Electrochimie et d'Electrométallurgie
38401 Saint Martin d'Heres
France

The plotting of the reoxidation potential curve of a point electrode shows a complex electronic structure for stabilized zirconia. A controlled reduction procedure, using oxygen pumps and gages, can be used to directly determine the relationship between the total conductivity and the deviation from stoichiometry of the oxide electrolytes under any reducing condition. A comparison of results obtained on doped ceria and stabilized zirconia confirms the complexity of the electronic structure of zirconia.

Severe electrochemical reductions of stabilized zirconia due to the passage of a high current density were investigated in the early 60's in connection with the use of this material as MHD electrodes.[1] The recent development of an electrically controllable and renewable oxygen getter based on reduced zirconia[2] gave additional impetus for such investigations.

Obviously the interval where zirconia can be used in galvanic cells for thermodynamic determinations is limited by the onset of a chemical reduction of the material, more precisely by the resulting increase in its electronic conductivity. This has prompted several studies[3,4] on more gentle reduction processes.

Basically a partial reduction of a solid oxide electrolyte is related to a valence change either of the basic cation (Zr^{4+} in the case of zirconia) or of a point defect (a dissolved chemical impurity such as a ferric ion or a structural defect such as an oxide-ion vacancy for instance). In the latter case, the relevant point defect can be looked upon as an electronic trap.

Such a reduction can obviously be obtained by equilibrating the material with a reducing system. It can also be produced by forcing a current higher than the limiting currents associated with the possible gas electrode reactions (involving species like O_2, CO_2, H_2O). This specific mode of reduction is made possible by the ionic conductivity of the oxide electrolytes. There is no difference between a chemically and an electrochemically reduced sample.

The main properties associated with the reduction are:

•A variation of the stoichiometry ratio due to oxygen release. This in turn implies a weight loss[5,6] and sometimes a modification of the lattice parameters.[7,8]

•A variation of the conductivities. The ionic conductivity changes due to

337

an increase in oxide-vacancy concentration and the electronic conductivity due to an increase in free-electron concentration.[1,9]

•A coloration and variation of other spectroscopic properties. A black coloration of reduced zirconia was first observed by Weininger and Zemany.[10] As a rule, coloration is due to entrapped electrons which can be excited to higher energy levels. Entrapped electrons can also be detected and characterized by other spectroscopic techniques such as ESR.[11-13]

•A variation of the electrode potential. An electrode in contact with zirconia reads a potential which varies as a function of the reduction degree as in aqueous solutions in the presence of electrochemical redox couples.[3,14]

Most of these properties were described in detail in a review paper.[15] We would like here to concentrate on typical experimental methods and to report some new results which confirm the complex electronic structure of stabilized zirconia. The methods described: (1) demonstrate the existence of reducible electronic traps in stabilized zirconia and determine their relative energy levels; (2) characterize the progression of the reduction front in the bulk of the material, considering two situations, depending on the relative magnitude of the additional electronic conductivity; and (3) measure the electronic conductivity in gradually reduced materials.

Determination of the Electronic Trap Levels

A rapid method of determining the electrode potentials (and the oxygen pressures) at which reduction reactions of point defects occur in solid electrolytes was proposed by Fabry et al.[3] This method is called the "point electrode technique" on account of the shape of the working electrode. The experimental cell is simply formed on the investigated zirconia pellet as sketched in Fig. 1. A large-surface counter electrode is deposited on one face and is maintained in contact with ambient air. A point metallic electrode is gently pressed against the other face. It is surrounded by an argon-oxygen mixture with low oxygen content. Basically the method consists, first, in electrochemically reducing a small amount of the investigated material around the point electrode by passing an appropriate direct current. After interrupting the current, the voltage of the cell E is recorded during the slow reoxidation of the material by the traces of oxygen contained in argon. Then, the cell works approximately like a titration cell, the slow supply of oxygen by the gas being analogous to the supply of the reactant from a buret. The diagram of E versus time exhibits waves as shown in Fig. 1. The voltages E^0 at the inflection points (i.e. at A, B, C, D in Fig. 1) are characteristic of dissolved redox systems, the valency change of which is responsible for the reduction of zirconia. This technique was extensively used by Fabry.[3,14,16] It was found that besides the main reduction of stabilized zirconia which occurs at approximately -2000 mV/air (wave A in Fig. 1), three point defects can be reduced in stabilized zirconia which result in a macroscopic chemical reduction of the material. The corresponding redox potentials vary as a function of temperature as shown in Fig. 2. The reduction which occurs near -1100 mV/air (wave D) has an extremely small amplitude and is likely to be limited to the surface or subsurface area. Redox reaction C seems to disappear above 1000 °C. Despite a large number of identification experiments, no definite conclusions could be drawn as to the nature of the point defects involved in all these redox reactions. They do not depend on the nature of the dopant.

When similar experiments are performed with working electrodes of large area, the response curve $E(t)$ is not as well structured and the waves are more or less smeared out. The initial reduction and reoxidation degrees are probably not equally distributed over the surface of contact and the working electrode averages the reoxidation process.

Progression of the Reduction Front in the Bulk of the Material
Sweep Voltammetry

When the electronic conductivity remains relatively small, the material behaves as an electrolyte and, in principle, traditional electrochemical methods can be applied. As an example, sweep voltammetry was used to characterize the change of valency of copper ions dissolved intentionally in a zirconia single crystal stabilized with 9 mol% yttria.[17] Copper was chosen because it is likely to be reduced within a redox stability interval of the electrolyte. The experimental cell is sketched in Fig. 3. A typical voltammogram is also shown in Fig. 3. Peak currents are observed. They are characteristic of slow diffusion processes of the electroactive species. A detailed analysis of the results showed that the diffusion activation energies of the oxidized and reduced species are similar: 1.6 and 1.5 eV, respectively. These activation energies are furthermore quite close to that of the electron-hole conductivity: 1.5 eV. Therefore, it was proposed that the limiting process of the reduction and subsequent reoxidation reaction is a diffusion of electron holes instead of copper ion diffusion which would be the usual mechanism in a conventional electrolyte.

Probes Technique

When the electronic conductivity in reduced zirconia is relatively important, the ohmic drop in the electrolyte decreases continuously and sometimes varies markedly. Electrochemical techniques cannot be easily used under these conditions.

The set of equations which describe the evolution of the material during its reduction is rather complex.[14,15] It comprises:

(1) The equations characterizing the local reactions or equilibria (defect creation or annihilation, association of defects, electron-hole recombination, redox reactions. . .).

(2) The equation giving the local magnitude of the particle fluxes as functions of their electrochemical potentials.

(3) The local balance equation for each point defect.

(4) The Poisson equation which defines the space charges in the material or the electroneutrality equation when these space charges are negligible.

(5) The limiting conditions.

This set of equations cannot strictly be solved. Two limiting simple cases could be treated[14,15] and gave the shape for the reduction front which moves from the cathode toward the anode:

(1) When the electronic conductivity in the reduced material is not too important (when the reduction is, for instance, produced under low current density), the solution obtained is quite similar to Fick's second law and the free-electron profile obeys an erfc function as shown in Fig. 4(a).

(2) When the electric field is high enough to be the predominant driving force for the mobile particles (electrochemical reduction under high current

density) the shape of the free-electron profile has the form of a step which is not altered when it moves from the cathode to the anode (Fig. 4(*b*)).

In the general case, the free-electron profile is intermediate between these two limiting shapes and the front zone becomes more and more diffuse as it progresses through the bulk of the material. An experimental verification of this conclusion was carried out by Fabry[14] with the experimental cell described in Fig. 5. Platinum probes were wrapped around a bar-shaped sample and pasted with platinum paint. The voltages of these probes with respect to the unpolarizable counter electrode of the cell measured the local free-electron concentration (on a logarithmic scale). The progression of the reduction zone could be observed during the electrochemical reduction by recording their time dependencies. Here again waves were observed indicating the existence of reducible point defects associated with successive reduction stages of stabilized zirconia. As predicted by the calculation, the waves gradually smeared out as the reduction progressed through the sample.

By referring to a simplified model called the "virtual cathode model,"[14,15] it is possible to estimate the order of magnitude of the relevant point defect concentration from the relationship between the current density and the progression rate of the reduction front. The point defects which are reduced at approx -1450 mV/air are only 3% as concentrated as the oxide ion vacancies. This result rules out the suggestion that F' centers could be formed at this stage of reduction.

Electronic Conductivity of Reduced Stabilized Zirconia

A quantitative use of the electrochemical reduction was proposed by Fouletier[18,19] to fix the stoichiometry ratio of the solid oxide electrolytes and to directly establish the nonstoichiometry-conductivity relationship.

Experimental Setup

The cell used was described in detail in a previous publication.[20] Its diagram is given in Fig. 6. The investigated sample is a cylinder approx 2 cm long and 1 cm in diameter, with metallic electrodes at both ends. These electrodes are connected either to a direct current supply for the reduction process or to an impedance meter for the conductivity measurements. An oxygen minigage with an enclosed Pd/PdO reference system[21] is used to measure the oxygen pressure in the vicinity of the sample. The gas circuit is represented in Fig. 7.

Procedure

Nonstoichiometry control: The argon gas carrier flows, at a constant flow rate D, according to marks 1 in Fig. 7. The oxygen content in the gas is fixed at approximately 0.1 ppm using the oxygen pump. It is monitored by the minigage located downstream in the experimental vessel and by the conventional oxygen gage shown in Fig. 7.

A constant current I is passed through the sample during a time τ in such a way that oxygen evolves from the downstream electrode and is flushed away by the gas carrier. The resulting variations in oxygen content of the gas are recorded by the gages. An example of such an oxygen content variation is shown in Fig. 8. The total amount Q of oxygen extracted from the sample is automatically calculated by integration according to equation

$$Q(\text{mol}) = \frac{D}{22.4} \int_0^\tau N_{O_2} \, dt \qquad (1)$$

and the change in the stoichiometry ratio according to

$$x = 2Q(M/m) \qquad (2)$$

M and m = sample molecular weight and weight.

This method enables us to control quite simply the stoichiometry ratio in the range 10^{-4}–10^{-1} by varying the parameters I and τ. It can be applied to any oxide exhibiting an ionic conductivity even if it is not predominant. It is much easier than equilibrating the sample with gas of variable compositions and enables extreme reducing conditions to be reached. The corresponding determination of the stoichiometry ratio x can also be very precise. It was demonstrated in a previous paper[2] that the accuracy of x can be better than 3%.

After the reduction process, the stopcocks are turned in such a way that the gas flows according to marks 2 in Fig. 7. Then the gas is purified on the zirconia-based oxygen getter before entering the experimental vessel. It was shown that all traces of oxygen are entrapped on the getter down to an oxygen pressure level smaller than 10^{-20} atm. The efficiency of the getter is maintained over several weeks. Under these conditions the gas carrier flowing around the sample is perfectly inert. It does not exchange any oxygen with it. This point was experimentally verified. Reduced samples were kept several days in such purified gas and no changes in conductivity, which is a sensitive parameter for reoxidations, were observed.

Conductivity Measurement: Before and after reduction of the sample, its conductivity is measured by impedance spectroscopy. The analysis of the impedance diagrams (Fig. 9) gives the bulk conductivity of the oxide according to the usual procedure.[22] With reduced samples, generally, the diagrams are simply concentrated around the real intercept, indicating a predominant electronic conductivity of the sample.

The fact that purified argon does not exchange any oxygen with the sample allows us to vary its temperature and directly determine the activation energy of its conductivity at constant stoichiometry ratio.

In the case of oxide electrolytes, the electronic conductivity $\sigma_e(x)$ can be easily deduced from the measured total conductivity $\sigma(x)$ when the oxide vacancies created by the reduction are negligible in concentration compared to the initial vacancy concentration. The ionic conductivity of the reduced material can simply be assumed to be equal to the initial conductivity σ_i. Consequently the electronic conductivity is calculated according to:

$$\sigma_e(x) = \sigma(x) - \sigma_i$$

Results

In a previous paper[20] results were presented for heavily and lightly doped ceria samples. With small deviations from stoichiometry, the main conclusions of interest here are:

- The electronic conductivity is proportional to the stoichiometry ratio.
- Therefore, all the "nonstoichiometry" electrons are likely to be located on cerium ions and to contribute equally to the conduction by jumping from cerium ion to cerium ion. The reduction process can be interpreted by the

reaction:

$$2\,Ce_{Ce} + O_O \rightarrow 2Ce'_{Ce} + V_{\ddot{O}} + \tfrac{1}{2}O_2$$

• The activation energy of the electronic conductivity is independent of the stoichiometry ratio, indicating the absence of any important interaction between the electrons.

Recent results obtained on yttria-doped zirconia (9 mol%) are reported in Figs. 10 and 11. Two temperature domains can be distinguished. At high temperature the results obtained previously by Casselton[1] and by us[19] are confirmed: In marked contrast with the results on ceria, the activation energy of the electronic conductivity varies from 1 to 0.2 eV as the deviation from stoichiometry increases. In the low temperature domain the activation energy diminishes very rapidly and is only 0.05 eV for a stoichiometry ratio of 1.41×10^{-3}.

This could be partly interpreted as follows: Some of the nonstoichiometry electrons injected in the material are entrapped on point defects according to reactions like:

$$2\,Fe'_{Zr} + O_O \rightarrow 2\,Fe''_{Zr} + V_{\ddot{O}} + \tfrac{1}{2}O_2$$

They do not contribute to the conductivity. Only a part of them, in equilibrium with those entrapped, are located on the basic zirconium cations and contribute to the electronic conductivity. This would explain why there is no simple proportionality between the stoichiometry ratio and the electronic conductivity and why the activation energy varies continuously.

This is consistent with the observation of successive reduction stages demonstrated by the other techniques previously described.

References

[1]R. E. W. Casselton, *J. Appl. Electrochem.*, **4**, 25 (1974).
[2]J. Fouletier and M. Kleitz, *Vacuum*, **25**, 307 (1975).
[3]P. Fabry, M. Kleitz, and C. Déportes, *J. Solid State Chem.*, **6**, 230 (1973).
[4]P. A. J. Swinkels, *J. Electrochem. Soc.*, **117**, 1267 (1970).
[5]G. Beranger, P. Desmarescaux, and P. Lacombe, *C. R. Hebd. Seances Acad. Sci.*, **259**, 1961 (1964).
[6]Ya. Gokhshstein and R. A. Khaikin, *High Temp. (Engl. Transl.)*, **8**, 965 (1969).
[7]R. E. W. Casselton, J. Penny, and M. J. Reynolds, *Trans. J. Br. Ceram. Soc.*, **76**, 115 (1971).
[8]R. E. W. Casselton, J. S. Thorp, and D. A. Wright, *Proc. Br. Ceram. Soc.*, **19**, 265 (1971).
[9]M. Jacquin, M. Guillou, and J. Millet, *C. R. Hebd. Seances Acad. Sci.*, **264**, 2101 (1967).
[10]J. L. Weininger and P. D. Zemany, *J. Chem. Phys.*, **22**, 1469 (1954).
[11]G. Bacquet and J. Dugas; p. 109 in Solid Electrolytes. Edited by P. Hagenmuller and W. Van Gool. Academic Press, London, 1978.
[12]J. S. Thorp, A. Aypar, and J. S. Ross, *J. Mater. Sci.*, **7**, 729 (1972).
[13]D. A. Wright, J. S. Thorp, A. Aypar, and H. P. Buckley, *J. Mater. Sci.*, **8**, 876 (1973).
[14]P. Fabry; Thesis, Grenoble, 1976.
[15]P. Fabry and M. Kleitz; p. 331 in Electrode Processes in Solid State Ionics. Edited by M. Kleitz and J. Dupuy. D. Reidel Publ. Co., Dordrecht, Holland, 1976.
[16]A. Pelloux, P. Fabry, and C. Déportes, *C. R. Hebd. Seances Acad. Sci.*, **276**, 241 (1973).
[17]P. Fabry and M. Kleitz, *J. Electrochem. Soc.*, **126**, 2183 (1979).
[18]J. Fouletier; Thesis, Grenoble, 1976.
[19]J. Fouletier and M. Kleitz, *J. Electrochem. Soc.*, **125**, 751 (1978).
[20]M. Levy, J. Fouletier, and M. Kleitz, *J. Phys. C*, **41** [6] 335 (1980).
[21]C. Déportes, M. Henault, F. Tasset, and G. Vitter, Fr. pat. 73 32671, Sept. 11, 1973.
[22]E. Schouler; Thesis, Grenoble, 1979.

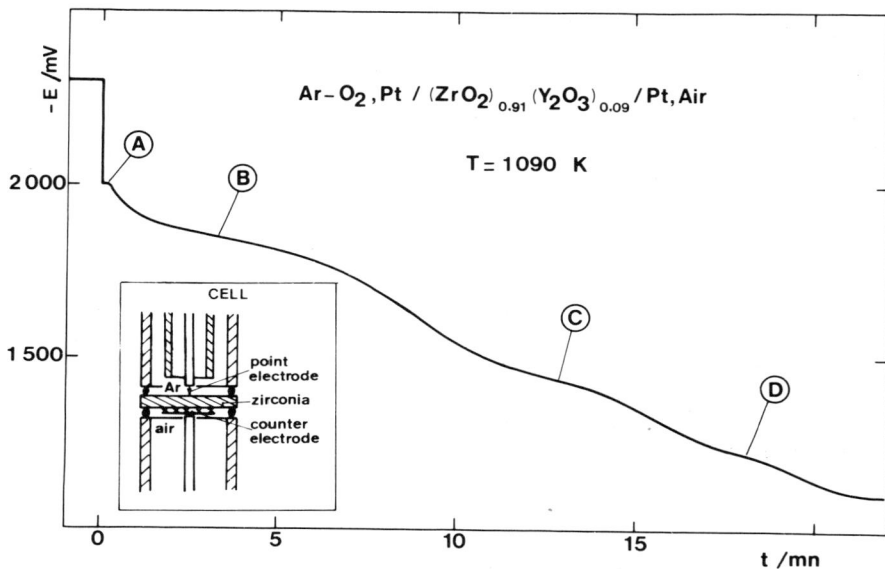

Fig. 1. Voltage variation of a point electrode after electrochemical reduction of yttria-stabilized zirconia.

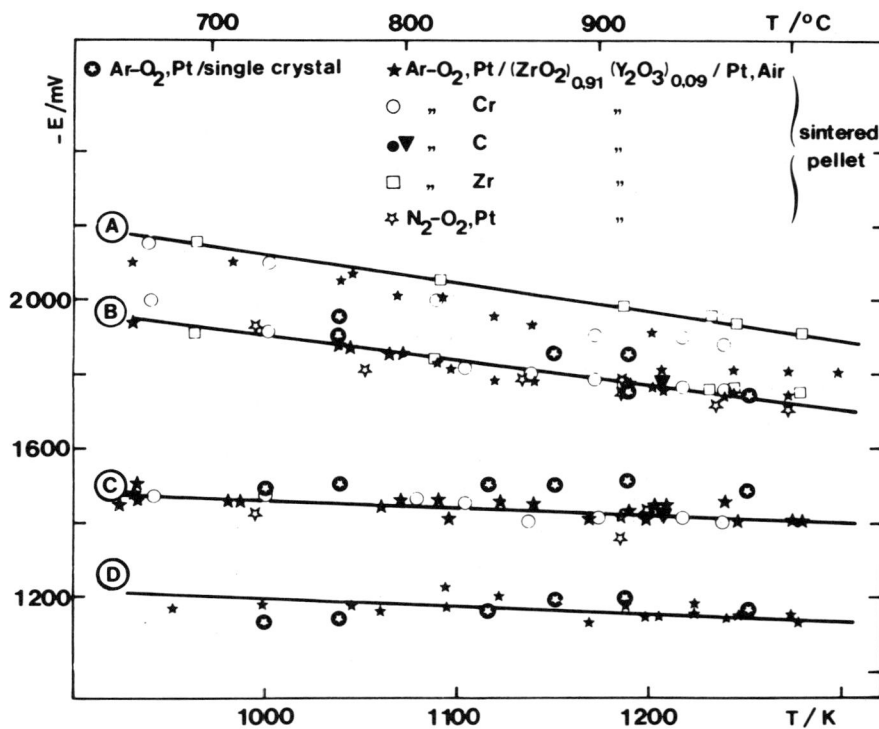

Fig. 2. Redox potentials of stabilized zirconia.

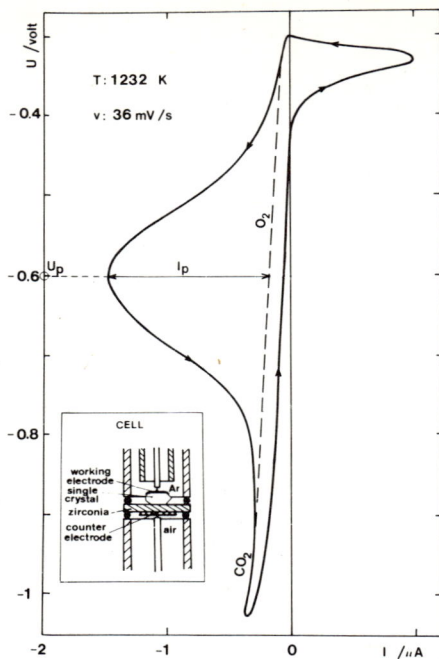

Fig. 3. Voltammogram of the electrode Pt/ZrO_2-Y_2O_3-Cu_xO (single crystal).

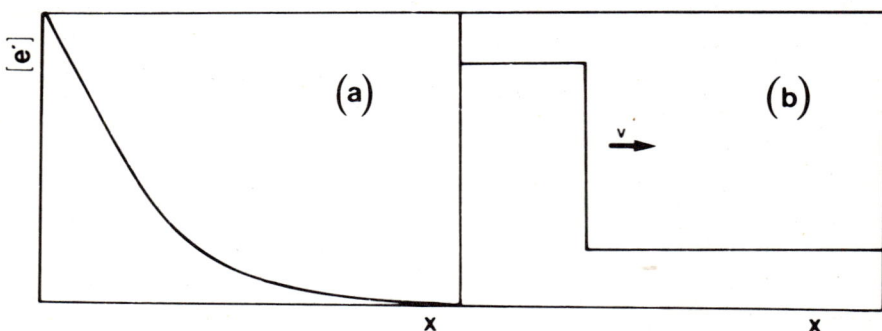

Fig. 4. Shape of the free-electron concentration profile in the reduction front of stabilized zirconia during electrochemical reduction. (a) Diffusion-controlled mechanism; (b) Electric-field-controlled mechanism.

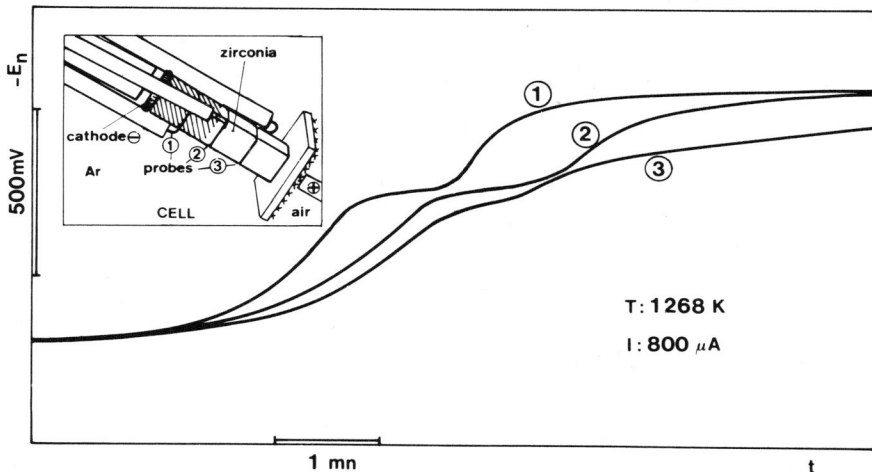

Fig. 5. Probe—potential variations during electrochemical reduction of a stabilized zirconia bar.

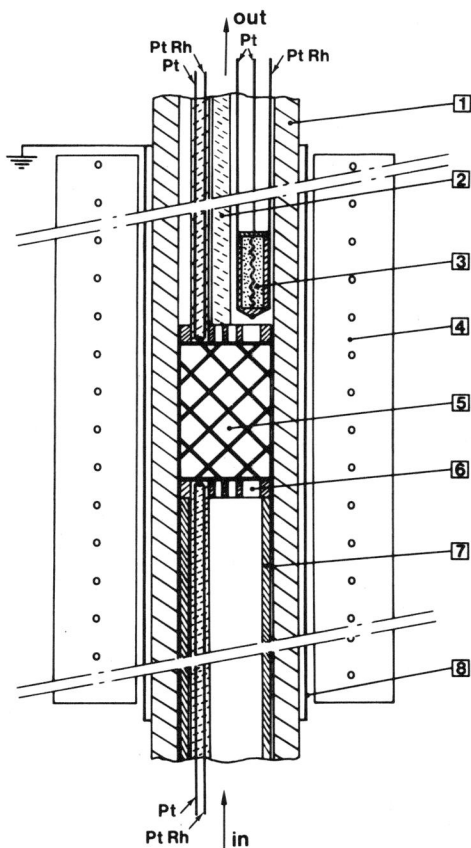

Fig. 6. Experimental cell (Ref. 20). (1) Alumina tube, (2) alumina rod, (3) oxygen minigage (Pd/PdO reference), (4) electric furnace, (5) investigated sample, (6) perforated alumina disks, (7) alumina tube, and (8) grounded screen.

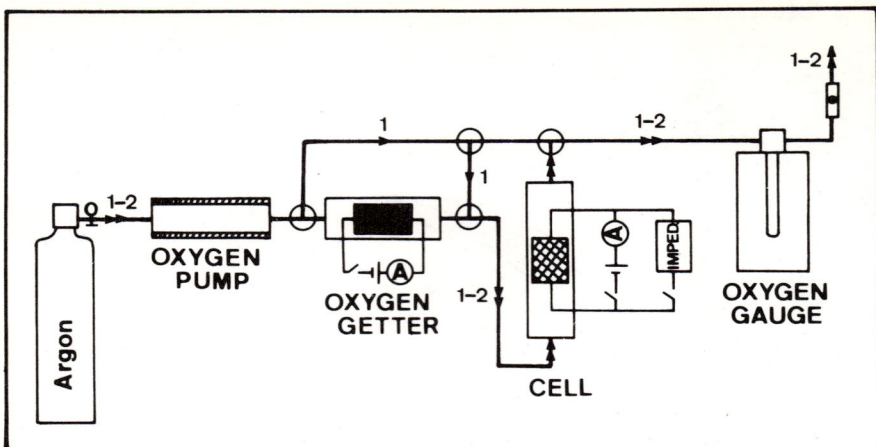

Fig. 7. Gas circuit. Way 1, reduction process; way 2, conductivity measurement (IMPED: impedancemeter).

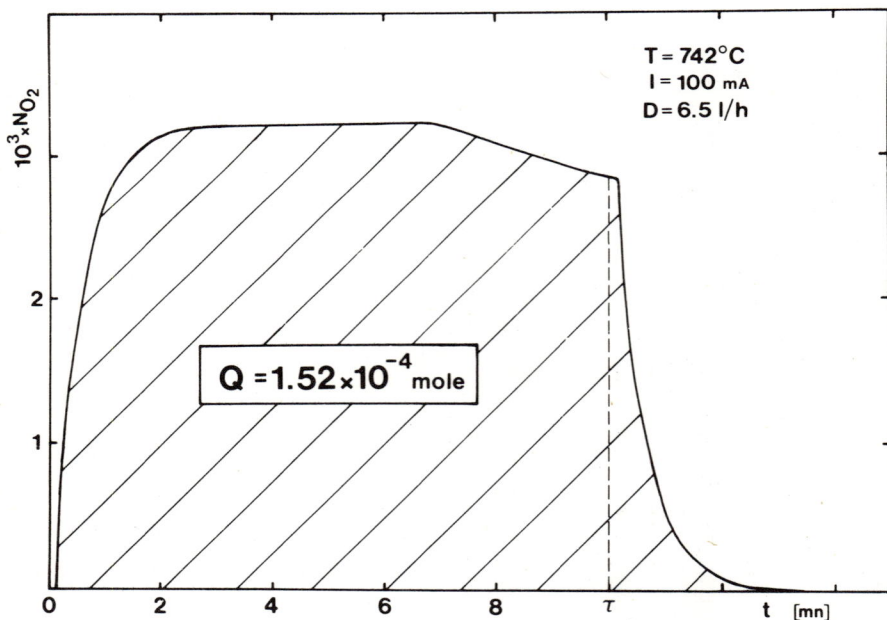

Fig. 8. Typical variation of the oxygen mole fraction N_{O_2} in the argon gas carrier during a reduction process.

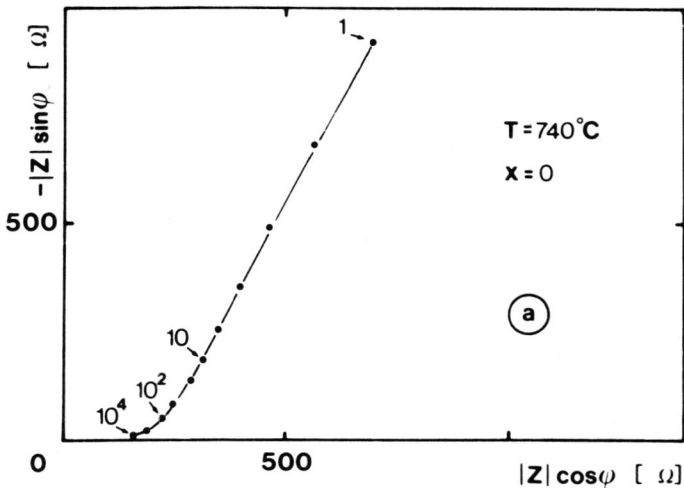

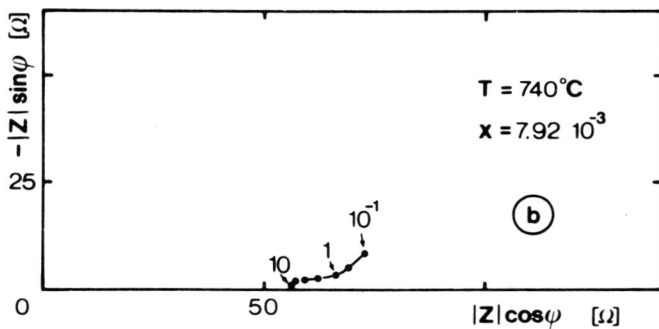

Fig. 9. Impedance diagrams of a stabilized zirconia sample (9 mol%). (a) Nonreduced sample, (b) reduced sample. Frequencies indicated in Hz.

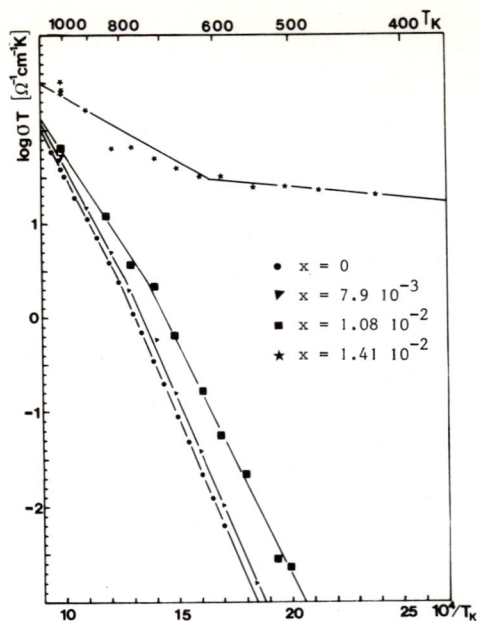

Fig. 10. Arrhenius plot of total conductivity of reduced and nonreduced yttria-doped zirconia (9 mol%).

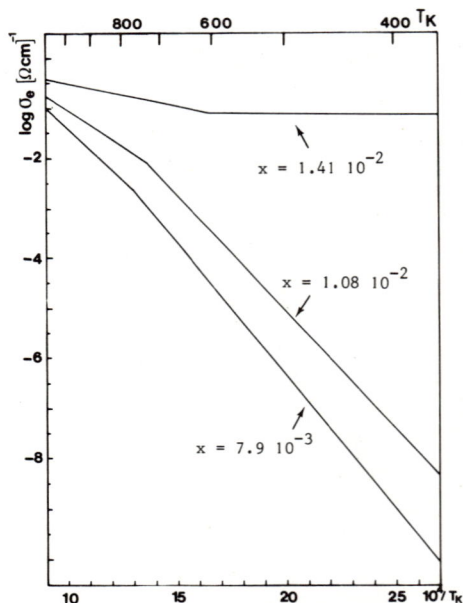

Fig. 11. Arrhenius plot of the electronic conductivity of reduced yttria-doped zirconia (9 mol%).

Determination of electronic conductivities and ionic domains of ZrO$_2$-Y$_2$O$_3$ by semipermeability measurements

M. Kleitz, E. Fernandez, J. Fouletier, and P. Fabry

Laboratoire d'Energétique Electrochimique
Institut National Polytechnique de Grenoble
E.N.S. d'Electrochimie et d'Electrométallurgie
38401 Saint Martin d'Heres
France

The measurement of the oxygen semipermeability flux through a pellet of solid oxide electrolyte provides the most accurate determination of its electronic conductivity. The basic equations are presented. Measurements were carried out with oxygen pumps and gages under various gases: pure oxygen, argon-oxygen, and hydrogen-water vapor mixtures. The free-electron conductivity was found important up to a rather high oxygen pressure at high temperature.

For most applications of zirconia-based solid electrolytes, a solely ionic conductivity is an essential requirement. However, these materials always exhibit an additional electronic conductivity. The characterization of this source of error, especially important at high temperature or under reducing conditions, is a necessity in any assessment of zirconia-based electrochemical devices.

Conventionally the electronic conductivity is viewed as arising from reactions:

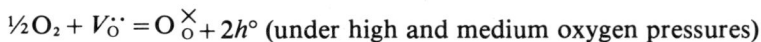

$$\tfrac{1}{2}O_2 + V_O^{\cdot\cdot} = O_O^X + 2h^\circ \text{ (under high and medium oxygen pressures)}$$

or

$$O_O^X = \tfrac{1}{2}O_2 + V_O^{\cdot\cdot} + 2e' \text{ (under low oxygen pressures)}$$

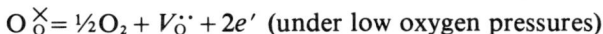

O^X is a normal oxide ion and $V_O^{\cdot\cdot}$ an oxide ion vacancy (Kroeger-Vink notations).

Under usual conditions, the oxide ion concentration and the oxide-ion vacancy concentration remain constant and therefore the concentration of free electrons e' and electron holes h° depend on the oxygen pressure P_{O_2} in equilibrium with the material as:

$$[e'] = k_e P_{O_2}^{-1/4} \tag{1}$$

$$[h^\circ] = k_h P_{O_2}^{1/4} \tag{2}$$

The parameters k_e and k_h are temperature-activated and may be written:

$$k_e = k_e^\circ \exp -\frac{E_e}{kT}$$

$$k_h = k_h^\circ \exp - \frac{E_h}{kT}$$

If the free-electron mobility and the electron-hole mobility are electron-concentration independent, the associated electronic conductivities vary as:

$$\sigma_e = \sigma_e^\circ P_{O_2}^{-1/4} \qquad (3)$$

$$\sigma_h = \sigma_h^\circ P_{O_2}^{1/4} \qquad (4)$$

The oxygen pressure dependencies were experimentally verified in various experiments.[1-6] However, other dependencies on oxygen partial pressure have been observed,[7-9] especially when the buffer effect of the gas is low. In a previous publication,[10] we demonstrated that, under these conditions, relation (4) is still applicable provided that the oxygen activity right on the electrolyte surface is taken instead of the oxygen partial pressure in the surrounding gas, which may be different.

The free-electron and electron-hole concentrations are evidently related through a recombination equation which is oxygen-pressure independent:

$$[e'][h^\circ] = K(T) \text{ with } K(T) = k_e k_h \qquad (5)$$

This means that the solid oxide electrolyte behaves as a simple semiconductor with respect to the electron concentrations. The unique property expressed by Eqs. (1) and (2) is that the Fermi level is oxygen pressure dependent.[26]

The existence of additional electronic conductivities has two main consequences in galvanic cells working under zero-current conditions. They are two aspects of a short-circuit effect.

•The measured voltage E is lower than the theoretical value E_{th}:

$$E = (1 - \bar{t}_e) E_{th}$$

where $\bar{t}_e$ is the average electronic number.

•A flux of electron holes or free electrons streams continuously through the electrolyte. It is electrically compensated by a counter migration of oxide ions. The net effect is a flux of oxygen through the material. The phenomenon is called oxygen electrochemical semipermeability. This oxygen flux disturbs the equilibria at the electrodes[10] and alters the oxygen activity in the surrounding systems.

The detrimental consequences of semipermeability can be very important. A resulting deviation of 100 mV from the theoretical voltage was observed in oxygen gages working at high temperature.

At present, the measurement of this oxygen electrochemical semipermeability flux provides the most sensitive determination of the parameters of the associated electronic conductivity. Recently, it was demonstrated[10] that such a measurement can be accurately carried out provided that appropriate experimental conditions are chosen.

Conventionally, this type of phenomenon is described within the framework of Wagner's theory.[2] Another way of handling the equation is proposed here which gives a more tangible description of the phenomena and leads to straightforward simplifications.

The reported experimental results are taken from a thesis.[12]

Theory

The following gas concentration cell is considered:

$$O_2(P')/\text{stabilized zirconia}/O_2(P'') \qquad \text{with } P' > P''$$
$$x = 0 \qquad\qquad\qquad x = L$$

Predominant Electron-Hole Conductivity

When the experimental conditions are such that the free-electron contribution is negligible, the electrochemical semipermeability can be described as follows.

The flux of electron holes is related to their electrochemical potential gradient $\partial\tilde{\mu}_h/\partial x$ by:

$$J_h = -\tilde{u}_h\,[h^\circ]\,\frac{\partial\tilde{\mu}_h}{\partial x} \tag{6}$$

with

$$\tilde{\mu}_h = \mu_h + F\Phi \tag{7}$$

and

$$\mu_h = \mu_h^\circ + RT\ln\,[h^\circ] \tag{8}$$

where $\tilde{u}_h$ is the hole electrochemical mobility, $[h^\circ]$ their concentration at location x, μ_h their chemical potential ($\mu_h^\circ = cst$), and Φ the inner electrical potential in stabilized zirconia,

$$J_h = -\tilde{u}_h\,[h^\circ]\left(\frac{\partial\mu_h}{\partial x} + F\frac{\partial\Phi}{\partial x}\right)$$

We have demonstrated[10] that under open-circuit conditions, and when the electronic conductivity is small with respect to the ionic conductivity, the following condition is satisfied:

$$F\frac{\partial\Phi}{\partial x} \ll \frac{\partial\mu_h}{\partial x} \tag{9}$$

The flux is then simply:

$$J_h = -RT\tilde{u}_h\,\frac{\partial[h^\circ]}{\partial x} \tag{10}$$

It is mostly due to a chemical diffusion. Furthermore, under steady-state conditions J_h is conservative:

$$J_h = C^{st}$$

Consequently:

$$J_h = \frac{RT}{L}\,\tilde{u}_h\,([h^\circ]' - [h^\circ]'') \tag{11}$$

where $[h^\circ]'$ and $[h^\circ]''$ are the hole concentrations at $x = 0$ and $x = L$, respectively.

If the electron-hole mobility is concentration independent, Eq. (11) can also be written:

$$J_h = \frac{RT}{F^2 L}\,(\sigma_h' - \sigma_h'') \tag{12}$$

with the conductivity σ_h given by:

351

$$\sigma_h = \tilde{u}_h F^2 [h\,^\circ] \tag{13}$$

The electrical compensation of the electron-hole current by the oxide ion current implies the following equation between J_h and the semipermeability flux $\tilde{J}_{O_2}$:

$$J_h = 4\tilde{J}_{O_2}$$

Therefore the oxygen semipermeability flux is equal to (in mol s^{-1} cm^{-2}):

$$\tilde{J}_{O_2} = \frac{RT}{4F^2L} (\sigma_h' - \sigma_h'') \tag{14}$$

or

$$\tilde{J}_{O_2} = \frac{RT}{4F^2L} \sigma_h^\circ (P'^{\,\frac{1}{4}} - P''^{\,\frac{1}{4}}) \tag{15}$$

This law was accurately verified.[10]
When the oxygen pressure P' and P'' are markedly different:

$$P' \gg P''$$

the following condition is obeyed:

$$\sigma' \gg \sigma''$$

Equation (15) can then be simplified to:

$$\tilde{J}_{O_2} = \frac{RT}{4F^2L} \sigma_h' \tag{16}$$

Under these conditions, an oxygen flux measurement is a direct determination of the electron-hole conductivity of the electrolyte in equilibrium with the higher oxygen pressure P'.

Predominant Free-Electron Conductivity

The same type of derivation can be developed and the oxygen semipermeability flux can be expressed as:

$$\tilde{J}_{O_2} = \frac{RT}{4F^2L} (\sigma_e'' - \sigma_e') \tag{17}$$

where σ_e'' and σ_e' are the free-electron conductivities of the material in equilibrium with the oxygen pressures P'' and P', respectively.

As above, when the oxygen pressures are markedly different, this equation can be simplified to:

$$\tilde{J}_{O_2} = \frac{RT}{4F^2L} \sigma_e'' \tag{18}$$

Predominant Free-Electron Conductivity on One Side and Predominant Electron-Hole Conductivity on the Other

In this case, Fabry[11] proposed that three zones be considered in the material as shown in Fig. 1:

The $x = 0$ side: Electron-hole conductivity is predominant and electronic recombination can be neglected because of extremely small concentration of free electrons.

The x = L side: Free-electron conductivity is predominant and electronic recombination can again be neglected.

The middle zone (λ in Fig. 1): The electron-hole and free-electron concentrations are of the same order of magnitude. An important recombination reaction takes place:

$$e' + h° = zero$$

Under usual conditions, in an air, H_2-H_2O cell for instance, the oxygen pressure variation interval is extremely broad and the concentrations $[e']$ and $[h°]$ vary over some orders of magnitude. It therefore seems reasonable to assume that the middle zone is narrow and can be assimilated to a plane (located at $x = \alpha$ in Fig. 1) as a first approximation.

Accordingly, at $x = \alpha$:

$$[e']^\alpha = [h°]^\alpha = \sqrt{K(T)}$$

and the semipermeability flux in the other zones obeys the equations:

For the $x = 0$ side: $\qquad J_h = 4\tilde{J}_{O_2} = \dfrac{RT}{F^2\alpha}\ (\sigma'_h - \sigma^\alpha_h)$ $\qquad\qquad$ (19)

For the $x = L$ side: $\qquad J_e = -4\tilde{J}_{O_2} = \dfrac{RT}{F^2(L-\alpha)}(\sigma^\alpha_e - \sigma''_e)$ $\qquad$ (20)

When the following conditions are satisfied:

$$\sigma'_h \gg \sigma^\alpha_h \tag{21}$$

$$\sigma''_e \gg \sigma^\alpha_e \tag{22}$$

these equations can be simplified to:

$$J_h = 4\tilde{J}_{O_2} = \frac{RT}{F^2\alpha}\ \sigma'_h \tag{23}$$

$$J_e = -4\tilde{J}_{O_2} = -\frac{RT}{F^2(L-\alpha)}\ \sigma''_e \tag{24}$$

Here again, under steady state conditions, the oxide ion flux is always conservative:

$$\tilde{J}_{O_2} = C^{st}$$

therefore:

$$J_h = -J_e$$

This condition allows us to calculate α by equalizing expressions (23) and (24):

$$\alpha = L\ \frac{\sigma'_h}{\sigma'_h + \sigma''_e} \tag{25}$$

With this value of α:

$$\tilde{J}_{O_2} = \frac{RT}{4F^2L}\ (\sigma'_h + \sigma''_e) \tag{26}$$

A similar expression was obtained by Iwase and Mori[5] following a different deviation.

A more accurate expression can be obtained in the same way from Eqs. (19) and (20) when conditions (22) are not satisfied.

Experimentation

The semipermeability measurements summarized in the following were taken using the cell designed by Fabry.[10,11,13] A pellet of oxide electrolyte was pressed between two alumina tubes and two metallic rings of platinum (Fig. 2). Oxygen permeating through the pellet was carried away by an appropriate gas circulating in the upper chamber and measured outside the cell. A vacuum was maintained outside the alumina tubes to minimize local short circuit effects at the platinum rings. With this cell assembly the measurements can be carried out under isothermal conditions.

The investigated zirconia pellets were yttria-doped (9 mol%). The preparation procedure was identical to that previously described.[10] The pellets were 2.2 cm in diameter and 3 to 3.6 mm thick. The investigated temperature range was 1170–1550 °C.

With each sample, two sets of measurements were performed with the following gas systems:

$$Air/ZrO_2-Y_2O_3/H_2-H_2O \qquad \text{(gas system I)}$$

$$Air/ZrO_2-Y_2O_3/Ar-O_2 \qquad \text{(gas system II)}$$

Air was in contact with the lower face of the sample. The other gas mixture was circulated in the upper chamber. The corresponding gas circuit is sketched in Fig. 2. It comprises a gas cylinder, a regulator, and an accurate needle valve to maintain the gas flow rate strictly constant, two electrochemical pumps, two oxygen gages, and an accurate flowmeter. With gas system I, argon-hydrogen (5%) mixtures were used and oxidized to a water vapor-hydrogen ratio of 5×10^{-4} by the pump EP1.

With argon-oxygen mixtures and oxygen contents greater than 1 ppm, the accuracy and reproducibility of the oxygen pressure measurements with gages OG1 and OG2 were within a few percent. With the $Ar-H_2-H_2O$ mixtures it was verified in a preliminary experiment[12] that approximately the same accuracy was reached provided that the temperatures of the gages were ≈ 700 °C.

To eliminate, as much as possible, the difficulties encountered in a previous study[10] and due to disturbances of the surface equilibria, measurements were performed with gas ensuring enough buffer effect on the side where the electronic conductivity plays a predominant part (H_2-H_2O with the first gas system and air with the second).

Two measurement procedures were used. According to the routine procedure, the oxygen semipermeability flux was determined from the voltages of gages OG1 and OG2. Gage OG1 read the oxygen pressure P_1 in the circulating gas supplied to the cell and gage OG2 read the pressure P_2 after it passed in the upper chamber. The difference:

$$\Delta P = P_2 - P_1 \qquad (27)$$

resulted from the oxygen semipermeability flow. The corresponding flux was simply calculated according to:

$$\tilde{J}_{O_2} = \frac{D}{22.4} \frac{\Delta P}{P_2} \frac{1}{S} \tag{28}$$

where D is the gas flow rate and S is the area of the sample.

To eliminate any source of error in the oxygen pressure measurement, especially with Ar-H_2-H_2O mixtures, another method was developed by Fernandez.[12] After passing through the upper chamber, the gas was reduced by the pump EP2. The reduction current I was controlled so that the voltage of gage OG2 was exactly equal to that of gage OG1. The quantity of oxygen extracted by the pump is then equal to the semipermeability flow. It was calculated according to Faraday's law:

$$J_{O_2} = \frac{I}{4F}$$

This method is very precise; however it proved to be rather tedious due to sluggish equilibration along the gas circuit. It was only used to check the reliability of the routine procedure.

Results

With the second gas system, Eqs. (16) and (26) were a priori assumed to be suitable. With the first gas system, Eq. (26) was the only reasonable choice according to published data.

Free-Electron Conductivity

The comparison of the results obtained with both gas systems showed that with the first gas system the electron-hole conductivity on the air side is much smaller than the free-electron conductivity on the H_2-H_2O side. It was found that:

$$\sigma_e^{H_2\text{-}H_2O} > 10^2 \, \sigma_h^{air}$$

Equation (26) can therefore be simplified to an expression similar to Eq. (18). This means that the oxygen semipermeability flux $\tilde{J}_{O_2}$ in this cell is mainly determined by the free-electron conductivity σ_e in stabilized zirconia in equilibrium with the H_2-H_2O mixture. It is related to this parameter by the equation:

$$\tilde{J}_{O_2} = \frac{RT}{4F^2L} \, \sigma_e \tag{29}$$

or

$$\tilde{J}_{O_2} = \frac{RT}{4F^2L} \, \sigma_e^\circ \, P_{O_2}^{-1/4} \tag{30}$$

The Arrhenius diagrams of the σ_e° values obtained with two samples are shown in Fig. 3.

The mean values of σ_e° were found to obey the following equation:

$$\sigma_e^\circ = 5.5 \times 10^5 \exp -\frac{3.72}{kT} \text{ (eV)} \tag{31}$$

Therefore, the proposed equation of the free-electron conductivity is:

$$\sigma_e = 5.5 \times 10^5 \, P_{O_2}^{-1/4} \exp -\frac{3.72}{kT} \text{ (eV)} \tag{32}$$

355

Numerical values are compared to published data in Table I.

Electron-Hole Conductivity

In contrast to the usual assumption, the results obtained clearly showed (cf. Fig. 6) that the electronic conductivity of stabilized zirconia is not necessarily due to a predominant electron-hole contribution under inert gas-oxygen mixtures at high temperature. Therefore, Eq. (26) must be used instead of Eq. (16) to express the experimental results obtained with the second gas system (air,Ar-O$_2$). In other words, the oxygen semipermeability flux must be corrected for the free-electron contribution to permit an accurate determination of the electron-hole conductivity. This correction was done by using Eq. (32). Figure 4 compares the corrected results to the results obtained without taking into account the free-electron contribution (by using Eq. (16)).

After correction, the equation obtained for σ_h was:

$$\sigma_h = 14 P_{O_2}^{\frac{1}{4}} \exp - \frac{1.5}{kT} \text{ (eV)} \tag{33}$$

Numerical values are again compared to published data in Table I. For the sake of comparison, the electronic conductivities σ_e and σ_h are plotted in Fig. 5 for different oxygen pressures in equilibrium with the material. For example, at 10^{-5} atm of oxygen, σ_e is greater than σ_h above 1270°C.

The oxygen pressure, $P\oplus$, where:

$$\sigma_e = \sigma_h$$

is given by:

$$P\oplus = 1.5 \times 10^9 \exp - \frac{4.36}{kT} \text{ (eV)} \tag{34}$$

(with $P\oplus$ in atm).

Ionic Domain

The ionic domain is traditionally defined as the oxygen pressure range where the ionic transport t_i is >0.5. The oxygen pressure limits defined by Patterson[14] as $P\oplus$ and $P\ominus$ correspond to the conditions $\sigma_h = \sigma_i$ and $\sigma_e = \sigma_i$, respectively. These parameters can be calculated by comparing the electronic conductivity equations to the published ionic conductivity data.

Strickler and Carlson's data,[15] which is in good agreement with recent results by Schouler,[16] was selected.

$$\sigma_i = 115 \exp - \frac{0.78}{kT} \text{ (eV)}$$

The calculated pressure limits vary as:

$$P\oplus = 4.55 \times 10^3 \exp \frac{3.04}{kT} \text{ (eV)} \tag{35}$$

$$P\ominus = 5.23 \times 10^{14} \exp - \frac{11.8}{kT} \text{ (eV)} \tag{36}$$

Arrhenius plots of these parameters are shown in Fig. 6 together with published data.

356

Discussion

The scatter of the published data on electronic conductivity of stabilized zirconia shown in Table I can be attributed to different experimental conditions and also to questionable interpretations.

•Different experimental determination techniques were used for the various reported data: oxygen semipermeability flux measurements,[1,3-6,10,12,21,22] dc polarization,[2,20] coulometric titration,[17-19] and emf measurements.[24,25]

•Most of the investigated samples for semipermeability were in the form of tubes resulting in nonisothermal conditions.[10,12]

•The compositions of the samples were very different: the dopant was most frequently CaO and sometimes Y_2O_3. In principle this should not matter as experimental studies[18,23] have shown that addition of various dopes in different concentrations does not change the oxygen semipermeability of the electrolyte. However, the presence of some monoclinic zirconia[20] probably has a marked influence on the electronic conductivity.

•The difference in impurity levels between commercial tubes[1,20,22] and tubes prepared from high purity materials[2] may be a key factor.

•A discussion of the results by Iwase and Mori[5] demonstrated that, in several cases, the calculations of the electronic conductivity were not correct because of the use of an oversimplified equation. In other cases, steady-state equations were used to interpret transient experimental results.[7]

•A basic source of error which has been stressed in this report is that the free-electron contribution was generally not allowed for in calculating the electron-hole conductivity.

•Lastly, note that the discrepancies in the $P_\oplus$ and $P_\ominus$ data shown in Fig. 5 can also be attributed to differences in the ionic conductivity of the electrolyte.

References

[1]A. W. Smith, F. W. Meszaros, and C. D. Amata, *J. Am. Ceram. Soc.*, **49** [5] 240 (1966).

[2]J. W. Patterson, E. C. Bogren, and R. A. Rapp, *J. Electrochem. Soc.*, **114**, 752 (1967).

[3]W. A. Fischer; p. 503 in Fast Ion Transport in Solids. Edited by W. Van Gool. North Holland, Amsterdam, 1973.

[4]S. F. Pal'guev, V. K. Gil'derman, and A. D. Neuimin, *J. Electrochem. Soc.*, **122**, 745 (1975).

[5]M. Iwase and T. Mori, *Met. Trans.*, **9B**, 365 (1978).

[6]M. Iwase and T. Mori, *ibid.*, p. 653.

[7]C. B. Alcock and J. C. Chan, *Can. Met. Quart.*, **11**, 559 (1972).

[8]L. Heyne and N. M. Beekmans, *Proc. Br. Ceram. Soc.*, **19**, 229 (1971).

[9]K. Kitazawa and R. L. Coble, *J. Am. Ceram. Soc.*, **57** [8] 360 (1974).

[10]J. Fouletier, P. Fabry, and M. Kleitz, *J. Electrochem. Soc.*, **123**, 204 (1976).

[11]P. Fabry; Thesis, Grenoble, 1976.

[12]E. Fernandez; Thesis, Grenoble, 1980.

[13]P. Fabry, M. Kleitz, and C. Déportes, *J. Solid State Chem.*, **5**, 1 (1972).

[14]J. W. Patterson, *J. Electrochem. Soc.*, **118**, 1033 (1971).

[15]D. W. Strickler and W. G. Carlson, *J. Am. Ceram. Soc.*, **47** [3] 122 (1964).

[16]E. Schouler, G. Giroud, and M. Kleitz, *J. Chim. Phys.*, **70**, 1309 (1973).

[17]D. A. J. Swinkels, *J. Electrochem. Soc.*, **117**, 1267 (1970).

[18]P. H. Scaife, D. A. J. Swinkels, and S. R. Richards, *High Temp. Sci.*, **8**, 31 (1976).

[19]T. H. Etsell and S. N. Flengas, *J. Electrochem. Soc.*, **119**, 1 (1972).

[20]L. M. Friedman, K. E. Oberg, W. M. Boorstein, and R. A. Rapp, *Met. Trans.*, **4**, 69 (1973).

[21]V. K. Gil'derman, L. Gil'mizyanov, A. D. Neuimin, and S. F. Pal'guev, *Sov. Electrochem. (Engl. Transl.)*, **15**, 1042 (1979).

[22]R. Hartung and H. H. Moebius, *Z. Phys. Chem.*, **243**, 133 (1970).

[23]V. K. Gil'derman, A. D. Neuimin, S. F. Pal'guev, and Yu. S. Toropov, *Sov. Electrochem. (Engl. Transl.)*, **12**, 1445 (1976).

[24]P. Fabry, M. Kleitz, and C. Déportes, *J. Solid State Chem.*, **5**, 1 (1972).

[25]A. V. R. Rao and V. B. Tare, *Scr. Metall.*, **6**, 141 (1972).

[26]M. Kleitz, P. Fabry, and E. Schouler; p. 1 in Electrode Processes in Solid State Ionics. Edited by M. Kleitz and J. Dupuy. D. Reidel Publ. Co., Dordrecht, Holland, 1976.

Table I. Values of the electronic conductivities and activation energies calculated from published data (OSFM: oxygen semipermeability flux measurement)

Ref.	Electrolyte ZrO_2	Method	Temp. (°C)	$\sigma_h^\circ (\Omega\cdot cm)^{-1}$ 900°C	1200°C	1600°C	E_a^h (eV)	$\sigma_e^\circ (\Omega\cdot cm)^{-1}$ 900°C	1200°C	1600°C	E_a^e (eV)
2	+ 15 mol% CaO	Polar techn.	800–1000	$6.0\ 10^{-5}$			2.3	$1.7\ 10^{-11}$			4.7
22	+ 11 mol% CaO	OSFM	800–1400	$1.0\ 10^{-6}$	$7.6\ 10^{-5}$		2.25				
8	+ 7.5 wt% CaO	OSFM	600–1000	$4.2\ 10^{-6}$	$2\ 10^{-4}$		1.9	$4.1\ 10^{-10}$	$7.3\ 10^{-7}$		3.7
3	+ CaO	OSFM	1342–1627			10^{-2}				$3.7\ 10^{-5}$	3.2
19	+ 10 mol% CaO	Coulom. titr.	600–1400				2.4	$1.2\ 10^{-10}$	$2.5\ 10^{-2}$		3.8
20	+ 3–4 wt% CaO	Polar techn.	1040–1250		$3.4\ 10^{-4}$		2.45		$1.5\ 10^{-7}$		
4	+ 15 mol% CaO	OSFM	1050–1250	$(7.8\ 10^{-6})$	$1.6\ 10^{-4}$		1.6				
	+ 10 mol% Y_2O_3			$(2.5\ 10^{-5})$	$8.9\ 10^{-5}$		1.9				
10	+ 9 mol% Y_2O_3	OSFM	950–1650	$3.1\ 10^{-6}$	$1.4\ 10^{-4}$	$3.4\ 10^{-3}$	1.9				
5	+ 11 mol% CaO	OSFM	1400–1550		$(6.6\ 10^{-4})$	$(3.6\ 10^{-3})$	1.0				
6	+ 5.8 wt% CaO	OSFM	1400–1550		$(7.8\ 10^{-4})$	$(9.8\ 10^{-3})$	1.5		$(6.1\ 10^{-8})$	$(1.6\ 10^{-4})$	4.7
21	+ 15 mol% CaO	OSFM	850–1450					$7.8\ 10^{-11}$	$1.3\ 10^{-7}$	$6.2\ 10^{-5}$	3.8
12	+ 9 mol% Y_2O_3	OSFM	1170–1550	$(3.5\ 10^{-6})$	$7.7\ 10^{-5}$	$(1.0\ 10^{-3})$	1.5	$(6.0\ 10^{-11})$	$1.1\ 10^{-7}$	$5.6\ 10^{-5}$	3.7

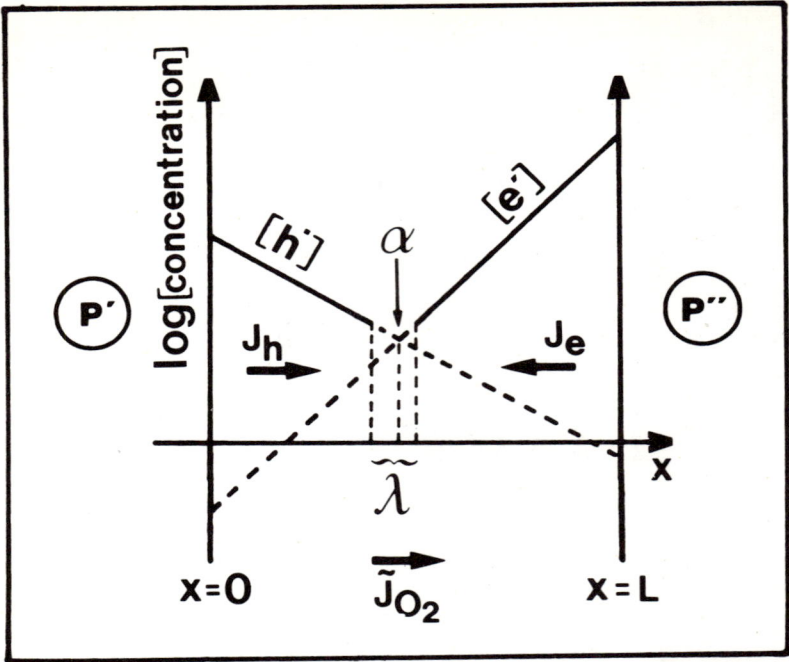

Fig. 1. Concentration profile of the electronic carriers in stabilized zirconia ($P' > P''$). λ is the recombination zone.

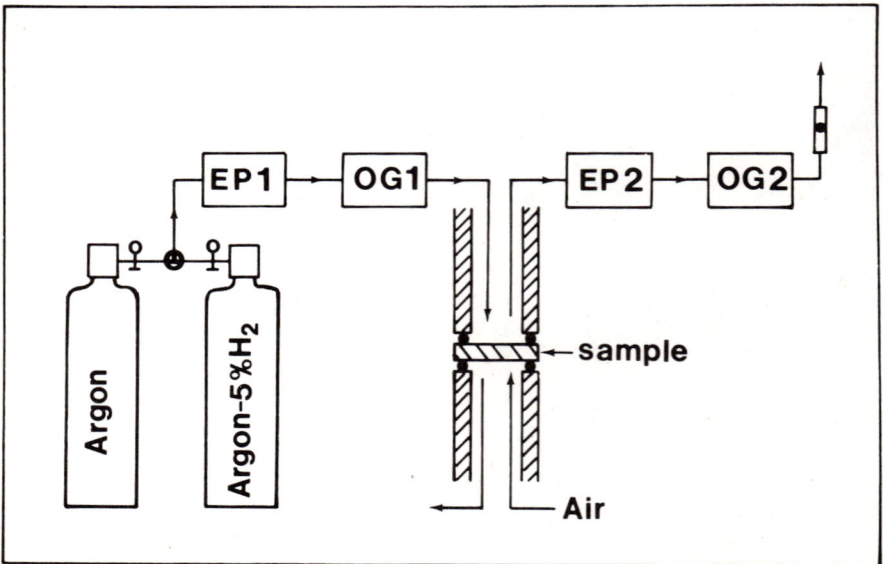

Fig. 2. Gas circuit. EP1 and EP2 are electrochemical pumps. OG1 and OG2 are oxygen gages.

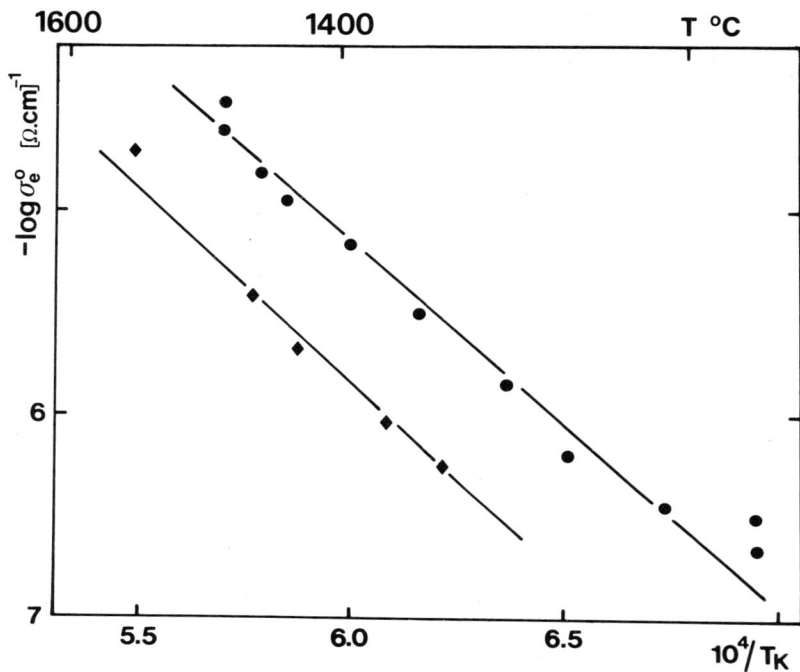

Fig. 3. Arrhenius plots of the free-electron conductivity constants of two yttria-doped (9 mol%) samples.

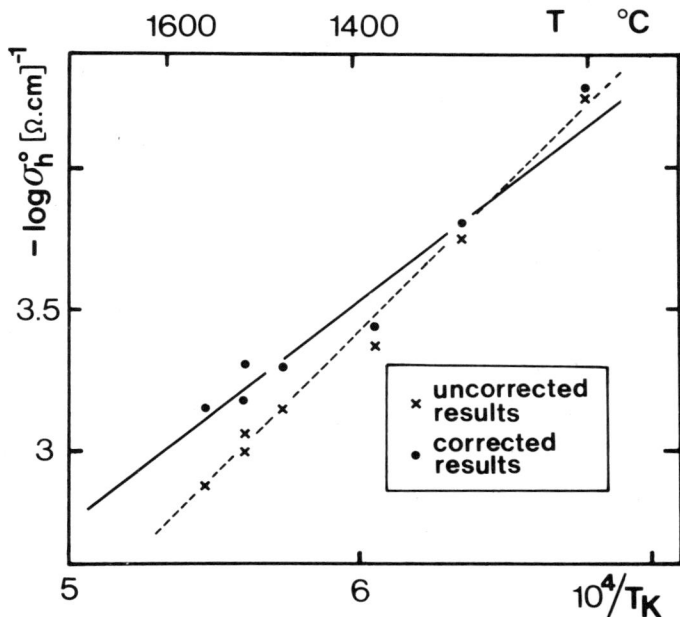

Fig. 4. Arrhenius plots of the electron-hole conductivity constant of yttria-doped zirconia (see text).

361

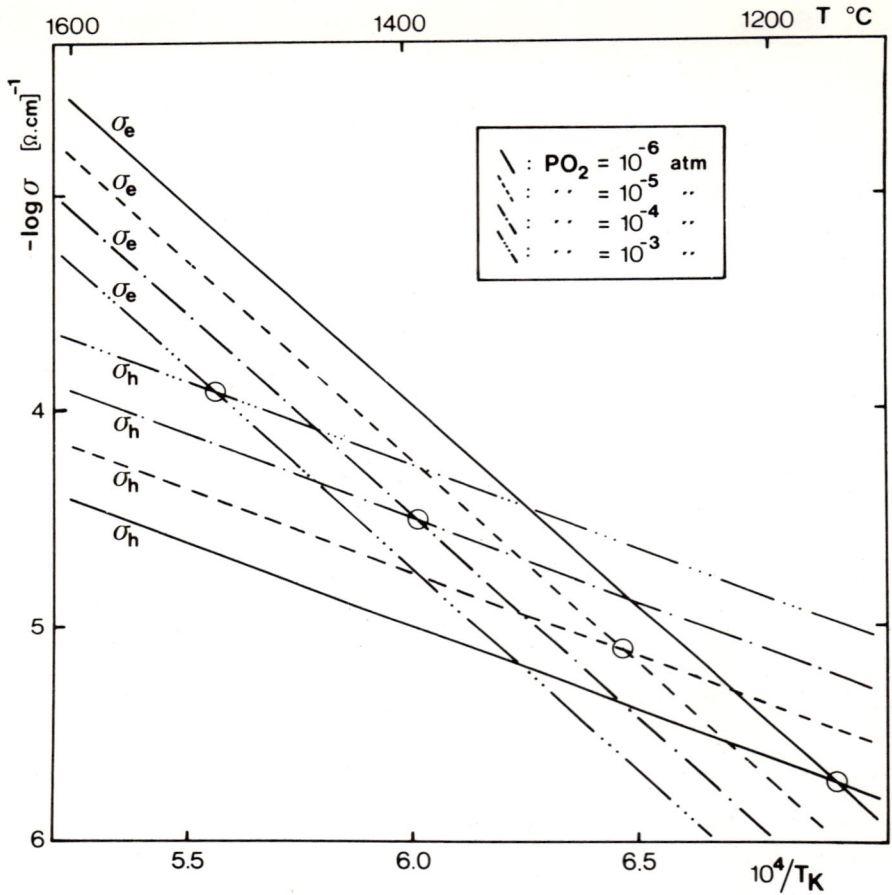

Fig. 5. Arrhenius plot of the electronic conductivities of yttria-doped stabilized zirconia (9 mol%) in equilibrium with the various oxygen pressures.

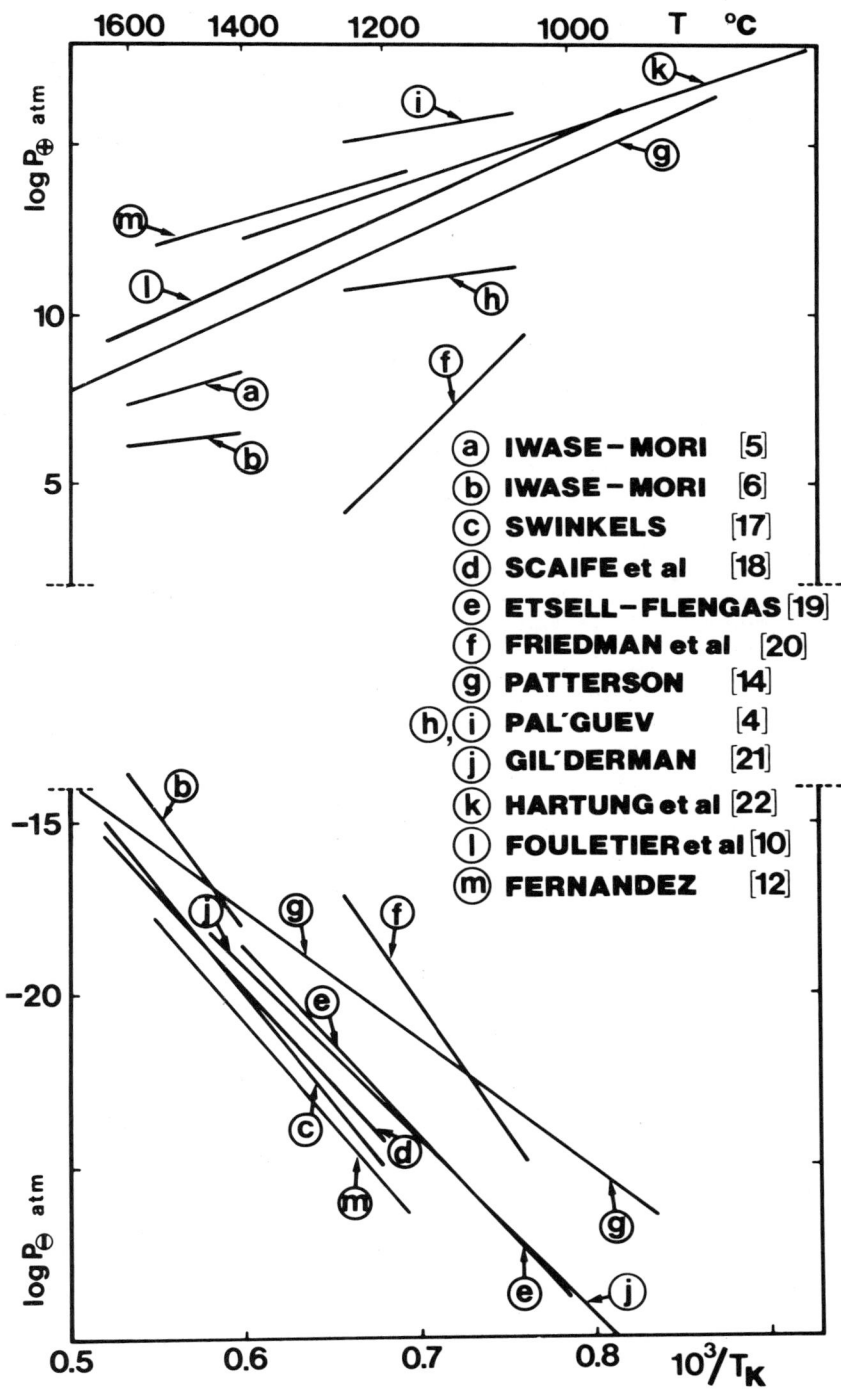

Fig. 6. Temperature dependence of parameters $P\oplus$ and $P\ominus$.

Characterization of yttria-stabilized zirconia oxygen solid electrolytes

F. K. MOGHADAM, T. YAMASHITA, AND D. A. STEVENSON

Department of Materials Science and Engineering
Stanford University
Stanford, Calif. 94305

Microstructural and conductivity changes occur when partially stabilized yttria-zirconia solid solutions, $ZrO_2 + 8$ wt% Y_2O_3 are annealed in the range 400–1000 °C in air. The conductivity was measured by the complex admittance technique and analyzed to separate out the bulk, grain boundary, and electrode contributions. A significant decrease was observed for anneals in the 800–1000 °C range; for example, a 23% decrease was observed after annealing for one week at 1000 °C. Concurrent observations of microstructural changes were made using scanning and transmission electron microscopy with associated energy-dispersive X-ray spectrometry. The decrease in conductivity during postsinter annealing at 800, 900, and 1000 °C is attributed to the following microstructural changes: (1) segregation of an yttria-rich layer near the grain boundaries and triple points (this latter phase is believed to have an adverse influence on the electrical properties of zirconia); (2) formation of tetragonal ZrO_2 at temperatures above 600 °C and retention of this metastable form along with the stabilized fluorite phase on cooling to lower temperatures; and (3) ordered domains forming within the disordered fluorite phase.

Solid electrolytes, which conduct by oxygen-ion transport, have been applied extensively for fundamental studies of thermodynamic and kinetic properties of systems containing oxygen. They are promising electrolytes in high temperature fuel cells and are presently receiving great interest in a variety of applications as oxygen sensors for medical applications and for combustion control. The most promising electrolytes are solid solutions of the fluorite modification of the oxides ZrO_2, HfO_2, ThO_2, and CeO_2. In all cases, the host oxide is doped with divalent or trivalent cations (e.g., alkaline earth ions, rare earth ions, or Y^{3+}) which introduces oxygen vacancies as charge-compensating defects. A satisfactory oxide electrolyte must meet three criteria[1]: (1) it must have a high ionic conductivity and a negligible electronic conductivity; (2) the ionic conductivity should be stable at cell operating temperatures; (3) the permeability of the oxide to gas must be very low and remain so during the entire life of the cell, including many cooling and heating cycles. The first two properties are influenced primarily by the electrolyte composition, while the third property is influenced primarily by the fabrication methods that are used and the resulting density and mechanical soundness.

Although fully stabilized calcia-stabilized zirconia (stabilized in the fluorite phase) has historically received the greatest attention, partially and fully stabilized solid solutions of $ZrO_2:Y_2O_3$ (YSZ) are currently most widely used. This system appears, at present, to have the best combination of electrical and mechanical properties. The use of partially stabilized compositions arose from an interest in improving the mechanical properties by precipitation hardening a fluorite matrix phase. The presence of a second phase, however, might be expected to compromise the desired electrolyte conductivity, especially during use due to aging effects. Some difficulties in using commercially available $ZrO_2 + 4.5$ mol% Y_2O_3 arising from aging have been reported.[2] We present here an experimental study of the changes in electrolyte properties that occur on aging and the concurrent changes in microstructure.

Experimental Procedures

The study was carried out on commercially available YSZ.* The specimen composition was $ZrO_2 + 8$ wt% (4.5 mol%) Y_2O_3, and the material was in the form of pellets 12.7 mm in diameter and 1.6 mm long. Specimen densities were calculated from weight and geometric measurements and the apparent densities of all specimens were 95 to 98% of the theoretical density. Theoretical densities were computed from measured cell dimensions and the anion vacancy model. For this particular composition the theoretical density was 6.05 g/cm³. Electrical conductivity measurements, scanning electron microscopy, transmission electron microscopy, and X-ray powder diffraction were used to characterize the material.

Ac Electrical Conductivity Measurements

The $ZrO_2 + 8$ wt% Y_2O_3 sample was in the form of a disk and was polished with diamond paste (6 μm) to produce optically flat, plane-parallel faces prior to the application of the platinum electrodes. Unfluxed, platinum paste† was applied to the polished surfaces of the electrolyte disk on both sides and fired at 800–850 °C in open air for 8–9 h to drive off the binder and sinter the platinum. This was done at least three times in order to obtain an adherent porous platinum electrode on each face.

The electrolyte disk was placed in a conductivity cell, with the specimen spring-loaded between two platinum foil contacts supported by ceramic disks. To prevent a short circuit between top and bottom platinum plates, the edges of the pellets were carefully polished. Stray voltage effects from the furnace windings were minimized by wrapping grounded nickel sheets (0.0051 cm thick) around the quartz jacket. The whole assembly was heated in a Kanthal Marshall tube furnace with a flat temperature zone of 43.2 cm for temperatures between 400 and 1000 °C and controlled with ±1 °C. The temperature was programmed for a heating and cooling rate of 1 °C/min between different isotherms and the conductivity measurements were made at 50 °C intervals from 400 to 1000 °C.

The ac conductivity measurements were made using Lissajous figures which allowed the cell admittance to be measured down to frequencies as low as 2 Hz.[3] The specimen was placed in series with a standard decade resistance resistance box‡ and connected to a variable amplitude and frequency sine wave generator§ as power source. The voltage waveform developed across the resistance box and that across a 100 Ω resistor in the reference leg of the bridge were displayed in the form of Lissajous figures on the CRT of an oscilloscope. Suitable adjustment of the bridge caused the long axis of the

Lissajous curve to lie along the 45° slope. The phase lag due to capacitance components of the specimen was obtained from the ratio of vertical intercepts to amplitudes. (This ratio is equal to the sine of the phase angle.) The results were plotted in complex admittance planes with conductance as real part (abscissa), and susceptance as imaginary part (ordinate).

The optimum output resistance of the oscillator is 600 Ω. To maintain this constant resistance across the terminals of the oscillator, a voltage divider was used. The voltage divider was designed also to provide an ac amplitude between 10 and 15 mV to avoid electrolysis of the electrolyte.

Scanning Electron Microscopy (SEM) Studies

The SEM studies were done on YSZ aged and unaged specimens. Since these materials are insulators, they were coated with thin layers (10 nm) of gold by sputtering and examined with an AMR 1000 SEM which had the capability of microchemical analysis with the EDS system. The microscope was operated under the optimum condition of 30 keV accelerating voltage and the specimen was tilted 45° from the optical axis of the microscope.

Transmission Electron Microscopy (TEM) Studies

Specimens were prepared for electron microscopy by breaking YSZ tubes** into large pieces. Disks 3 mm in diameter were cut from these pieces by an ultrasonic cutting machine and thinned to $\approx 100 \ \mu m$ by a mechanical milling machine with 3 μm and 1 μm diamond paste. Final thinning was done by ion beam milling at 5 kV.

The specimens were coated with carbon to eliminate charging by the electron beam. An electron microscope†† equipped with the 60° high tilt stage and an energy dispersive X-ray analyzer was used throughout. The STEM mode on the microscope and the LaB_6 filament allowed elemental analysis from a very small region of the specimen (10 nm diam. area). The operating voltage of the microscope was 120 kV.

Results and Discussion

Ac Admittance Measurements

Ac admittance measurements used the symmetrical cell:

$$Air, Pt \,|\, 8wt\% \, Y_2O_3 : ZrO_2 \,|\, Pt, Air$$

The porous Pt-paste electrodes act as reversible electrodes. Figure 1 shows complex admittance plots for an yttria-stabilized zirconia at two temperatures. Three distinctive features are noticed in this figure: (1) the presence of two well defined semicircles at low temperature (< 700 °C) indicating the presence of grain boundary effects; (2) the depression of the center of the semicircles by quite large angles of 20° to 40° below the real axis (abscissa) which suggests the existence of strong frequency dependent equivalent circuit elements; (3) a finite value for dc conductance which reveals the reversibility of the platinum paste electrodes.

Figure 2 shows a schematic complex admittance plot with the equivalent circuit proposed by Bauerle[4] as inset. The left arc is ascribed to the admittance of the electrode and in the equivalent circuit is shown by the network consisting of a charge transfer resistance, R_{CT}, in parallel with a double-layer capacitance, C_{DL}. This network is in series with the network composed of a parallel grain boundary resistance and capacitance (R_{GB} and C_{GB}) in series with the bulk resistance, R_B. This latter network accounts for the arc on the right; the geometric capacitance is neglected. A displacement of the semicircle centers from the abscissa is a measure of the frequency dependence of the circuit elements; the large displacement of the first arc is ascribed to the frequency dependence of C_{DL} arising in turn from the frequency dependence of the dielectric constant. The bulk resistance, R_B, the grain boundary resistance, R_{GB}, and the total electrolyte resistance, $R_T = R_B + R_{GB}$, may be obtained by the formula given in Fig. 2. The temperature dependence of total conductivity, σ_T, contains the contribution of lattice conductivity as well as intergranular conductivity.

Assuming, in accord with Bauerle and Hrizo,[5] that both R_B and R_{GB} show exponential dependence on temperature, the total electrolyte resistance, R_T, would not be expected to show a simple exponential dependence. Rather R_T is given by the relation:

$$\frac{R_T}{T} = \frac{R_B}{T} + \frac{R_{GB}}{T} = R_{0B}\exp\left(\frac{Q_{act}^B}{kT}\right) + R_{0GB}\exp\left(\frac{Q_{act}^{GB}}{kT}\right) \qquad (1)$$

where Q_{act}^B and Q_{act}^{GB} are the activation energies for oxygen ion immigration through the intra- and intergrain regions, respectively, and the pre-exponentials can be determined by the appropriate plot to $1/T = 0$.

It is difficult to compare the present conductivity results with prior work[6-9] on the electrical conductivity of ZrO_2-Y_2O_3 systems because the complex-plane technique used in this investigation was not used in any of the earlier studies except that of Bauerle and Hrizo.[5] In those studies, the results were obtained by monofrequency ac conductivity measurements (at $f = 1$ kHz) or a four-probe dc technique. At low frequencies, like 1 kHz, there may be an appreciable effect of electrode polarization; electrode polarization in some cases dominates at intermediate temperatures and always does at elevated temperatures. Except for the studies of Bauerle[4] and Bauerle and Hrizo,[5] none of the earlier investigations separated the bulk and grain boundary contribution to the total conductivity, but rather measured R_T only. Starting with Eq. (1) for R_T/T, it is easy to demonstrate that ($\sigma_T \cdot T$) or (R_T/T) should show some curvature when displayed on an Arrhenius plot; such curvature is readily apparent in the Arrhenius plot of the present R_T result (Fig. 3). The activation energies quoted in most of the earlier reports[6-9] may represent the best linear fit to conductivity-temperature plots that show similar curvature. The reported activation energy for conduction in 6 mol% Y_2O_3 is 0.8 eV and for 4.5 mol% Y_2O_3 (8 wt%) is 1.07 eV.[6] The activation energies of 1.167 eV and 0.682 eV were reported for high purity $ZrO_2 + 10$ mol% Y_2O_3 for grain boundary and bulk conductivity, respectively.[5] In the same plot R_B and R_{GB} are shown. At 550 °C the contributions to R_T caused by bulk and grain boundary conduction are the same magnitude. A break point in R_B at 600 °C is observed and coincides with the eutectoid temperature for the ZrO_2-Y_2O_3 system.[10,11] The eutectoid transition produces the less conductive phase with monoclinic symmetry through the following reaction:

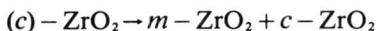

$$(c) - ZrO_2 \rightarrow m - ZrO_2 + c - ZrO_2$$

Although the activation energy for $T > 600$ °C is smaller than that for $T < 600$ °C, the corresponding pre-exponential factors show the opposite trend. (The pre-exponentials in resistivity or conductivity Arrhenius expressions are believed to be more structure and composition-sensitive than the activation energies.)

The activation energy values for bulk conduction vary only slightly from specimen to specimen, whereas those for grain boundary conduction (which depends on such factors as microstructure and grain boundary composition) show greater variation.

The most plausible explanation for the electrolyte polarization behavior reported here for YSZ appears to be that oxygen ion conduction of the electrolyte is partially blocked at the grain boundaries by an impurity phase. The existence of a thin grain boundary layer of an impurity phase is postulated to explain the experimental observations. The TEM results are used as direct evidence to support the idea that such a "second phase" exists as a thin layer at the grain boundaries.

One of the primary requirements for high temperature applications is that they should have a stable performance for a long service life. A phenomenon, first investigated by Carter and Roth[12] for $ZrO_2 + 15$ mol% CaO, was the steady increase in resistivity due to annealing at $\approx 1000 °C$ and this seems to be a general problem with stabilized zirconia electrolytes. The observed aging is a cause for concern and it was necessary to investigate the magnitude of the effect for $ZrO_2 + 8$ wt% Y_2O_3. The total conductivity of YSZ in air at $1000 °C$ was obtained from a complex plot analysis, with the results shown in Fig. 4. A rapid decrease in conductivity occurs during the first 24 h (a 10% decrease) and between 80 and 100 h (a 9.7% decrease), with a very gradual decrease during other time intervals. This behavior suggests that there may be two processes contributing to the conductivity decrease. We propose that the first abrupt decrease arises from the precipitation of an yttria-rich second phase at the grain boundaries and simultaneous formation of intragranular second phase (most likely, according to the phase diagram, the tetragonal ZrO_2 phase). The second abrupt decrease in conductivity we attribute to an order-disorder transformation, with slower kinetics. A total conductivity decrease of 23% was observed after annealing at $1000 °C$ for one week; half of the conductivity change occurred in the first 24 h and the other half after ≈ 100 h.

The bulk, grain boundary, and total resistance were calculated from appropriate Arrhenius plots for two representative temperatures and are given in Table I. The increase in R_{GB} is explained by a second phase at the grain boundaries and the increase in R_B is attributed to formation of tetragonal ZrO_2 in the form of isolated regions or particles dispersed within the grains. Micrographic sections in SEM and TEM show this clearly.

X-Ray Powder Diffraction

The room temperature X-ray diffraction patterns for the unaged $ZrO_2 + 8$ wt% Y_2O_3 were indexed using published data.[13] The positions of the peaks change only slightly with composition, but their relative intensities change markedly. The X-ray spectrum on the powder, which was obtained by crushing the aged samples, showed no significant difference from that of the unaged one and showed that the only phase present was the cubic, fcc, phase with a lattice parameter of $a = 0.5112$ nm, in good agreement with Stubican et al.[11] This was also noted in the TEM study of this material, which is described in the following section.

The presence of tetragonal ZrO_2 could not be ruled out because it is extremely difficult to distinguish from cubic ZrO_2, since the reflections of cubic and tetragonal phases overlap. With the resolution of $2\theta = 2°/2.54$ cm and peak spreading, which we observed in the present case, overlapping pairs could not be readily distinguished.

SEM Studies on Aged and Unaged Specimens

The grain size obtained from scanning electron micrographs, as well as optical micrographs, for as-received material is 10–20 μm and no appreciable change in grain size was noticed after aging at 1000 °C. Energy dispersive X-ray spectra (EDS) of the grains and grain boundaries revealed the enrichment of yttria at the grain boundary. The problem encountered in EDS was the overlap of the Zr-L line with the Y-L line and with the Au-M line (Au is used as a conductive layer to prevent charging). The only way to distinguish them is to ignore the low energy regime (< 10 keV) and observe the high energy peaks, i.e. K lines of Zr and Y which were located between 10 and 20 keV and could be distinguished from Au-L lines ($L\alpha_1$ and $L\beta_1$) which occur at 9.7 and 11.5 keV, respectively. Another disadvantage with energy dispersive spectroscopy in the SEM is the resolution limit of 1 μm. It is not possible to observe particles of phases smaller than this limit; therefore, TEM was used to resolve inhomogeneities at this level.

TEM Studies on Aged and As-Received Specimens

TEM studies were done on both as-received (AR) and aged specimens. Figure 5(A) is a bright field transmission electron micrograph of a typical grain found in both AR and aged specimens. Grains of this type constitute well over 90% of all the grains in both samples. The appearance of the dark contrast features within the grain change noticeably when the specimen is tilted. Local change in contrast can also be observed when a particular region is heated by the electron beam. In some low index zone axis orientations, the dark features become so dense as to make the grain appear completely black in the bright field image. These observations along with the fact that no Kikuchi pattern could be obtained from these grains indicated the presence of considerable lattice strain within the grain. The diffraction pattern (DP) corresponding to the BF image is shown in Fig. 5(B). The DP contains reflections forbidden by the symmetry of the cubic fluorite structure and, furthermore, spot splitting in the high index reflections can be observed. Careful measurements of the diffraction pattern indicate that these effects are due to the presence of tetragonal phase within the cubic matrix. A dark field image taken with the fluorite forbidden (100) reflection is shown in Fig. 5(C). It clearly shows that the tetragonal phase exists as finely dispersed particles which are ≈ 10 nm wide and 100 nm long. The yttria composition within the grain was determined by the EDS analysis and it was found to be close to the expected overall composition of 8 wt%. The corresponding EDS spectrum is shown in Fig. 5(D).

Dark field imagings in several crystal orientations were done in order to determine the morphology and the habit plane of the tetragonal particles. The possible habit planes can be inferred from the diffraction pattern taken from a grain oriented to the [001]$_f$ zone axis, an example of which is shown in Fig. 6. The higher order reflections are split into three spots, two of which are associated with the tetragonal precipitates in three orientations. (The third tetragonal precipitate reflections overlap the cubic matrix.) This indicates

369

that the (100), plane could be parallel to any of the three {100}, planes. This is to be expected, since the difference in the lattice parameter between the cubic and the tetragonal phase is < 1%. Previous work on YSZ of similar composition by Hannink[14] also reports the presence of tetragonal particles, with similar orientation relationship with the cubic matrix. The morphology of the tetragonal particles has been described as rectangular plate-like structure. The presence of the tetragonal phase can be explained in terms of the phase diagram.[10,11] The tetragonal ZrO_2 is metastable at room temperature and it is retained in the microstructure from the sintering temperature of 1600–1800 °C where the two phases, (c) ZrO_2 + (t) ZrO_2, coexist. A two-phase structure has an advantage over a homogeneous one from a mechanical strength point of view because of dispersion strengthening.

The following important differences are observed for aged samples (1000 °C for one week) in contrast to the AR samples:

(1) The presence of intergranular phase, which concentrates yttria and other impurities.

(2) More frequent occurrence of grains with enriched tetragonal phase and the presence of completely clear strain-free grains which have been identified as being tetragonal.

Figure 7(A) shows a bright field micrograph of a triple point with an amorphous second phase, as indicated by the diffuse diffraction ring (shown in the inset). Aged specimens contained numerous triple points and gaps between the grains which were sometimes filled with an amorphous material. Microchemical analysis of the triple points indicated a high yttria concentration as seen in the spectrum shown in Fig. 7(B). High concentration of other impurities such as Al, Ca, Fe, and Cu were often detected as well. These impurities could have been already present in the ZrO_2 and Y_2O_3 powders, or else they became incorporated into the material during the fabrication process.

Aged specimens contained grains with higher content of the tetragonal phase as indicated by the greater intensity of the tetragonal reflections in the diffraction pattern. This effect may be due either to growth[14] or to nucleation of tetragonal particles with aging. Completely clear grains were often observed in the aged specimens which were free of lattice strain, and they were usually found to be pure tetragonal ZrO_2. One example of such a grain is shown in Fig. 8(A). The diffraction pattern (shown in Fig. 8(B)) was indexed as (010) tetragonal orientation. The streaking in the DP is due to the presence of lath-like defects seen in the BF image, which are probably stacking faults. The c/a ratio was 1.02 which is reasonably close to the reported ratio of 1.017 for pure tetragonal ZrO_2. It is important to note that tetragonal ZrO_2, in contrast to cubic ZrO_2, has a very small solubility for Y_2O_3 even at high temperatures because of the dissimilarity in the structure. The lower yttria content of the tetragonal phase is clearly shown by the EDS spectrum shown in Fig. 8(C). The measured yttria concentration was 1–2 wt%.

In addition to the points mentioned above concerning the aged specimens, diffuse scattering was also observed in the DP for certain orientations. In previous studies of the ZrO_2-CaO system, similar observations of diffuse scattering were related to the existence of very small (5 nm) ordered microdomains of the $CaZr_4O_9$ structure type.[15,16] An ordered structure was previously detected in CSZ by Tien and Subbarao[17] using powder X-ray diffraction analysis of superlattice lines. In YSZ, ordering cannot be studied by

X rays, because the scattering factors of Zr^{4+} and Y^{3+} are essentially the same.[18] The atomic scattering factor for electrons for Y^{3+} and Zr^{4+}, however, are different and ordering in YSZ may be studied by TEM in order to reveal the origins of diffuse scattering in some of the diffraction patterns. Normally, for aging time of one week at 1000 °C, the extent of ordering should be very small for 8 wt% YSZ due to slow kinetics. Much higher yttria concentration is required before ordering becomes significant. When the tetragonal particles form within the cubic matrix, however, excess yttria that is insoluble in the tetragonal phase will be rejected into the cubic matrix. Consequently, the yttria concentration would be much higher than average within the cubic matrix. Such inhomogeneity in the concentration of yttria may allow ordering to occur.

Summary and Conclusions

•Ac electrical conductivity measurements were made on commercially available YSZ, and the bulk, grain boundary, and electrode resistances were separated out by the analysis of the complex admittance plots.

•A decrease in isothermal conductivity on annealing of YSZ at 1000 °C was observed and was studied in conjunction with microstructural studies. A 50% decrease in conductivity was observed due to aging at 1000 °C for one week. (The conductivity was measured at 750 °C.)

•X-ray powder diffraction showed no difference between aged and AR material.

•SEM studies showed intergranular second phases which are thought to be responsible for degradation of conductivity upon high temperature annealing.

•TEM studies showed the following important structural features of aged specimens:

(a) Fine, dispersed intragranular tetragonal ZrO_2 which is believed to strengthen the material.

(b) An intergranular boundary phase with a high yttria content and with other impurities such as Ca, Fe, Cu, and Al concentrated at the grain boundaries.

(c) Tetragonal grains within the cubic grains which are believed to contribute to the decrease of conductivity for high temperature annealing.

(d) Evidence of ordered microdomains from observations of diffuse scattering in certain crystallographic orientations.

Acknowledgments

The support of this work by the Department of Energy under Contract No. DE–AT03–76 ER 70037 is greatly appreciated. The authors are grateful to Robert Sinclair for his role in the development of facilities and expertise in electron microscopy at Stanford University and to the Center for Materials Research (Stanford University) for microscopes and analytical facilities.

References

[1]T. L. Markin, R. J. Bone, and R. M. Dell; pp. 15–35 in Superionic Conductors. Edited by G. D. Mahan and W. L. Roth. Plenum, New York, 1976.

[2]D. A. Stevenson, B. Heshmatpour, and F. K. Moghadam; pp. 117–20 in Proc. Int. Conf. Fast Ion Transport in Solids, Electrodes, and Electrolytes. Edited by P. Vashishta, J. N. Mundy, and G. K. Shenoy. Lake Geneva, Wis., May 21–25, 1979.

[3]P. H. Bottelberghs; pp. 145–72 in Solid Electrolytes: General Principles, Characterization, Materials, Applications. Edited by P. Hagenmuller and W. Van Gool. Academic, New York, 1978.

[4]J. E. Bauerle, "Study of Solid Electrolyte Polarization by a Complex Admittance Method," *J. Phys. Chem. Solids*, 30, 2657 (1969).

[5]J. E. Bauerle and J. Hrizo, "Interpretation of the Resistivity Temperature Dependence of High Purity $(ZrO_2)_{0.90}(Y_2O_3)_{0.10}$," *ibid.*, p. 565

[6]D. W. Strickler and W. G. Carlson, "Ionic Conductivity of Cubic Solid Solutions in the System $CaO-Y_2O_3-ZrO_2$," *J. Am. Ceram. Soc.*, 47 [3] 122 (1964).

[7]D. W. Strickler and W. G. Carlson, "Electrical Conductivity in the ZrO_2-Rich Region of Several $M_2O_3-ZrO_2$ Systems," *ibid.*, 48 [6] 286 (1965).

[8]J. M. Dixon, L. D. LaGrange, U. Merten, C. F. Miller, and J. T. Porter, "Electrical Resistivity of Stabilized Zirconia at Elevated Temperatures," *J. Electrochem. Soc.*, 110 [4] 276 (1963).

[9]M. M. Nasrallah and D. L. Douglas, "Ionic and Electronic Conductivity in Y_2O_3-Doped Monoclinic ZrO_2," *ibid.*, 121 [2] 255 (1974).

[10]K. K. Srivastava, R. N. Patil, C. B. Choudhary, K. V. G. K. Gokhale, and E. C. Subbarao, "Revised Phase Diagram of the System $ZrO_2-YO_{1.5}$," *Trans. J. Br. Ceram. Soc.*, 73, 85 (1974).

[11]V. S. Stubican, R. C. Hink, and S. P. Ray, "Phase Equilibria and Ordering in the System $ZrO_2-Y_2O_3$," *J. Am. Ceram. Soc.*, 61 [1-2] 17 (1978).

[12]R. E. Carter and W. L. Roth; p. 125 in Electromotive Force Measurements in High Temperature Systems. Edited by C. B. Alcock. Elsevier, New York, 1968.

[13]T. K. Gupta, J. H. Bechtold, R. C. Kuznicki, L. H. Cadoff, and B. R. Rossing, "Stabilization of Tetragonal Phase in Polycrystalline Zirconia," *J. Mater. Sci.*, 12 [12] 241 (1977).

[14]R. H. J. Hannink, "Growth Morphology of the Tetragonal Phase in Partially Stabilized Zirconia," *ibid.*, 13 [11] 2487 (1978).

[15]J. G. Allpress and H. J. Rossell, "A Microdomain Description of Defective Fluorite-Type Phases $Ca_xM_{1-x}O_{2-x}$ (M = Zr,Hf; x = 0.1–0.2)," *J. Solid State Chem.*, 15, 68 (1975).

[16]B. Hudson and P. T. Moseley, "On the Extent of Ordering in Stabilized Zirconia," *ibid.*, 19, 383 (1976).

[17]T. Y. Tien and E. C. Subbarao, "X-ray and Electrical Conductivity Study of the Fluorite Phase in the System ZrO_2-CaO," *J. Chem. Phys.*, 39 [4] 1041 (1963).

[18]International Tables for X-ray Crystallography, Vol. IV. Kyonoch Press, England, 1974; pp. 81, 82, 158, and 167.

---

*Sources of specimens were Corning Glass Works, Zircoa Div., Solon, Ohio and Zircar Products, Florida, New York.

†No. 6926, Engelhard Minerals & Chemicals Corp., East Newark, N.J.

‡Model 1433-X; GENRAD, Inc., Concord, Mass.

§Model 1310B; GENRAD, Inc.

‖Model 475; Tektronix, Inc., Beaverton, Ore.

**Corning Glass Works, Zircoa Div.

††Model EM400; Philips Electronic Instruments, Mahwah, N.J.

Table I. Effect of annealing on bulk, grain boundary, and total resistance of $ZrO_2 + 8$ wt% Y_2O_3

	Before annealing		After annealing at 1000 °C for 1 week	
	$T = 500\,°C$	$T = 800\,°C$	$T = 500\,°C$	$T = 800\,°C$
$R_B(\Omega)$	301.016	10.24	397.93	22
$R_{GB}(\Omega)$	339.2	4.28	1287.7	5.72
$R_T(\Omega)$	652.10	15.69	1870.49	25.12

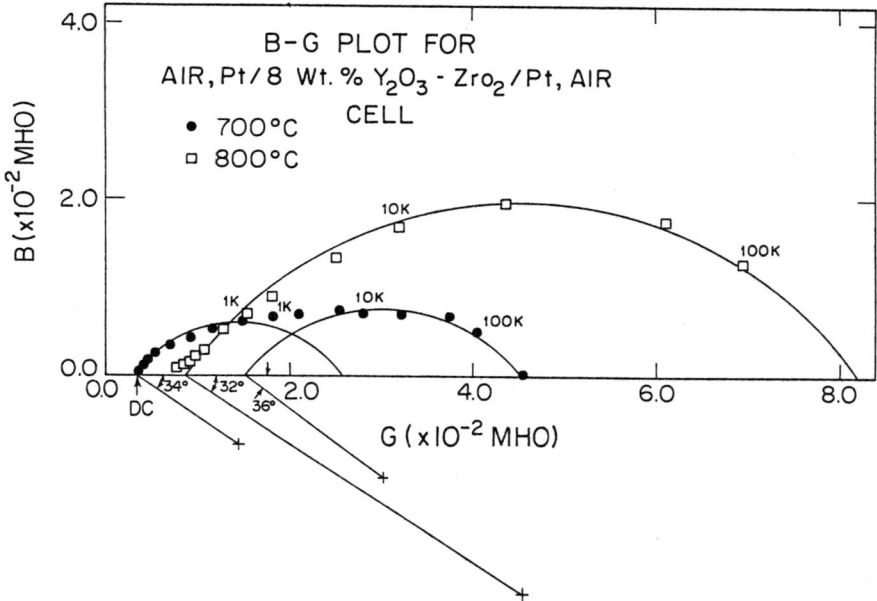

Fig. 1. Complex plane admittance diagram of a YSZ sample at two indicated temperatures; electrode area = 1.37 cm², electrode distance = 0.16 cm, frequency range if 2 Hz to 400 kHz (numbers are frequencies in Hz).

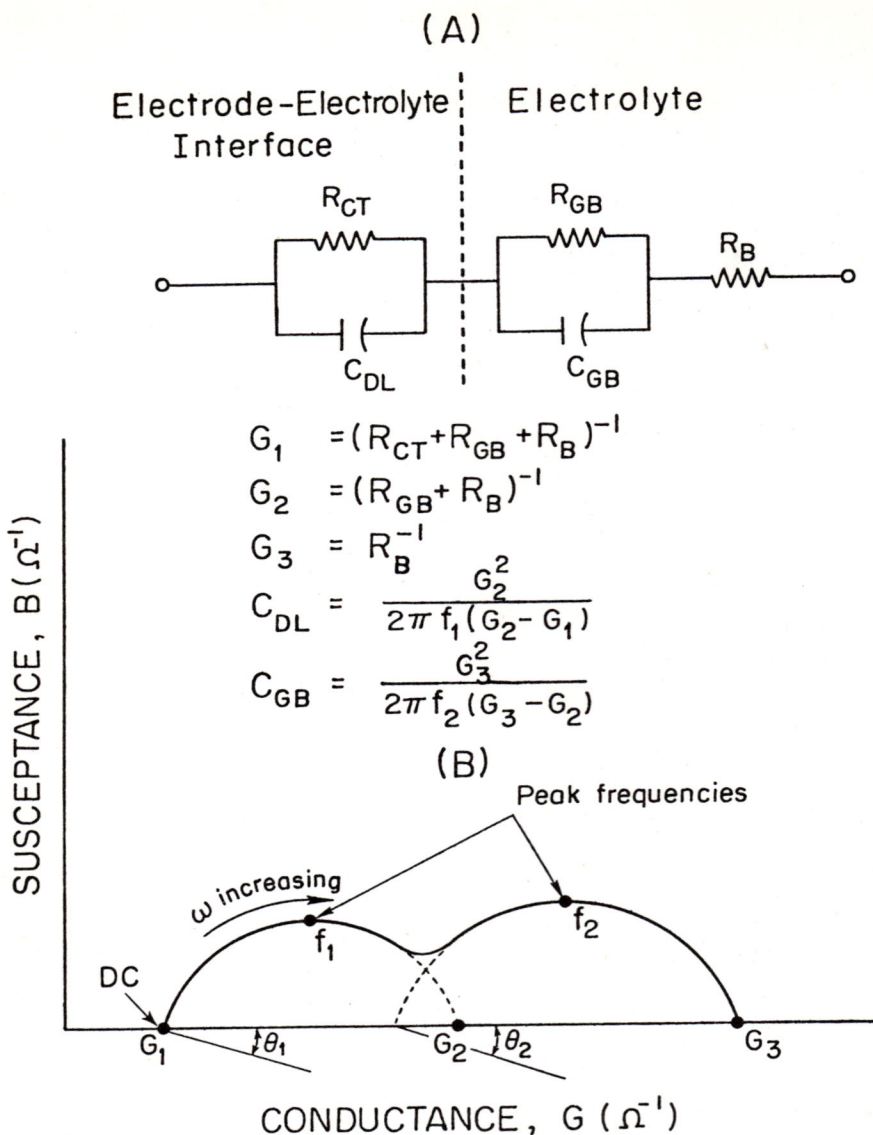

(A)

Electrode-Electrolyte Interface | Electrolyte

$$G_1 = (R_{CT} + R_{GB} + R_B)^{-1}$$
$$G_2 = (R_{GB} + R_B)^{-1}$$
$$G_3 = R_B^{-1}$$
$$C_{DL} = \frac{G_2^2}{2\pi f_1 (G_2 - G_1)}$$
$$C_{GB} = \frac{G_3^2}{2\pi f_2 (G_3 - G_2)}$$

(B)

Peak frequencies

ω increasing

Fig. 2. (A) Equivalent circuit for porous Pt electrode/zirconia solid electrolyte system; R_B = bulk resistance, R_{GB} = grain-boundary resistance, R_{CT} = charge-transfer or boundary layer resistance, C_{GB} = grain-boundary capacitance, and C_{DL} = boundary or double layer capacitance. (B) Schematic of complex admittance plot for equivalent circuit shown in (A).

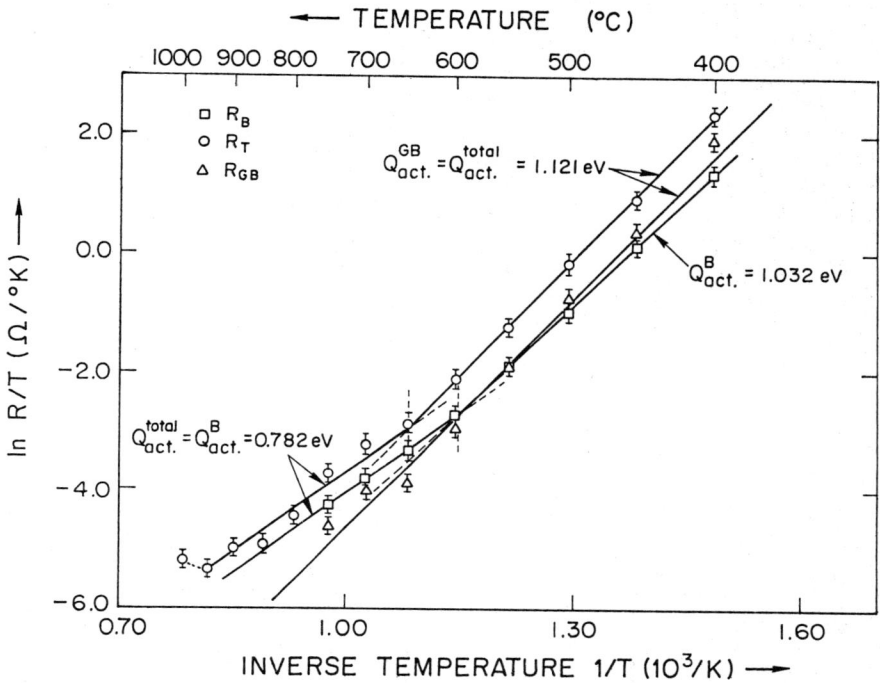

Fig. 3. Arrhenius plots for the resistance of a $ZrO_2 + 8$ wt% Y_2O_3 sample with Pt-paste electrodes.

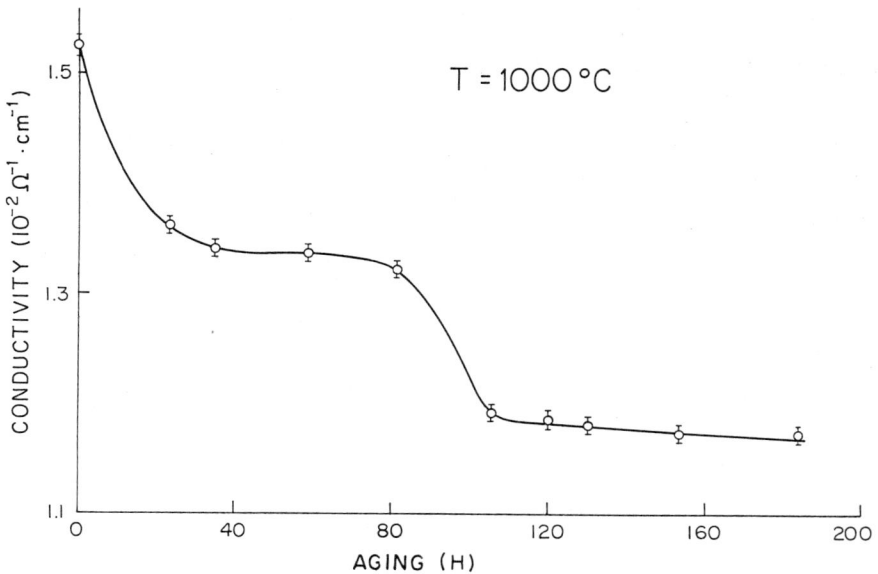

Fig. 4. Isothermal change of the total conductivity of a $ZrO_2 + 8$ wt% Y_2O_3 sample at 1000°C in air.

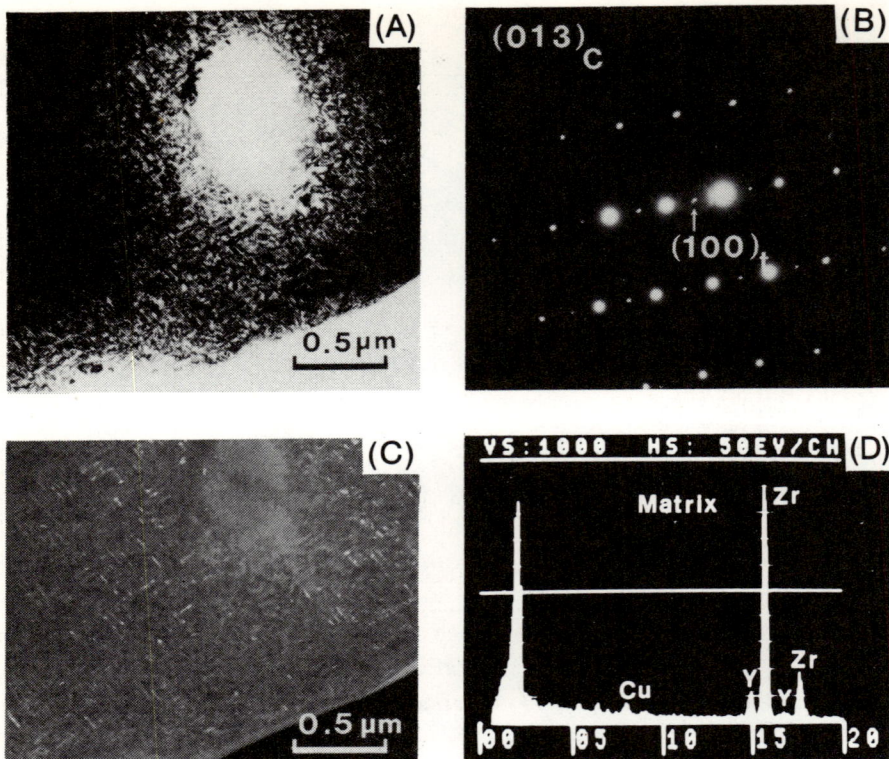

Fig. 5. (*A*) Bright-field transmission electron micrograph of an aged YSZ specimen. Dark contrast features are tetragonal precipitates. (*B*) Selected area diffraction patterns (SADP) from the region shown in Fig. 5(*A*). The (013) cubic pattern contains fluorite forbidden reflections which come from the tetragonal precipitates within the matrix. (*C*) Dark-field image formed by using the (100)ₜ spot indicated in Fig. 5(*B*). Rod-like tetragonal precipitates which give rise to the fluorite forbidden reflections appear bright in the image. (*D*) Energy dispersive X-ray spectrum (EDS) from the grain shown in Fig. 5(*A*). (Oxygen X rays are too soft to permit detection in the EDS system.) The measured yttria content is close to the expected bulk composition.

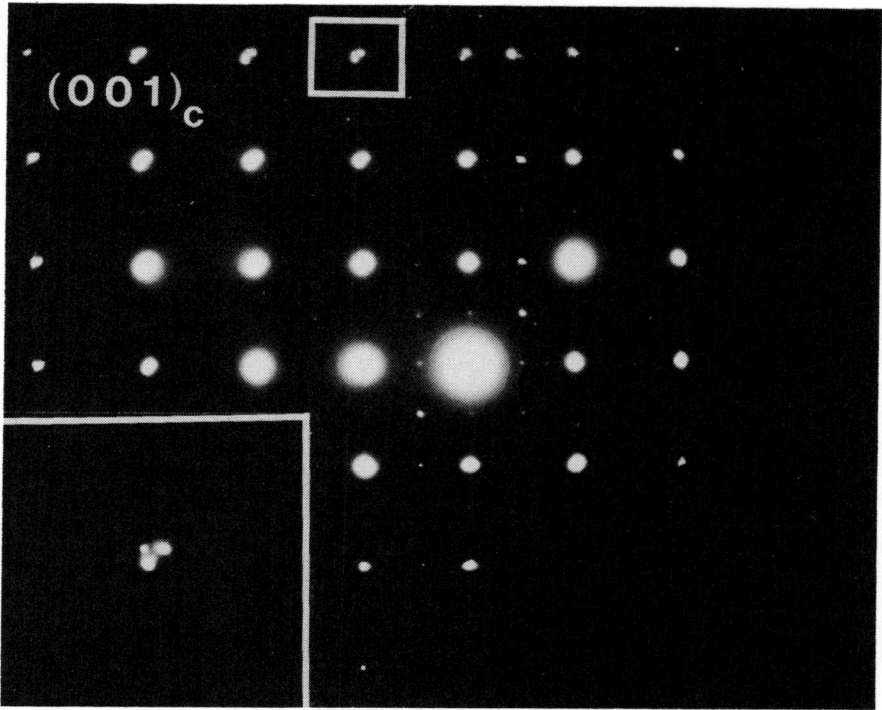

Fig. 6. [001] diffraction pattern of YSZ (8 wt% Y_2O_3). Tetragonal satellites around fluorite reflections at high order reflections are shown. There are three tetragonal variants for each fluorite reflection.

377

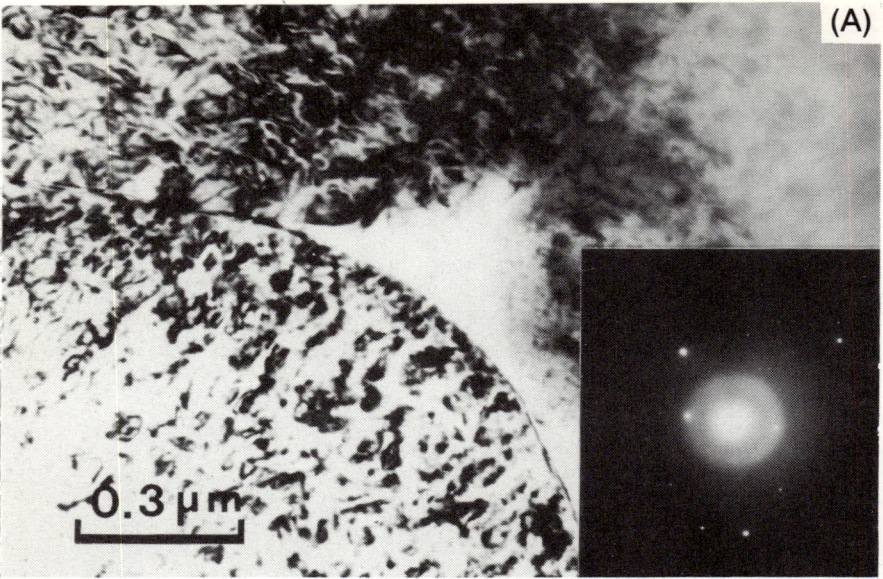

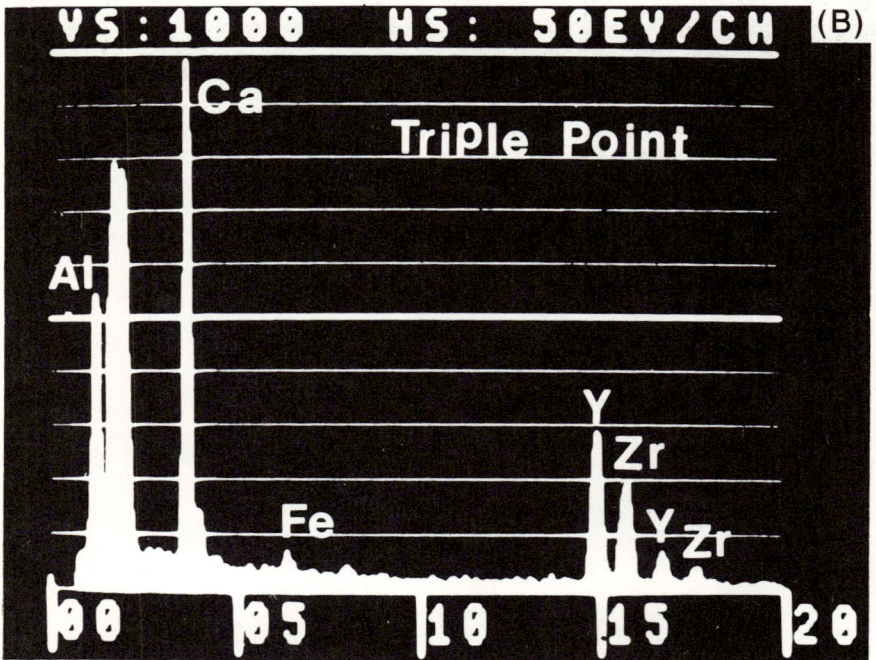

Fig. 7. (*A*) Bright-field transmission electron micrograph showing a second phase at the triple point. Corresponding microdiffraction pattern is shown in the inset. The diffuse diffraction ring indicates the presence of a glassy phase. (*B*) Fine-probe EDS from the amorphous phase at the triple point. High concentrations of yttria and other impurities are found at the triple point.

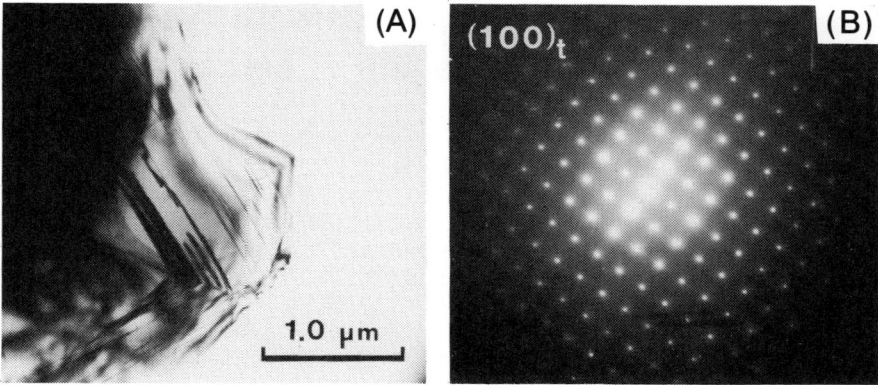

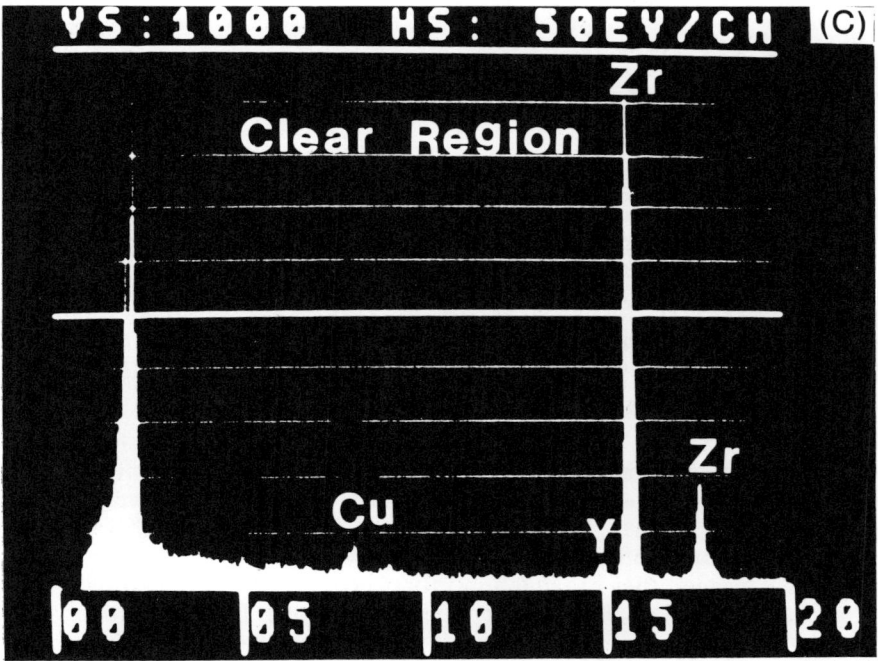

Fig. 8. (*A*) Bright-field TEM image of a tetragonal grain. The dark lines are stacking faults. (*B*) (100), diffraction pattern from a circular region shown in Fig. 8(*A*). The streaking in the diffraction pattern comes from the stacking faults. (*C*) EDS from Fig. 8(*A*). A comparison with the EDS from more typical grains (Fig. 5(*A*)) shows that clear regions contain less yttria than the matrix (1–2 wt% Y_2O_3).

Study of electronic minority defects in stabilized zirconia by thermal emission of electrons

P. ODIER

Centre de Recherche sur la Physique des Hautes Températures
Centre National de la Recherche Scientifique
1 D Avenue de la Recherche Scientifique
45045 Orléans Cedex
France

J. P. LOUP

Laboratoire de Physique des Matériaux Solides
Faculté des Sciences
Parc de Grandmont
37200 Tours
France

The thermal emission of electrons has been used to study stabilized zirconia in which the vacancy concentration is fixed by the stabilizing concentration. As a result, the electron concentration is proportional to $Po_2{}^{-1/4}$. The variations of the density of the emitted current follow this law exactly, showing that the thermal emission of electrons can be used to study electronic minority defects. Evidence is also offered that the electron affinity of an oxide is lowered by additions of low electron affinity compounds.

Stabilized zirconia has been used for many years as an electrolyte. Its ionic conductivity is well known and does not change over a very wide range of oxygen pressure. However, there are some mobile electrons which can cause short-circuit conditions. Information about these electrons is very sparse, essentially because they are in a minority. The thermal emission of electrons provides information about the conduction electrons. In a first attempt, we have obtained data on the variations of the electron concentration in the conduction band with the oxygen pressure, equilibrating the oxide. Further work will attempt to establish absolute values.

The thermal emission of electrons is also very sensitive to the structure of the surface area via the electron affinity. Results obtained on pure zirconia are compared[1] with those on calcia-stabilized and yttria-stabilized zirconia and discussed.

Thermal Emission of Electrons as a Probe of Electronic Minority Defects

The work function Φ is generally used in the expression of the density of the emitted current j_0.

$$j_0 = A_0(1 - \bar{r})\, T^2 \exp\left(-\frac{\Phi}{kT}\right) \qquad (1)$$

380

Φ is defined as the difference in energy between the Fermi level and the energy level of an electron in the vacuum, outside the solid (Fig. 1). A_0 is a constant (1.2 MA m^{-2} K^{-2}), $\bar{r}$ is the mean reflection coefficient of the potential barrier at the surface (which is subsequently neglected), T is the absolute temperature, and j_0 is obtained at zero electrical field. Φ depends on T and, among others, parameters of the position of the Fermi level in the band gap for oxides or semiconductors. It is thus convenient to introduce the electron affinity, χ, defined in Fig. 1 as $\Phi = \chi + \epsilon_c - \epsilon_F$, ϵ_c and ϵ_F being the energies of the bottom of the conduction band and of the Fermi level. In the case of a nondegenerated semiconductor, the density of electrons in the conduction band may be expressed as a function of $\epsilon_c - \epsilon_F$ by

$$n = N_c \exp \left(-\frac{\epsilon_c - \epsilon_F}{kT} \right) \qquad (2)$$

Equation (1) can thus be rewritten

$$j_0 = A_0 T^2 \, \frac{n}{N_c} \, \exp \left(-\frac{\chi}{kT} \right) \qquad (3)$$

Equation (3) shows that j_0 is directly proportional to n. This provides a simple method to investigate electronic defects, even minority electronic defects. The great advantage of the technique is that it is sensitive only to electrons and not to all the charge carriers, as is electrical conductivity. The electronic contribution of the electrical conductivity is $\sigma_e = ne\mu_n + pe\mu_p$. This is true as long as no additional band bending has occurred at the surface, which would add additional terms in the exponential of Eq. (3). Such effects may occur by oxygen chemisorption at low temperature but cannot explain the general features observed at high temperature for oxides.[1-3]

This method has been applied to calcia-stabilized and yttria-stabilized zirconia. These electrolytes are nearly pure ionic conductors and no experiments were able to give directly the variations of n with the oxygen pressure P_{O_2} in the range 1–10^{-10} atm, mainly because the electron concentration is small with respect to the oxygen vacancy concentration. In the present experiment, the density of conducting electrons is varied by changing P_{O_2}, equilibrating the oxide at high temperature. The results are analyzed in terms of point defects. Activation energies have also been obtained and are discussed, with the aim of obtaining data on the electron affinity.

Experimental Results

The experimental procedure consists in following either the variations of j_0 with P_{O_2} at constant T or the variations of j_0 with T at constant P_{O_2}. A high vacuum system was used in which pure oxygen is introduced, the pressure range covered being 10^{-5}–10^{-11} atm, and the gas composition being controlled by a mass spectrometer. The emitted current was measured by the diode technique, the oxide constituting the cathode and the anode made of platinum. A cylindrical geometry was used with guard anodes insuring that the collected electrons came only from the sample. The oxide was deposited on a metallic wire (pure iridium) by electrophoresis of a suspension of the oxide in isopropyl alcohol (the thickness of the sample is in the range of 50 μm). The cathode was then Joule-heated. The temperature was measured on the iridium wire between the central anode and the guard. The emitted current was measured in the central anode with an electrometer for various voltages and extrapolated to zero electrical field. The density of the emitted current was then calculated taking account of the geometric surface.

Results for ZrO₂-7.5 mol% Y₂O₃ and ZrO₂-10 mol% CaO

It was established that the zirconia was stabilized in both cases, even with ZrO_2-7.5% Y_2O_3, which is at the limit of the stabilizing composition. Before measurements, the sample was heated at a high temperature in the high oxygen pressure range, then isotherms are made by reducing P_{O_2}. Measurements were reversible and the data are shown in Figs. 2 and 3. The total pressure had no influence, as demonstrated by an experiment made with N_2-20% O_2 where the oxygen partial pressure was the controlling parameter. Both oxides present the same features, namely j_0 is proportional to $P_{O_2}^{-1/4}$. Departures from this law were small for ZrO_2-10% CaO and almost insignificant for ZrO_2-7.5% Y_2O_3 and may originate in small temperature variations.

Thermal Emission of Electrons versus Temperature

Isobars for both compounds were also obtained and plotted in Figs. 4 and 5. They are straight lines, over a wide range of temperature, whose slopes do not change with the oxygen pressure. The slopes of the isobars lead to the following energies of activation:

$$E_{ZrO_2 - xY_2O_3} = 6.5 \text{ eV}$$

$$E_{ZrO_2 - yCaO} = 5.6 \text{ eV}$$

Discussion

The Oxygen Dependency of n

As observed in Figs. 3 and 4, j_0 is proportional to $P_{O_2}^{-1/4}$; the slope is independent of the temperature in the investigated range 1500–1800 K. At the same time, isobars (Figs. 5 and 6) are straight lines whose slopes are independent of the oxygen pressure. As assumed elsewhere,[1-3] it is unlikely that such variations are due to variations in the band curvature, if it exists, with the oxygen pressure. At such a high temperature, diffusion is probably fast enough to ensure equilibration of the bulk, and the work function reflects the Fermi level position.[4] Accordingly, a simple point-defect model can explain the oxygen dependency. We use the classic description of Kroeger.[5] The incorporation of yttria or calcia in the lattice of the zirconia may be written

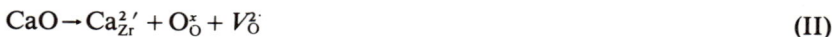

$$Y_2O_3 \rightarrow 2Y'_{Zr} + 3O^x_O + V^{\cdot\cdot}_O \tag{I}$$

$$CaO \rightarrow Ca^{2'}_{Zr} + O^x_O + V^{\cdot\cdot}_O \tag{II}$$

Thus incorporation of Y_2O_3 or CaO creates oxygen vacancies responsible for the high ionic conduction and charge compensation is effected by the substitution cations:

$$2[V^{\cdot\cdot}_O] = [Y'_{Zr}] = [Y] \text{ tot} \tag{4}$$

$$[V^{\cdot\cdot}_O] = [Ca^{2'}_{Zr}] = [Ca] \text{ tot} \tag{5}$$

The equilibration of the oxide with the oxygen pressure is controlled by the equilibrium reaction

$$O_O^x \rightleftharpoons \tfrac{1}{2}O_2 + V_O^{\cdot\cdot} + 2e \qquad (III)$$

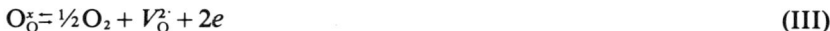

Since the activity of the oxygen vacancies is fixed by the dopant, the electron concentration n is fixed by the oxygen pressure, e.g. in the case of doping by CaO

$$n = \frac{K^{1/2}}{\gamma^{1/2}[Ca]^{1/2}_{tot}} P_{O_2}^{-1/4} \qquad (6)$$

where K is the constant of mass associated with equilibrium (III):

$$K = K_o \exp\left(-\frac{H_{V_O^{\cdot\cdot}}}{kT}\right) \qquad (7)$$

and $H_{V_O^{\cdot\cdot}}$ is the enthalpy for the creation of $V_O^{\cdot\cdot}$. Since the concentration of $V_O^{\cdot\cdot}$ is high, activity coefficients γ should be considered. It is thus found that n is proportional to $P_{O_2}^{-1/4}$, and this explains the observations of Figs. 3 and 4 because j_0 is simply related to n (Eq. (3)). The model assumes the predominance of point defects, which may be a questionable assumption in such highly defective materials. As suggested by Kroeger[6] association of defects in neutral form will probably occur, decreasing the $V_O^{\cdot\cdot}$ concentration but having no influence on the oxygen-pressure dependency of n. Absolute measurements of n would be very helpful in confirming this interpretation.

Until now, there have been no direct determinations of the variations of n in our range of oxygen pressure, mainly because the electronic transport numbers are too small. Indirect experiments based on electrochemical measurements[7,8] provide only some data in the available oxygen pressure range of the electrolyte. Recently measurements of permeability gave the variations of σ_P[9,10] and σ_n[11] with P_{O_2}. They are in agreement in finding that $\sigma_P \propto P_{O_2}^{1/4}$ and $\sigma_n \propto P_{O_2}^{-1/4}$. Our experiment on thermal emission of electrons is, however, more direct in principle and is able to give absolute values of n if exponential terms of Eq. (3) are known.

Activation Energies

From Eqs. (3), (6), and (7), one sees that the exponential term of j_0 is in fact

$$E = \chi + \tfrac{1}{2}H_{V_O^{\cdot\cdot}} \qquad (8)$$

The plot of $\log j_0$ as a function of $1/T$ gives approximately the energy E. Instead of plotting $\log j_0$ vs $1/T$, it would be more correct to plot $\log j_0/T^m$, m being the power of T in the preexponential term of Eq. (3) which is different from zero ($m = 2$ in the case of $V_O^{\cdot\cdot}$), but this contribution can be neglected to a first approximation. As a result of such plots, values of activation energies are obtained at zero Kelvin.

An order of magnitude $H_{V_O^{\cdot\cdot}}$ was deduced by Loup[1] from measurements on pure zirconia, viz. $H_{V_O^{\cdot\cdot}} \cong 9$ eV. Fabry[10] and Heyne and Beekmans[9] give for the hole conduction of 9% yttria-stabilized zirconia $\sigma_p \cong AP_{O_2}^{1/4}$ exp $(-(1.9/kT))$, in which the exponential term is $E_g + E_{\mu_p} - \tfrac{1}{2}H_{V_O^{\cdot\cdot}}$ where E_g and E_{μ_p} are the band gap and hole mobility activation energy, respectively. Neglecting the term E_{μ_p} would lead to $H_{V_O^{\cdot\cdot}} \cong 6.2$ eV, since the band gap is in the range ≈ 5 eV.[12] This value is probably below the true value; on the other hand, 9 eV may be too high. Recently, Ounalli et al.[11] found an activation energy for σ_n

near 3.8 eV by electrochemical transport of oxygen in 10% calcia-stabilized zirconia. Using the same type of assumption for the mobility as above would lead to $Hv_O^{2\cdot} \cong 7.6$ eV. The differences in $Hv_O^{2\cdot}$ between pure zirconia and stabilized zirconia and between calcia-stabilized and yttria-stabilized zirconia are unexplained but may be due to several factors. One of these could be systematic errors in experiments, and more work should clarify this source. The values extracted from Refs. 9–11 may be in agreement if one considers $E_{\mu P} \cong 0.5$ eV and $E_{\mu n} \cong 0.0$ eV. If associates of defects are present, a negative energy will enter the activation energy of the electron conductivity, thus leading to a higher value of $Hv_O^{2\cdot}$, closer to that calculated by Loup and Anthony.[1] In compounds with such high defect concentrations, association may be important even at high temperature. The decrease in $Hv_O^{2\cdot}$ from the pure zirconia value to the value for stabilized zirconia may also be related to the well-known decrease in activation energy for the electrical conductivity observed when an increasing amount of yttria or calcia is added to the zirconia. Whatever the real reason, it remains that $Hv_O^{2\cdot}$ is between 6.2 and 9 eV which has been assumed in the following.

From the activation energy obtained by thermal emission of electrons, the electron affinity can be calculated. In pure zirconia[1-13] two methods lead to approximately $\chi_{ZrO_2} \cong 3.5$ eV, in good agreement with empirical correlations.[14] From the above considerations and Eq. (8), it is thus deduced that, for stabilized zirconia, 3.5 eV $> \chi_{ZrO_2 - xY_2O_3} \geq 2$ eV and 2.5 eV $> \chi_{ZrO_2 - yCaO} > 1.1$ eV. It is then obvious that adding yttria or calcia to the zirconia decreases its electron affinity. In fact, as we observed,[2] the electron affinity of yttria is $\chi_{Y_2O_3} \cong 2$ eV, again in agreement with empirical correlations[15]; the electron affinity of CaO is ≈ 0.8 eV.[16] It is known that the addition of low-work-function compounds to a material decreases its own work function.[17] Our measurements are in agreement with that rule but additionally, at least in our case, it is the electron affinity which is involved. This is consistent with the prediction of empirical correlations,[14,15] expressing that the electron affinity depends mostly on the electronegativity of the compound.

Summary

We have followed, by thermal emission of electrons, the variations of the concentration of conducting electrons in stabilized zirconia. This cannot be done by conventional techniques, such as electrical conductivity. The electron affinity is lowered by addition of low-electron-affinity compounds.

References

[1]J. P. Loup and A. M. Anthony, *Phys. Status Solidi A*, **38** [1] 499 (1970).
[2]P. Odier and J. P. Loup, *J. Solid State Chem.*, **34** [1] 107 (1980).
[3]P. Odier, J. C. Rifflet, and J. P. Loup; to be published in the Proceedings of the 9th International Symposium on the Reactivity of Solids, Krakow, 1980.
[4](a) J. Nowotny and I. Sikora, *J. Electrochem. Soc.*, **125** [5] 781 (1978).
  (b) J. P. Bonnet, J. Nowotny, M. Onillon, and I. Sikora, *Oxid. Met.*, **13** [3] 273 (1979).
[5]F. A. Kroeger, The Chemistry of Imperfect Crystals. North Holland, Amsterdam, 1974.
[6]F. A. Kroeger, *J. Am. Ceram. Soc.*, **49** [4] 215 (1966).
[7]H. Schmalzreid, *Z. Elektrochem.*, **66** [7] 572 (1962).
[8]L. M. Friedman, K. E. Oberg, W. M. Boorstein, and R. A. Rapp, *Metall. Trans.*, **4** [1] 69 (1973).
[9]L. Heyne and W. M. Beekmans, *Proc. Br. Ceram. Soc.*, **19**, 265 (1971).
[10](a) P. Fabry; Thesis, Grenoble, 1976.

(b) J. Foultier, P. Fabry, and M. Kleitz, *J. Electrochem. Soc.*, **123**, 204, (1976).

[11]A. Ounalli, B. Cales, and J. F. Baumard; private communication, 1980.

[12]W. H. Strehlow and E. L. Cook, *J. Phys. Chem. Ref. Data*, **2** [1] 163 (1973).

[13]J. P. Loup and A. M. Anthony, *C. R. Hebd. Seances Acad. Sci.*, **269B**, 918 (1969).

[14]M. A. Butler and D. S. Ginley, *J. Electrochem. Soc.*, **125** [2] 228 (1978).

[15]P. Odier and J. C. Rifflet; to be published in the *International Journal of Thermophysics.*

[16]K. Y. Tsou and E. B. Hensley, *J. Appl. Phys.*, **45** [1] 47 (1974).

[17]G. Hermann and S. Wagener, The Oxide Coated Cathode. Chapman and Hall, London, 1951.

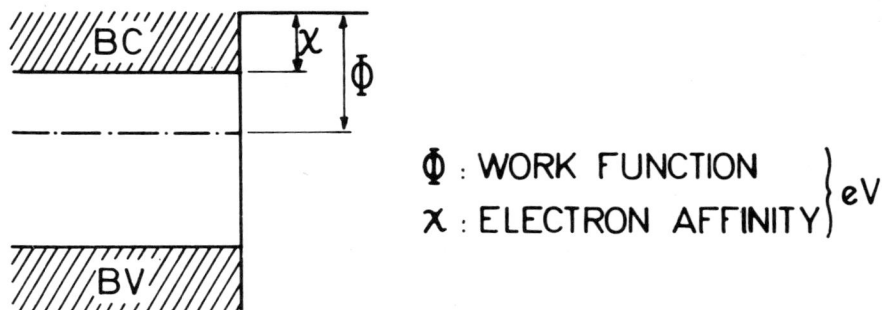

Fig. 1. Work function and electron affinity.

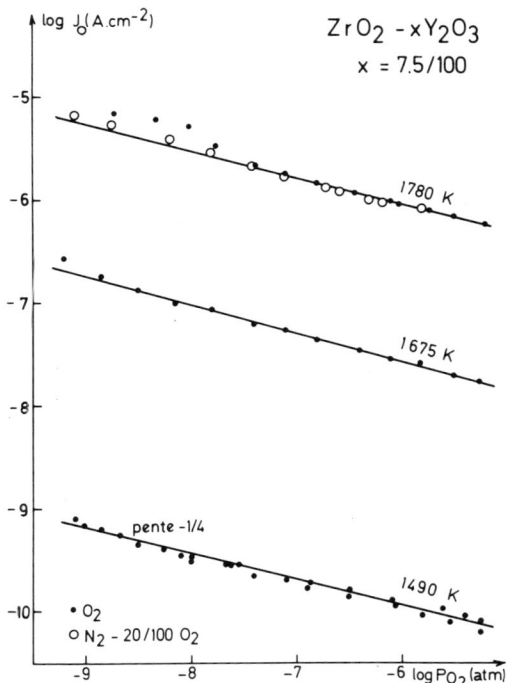

Fig. 2. Thermal emission vs oxygen pressure for ZrO_2-7.5%Y_2O_3.

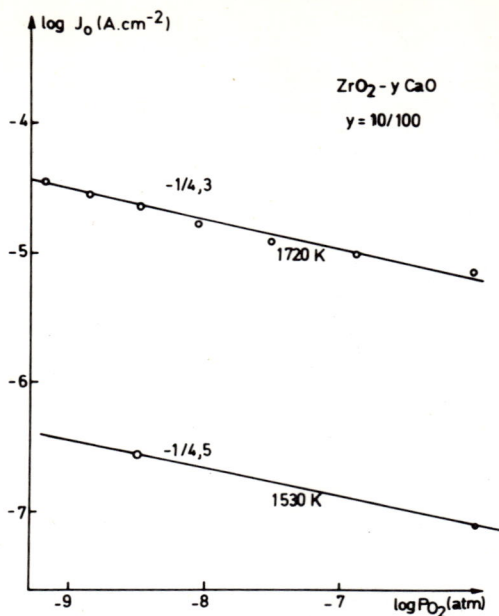

Fig. 3. Thermal emission vs oxygen pressure for ZrO_2-10%CaO.

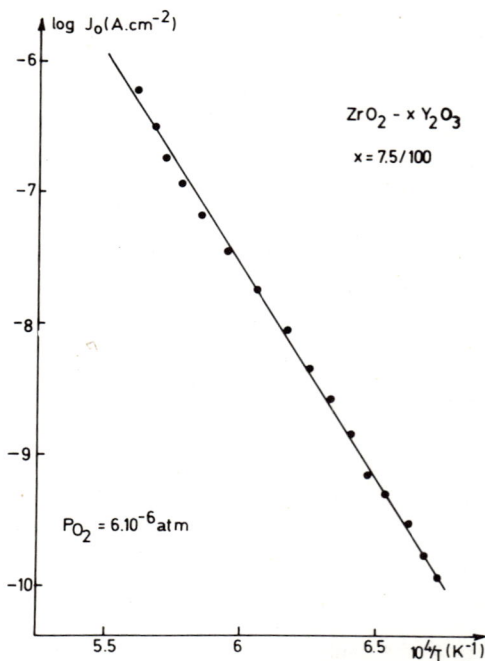

Fig. 4. Thermal emission vs temperature for ZrO_2-7.5%Y_2O_3.

386

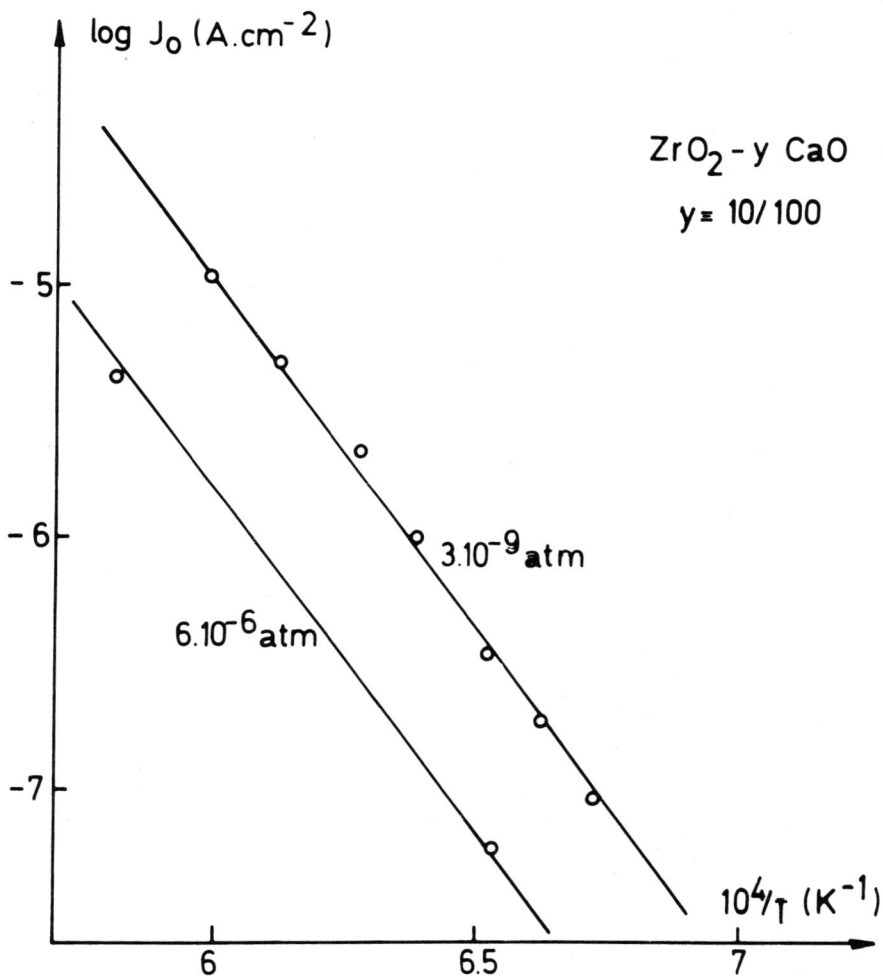

Fig. 5. Thermal emission vs temperature for ZrO_2-10%CaO.

ZrO$_2$ oxygen sensors in automotive applications

E. M. LOGOTHETIS

Staff, Research
Engineering and Research
Ford Motor Company
Dearborn, Michigan 48121

Electrochemical oxygen sensors based on ZrO$_2$ materials are finding increasing use as automotive engine control devices in engine systems designed to reduce emissions and fuel consumption. Their usefulness derives from the fact that the oxygen partial pressure in the exhaust gas of an engine is a function of the air-to-fuel ratio (A/F) of the mixture introduced into the cylinders of the engine. In this paper, automotive ZrO$_2$ oxygen sensor designs, their characteristics, and their performance in the auto exhaust will be reviewed. The harsh conditions prevailing in the automotive environment impose severe requirements on the electrical, chemical, and mechanical properties of the sensor materials. Additional complications arise from the fact that the auto exhaust consists of many gaseous components that are not in thermodynamic equilibrium. Consequently, ideal Nernst-type behavior is not usually observed; departures from ideality are found in both the steady-state and the transient response of the sensors. Examples to be discussed include shifts in the emf vs air-to-fuel ratio characteristic and response-time asymmetries. Models developed to explain these phenomena, especially those involved in the exchange of oxygen at the solid/gas interface, will be discussed briefly and compared with experimental results. Finally, some results of the performance of systems using ZrO$_2$ sensors will be presented.

In the past few years, the need to improve the performance of internal combustion engines with respect to emissions and fuel economy has led to the investigation of various engine systems and control strategies, some of which are based on feedback control of the air-to-fuel ratio, A/F. Air-to-fuel ratio is defined as the ratio of the mass of air to the mass of fuel that makes up the mixture in the cylinders of an engine before combustion. Figure 1 shows the dependence of the concentrations of various exhaust gas constituents on A/F.[1] $(A/F)_{stoich}$ is the stoichiometric air-to-fuel ratio, i.e. the one that contains just enough oxygen to convert all the hydrocarbons (HC) of the fuel to CO$_2$ and H$_2$O. The region to the left of $(A/F)_{stoich}$ is the so-called fuel-rich region, whereas that to the right of $(A/F)_{stoich}$ is the fuel-lean region. In the fuel-rich region, the insufficiency of O$_2$ causes the formation of CO and H$_2$; in the lean region, the excess oxygen appears as free oxygen molecules. Incomplete combustion results in unburned HC in the exhaust gas; high engine temperatures result in the formation of oxides of nitrogen (NO$_x$).

Of the three exhaust gas constituents for which there are federal standards, HC and CO are found in high concentrations in rich A/F mixtures but decrease significantly as the A/F ratio increases; the concentration of NO$_x$, on the other hand, has a maximum in the lean region at ≈ 16 and decreases for richer and leaner A/F mixtures. The concentration of H_2 is $\approx \frac{1}{3}$ that of CO and the concentration of H_2O has a variation similar to that of CO_2. In addition to A/F ratio, the exhaust gas composition is also influenced by parameters such as engine design, fuel metering system, cylinder-to-cylinder maldistributions, ignition timing, etc. It must be emphasized from the beginning that the exhaust gas composition is variable and far from thermodynamic equilibrium. For conventional engines, fuel economy and the power generated by the engine are also functions of the A/F ratio: Fuel economy is higher for lean A/F mixtures, whereas power is higher for rich A/F mixtures.

One engine control system that can achieve present federal emissions and fuel economy requirements utilizes a three-way-catalyst (TWC) and an A/F sensor to control the A/F ratio at its stoichiometric value.[2-6] The need for A/F control in this type of system is indicated in Fig. 2, where the efficiency of a TWC for the oxidation of CO and HC and for the reduction of NO_x is plotted as a function of the A/F ratio. It is apparent that high overall efficiency for the simultaneous removal of all three gases with a TWC is achieved only within a very narrow A/F ratio range near the stoichiometric value. This then establishes the need for an A/F sensor that can accurately maintain the A/F near stoichiometry in this type of engine control system. Control of the A/F ratio away from stoichiometry has also been considered.[7,8] Although such control systems have not yet been developed, the continuing thrust for fuel economy may eventually lead to the development of systems based on feedback control of the A/F ratio at lean A/F values.

Presently, the only automotive A/F sensors in existence are oxygen sensors that sense the oxygen partial pressure P_{O_2} in the exhaust gas. Figure 3 shows the dependence of P_{O_2} of the exhaust gas on the A/F ratio. The solid line corresponds to the case that the exhaust gas is in thermodynamic equilibrium at a given temperature (700 °C in Fig. 3). The main feature is the abrupt and large change in P_{O_2} at the stoichiometric A/F ratio; for A/F ratios larger than $(A/F)_{stoich}$, P_{O_2} is determined by the excess O_2 and is high; for A/F ratios smaller than $(A/F)_{stoich}$, the exhaust gas is reducing, thereby establishing very low P_{O_2}. In contrast, the partial pressure of the free oxygen depicted by the dotted line in Fig. 3 does not change significantly at the stoichiometric A/F ratio. Therefore, in order to sense the large change in the thermodynamic equilibrium P_{O_2} at $(A/F)_{stoich}$, the sensor must have the ability, i.e. the catalytic activity, to equilibrate the exhaust gas near the sensor surface. It is noted that, in the lean region, P_{O_2} depends only weakly on A/F ratio, a fact that makes difficult the use of oxygen sensors for the measurement of lean A/F ratios.

ZrO₂ Oxygen Sensors

The device most widely used at the present time as an A/F sensor is a ZrO_2 galvanic cell.* This device, shown schematically in Fig. 4, is essentially an oxygen concentration cell.[11,12] It consists of a ZrO_2 ceramic, usually in the form of a tube, with appropriate electrodes (e.g. platinum) deposited on both sides of the ceramic. ZrO_2 doped with ions such as Ca or Y is an ionic conductor with very high conductivity for oxygen ions. The properties of ZrO_2 materials have been discussed in great detail in other papers in this volume. When the two sides of the ZrO_2 ceramic are exposed to different oxygen par-

389

tial pressures $P_{O_2}(1)$ and $P_{O_2}(2)$, an electromotive force develops across the electrodes given by the Nernst equation[12]:

$$emf = (RT/4F) \ln [P_{O_2}(1)/P_{O_2}(2)] \tag{1}$$

where R and F are the gas and Faraday constants and T is the temperature. This equation corresponds to the following electrochemical reaction

$$O_2(gas) + 4e^-(electrode) \rightleftharpoons 2O^{2-}(electrolyte) \tag{2}$$

If CO is also present, another electrochemical reaction is possible

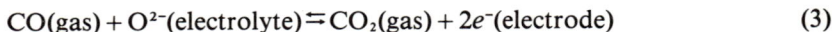

$$CO(gas) + O^{2-}(electrolyte) \rightleftharpoons CO_2(gas) + 2e^-(electrode) \tag{3}$$

In addition, the following reaction occurs catalytically on the electrode (platinum) surface

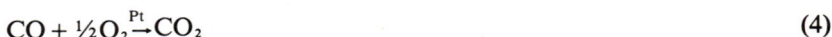

$$CO + \tfrac{1}{2}O_2 \xrightarrow{Pt} CO_2 \tag{4}$$

It is important to note that Eqs. (2)–(4) represent overall reactions which mask intermediate reaction steps involving surface adsorption and desorption and reactions between adsorbed species. It is also pointed out that similar reactions occur with the H_2 and H_2O present in the exhaust gas and more complicated ones involving HC and NO_x.

Oxygen sensors based on ZrO_2 concentration cells have a long history. For many years, various types of ZrO_2 cells have been developed and used in scientific studies and applications.[11-15] However, the application of these devices to engine control has not been straightforward. The reason is that automotive oxygen sensors must be not only simple and inexpensive but also able to operate and survive in the harsh environment of the exhaust gas, namely at temperatures $>900\,°C$ and temperature changes of $\leq 50\,°C/s$, in oxidizing and very reducing environments and in the presence of strong vibrations and many chemical compounds and elements from gasoline, oil, and various automotive materials.

At the present time, automotive ZrO_2 oxygen sensors have been developed only for stoichiometric control for the TWC applications.[16-19] Experimental ZrO_2 devices intended for operation in lean A/F mixtures where P_{O_2} does not change significantly have been reported but will not be discussed here. These are, generally, heated or temperature-compensated devices, either Nernst-type concentration cells or oxygen-pumping devices.[8,15,20-23]

Since stoichiometric ZrO_2 sensors are required to sense only the large change in P_{O_2} at the stoichiometric A/F ratio, they are simply made and do not have an external heater. The device is inserted in the exhaust manifold (or the exhaust pipe) and is heated by the exhaust gas.[†] Figure 5 shows a typical configuration of this type of sensor. The housing is essentially an 18 mm spark plug shell. The sensor body is made from fully or partially stabilized ZrO_2. Yttria is exclusively used as a stabilizer[11] because it results in a higher conductivity material and thus a lower impedance device at low temperatures. The inner and outer electrodes are generally made from platinum. The physical and chemical structure of these electrodes is a compromise between the desire for performance (thin, high surface area, porous electrodes for easy gas transport and high catalytic activity) and the desire for durability (thick, well-sintered, and stable electrodes).[24] The outer electrode is protected from erosion with a porous oxide layer such as spinel[16,18,24]; the presence of this layer also helps to bring the exhaust gas closer to thermodynamic equilibrium near the electrode by restricting the amount of exhaust gas reaching the electrode.[25] The microstructure and the thickness of this layer, however, can also affect the response time of the sensor and introduce asymmetries in the rich-to-lean and the lean-to-rich response times of the sensor; this will be discussed later. Gastight sealing of the zirconia ceramic to the metal housing is achieved with a graphite or a metal gasket.

Figure 6 shows some data on the dependence of the emf of a commercial ZrO_2 sensor on A/F ratio. These results were obtained in laboratory tests with the sensor located in the exhaust ($T = 700\,°C$) from a 351 CID engine. The data of Fig. 6 are typical of most ZrO_2 sensors at the higher exhaust gas temperatures. The large steep change in the sensor emf at the stoichiometric A/F ratio is, of course, a reflection of the similar behavior of the thermodynamic equilibrium P_{O_2} in the autoexhaust. At low temperatures, however, the emf vs A/F dependence of ZrO_2 sensors shows departures from ideal behavior. This will be discussed in the next section.

A very important sensor property is the response time of the sensor to changes in the A/F ratio. For feedback control purposes, the important response time is the lean-to-set point (or the rich-to-set point) response time defined as the time required for the sensor voltage to change from its lean value V_L (rich value V_R) to the control set point value (frequently taken as $(V_L + V_R)/2$), when the A/F ratio is changed from lean (rich) to rich (lean) values. The response times of well-made sensors are generally < 200 ms above $350\,°C$. Available data[11,19,24] indicate that response times are determined by mass transport through the protective spinel coating and the porous electrode and by gas reaction kinetics on the catalytic electrode.

Limit-Cycle Feedback Control

Because of the singularity in the sensor response at $(A/F)_{stoich}$ (Fig. 6), the feedback control of the A/F ratio is not directly proportional to the A/F ratio but is based on a limit-cycle principle. Although this type of control can be implemented in a number of ways, the basic principles are nevertheless the same and may be understood with the help of Figs. 7 and 8. In Fig. 7, the system is controlled by sensor No. 1 which is assumed to have zero response time. The upper curve shows the waveform of the fuel metering system (e.g. fuel injection or feedback carburetor) which is driven toward increasing (lean) or decreasing (rich) A/F ratios, depending on the sensor signal. The second curve represents the corresponding waveform of sensor No. 1.‡ The output from the sensor goes to a comparator with a set-point value between V_L and V_R, e.g. $(V_L + V_R)/2$. The output of the comparator (third curve in Fig. 7) is 5.0 V if emf $> (V_L + V_R)/2$ indicating a rich A/F mixture or 0.0 V if emf $< (V_L + V_R)/2$ indicating a lean A/F mixture.

To explain the waveforms of Fig. 7, suppose that at some specific time the A/F ratio, as it decreases, passes through the stoichiometric value. The sensor will not immediately sense this change but will continue showing a low emf (lean indication) because it takes some time τ for the gas to travel from the engine cylinders to the sensor location. After this time τ has elapsed, the stoichiometric exhaust gas arrives at the sensor and the sensor emf will change to a high value (e.g. 800 mV) indicating rich A/F mixtures. This change in the sensor signal and the corresponding change in the output of the comparator will direct the fuel metering system to change direction toward leaner mixtures, i.e. increasing A/F ratios. Because of symmetry, after time

τ, the A/F ratio in the cylinders will again pass through stoichiometry, but the sensor will not indicate transition to lean A/F mixtures unless an additional time τ has elapsed. At this moment, the fuel metering system will be directed by the sensor/comparator to change direction toward decreasing (rich) A/F ratios completing a full cycle. It is apparent that the limit-cycle period is $T = 4\tau$. In closed-loop operation the time-averaged output from the comparator is constrained to be 2.5 V; the fuel metering system spends then equal times in the rich and in the lean A/F regions and the time-averaged A/F ratio is stoichiometric as required. If the sensor rich-to-lean and lean-to-rich response times are equal but nonzero, τ_{sensor}, the limit cycle period becomes $T = 4\tau + 4\tau_{sensor}$.

Suppose now that another sensor, sensor No. 2, has an asymmetric response time, $\tau_{LR} < \tau_{RL}$. The response of this sensor when sensor No. 1 is in control is shown by the fourth curve in Fig. 7. It is apparent that the time average of the output from a comparator connected to sensor No. 2 will be > 2.5 V, indicating, quite wrongly, an overall rich A/F mixture. When sensor No. 2 is in control, the value of the time-averaged A/F ratio changes. This is shown in Fig. 8. The time-averaged output from the comparator is again constrained to be 2.5 V, i.e. sensor No. 2 spends equal times with high (rich) and zero (lean) emf values. The fuel metering system, however, spends longer times in the lean than in the rich A/F region.

Therefore, the average A/F ratio is no longer stoichiometric but lean. The time-averaged output from the comparator of sensor No. 1 will, of course, be < 2.5 V, correctly indicating a lean mixture. Thus, an asymmetry in the lean-to-rich and rich-to-lean response times results in a dynamic shift in the control point away from the stoichiometric A/F ratio. This effect can be substantially compensated by system control design (e.g. introducing asymmetric sweep of the A/F ratio) provided that the sensor properties are known and reproducible.

Nonideal Behavior of ZrO₂ Sensors

The ideal response of a ZrO_2 sensor, in the exhaust gas indicated in Fig. 6, is not usually observed, especially at low temperatures. The reason is that the auto exhaust is an open system consisting of a continuously flowing mixture of many gaseous components that are not in thermodynamic equilibrium. Figure 9 compares the expected ideal behavior with some results of engine studies by Fleming[26] on a commercial sensor in the range 370–500 °C. In contrast to an ideal sensor, this sensor does not always make its step-like emf change at the stoichiometric A/F ratio but at an A/F ratio that becomes progressively leaner as the temperature decreases. The magnitude of these results appears to be rather extreme; nevertheless all ZrO_2 sensors show to some degree this type of behavior at low enough temperatures.

The nonideal behavior shown in Fig. 9 may be understood in terms of nonequilibrium chemical reactions and electrochemical processes occurring on the sensor electrodes. In this brief review, a simplified model will be used to indicate how this behavior arises; more complete analysis can be found elsewhere.[26-29]

First consider, however, the results of some laboratory experiments[30] using a simpler and better-controlled system than the auto exhaust. In these experiments, a ZrO_2 oxygen cell was exposed to a continuous and uniform flow of a gas mixture consisting only of CO and O_2 in N_2 at a total pressure of 1 atm. The ratio R_{FG} of CO to O_2 was changed and the corresponding emf of the ZrO_2 cell was measured. Figure 10 shows some results on the dependence of the emf on R_{FG} for two temperatures. The stoichiometric ratio of this mixture is at $R_{FG} = 2$. Gas mixtures with R_{FG} values >2 correspond to simulated rich A/F mixtures and those with R_{FG} values <2 correspond to simulated lean A/F mixtures. For temperatures $>450\,°C$, near-thermodynamic-equilibrium behavior is observed; for example, at $500\,°C$, the experimental data in Fig. 10 are in agreement with emf values (curve a) calculated from Eqs. (1) and (4) assuming thermodynamic equilibrium. As the temperature decreases, the system begins to depart from thermodynamic equilibrium as one of the intermediate reaction steps becomes the rate-limiting process. For example, at $350\,°C$ the observed ZrO_2 cell response (curve b) deviates substantially from the calculated thermodynamic equilibrium response (curve c). Figure 10 also shows the average CO_2 production rate as a function of R_{FG} at $350\,°C$. The peaked response is similar to the one commonly observed in the catalytic oxidation of CO and hydrocarbons.[31]

Some insight as to how this behavior arises may be obtained by an analysis of the simple model described by the following reactions:

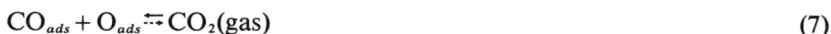

$$O_2(gas) \rightleftharpoons 2O_{ads} \tag{5}$$

$$CO(gas) \rightleftharpoons CO_{ads} \tag{6}$$

$$CO_{ads} + O_{ads} \rightleftharpoons CO_2(gas) \tag{7}$$

This model considers only adsorption/desorption of oxygen and CO and reaction between these two species on the surface of the platinum electrode, neglecting charge transfers altogether. From these equations one obtains the following differential equations for the concentrations of adsorbed oxygen and adsorbed CO and for the production of CO_2:

$$\frac{d[O_{ads}]}{dt} = K_O^a P_{O2} - K_O^d[O_{ads}]^2 - r[O_{ads}][CO_{ads}] \tag{8}$$

$$\frac{d[CO_{ads}]}{dt} = K_{CO}^a P_{CO} - K_{CO}^d[CO_{ads}] - r[O_{ads}][CO_{ads}] \tag{9}$$

$$\frac{dCO_2}{dt} = r[O_{ads}][CO_{ads}] \tag{10}$$

K_O^a, K_O^d, K_{CO}^a, K_{CO}^d are the adsorption and desorption rate constants for oxygen and CO, and r is the rate constant for the reaction between the adsorbed oxygen and CO.

The solution of Eqs. (8)–(10) reveals a variety of behavior, depending on the relative magnitude of the various rate constants. Some of the results of this analysis are presented in Figs. 11 and 12. Figure 11 shows some results for the case that $\frac{1}{2}K_O^a = K_O^d = K_{CO}^a = K_{CO}^d$ and the ratio r/K_j^i is changed. The main feature in this figure is the large and step-like change in the concentration of adsorbed oxygen when r is much larger than K_j^i. As the value of the reaction rate constant r decreases toward the values of the K_j^i's, that is, as the catalytic activity for the oxygen/CO reaction decreases, the step-like change in the concentration of O_{ads} becomes smaller and finally disappears. If the assumption is made that the oxygen activity is in one-to-one correspondence with the concentration of O_{ads}, then the sensor emf will exhibit a similar dependence on R_{FG}. Figure 12 shows results for the case that r/K_j^i is large and (approximately) constant and the relative magnitude of the oxygen and CO adsorption and desorption rate constants varies. The main result in this case is that the position of the step-like change in the concentration of O_{ads} depends on the relative magnitude of the K_O^i and K_{CO}^i. For example, when $\frac{1}{2}K_O^a = K_O^d = K_{CO}^a = K_{CO}^d$, the step occurs at the stoichiometric ratio $R_{FG} = 2$. However, as the value of K_O^a changes from larger values to smaller values, the step moves continuously from $R_{FG} > 2$ to $R_{FG} < 2$. The computed values of the concentrations of O_{ads} and CO_{ads} can be used to calculate the rate of CO_2 production from Eq. (7). The results of this calculation are also in qualitative agreement with the measured behavior in Fig. 10.

The analysis of this simple model leads to the following main conclusions for the behavior of the ZrO_2 oxgyen sensors in the exhaust gas: (1) The magnitude of the step-like change in the sensor emf depends on the catalytic activity of the electrode (i.e. the reaction rate constant r) and increases with increasing catalytic activity. (2) The position of the step-like change in the emf depends on the adsorption and desorption characteristics of the electrode and shifts to leaner A/F ratios as the CO adsorption rate becomes larger than that of oxygen. These conclusions are in agreement with those reached with other models.[26-28]

Another interesting example of nonequilibrium behavior observed in the experiment of Fig. 10 is the appearance, under certain conditions, of oscillations in the emf of the ZrO_2 cell.[30] These oscillations were accompanied by simultaneous oscillations in the CO_2 production and in the surface temperature. The electrical oscillations arise from oscillations in the reaction rate and in the concentration of adsorbed oxygen and CO. This phenomenon is an example of oscillating chemical reactions found in many open chemical and biological systems (e.g. biological clocks) far from thermodynamic equilibrium.[32]

Other related nonideal effects observed with ZrO_2 sensors include dependence of the sensor characteristics on the concentration of NO_x in the exhaust gas,[33] dependence of the sensor response time on the magnitude of the change in the A/F ratio,[33] and a hysteresis in the static response in the sense that the step-like emf change occurs at different A/F values depending on whether the A/F ratio changes from rich-to-lean or from lean-to-rich.[19]

The degree of nonideal behavior exhibited by a given ZrO_2 sensor depends on the materials used and the details of the sensor structure. For example, improvement toward achieving thermodynamic equilibrium can be provided by the porous spinel coating. This coating decreases the gas flow from the bulk of the gas to the electrode surface, enabling the gas mixture in the vicinity of the electrode to come closer to thermodynamic equilibrium even when the reaction rates are not very high.[24,25,27,28] On the other hand, the presence of the spinel layer can cause shifts in the position of the emf step away from the stoichiometric A/F ratio because of differences in the diffusion coefficient of the various gas species. Specifically, the larger diffusion coefficient of H_2 creates a diffusion enrichment of hydrogen relative to oxygen at the electrode surface. This causes the step-like emf change to occur at more oxygen-containing A/F mixtures, i.e. lean mixtures. This phenomenon has been analyzed in detail by Takeuchi et al.[34] In addition to this static shift, the higher diffusion of H_2 influences also the dynamic response of the sensor; in particular, the lean-to-rich response time becomes shorter than the rich-to-lean response time.[24] According to the discussion of the limit-cycle operation, this time asymmetry causes a dynamic shift of the control point to the lean side of the stoichiometric A/F ratio.

Additional complications can arise from changes in sensor properties due to aging or due to abnormal conditions during the operation of the sensors on vehicles. For example, use of leaded gasoline poisons the Pt electrode and decreases its catalytic activity. Some of these effects will be discussed in the following section.

System Performance of ZrO₂ Sensors

Because of their construction and the environment in which they operate, automotive ZrO_2 oxygen sensors do not behave ideally and their properties change with changes in operating conditions and with aging. Considering the narrowness of the TWC efficiency window, these changes would normally be expected to lead to system failure. Fortunately, this is not the case, because the efficiency window of a TWC is wider when the A/F ratio is modulated at some low frequency as is, in fact, the case during the limit-cycle operation. Figure 13 shows some results obtained by Gandhi *et al.*[6] on the efficiency of a TWC for steady state and for the case that the A/F ratio is modulated by ± 1 A/F unit at a frequency of 1 Hz. As the figure shows, under modulation, the efficiency of a TWC decreases somewhat but the window opens up considerably. As a result, the system can reasonably well accommodate relatively small changes in the sensor properties.

Considering the harsh environment of the automobile, the ZrO_2 oxygen sensors have been performing remarkably well. In the last three years, many feedback-controlled vehicles with ZrO_2 sensors have been produced by the automobile manufacturers with positive results. In regard to durability, detailed data on sensor aging have been presented by Gruber and Wiedenmann[24] and by Young and Bode.[19] Under normal conditions, the main effects of aging are an increase in the response time, a decrease in the sensor emf, and a shift in the position and some loss in the sharpness of the step-like emf change. As expected, these changes are more pronounced at the lower temperatures, but generally are not so severe as to lead to system failure. Abnormal operating conditions, however, can lead to severe degradation in sensor properties. Leaded gasoline, for example, poisons the catalytic electrode to a degree that depends on the sensor temperature. The main effects of lead poisoning are a decrease in the rich voltage and an increase in the lean-to-rich response time that causes a dynamic rich shift in the control point.[19,24] Abnormal consumption of oil may cause the formation of heavy deposits on the sensor. If so, the resulting plugging leads to an increase in the rich-to-lean and the lean-to-rich response times; the latter, however, is influenced less since the hydrogen can still diffuse easier than oxygen. The net result is a dynamic shift of the control point to lean A/F mixtures.

Conclusions

The application of ZrO_2 sensors to engine control systems involves the simultaneous satisfaction of many conflicting requirements. These include cost, simplicity, extended durability, and performance in the harsh and complicated exhaust gas environment. The single most important difficulty is the fact that the exhaust gas is an open chemical system far from thermodynamic equilibrium. As a result, present sensors exhibit a variety of nonideal behavior, especially at low temperatures. In spite of the difficulties, automotive ZrO_2 sensors have been developed to such a degree that their use for stoichiometric A/F ratio control has been successful. This success has been aided by the fact that limit-cycle operation broadens the TWC efficiency window.

Further work on improving sensor properties, such as low temperature operation, response time, and light-off characteristics, appears warranted. In addition, research aiming at the development of sensors that can operate in the fuel-lean region where fuel consumption is lower would be desirable.

References

[1] J. Harrington and R. C. Shishu, Paper No. 730 476, SAE Congress, Detroit, May 1973.

[2] R. Zechnall, G. Baumann, and H. Eisele, Paper No. 730 566, SAE Congress, Detroit, May 1973.

[3] J. G. Rivard, Paper No. 730 005, SAE Congress, Detroit, January 1973.

[4] G. T. Engh and S. Walman, Paper No. 770 295, SAE Congress, Detroit, February 1977.

[5] (a) R. A. Spilski and W. D. Creps, Paper No. 750 371, SAE Congress, Detroit, February 1975.
(b) R. Seiter and R. Clark, Paper No. 780 203, SAE Congress, Detroit, February 1978.

[6] H. S. Gandhi, R. G. Delosh, A. G. Piken, and M. Shelef, Paper No. 760 201, SAE Congress, Detroit, February 1976.

[7] T. Y. Tien, H. L. Stadler, E. F. Gibbons, and P. J. Zacmanidis, *Am. Ceram. Soc. Bull.*, **54**, 280 (1975).

[8] M. Hubbard, J. J. Bonilla, K. W. Randall, and J. D. Powel, Paper No. 760 287, SAE Congress, Detroit, February 1976.

[9] E. M. Logothetis, *Ceram. Eng. Sci. Proc., 8th Automotive Mater. Conf.*, **1**, 281 (1980).

[10] M. J. Esper, E. M. Logothetis, and J. C. Chu, Paper No. 790 140, SAE Congress, Detroit, February 1979.

[11] H. Dietz, W. Haecker, and H. Jahnke; p. 1 in Advances in Electrochemistry and Electrochemical Engineering, Vol. 10. Edited by H. Gerischer and C. Tobias. Wiley, New York, 1977.

[12] P. Kofstad, Nonstoichiometry, Diffusion, and Electrical Conductivity in Binary Metal Oxides. Wiley Interscience, New York, 1972; Chapters 6 and 8.

[13] J. Weissbart and R. Ruka, *Rev. Sci. Instrum.*, **32**, 593 (1961).

[14] (a) H. Schmalzreid, *Z. Electrochem.*, **66**, 572 (1966).
(b) H. Schmalzreid, *Z. Phys. Chem. (Frankfurt)*, **38**, 87 (1963).

[15] N. M. Beekmans and L. Heyne, *Philips Tech. Rev.*, **31**, 112 (1970).

[16] H. Dueker, K. H. Friese, and W. D. Haecker, Paper No. 750 223, SAE Congress, Detroit, February 1975.

[17] W. J. Fleming, D. S. Howarth, and D. S. Eddy, Paper No. 730 575, SAE Congress, Detroit, May 1973.

[18] E. Hamann, H. Manger, and L. Steinke, Paper No. 770 401, SAE Congress, Detroit, February 1977.

[19] C. T. Young and J. D. Bode, Paper No. 790 143, SAE Congress, Detroit, February 1979.

[20] D. S. Howarth and R. V. Wilhelm, Jr., Paper No. 780 212, SAE Congress, Detroit, February 1978.

[21] J. A. Riedijk, 2nd Automotive Emissions Conference, University of Michigan, Ann Arbor, September 13–14, 1973.

[22] R. E. Hetrick and A. Fate; unpublished work.

[23] R. E. Hetrick, E. M. Logothetis, and D. K. Hohnke, *AIP Conf. Proc. Physics in the Automotive Industry*, Detroit, May 1980.

[24] See, for example, H. U. Gruber and H. M. Wiedenmann, Paper No. 800 017, SAE Congress, Detroit, February 1980.

[25] (a) R. Bosch GmbH, Ger. Offenl. 2 416 629 (1975).
(b) W. H. McIntyre, R. W. Wallace, and M. J. Troha, U.S. Pat. 3 928 161 (1975).

[26] W. J. Fleming, *Ceram. Eng. Sci. Proc., 8th Automotive Mater. Conf.*, **1**, 272 (1980); and Paper No. 800 020, SAE Congress, Detroit, February 1980.

[27]W. J. Fleming, Paper No. 770 400, SAE Congress, Detroit, February 1977.

[28]L. Heyne, Paper No. 78, Electrochemical Society Meeting, Seattle, Wash., May 1978.

[29]Similar nonequilibrium behavior is observed with resistive-type oxygen sensors. For discussion of this topic see Ref. 9.

[30]R. E. Hetrick and E. M. Logothetis, *Appl. Phys. Lett.*, **34**, 117 (1979).

[31]See, for example, E. McCarthy, J. Zahradnik, G. C. Kuczynski, and J. J. Carberry, *J. Catal.*, **39**, 29 (1975).

[32]Faraday Symposia of the Chemical Society, No. 9, The Physical Chemistry of Oscillating Phenomena, The Faraday Division, Chemical Soc., London, 1974.

[33]M. A. Shulman and D. R. Hamburg, Paper No. 800 018, SAE Congress, Detroit, February 1980.

[34]T. Takeuchi, K. Saji, and I. Igarashi, Paper No. 74, Electrochemical Society Meeting, Seattle, Wash., May 1978; Extended Abstracts **78-2**, No. 74 (1978).

---

*Another *A/F* sensor is a resistive-type oxygen sensor consisting of a metal oxide (e.g. TiO_2) having a resistance that depends on Po_2 (Refs. 7, 9, 10).

†The temperature of the exhaust gas depends on many parameters, e.g. type of engine, vehicle weight, engine rpm, load, timing, etc. Normal temperatures (in the exhaust manifold) can be as low as 250–300 °C and as high as 900–950 °C.

‡For the purpose of explaining the feedback control, the sensor waveforms are idealized in the sense that the lean emf value V_L and the rich emf value V_R are taken as exactly 0 and 800 mV, respectively. In reality, V_L and V_R vary, depending on the *A/F* ratio and the temperature.

Variation of Exhaust Emissions with Air/Fuel Ratio for Single Cylinder Engine using Indolene 30 Fuel (#1)

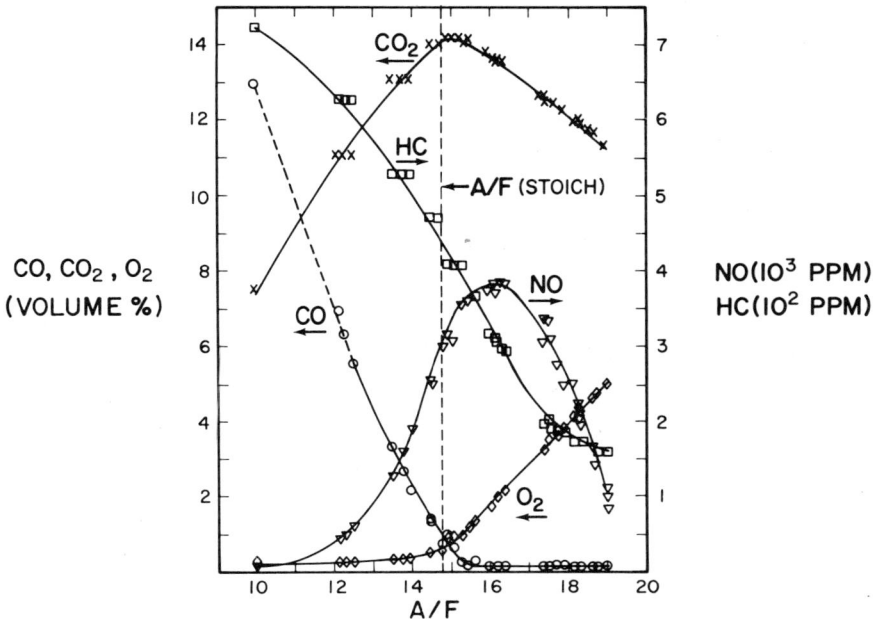

Fig. 1. Dependence of the concentration of some exhaust gas constituents on the air-to-fuel ratio, *A/F*.

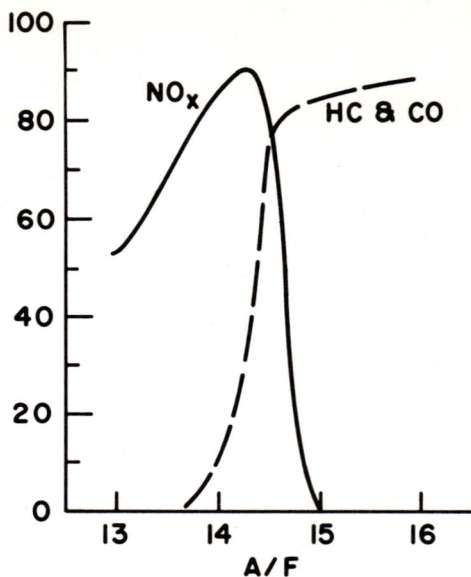

Fig. 2. Conversion efficiency of a three-way catalyst for NO_x, HC, and CO as a function of A/F.

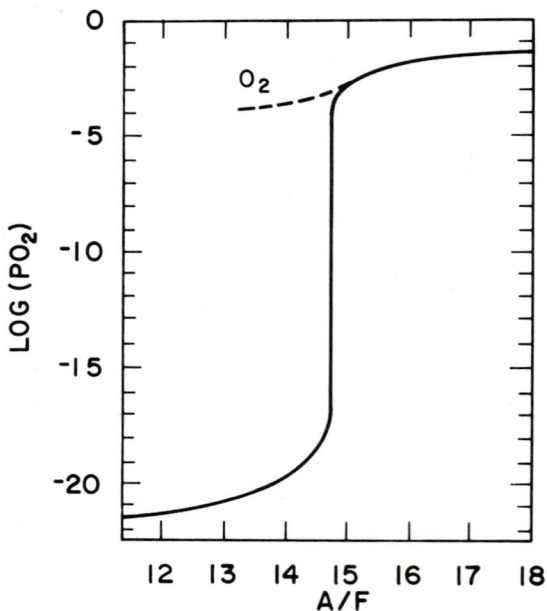

Fig. 3. Dependence of the P_{O_2} of the exhaust gas on A/F. The solid line represents the thermodynamic equilibrium P_{O_2} at 700 °C and the dotted line represents the P_{O_2} of the free oxygen.

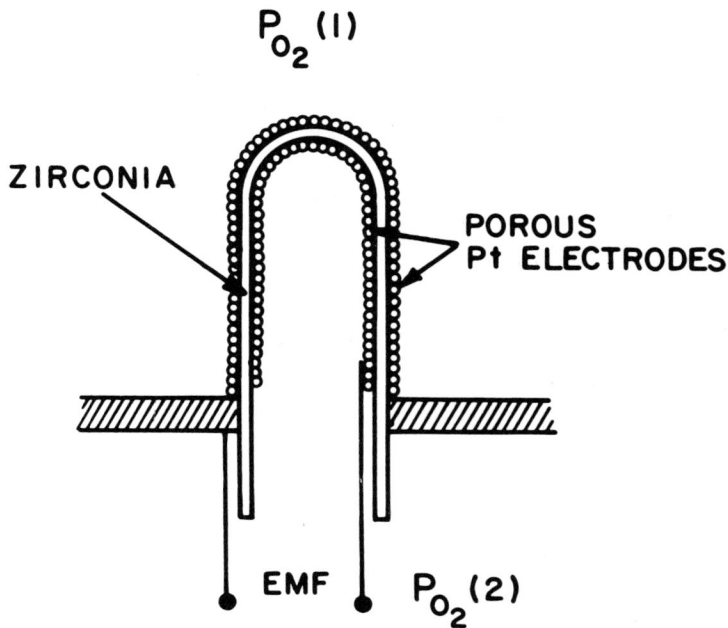

$P_{O_2}(1)$

ZIRCONIA

POROUS Pt ELECTRODES

EMF $P_{O_2}(2)$

Fig. 4. Schematic diagram of a ZrO_2 oxygen concentration cell.

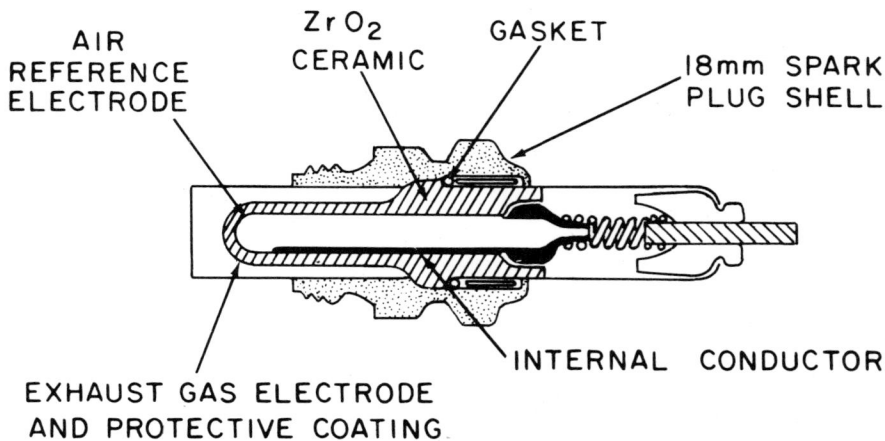

AIR REFERENCE ELECTRODE

ZrO_2 CERAMIC

GASKET

18mm SPARK PLUG SHELL

INTERNAL CONDUCTOR

EXHAUST GAS ELECTRODE AND PROTECTIVE COATING

Fig. 5. Cross section of a typical automotive ZrO_2 oxygen sensor.

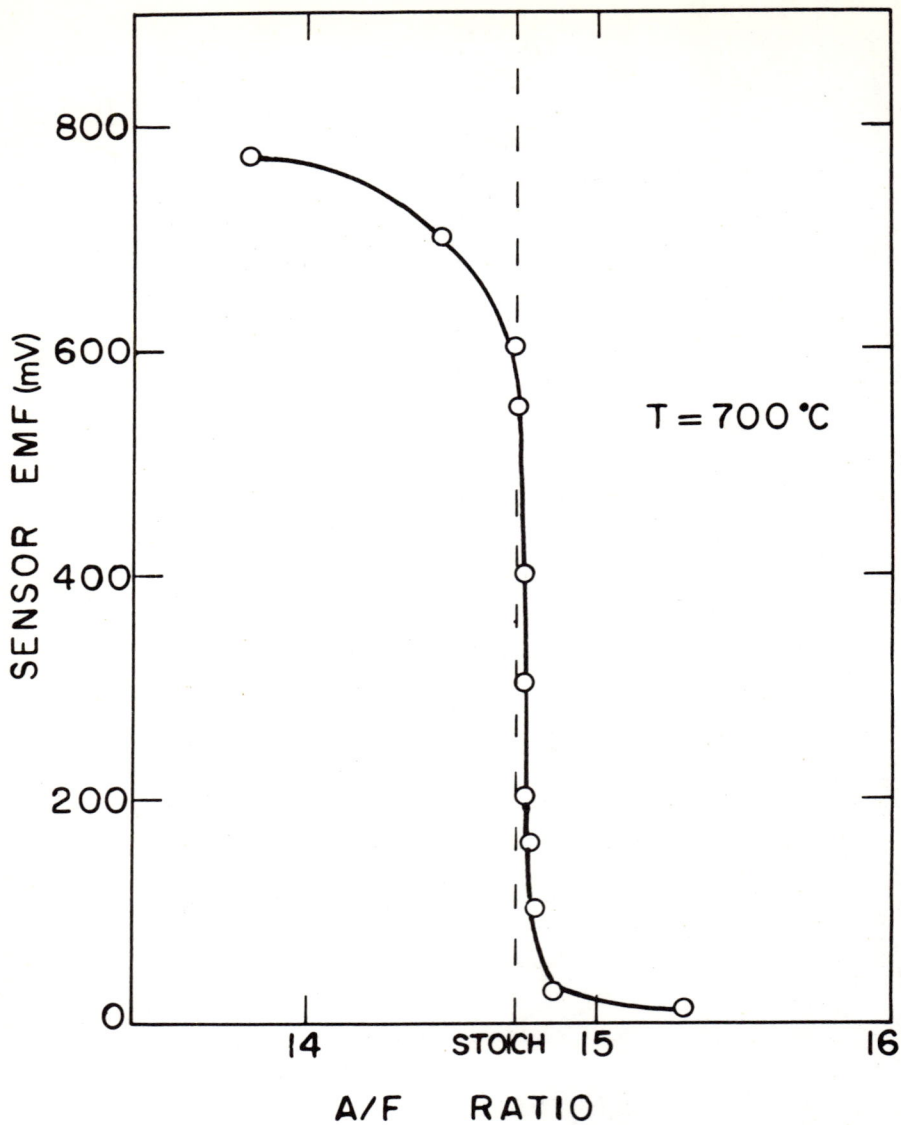

Fig. 6. Dependence of the emf of a ZrO$_2$ oxygen sensor on A/F at 700°C.

SENSOR #1 IN CONTROL

AIR/FUEL RATIO

LEAN
STOICH
RICH

τ τ

TIME

SENSOR #1 EMF

800 mV
0

2τ

$\tau_{SENSOR\#} = 0$ $\quad T_{LC} = 4\tau$

(IF $\tau_{SENSOR} > 0 \rightarrow T_{LC} = 4\tau + 4\tau_{SENSOR}$)

COMPARATOR OUTPUT

5.0 V
0

SENSOR #2 EMF

800 mV
0

τ_{LR} τ_{RL}

$2\tau - \tau_{LR} + \tau_{RL}$ $\quad 2\tau + \tau_{LR} - \tau_{RL}$

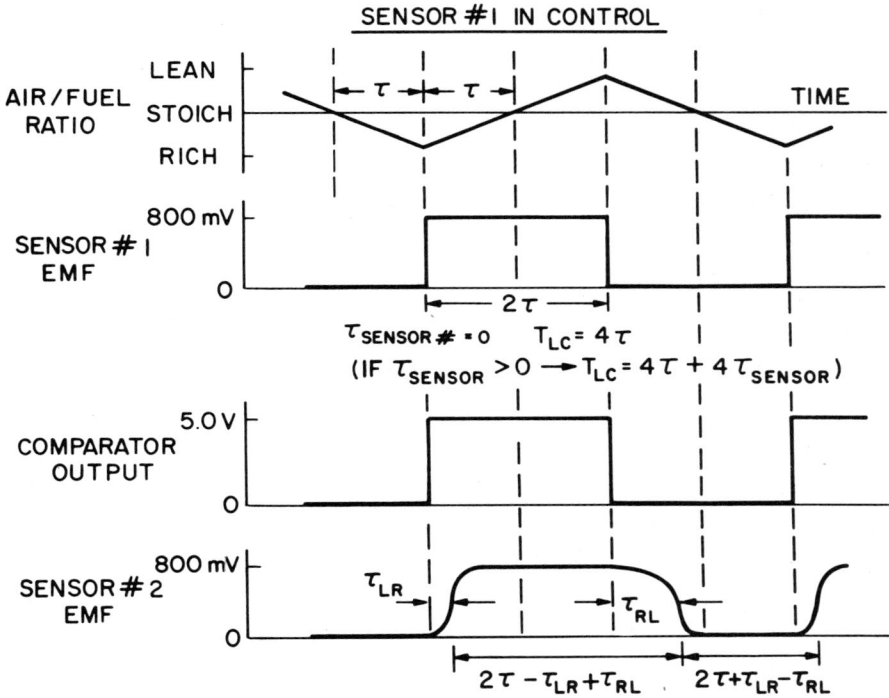

Fig. 7. Limit-cycle feedback control using sensor No. 1. See text for discussion of the waveforms.

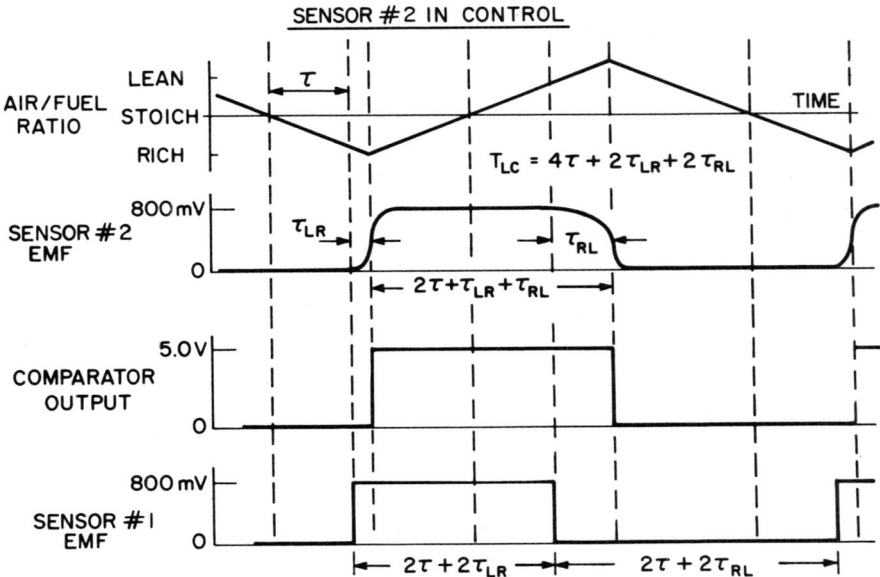

SENSOR #2 IN CONTROL

AIR/FUEL RATIO

LEAN
STOICH
RICH

τ

TIME

SENSOR #2 EMF

800 mV
0

τ_{LR} τ_{RL}

$T_{LC} = 4\tau + 2\tau_{LR} + 2\tau_{RL}$

$2\tau + \tau_{LR} + \tau_{RL}$

COMPARATOR OUTPUT

5.0 V
0

SENSOR #1 EMF

800 mV
0

$2\tau + 2\tau_{LR}$ $\quad 2\tau + 2\tau_{RL}$

Fig. 8. Limit-cycle feedback control using sensor No. 2. See text for discussion of the waveforms.

401

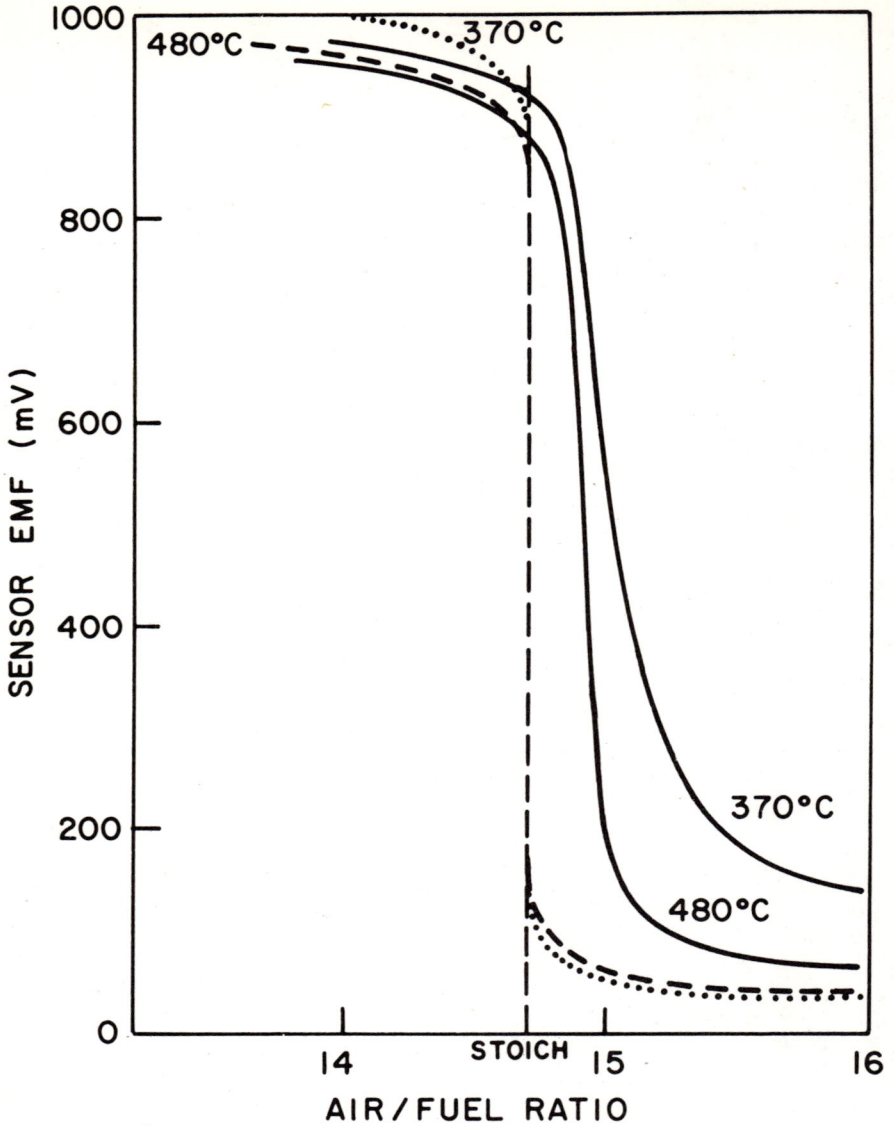

Fig. 9. Comparison between the measured emf of an automotive ZrO_2 oxygen sensor (solid lines) and the calculated emf of an ideal sensor (dotted lines) in the range 370–500 °C. The experimental data are from Ref. 26.

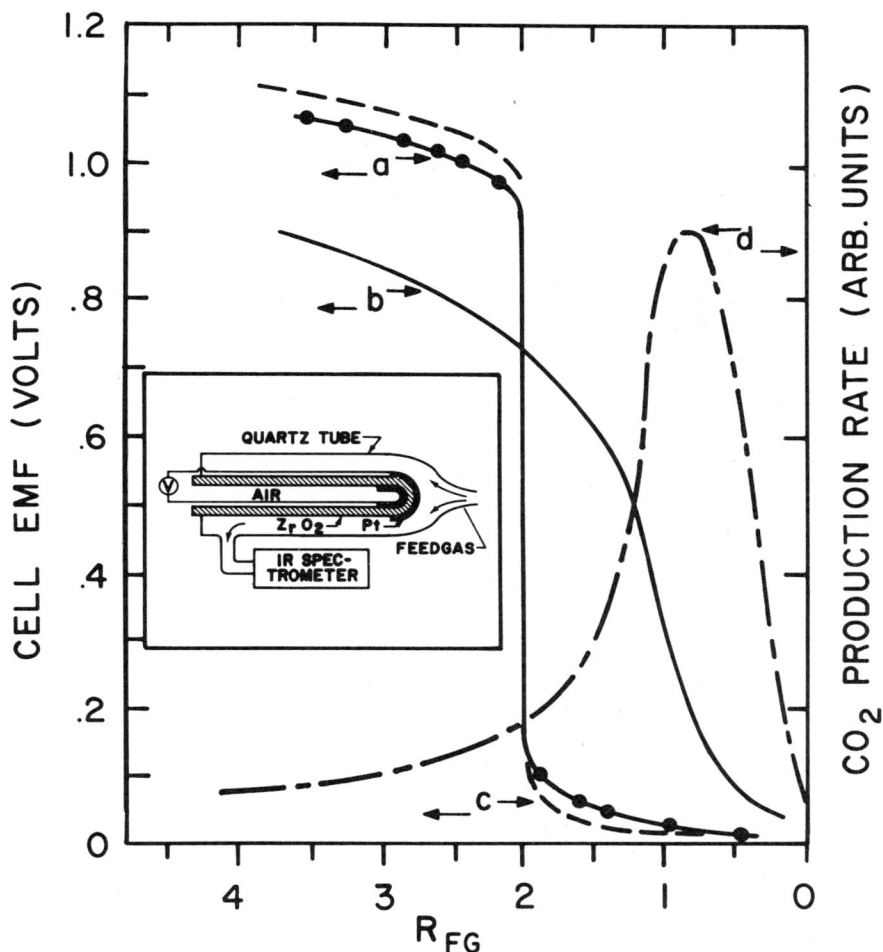

Fig. 10. Dependence of the emf of a ZrO_2 oxygen cell on R_{FG}, the ratio of P_{CO} to P_{O_2} in the feed gas. The measured values are given by the data points for 500 °C and by curve (b) for 300 °C. Curves (a) and (c) show the calculated thermodynamic equilibrium response at 500° and 350 °C, respectively. Curve (d) represents the average CO_2 production rate at 350 °C. The inset shows a schematic diagram of the apparatus.

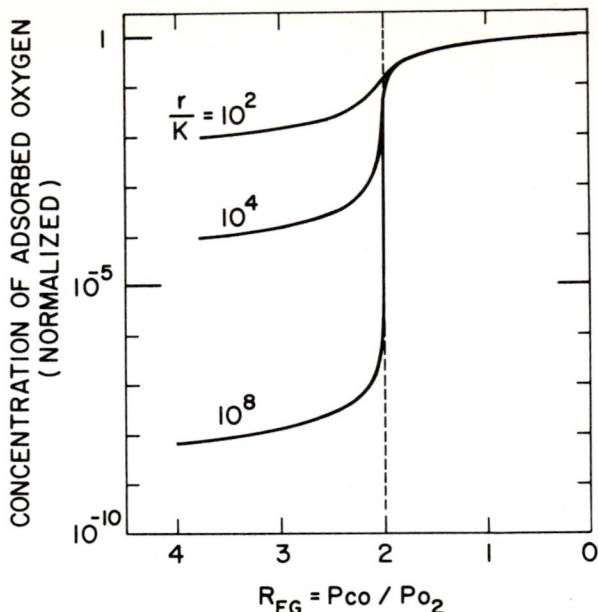

Fig. 11. Concentration of adsorbed oxygen vs R_{FG} calculated from the simple model discussed in the text for several values of the ratio r/K.

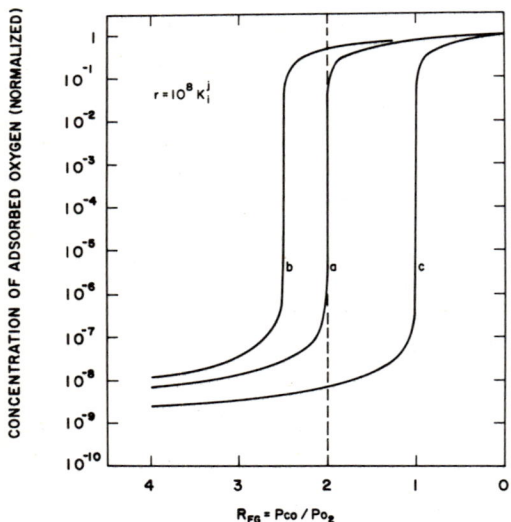

Fig. 12. Concentration of adsorbed oxygen vs R_{FG} calculated from the simple model discussed in the text for $r = 10^8 \, K_j^i$. The various curves were obtained for the following values of the parameters K_j^i. Curve (a): $K_O^a/2 = K_O^d = K_{CO}^a = K_{CO}^d$; curve (b): $K_O^a/3 = K_O^d = K_{CO}^a = K_{CO}^d$; curve (c): $K_O^a = K_O^d = K_{CO}^a = K_{CO}^d$.

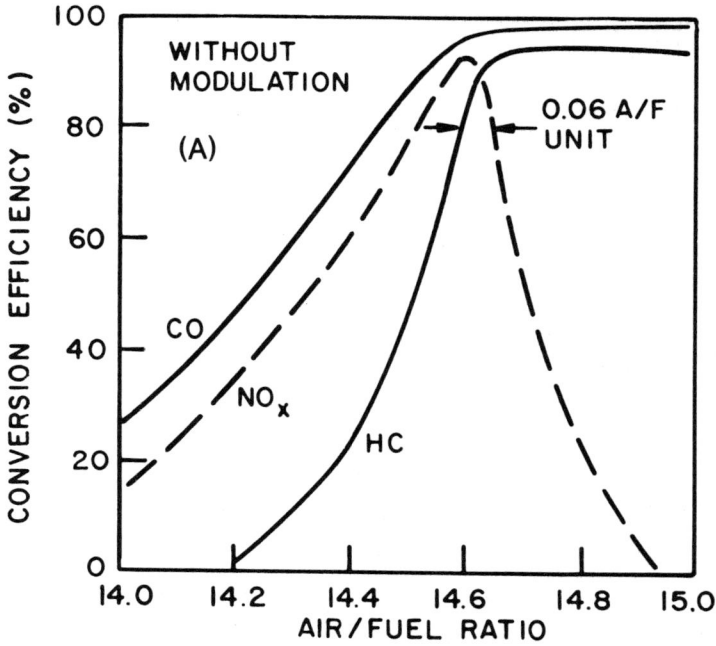

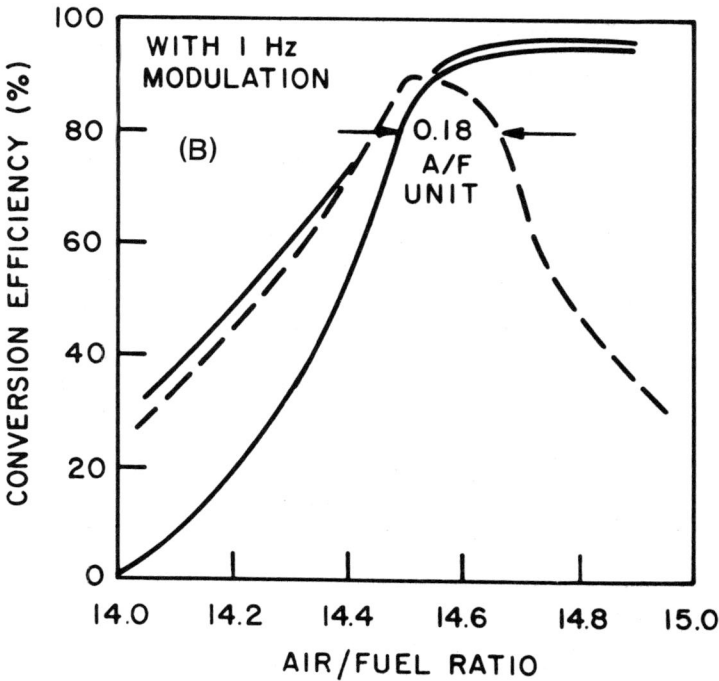

Fig. 13. Efficiency of a TWC as a function of *A/F* for steady state (*A*) and for modulated *A/F* ratio (*B*).

Zirconia fuel cells and electrolyzers

Hugh S. Isaacs

Department of Energy and Environment
Brookhaven National Laboratory
Upton, New York 11973

The thermodynamic principals and the development and application of fuel cells and steam electrolyzers using zirconia electrolytes are presented. The effects of overvoltage losses on the efficiencies and factors determining both the physical and electrochemical parameters are discussed. The rate-controlling process for the oxygen electrochemical reaction is considered to be the adsorption or desorption of oxygen on the electrode.

High temperature electrochemical cells with zirconia electrolytes[1] can be used either as electrolyzers or as fuel cells without modification in design. In contrast, low temperature phosphoric acid fuel cells[2] or alkaline water electrolyzers[3] can be operated in only one mode because of the different methods required for adding or removing the reactants or products when liquid electrolytes are used. Liquid electrolytes may also flood the porous electrodes, evaporate, change in composition, decompose, and lead to poor performances, whereas the composition of the solid zirconia electrolyte is invariant.

In an analysis of four fuel cell systems[4,5] (phosphoric acid, alkaline, molten carbonate, and zirconia), the capital and electricity costs and overall system efficiencies of fuel cell power plants were projected. The zirconia fuel cells were the most beneficial in all these parameters.

Disadvantages in the use of solid zirconia systems are the difficulties and costs of producing thin layer cell stacks, high temperature sealing, and the fact that the construction of modules representative of large-scale plants must still be demonstrated.[6,7] However, there have been major developments in the production of fuel cells[8,9] and the long-term stability of single cells has been demonstrated.[6,7] Thus, there are no major technical obstacles to the development of high temperature fuel cells or steam electrolyzers.

Development of Solid Oxide Fuel Cells

A solid oxide was first used as an electrolyte by Nernst[10] in his glower. The electrolyte was a mixture of zirconia (85%) and yttria (15%). During the passage of an electrical current, oxygen evolution occurred at the anode and oxygen reduction at the cathode. Baur and Preis[11] were the first to use a solid oxide as an electrolyte in a fuel cell operating at $> 1000\,°C$. Coke was used as the anode. The electrolyte was zirconia stabilized with yttria or magnesia. Only partial oxidation of the carbon occurred which resulted in a low fuel cell efficiency and current outputs of < 1 mA/cm². Later single cell work by Weisbart and Ruka[12] with calcia-stabilized zirconia demonstrated that current densities well above 100 mA/cm² could be achieved with oxygen and a hydrogen/water fuel.

The pioneering work for the development of solid electrolyte cell stacks has been mainly at two research laboratories—Westinghouse Research and Development Center[8,9,13-16] and Brown, Boveri, and Cie. (BBC), Heidelberg.[6,7] Various configurations have been investigated: flat plate types with the electrolyte in the form of thin disks, segmented tube designs in which short cylinders of the electrolyte are joined together with conducting seals (bell and spigot type) and thin layer electrolytes.[11-20]

Cells with tubular electrolytes, fabricated and tested at BBC,[7] had a diameter of 25 mm, height of 12 mm, and wall thickness of 0.5 mm. When operated in the fuel cell at 1000 °C, the maximum power density was 0.3 W/cm². Open circuit potentials and power densities in single cells remained constant over a period of 50 000 h. BBC has also constructed and tested multicell stacks; the number of cells in stacks varied from 20 to 100. Several types of materials have been tested for interconnection of cells in series. These include cobalt chromite, lanthanum chromite doped with strontium or nickel, and lanthanum manganite doped with strontium. A 10-cell stack was tested for > 10 000 h. Figure 1 shows the potential-current relation for this cell and factors contributing to the polarization.[7]

Accorsi and Bergman[18] fabricated cells with ytterbia-stabilized zirconia disks ≈ 0.4 mm thick. The oxygen electrodes were tin-doped indium oxide on a sublayer of urania-doped zirconia. The cathodes were sputtered nickel-zirconia cermets. The cells were operated below 1.5 V at 1 A/cm² with a 50% H_2 to H_2O mixture for > 1000 h at 900 °C. The measurements allowed separation of the voltage losses from various components and also the operation of the units as fuel cells. A major problem was separation and volatilization of the nickel cermet electrode, which increased the potential from 1.45 to 1.49 V.

Doenitz et al.[19] developed a tubular electrolyzer cell stack with cylindrical electrolyte sections, which were sintered to the interconnection materials. The electrolyte was yttria-stabilized ZrO_2; perovskite-structure oxides were used as anodes and nickel cermets as cathodes. The interconnections between the electrolyte cylinder were again perovskite of unspecified composition. The cells operated at 900 °C at 0.4 A/cm² and at 1.3 V.

Westinghouse Research and Development Center designed, fabricated, and tested solid electrolyte fuel cells during the 1960's.[13-16] A fuel cell battery, with the bell and spigot design consisting of 400 cells, was constructed and tested.[14] The open circuit potential was 200 V with a maximum power output of 102 W. At this power output, the voltage efficiency was nearly 60%. In these cells, the electrolyte resistance was a major contribution to the overpotential losses.

In order to reduce the resistance of the electrolyte, Westinghouse deposited thin layers of fuel electrode, electrolyte, air electrode, and interconnections on a porous support structure.[13,14] Five-cell, solid electrolyte fuel cell batteries were constructed. The major difficulty of cracking in these cell stacks was caused by stress resulting from thermal expansion mismatch between the Cr_2O_3 interconnection material and zirconia.

The requisite properties of the interconnection material are: (1) moderate material costs, (2) nearly invariant composition in both fuel and air atmospheres, (3) no reaction with other cell components at 1000 °C, (4) resistivity of < 50 ohm-cm and nearly 100% electronic conduction at 1000 °C

407

(5) negligible metal ion conductions, (6) thermal expansion characteristics compatible with other cell components, (7) absence of destructive phase transformations in the region from room temperature to 1000 °C, (8) low volatility of oxide components in the working atmosphere, and (9) fabricability as a thin, gas-impervious layer.[15]

Lanthanum chromite doped with magnesium and aluminum was found to have the requisite properties for the interconnection material and was incorporated in a Westinghouse 5-cell thin film solid electrolyte fuel cell.[15,16] The total thickness of the cell sandwich is ≈ 100 μm and it is made up of the following layers[14,16]: (1) A slurry of nickel oxide and yttria-stabilized zirconia is sintered on the porous support tube. The nickel oxide is reduced with hydrogen leaving a porous cermet 20 μm thick. This nickel is dissolved electrochemically to achieve a band and ring structure for multicell fabrication. (2) Electrochemical vapor deposition of a gastight interconnection layer (doped lanthanum chromite) with a thickness of ≈ 20 μm. (3) Electrochemical vapor deposition of yttria-stabilized zirconia (electrolyte layer) 20 μm thick. During this deposition, interconnections are masked. (4) Chemical vapor deposition of tin-doped indium oxide, the air electrode current collector. The porous electrolyte layer under the air electrode current collector is then impregnated with praseodymium nitrate solution which is thermally decomposed to the oxide. Praseodymium oxide impregnation reduces the air electrode overpotential. A cross section of two interconnected solid electrolyte fuel cells is schematically represented in Fig. 2.[16]

The fuel cell developments reflect directly on the development of water electrolyzers, and the cell stacks of BBC and Westinghouse have also been operated as water electrolyzers.

Thermodynamic Principles

The overall reaction:

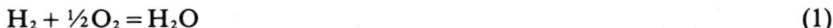

$$H_2 + \tfrac{1}{2}O_2 = H_2O \tag{1}$$

in the forward direction is representative of the overall fuel cell process and of the electrolyzer in the reverse direction. The enthalpy (ΔH^0) per mole of hydrogen for the reaction is

$$\Delta H^0 = \Delta G^0 - T\Delta S^0 \tag{2}$$

where the free energy ΔG^0 is related to the electric potential E^0 by the Gibbs equation

$$\Delta G^0 = 2FE^0 \tag{3}$$

The above equations are referred to the standard state, i.e. each gas in Eq. (1) is at 9.8066×10^4 Pa (1 atm) pressure. Under equilibrium conditions, E^0 is a measure of the electrical energy, while $T\Delta S^0$ represents the thermal energy and is the product of the absolute temperature T and the entropy ΔS^0. In the case of a fuel cell, the reaction is exothermic, and the heat and electricity are liberated. In the case of steam electrolysis, both electrical and thermal energy must be supplied to the cell for the endothermic reverse reaction. A detailed analysis of the thermodynamic consequences during steam electrolysis has been published.[21]

Equation (2) is shown diagrammatically in Fig. 3(A) under equilibrium

408

conditions; the electrical and thermal energy requirements for a particular temperature are also shown. However, when current is passed, there are voltage losses leading to the liberation of heat. In fuel cell operation, the losses reduce the amount of electrical energy taken from the electrochemical system. This is shown in Fig. 3(B) where the potential loss η (i.e. the overvoltage) increases the heat liberated by $2F\eta$ and reduces the electrical energy by an equivalent amount. During electrolysis, the losses necessitate an applied potential greater than E^0 and again the loss $2F\eta$ is liberated as heat. This heat, supplied electrically, reduces the amount to be added thermally for the endothermic reaction, as shown in Fig. 3(C). When

$$2F\eta = T\Delta S^0 \qquad (4)$$

all the energy for the reaction is supplied electrically and none thermally, and

$$E_{TN} = E^0 + 2F\eta = \Delta H^0 \qquad (5)$$

where E_{TN} is termed the thermoneutral potential.

The efficiency of the electrochemical process decreases with the magnitude of η and is a maximum when $|\eta| = 0$. In the case of fuel cell operation, the removal of heat is not a major difficulty, as this can be accomplished by increasing the flow of air normally used to supply oxygen for the cathodic reaction. During electrolysis at potentials below E_{TH}, heat must be supplied and with low η this requires repeated heating of the steam/hydrogen mixture, or cooling of the cell stack will be excessive.

The production of hydrogen from a heat source at temperatures $>900\,°C$ has been analyzed,[21] in particular for a thermonuclear reactor.[22] Other possible future heat sources are magnetohydrodynamic systems or solar furnaces. In the conversion process if only a heat source is considered, part of the heat is added directly to the electrochemical cell, but the major part must be used to produce electricity. Based on conventional electricity production methods, this has only a 40% efficiency, whereas the heat used directly approaches 100% efficiency for the production of chemical energy stored in the hydrogen. Hence, the greater the quantity of heat added directly, the more efficient is the overall conversion of thermal to chemical energy. This may be accomplished by keeping the overvoltage as low as possible.

An alternate way of increasing the ΔG^0 to $T\Delta S^0$ ratio or the overall efficiency is by increasing the temperature. An investigation with this objective was carried out by Yang and Isaacs[23] to determine the highest achievable operating temperature of a zirconia cell. The cell tested used doped lanthanum chromite as electrodes which were compatible with yttria-stabilized zirconia up to $1400\,°C$.[23]

Overvoltage Losses During Operation of Fuel Cells and Steam Electrolyzers

Physical Factors

The observed current-potential or overvoltage characteristics are shown in Fig. 1 for a 10-cell stack.[7] Included in this figure are the magnitudes of various components contributing to the loss in potential; Table I gives the percentage losses from each cause. The major loss is a result of polarization at the anode and cathode ($V_A + V_C$) because of slow electrochemical reaction kinetics. The small deviations from the theoretical (thermodynamic) potential resulted from pores in the interconnection between cells.[7] All other losses were ohmic arising from the bulk resistances of cell components and excluded all interfacial effects. The largest contribution was from the electrolyte IR_E followed by the cathode IR_C (air electrode), the interconnections IR_{CM}, and the anode IR_A (hydrogen or fuel electrode).

As discussed above, using thin-layer-supported fuel cells reduces the electrolyte thickness from ≈ 0.4 to 0.02 mm and hence its associated resistance. In other words, manufacturing techniques can reduce the electrolyte resistance a negligible amount. Changes in the physical dimensions of the electrode materials and interconnections can be calculated and optimized to reduce the resistances of these components.[24,25]

Electrochemical Factors

The electrochemical reactions and processes at the oxygen or air electrode are not well understood. Many possible rate-controlling steps have been proposed with little agreement between qualitative interpretations and less agreement between results.[26-27] There is a definitive lack of unambiguous information. The understanding of the fuel or hydrogen electrode processes is even more precarious as less information is available. Reducing the electrolyte resistance, as mentioned above, by more than an order of magnitude will further emphasize the polarization losses at the electrode. From Table I polarization would then account for $> 50\%$ of the voltage losses.

The electrochemical reactions during current flow may be limited by processes taking place in the gas, the solid phases, or at their interfaces. If it were known which process is limiting, then it would be possible to design electrodes to maximize the rates by changing the porosity and tortuosity of the electrode, its composition, surface purity, area of contact with the electrolyte, particle size, size distribution, or distance between electrode particles.

An example of the polarization behavior for a single platinum ball contact electrode measured in air at 1000 °C is shown in Fig. 4. Three electrodes were used in these measurements, the single-contact electrode under test, a counter electrode to carry current to or from this electrode, and a reference which carried no current and against which the potential of the test electrode was measured. At positive potentials anodic currents were a result of oxygen evolution and at negative potentials cathodic currents resulted from oxygen reduction. The full line shows the measured variations in current as the potential was changed at a rate of 5 mV/s. These results incorporated the effect of electrolyte resistance. The points in Fig. 4 show the variations of the interfacial potential with current after correcting for this electrolyte ohmic polarization. The electrolyte resistance was taken as 478 ohms. The current varied logarithmically at positive potentials; at negative potentials the electrode polarized more rapidly and also showed larger hysteresis with greater cathodic currents when the potential was increased. It should be pointed out that these results were obtained without contamination by a volatile bismuth oxide from any fluxed paste. Bismuth is often added to platinum pastes and has a very marked effect on the electrode behavior.[29]

The overall oxygen electrochemical reaction

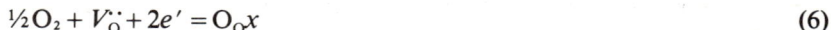

$$\tfrac{1}{2}O_2 + V_O^{\cdot\cdot} + 2e' = O_O x \qquad (6)$$

requires oxygen from the gaseous phase, oxygen vacancies from the electrolyte, and electrons from the electrode. All these species may interact either simultaneously or sequentially to form an oxygen ion in the oxide, but the

410

former is less likely.

A simultaneous reaction would require that the reactants meet at the three-phase boundary line (TPBL) between gas, electrode, and electrolyte. The reaction rate then depends on the supply of oxygen. The charged species give ohmic polarization but do not control the rate, which depends on the arrival of oxygen at the boundary. Some dimension of this boundary is required even for the simplest calculation of the rate of oxygen arrival; a minimum thickness would be 1 at. diam. ($\approx 4.10^{-8}$ cm). The equivalent current for a surface in oxygen at 9.8×10^4 Pa pressure, assuming that each molecule impacting reacts at the surface, is $\approx 10^4$ A/cm^2.[26] The three-phase boundary current is then calculated to be $\approx 10^{-4}$ A/cm length of TPBL in air. This value can be compared to the experimental value obtained from Fig. 4 after determining the TPBL. The resistance of the contact, R, measured using ac techniques,[30] can be used to determine the radius of contact, r, and hence the length of the TPBL. Assuming the contact is circular, then

$$R = \varrho/4r \tag{7}$$

where ϱ is the resistivity of the electrolyte. The TPBL is then equal to the circumference of the contact, i.e. ≈ 0.1 cm. The specific current at -0.33 V is then $\approx 10^{-2}$ A/cm. This value is approximately two orders of magnitude greater than the calculated value and shows that the active surface must be at least 100 at. diam. The calculation is highly simplified, but it clearly demonstrates an area is required for the reaction to take place with part of it removed from the TPBL. Therefore the simultaneous combination of the three reacting species is highly unlikely. The extended area of gas reacting at the surface also necessitates that the reaction takes place sequentially, and areas of both the electrode and the electrolyte must be involved.

The rate-controlling step may be either a bulk or surface reaction or process. The schematic in Fig. 5 shows some possible controlling steps[26]: (1) gas diffusion control external to the electrode or within pores, (2) adsorption and dissociation on the electrode surface of the electrolyte, (3) diffusion of dissolved oxygen in the electrode or electrolyte, (4) charge transfer, (5) diffusion of electron holes in the electrolyte, and (6) diffusion of adsorbed oxygen on the electrode, electrolyte to the TPBL, or into the electrode/electrolyte interface.

Generally, diffusion processes are unlikely, although for very thick or low porosity electrodes, this process may be rate controlling. Calculated rates for gas diffusion are ≈ 25 A/cm^2.[21] However, the current densities for most electrodes are usually lower than these values, even though the electrodes have relatively high porosity. The absence of diffusion processes is also demonstrated by the results in Fig. 4, as the cathodic currents are larger on increasing the potential. If diffusion were the dominant process, then any concentration gradients generally would decrease with time and lead to lower currents rather than the larger currents observed on increasing the potential.

In addition, impedance measurements[32-34] do not show diffusion-controlled (Warburg impedance) characteristics with solid electrodes on zirconia, although they are clearly observed with liquid electrodes[34] and in many other diffusion-controlled electrochemical processes.[35] The absence of any Warburg impedance eliminates the diffusion of oxygen in the gas, pores, or solid phase, holes in the electrolyte, or adsorbed oxygen on the electrode or electrolyte as rate-controlling mechanisms.

The interfacial impedance measurements in air or oxygen at high temperatures show high capacitance of hundreds to thousands of microfarads which can only be accounted for by chemical and not physical models. The most consistent interpretation of these curves, and their dependence on potential[36] or oxygen pressure,[37] is the presence of adsorbed oxygen.[38] Olmer and Isaacs[29,30] have shown that the adsorption of oxygen on platinum accounted for the observed variations of the impedance with potential and changes in the number of oxygen adsorption sites. Kleitz *et al.*[38] considered the adsorption to be present at the zirconia/electrode interface. Fabry and Kleitz[37] have shown both the metal and the electrolyte influence the temperature dependence of single contact electrodes and consider both to be important factors. They measured the pressure dependence of the resistance of the single contacts and analyzed the temperature changes of this dependence. The measurements depend on both the interfacial or polarization resistance and electrolyte resistances. With changes in temperature and oxygen partial pressure, the size of the microsystem can change and vary the electrolyte constriction resistance.[20,26] The change in the microsystem dimension could therefore lead to an apparent dependence of the rate-controlling process on the electrolyte conductivity, whereas the reaction may only be controlled by adsorption of oxygen on the electrode material. The polarization resistance varied with the metal electrodes tested and other work by the authors indicated that the rates increased as $Pd < Rh < Au < Pt < Ag$.[39]

The comparison of the relative behavior of electrode materials using single-point contacts is difficult because of the unknown size of contact with the electrolyte. Isaacs and Olmer[30] also studied single-contact electrodes and used the initial voltage response to square current waves to determine the electrolyte resistance, as described above. The initial response is effectively a very high frequency measurement and the impedance of the entire electrode/electrolyte interface is low compared to the electrolyte path. In contrast, only the dc current flows to the TPBL, as negligible current flows to the electrode/electrolyte interface. The comparison was made on the basis of currents per unit length of TPBL at a negative interfacial potential of -100 mV in air at 1000 °C. These specific currents were determined for noble metal and oxide electrodes and varied by approximately three orders of magnitude depending on the electrode material. The highest specific current was with rhodium (> 20 mA/cm) and the lowest with gold (≈ 0.03 mA/cm). Of the oxides tested, $La_{0.5}Sr_{0.5}FeO_3$ gave the highest specific currents (14 mA/cm). The order of increasing currents was $Au < In_2O_3 < NiO < La_{0.95}Mg_{0.05}Cr_{0.85}Al_{0.15}O_3 < Co_3O_4 < Pt < Cr_2O_3 < Pd < PrCoO_3 < La_{0.8}Ba_{0.2}CoO_3 < La_{0.5}Sr_{0.5}FeO_3 < Rh$. These measurements clearly showed that the electrode material played a dominant role in the electrochemical reaction rates.

These reaction rates were dependent on the surface conditions and not on the bulk, e.g. diffusion of oxygen within the electrode. This was shown by doping the surface of platinum with small amounts of gold, praseodymium oxide, or bismuth oxide, which markedly changed the specific current but could not alter bulk diffusion rates.

The rate-controlling step for the oxygen reaction therefore depends on the surface of the electrode material in close proximity to the electrolyte and does not involve diffusion control of the adsorbed oxygen. The slow reaction is probably

$$O_2 \rightleftharpoons 2O_{ads} \tag{8}$$

where the O_{ads} determines the electrode potential by the fast electrochemical or charge transfer reaction

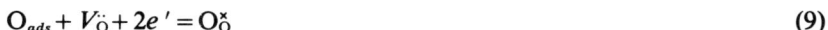

$$O_{ads} + V_{\ddot{O}} + 2e' = O_O^x \qquad (9)$$

The rate of the reaction depends on the number of sites available for adsorption and their occupancy or coverage. This model, as mentioned above, is consistent with the impedance measurements, and also offers an explanation for the current hysteresis in Fig. 5. The higher currents on increasing the potential result from an increase in the number of sites available for oxygen adsorption which are released after reduction at negative potentials.

The model of adsorption on the metal phase has also been used to account for the behavior of oxygen meters. For example, Vayenas and Satsburg[40] found that SO_3 chemisorbed on noble metals, at low temperatures, dominated the oxygen activity of the adsorbed oxygen and poisoned the equilibration with gaseous oxygen. This behavior dominates at low temperatures, and adsorption of impurities could account for the poor performance of oxygen meters generally observed below 500 °C.

Acknowledgments

Work was performed under the auspices of the United States Department of Energy, contract No. DE-ACO2-76CH00016.

References

[1]Workshop High Temperature Solid Electrolyte Fuel Cells. Edited by H. S. Isaacs, S. Srinivasan, and I. L. Harry. Brookhaven National Laboratory, Upton, New York, 1978 (BNL 50756).

[2]From Electrocatalysis to Fuel Cells. Edited by G. Sandstede. University of Washington Press, Seattle, 1972.

[3]Industrial Water Electrolysis. Edited by S. Srinivasan, F. J. Salzano, A. R. Landgrebe. The Electrochemical Society, Princeton, New Jersey, 1978.

[4]C. J. Warde, R. J. Ruka, and A. O. Isenberg; p. 67 in Fuel Cells for Public Utility and Industrial Power. Edited by R. Noyes. Noyes Data Corporation, 1977.

[5]M. Warshay; p. 89 in Ref. 1.

[6]F. J. Rohr; p. 122 in Ref. 1.

[7]F. J. Rohr; p. 431 in Solid Electrolytes. Edited by P. Hagenmuller and W. Van Gool. Academic Press, New York, 1978.

[8]A. O. Isenberg; p. 572 in Electrode Materials and Processes for Energy Conversion and Storage. Edited by J. D. E. McIntyre, S. Srinivasan, and F. G. Will. The Electrochemical Society, Princeton, New Jersey, 1977.

[9]A. O. Isenberg, National Fuel Cell Seminar Abstracts, San Diego, California, 1980; p. 135.

[10]W. Nernst, Z. Elektrochem., 6, 41 (1900).

[11]E. Baur and H. Preis, ibid., 43, 727 (1937).

[12]J. Weisbart and R. J. Ruka; p. 37 in Fuel Cells, Vol. 2. Edited by G. J. Young. Reinhold, New York, 1963.

[13]E. F. Sverdrup, Project Fuel Cell, Final Report, No. 57, Office of Coal Research, 1970.

[14]D. H. Archer et al.; pp. 332, 343 in Fuel Cell Systems Advances in Chemistry Series. Edited by G. J. Young and H. R. Linden. American Chemical Society, Washington, D.C., 1965.

[15]R. J. Ruka; p. 56 in Ref. 3.

[16]A. O. Isenberg; abstract 141 in Extended Abstracts, Fall Meeting of the Electrochemical Society, Los Angeles, Calif. The Electrochemical Society, Princeton, New Jersey, Vol. 79-2, 1979.

[17]W. Baukal and W. Kuhn, Int. Power Sources, 1, 91 (1976/77).

[18]R. Accorsi and E. Bergman, J. Electrochem. Soc., 127, 804 (1980).

[19]W. Doenitz, R. Schmitberger, and E. Steinheil; p. 266 in Ref. 3.

[20]H. S. Spacil and C. S. Tedmon, J. Electrochem. Soc., 116, 1170 (1969).

[21]H. S. Spacil and C. S. Tedmon, ibid., pp. 1618, 1627.

[22]H. S. Isaacs, J. A. Fillo, V. Dang, J. R. Powell, M. Steinberg, F. J. Salzano, and R. Benenati; p. 249 in Ref. 3.

[23]C. Y. Yang and H. S. Isaacs; unpublished work.

[24]E. F. Sverdrup, C. J. Warde, and R. L. Eback, *Energy Convers.*, **13**, 129 (1973).
[25]T. Kudo and H. Obayashi, *ibid.*, **15**, 121 (1976).
[26]S. Pizzini; p. 461 in Fast Ion Transport in Solids. Edited by W. Van Gool. North Holland, Amsterdam, 1973.
[27]R. E. W. Casselton, *J. Appl. Electrochem.*, **4**, 25 (1974).
[28]B. C. H. Steel; p. 367 in Electrode Processes in Solid State Ionics. Edited by M. Kleitz and J. Dupuy. D. Reidel Publishing Co., Dordrecht, Holland, 1976.
[29]L. J. Olmer and H. S. Isaacs; unpublished work.
[30]H. S. Isaacs and L. J. Olmer; unpublished work.
[31]P. Fabry, M. Kleitz, and C. Deportes, *J. Solid State Chem.*, **5**, 1 (1972).
[32]J. E. Bauerle, *J. Phys. Chem. Solids,* **30**, 2657 (1969).
[33]E. J. L. Schouler and M. Kleitz, *J. Electroanal. Chem. Interfacial Electrochem.,* **64**, 135 (1975).
[34]E. J. L. Schouler; Thesis, Grenoble, France (1979).
[35]J. E. B. Randles and K. W. Somerton, *Trans. Faraday Soc.*, **48**, 937, 951 (1952).
[36]L. J. Olmer and H. S. Isaacs, *Extended Abstracts,* Fall Meeting of the Electrochemical Society, Hollywood, Fla., October 1980. The Electrochemical Society, Princeton, New Jersey, Vol. 80-2, 1980.
[37]P. Fabry and M. Kleitz, *J. Electroanal. Chem. Interfacial Electrochem.*, **57**, 167 (1974).
[38]M. Kleitz, P. Fabry, and E. J. L. Schouler; p. 1 in Ref. 28.
[39]M. Kleitz, P. Fabry, and E. Schouler; p. 451 in Ref. 26.
[40]C. G. Vayenas and H. M. Satsburg, *J. Catal.*, **57**, 296 (1979).

---

*Work was performed under the auspices of the United States Department of Energy, contract No. DE-ACO2-76CH00016.

Table I. Overvoltage losses in a solid oxide fuel cell*

Factors causing loss	Loss (%)
Polarization anode + cathode	39
Electrolyte resistance	26
Cathode resistance	16
Interconnection resistance	10
Open circuit deviation from theoretical	6
Anode resistance	3
Total	100

*From Ref. 7 and Fig. 1.

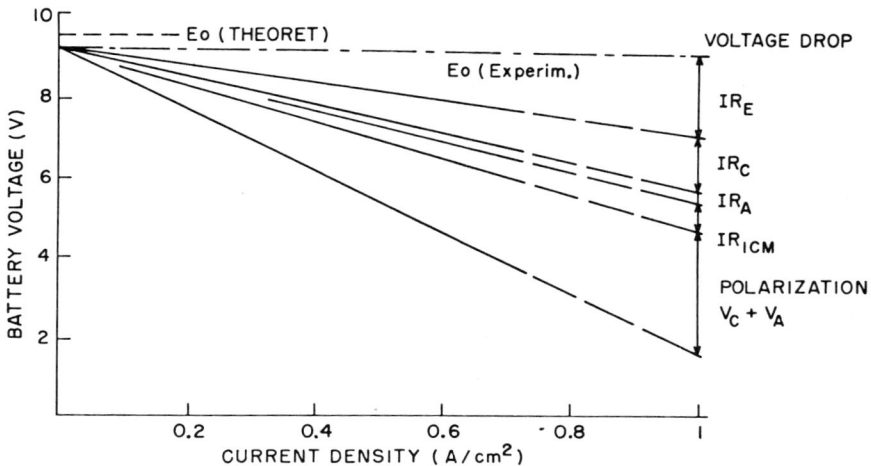

Fig. 1. Resistance and polarization losses of a 10-cell module as a function of current density at 1000 °C. Fuel, $H_2 + 31\%$ H_2O; oxidant, air (Ref. 7).

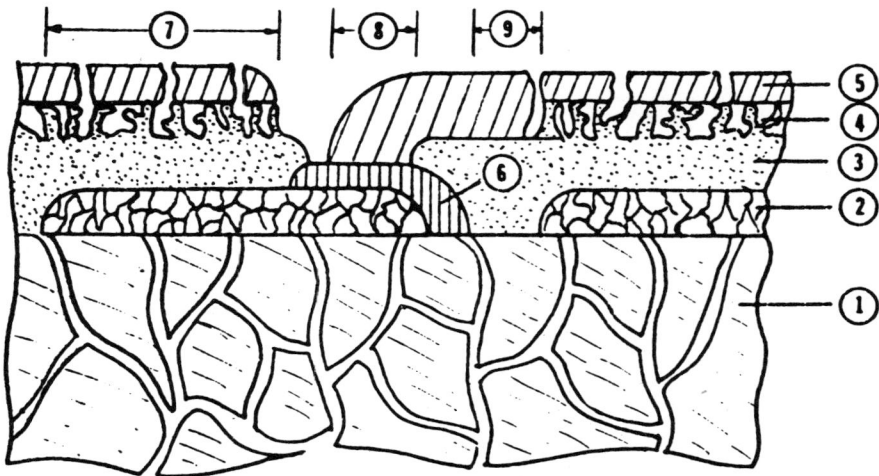

Fig. 2. Cross section through two interconnected solid oxide fuel cells deposited on a porous tube. (1) Porous support, (2) cermet fuel electrode, (3) electrolyte, (4) active fuel electrode, (5) air electrode current collector, (6) interconnection layer, (7) active cell area, (8) interconnection area, and (9) intercell area (Ref. 16).

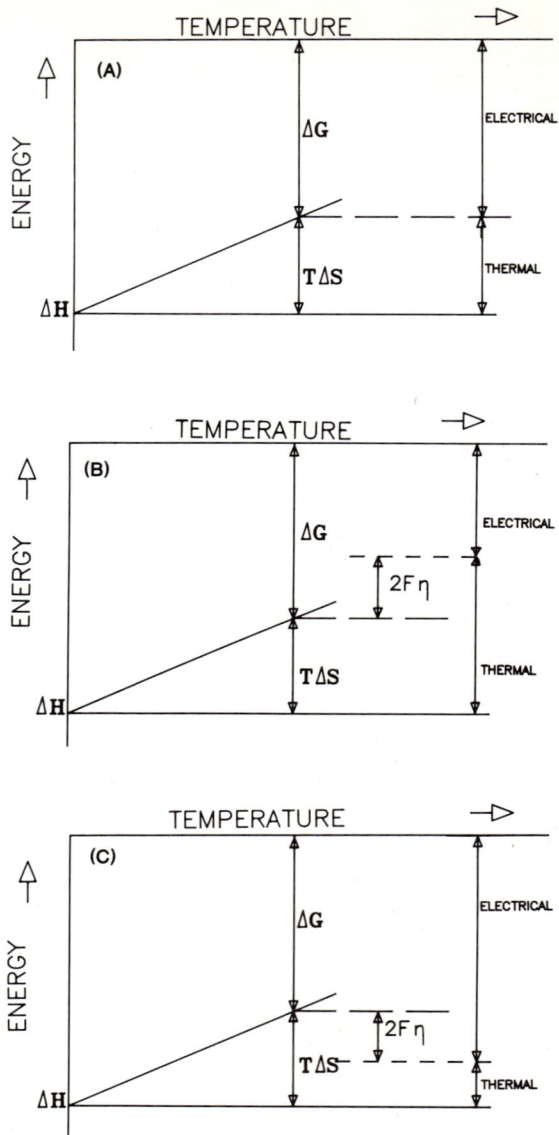

Fig. 3. (*A*) Schematic of the free energy of formation of water as a function of temperature, showing the thermal and electrical contributions to the heat of formation. (*B*) Schematic of the free energy of formation of water, showing the effect of overvoltage (η) on the thermal and electrical contributions during fuel cell operation. (*C*) Schematic of the free energy of formation of water, showing the effect of overvoltage (η) on the thermal and electrical contributions during water electrolysis.

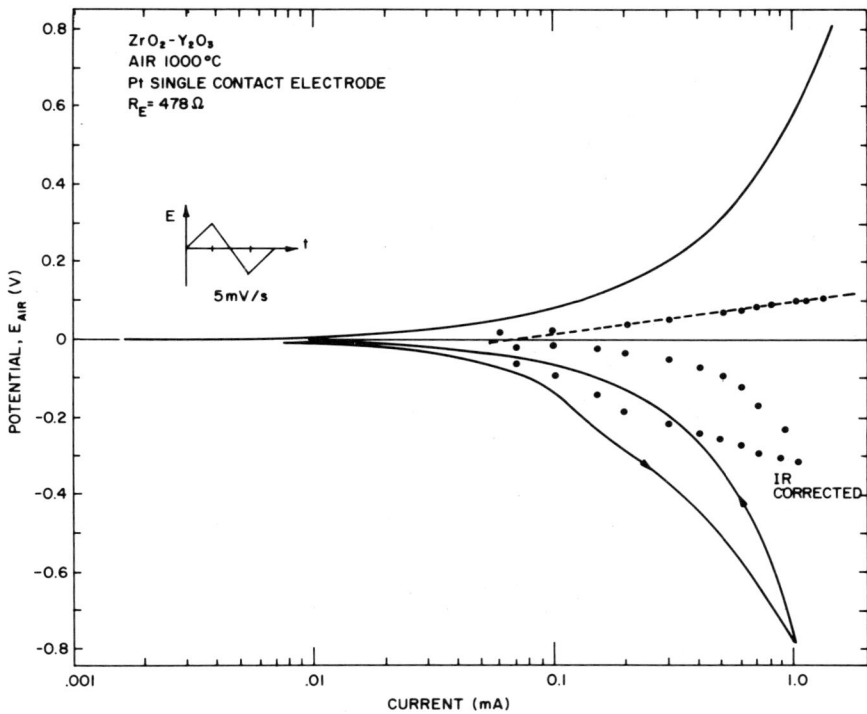

Fig. 4. The variation of current with potential for a single-contact platinum electrode on zirconia in air at 1000 °C measured at 5 mV/s. (−−) As measured; (•) after connecting for electrolyte resistance.

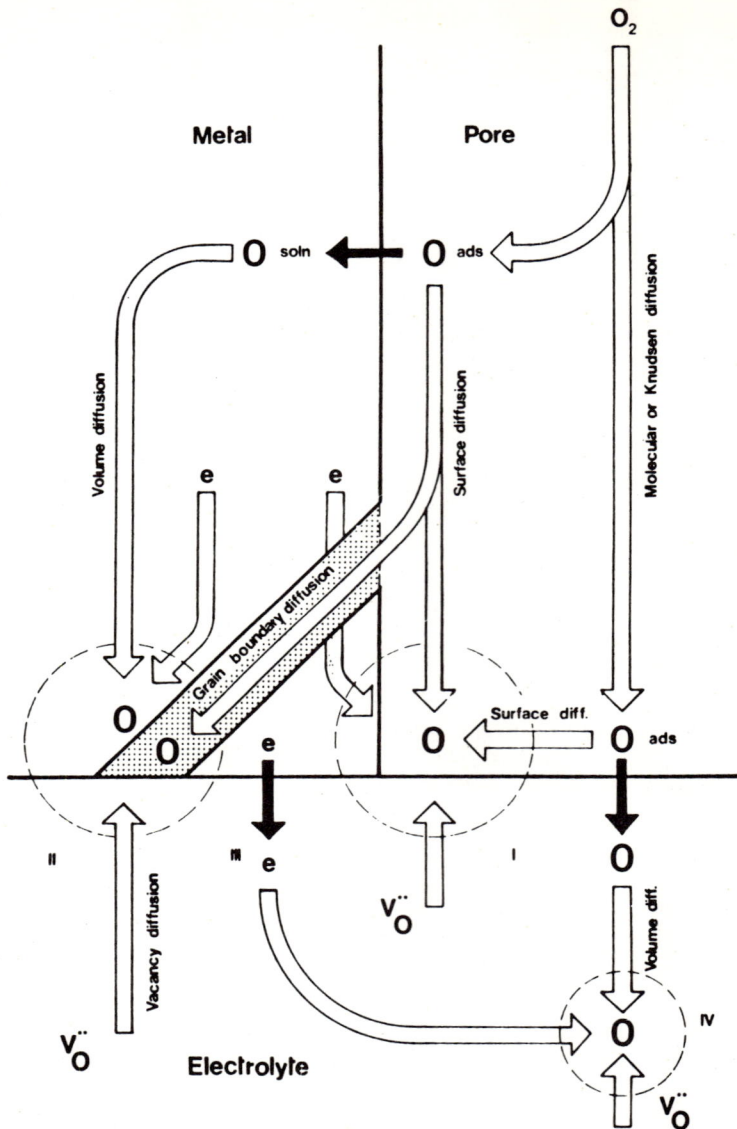

Fig. 5. Schematic of possible steps for oxygen reduction at a metal electrode on a zirconia electrolyte (Ref. 26).

Zirconia-, hafnia-, and thoria-based electrolytes for oxygen control devices in metallurgical processes

D. JANKE

Max-Planck-Institut für Eisenforschung GmbH
4000 Düsseldorf
Federal Republic of Germany

Oxygen probes with solid electrolytes are increasingly used to control melting, refining, and casting in steel and nonferrous metal making as well as exhaust gases in combustion processes. A survey is made of fundamentals of the electrochemical methods, the design of various types of oxygen sensors used in molten metals or gas atmospheres, calibration of the probes, accuracy and sources of error of the electrochemical method, and aspects of discontinuous and continuous oxygen measurements. Several properties of the ceramic electrolytes based on zirconia, hafnia, or thoria are outlined with respect to their application under special conditions such as high temperatures between 1000 and 1700 °C, low oxygen potentials down to $\approx 10^{-15}$ Pa, and rapid temperature changes causing high thermal stress in the ceramic parts of the probes. For the present, stabilized zirconia is exclusively applied in oxygen probes for industrial purposes, but it has been shown experimentally that doped hafnia or thoria exhibits higher thermochemical stability and lower partial electronic conductivity which favor these two materials for measuring extremely low oxygen potentials.

Oxygen is of great importance in the manufacture of liquid metals. The various metallurgical steps of production and refining of molten metals can essentially be considered as reduction and oxidation processes. Oxygen is present in metals both dissolved and precipitated as gaseous, liquid, or solid oxide. The dissolved oxygen characterizes the degree of oxidation of the liquid metal and has to be considered as an important quantity to control metal-making processes. This quantity may be expressed as the partial pressure or the chemical activity of the dissolved oxygen in the metal according to the reaction of dissolution

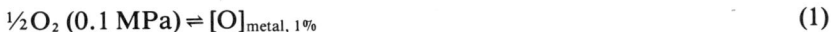

$$\tfrac{1}{2}O_2 \,(0.1\ \mathrm{MPa}) \rightleftharpoons [O]_{\mathrm{metal},\,1\%} \tag{1}$$

$$\ln K_O = \ln \frac{a_O}{p_{O_2}^{1/2}} = -\frac{\Delta G_O^\circ}{RT} = -\frac{\Delta H_O^\circ}{RT} + \frac{\Delta S_O^\circ}{T} \tag{2}$$

$$\ln p_{O_2} = 2 \ln a_O + \frac{2\Delta H_O^\circ}{RT} - \frac{2\Delta S_O^\circ}{R} \tag{3}$$

$$\ln a_O = \tfrac{1}{2} \ln p_{O_2} - \frac{\Delta H_O^\circ}{RT} + \frac{\Delta S_O^\circ}{R} \tag{4}$$

Equation (4) is graphically represented in Fig. 1 for liquid iron at temperatures between 1500 and 1700 °C. In Eqs. (1) to (4) Henry's law is assumed to be obeyed for the dilute solution range and the 1% solution is taken as the standard state which allows the oxygen activity to be set equal to the weight percent of chemically analyzed oxygen. In steelmaking processes a wide range of oxygen activities must be considered. Oxygen activities of oxidized steel in the BOF converter, for instance, range from 0.02 to 0.1. On the other hand, oxygen activities as low as 0.0002 are established in fully deoxidized steel in refining and casting ladles.

Remarkable progress has been achieved in controlling oxidation and deoxidation of molten metals by electrochemical methods. Their application is particularly in progress in the field of steelmaking throughout the world. An annual consumption of 200 000 one-reading probes for the determination of oxygen in liquid steel is roughly estimated for the present. The use of these probes is mainly concentrated in West European and Japanese steel plants.

On the other hand, electrochemical probes monitoring oxygen partial pressures of gas phases are receiving increasing attention. They are mainly used to control the waste gas composition in the refining process of stainless steel melts and the reducing atmosphere in solid metal annealing processes.

In this report, a survey is given of: (1) the fundamentals of the electrochemical method, (2) the design of various types of probes used in molten metals or gas atmospheres, (3) the calibration of the probes, (4) the accuracy and sources of error of the electrochemical method, and (5) aspects of discontinuous and continuous oxygen measurements.

Properties of the oxide ceramic electrolytes predominantly based on zirconia are outlined with respect to their application under special conditions such as: (1) temperatures between 1000 and 1700 °C, (2) low oxygen potentials down to $\approx 10^{-15}$ Pa, and (3) rapid temperature changes causing high thermal stress in the ceramic parts of the probes.

Fundamentals of the Electrochemical Method

Devices for the detection of oxygen in liquid metals or gases at elevated temperatures are based on oxygen concentration cells of the type

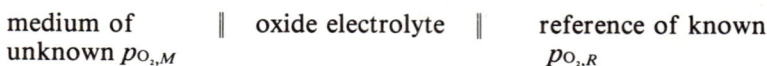

medium of	‖ oxide electrolyte ‖	reference of known
unknown $p_{O_2,M}$		$p_{O_2,R}$

The measured emf as a function of the ratio $p_{O_2,R}/p_{O_2,M}$ can be written as

$$E = \frac{RT}{4F} \int_{p_{O2,M}}^{p_{O2,R}} t_{ion}\, d \ln p_{O_2} \qquad \text{(C. Wagner)} \qquad (5)$$

in the case of a mixed ionic and electronic conductivity in the electrolyte or as

$$E = \frac{RT}{4F} \ln (p_{O2,R}/p_{O2,M}) \qquad \text{(W. Nernst)} \qquad (6)$$

in the case of a predominant ionic conductivity in the electrolyte.

For measurements in liquid metals the oxygen activity a_O is substituted for the unknown $p_{O_2,M}$ with regard to Eqs. (1)–(4) as follows:

(a) Ionic and electronic conduction in the electrolyte

$$E = \frac{RT}{F} \ln \frac{p_{e'}{}^{1/4} + p_{O_2,R}{}^{1/4}}{p_{e'}{}^{1/4} + p_{O_2,M}{}^{1/4}} = \frac{RT}{F} \ln \frac{p_{e'}{}^{1/4} + p_{O_2,R}}{p_{e'}{}^{1/4} + (a_O/K_O)^{1/2}} \tag{7}$$

(H. Schmalzried)

$$a_O = \exp\left(-\frac{\Delta G_O^\circ}{RT}\right)\left[(p_{e'}{}^{1/4} + p_{O_2,R}{}^{1/4}) \exp \frac{EF}{RT} - p_{e'}{}^{1/4}\right]^2 \tag{8}$$

(b) Predominant ionic conduction in the electrolyte

$$E = \frac{RT}{4F} \ln \frac{p_{O_2,R}}{p_{O_2,M}} = \frac{1}{2F}\left[\frac{1}{2}RT \ln p_{O_2,R} - \Delta G_O^\circ\right] - \frac{RT}{2F} \ln a_O \tag{9}$$

$$a_O = \exp\left[-\frac{\Delta G_O^\circ + 2FE}{RT}\right] p_{O_2,R}^{1/2}$$

For measurements in gases, relationships can be applied as written in Eq. (7) or (9).

The parameter $p_{e'}$, in the foregoing equation represents the oxygen partial pressure $p_{O_2,M}$ where ionic and electronic conductivity are identical, which means that the transference numbers are $t_{ion} = t_{e'} = 0.50$. It has been introduced and the mathematical expressions have been developed by H. Schmalzried.[1,2] For further detailed information, see Refs. 3–8.

Properties of Oxide Ceramic Electrolytes for Metallurgical Purposes
Composition and Chemical Stability

For metallurgical applications the following zirconia-based electrolyte materials are normally used:

(a) Partially stabilized ZrO_2 with 2.5 to 3.0% MgO (7 to 8.5 mol%) which is the preferred composition for commercial one-reading probes in metallic melts.

(b) Fully stabilized ZrO_2 with 5 to 7.5% CaO + MgO (10 to 25 mol%) which is mainly used for commercial gas analyzers.

There are other materials such as stabilized HfO_2 with 5% CaO (16.5 mol%) and yttria-doped ThO_2 with 5–10% Y_2O_3 (6 to 11 mol%) which have been studied under laboratory conditions and seem to be promising solid electrolytes, especially for the detection of extremely low oxygen potentials.

The oxides of ZrO_2, HfO_2, and ThO_2 exhibit high chemical stability. This is evident from the free energies of formation of the three oxides compiled in Table I.[9–12]

Ionic and Electronic Conductivity

In the solid solutions of ZrO_2-MgO, ZrO_2-CaO, HfO_2-CaO, and ThO_2-Y_2O_3 the ionic conductivity is predominant within certain limits of the oxygen partial pressure. For the study of metallurgical reactions, the lower limit is of particular interest in most cases. The concept of the parameter $p_{e'}$[1,2] has proved to be convenient and useful to consider contributions of electronic conductivity at extremely low oxygen partial pressures. Parameters $p_{e'}$ have been determined experimentally at elevated temperatures, using a polarization technique. Table II shows that hafnia- and thoria-based electrolytes exhibit a remarkably lower partial electronic conductivity than zirconia-based electrolytes.[13-15] Using these parameters $p_{e'}$, calibration functions can be calculated for oxygen concentration cells according to Eqs. (7) and (8). In Fig. 2 functions of this type are presented for the cell

$$Cr + Cr_2O_3 \| \text{solid oxide electrolyte} \| \text{iron melt}$$

at 1600 °C using different electrolyte compositions.

Structure and Thermal Stress

For the high-temperature use in the field of metal making, ceramic parts of ZrO_2, HfO_2, or ThO_2 are desired with special requirements such as high density and sufficient thermal shock resistance. When gaseous media are involved, the porosity of the electrolyte material should be as low as possible. The apparent porosity of the ZrO_2 electrolyte wall, for example, should be < 1% when oxygen measurements are made in liquid steel with a platinum-air reference system. But a somewhat higher porosity of the electrolyte can be tolerated when a solid metal-metal oxide reference is applied instead.

Thermal shock is a particular problem for the use of oxide electrolyte devices in metallurgical processes. At a fixed size and shape of the ceramic body, thermal shock resistance is determined by the conditions of thermal shock (temperature gradient and heat transfer coefficient) and the thermo-mechanical properties of the electrolyte material (thermal expansion coefficient, thermal conductivity, tensile strength, and modulus of elasticity).

ZrO_2-, HfO_2-, or ThO_2-based oxide materials exhibit poor thermal shock resistance, which is mainly due to high thermal expansivity and low thermal conductivity. In partially stabilized ZrO_2 the structure can be strengthened in a controlled manufacturing process making use of the different thermal expansivities of the monoclinic and cubic phase and the volume change in the monoclinic-tetragonal transformation. Partially stabilized ZrO_2 is a preferred electrolyte material for small tubes closed at one end which are used for rapid immersion measurements in liquid steel. Typical structures of partially stabilized ZrO_2 with 2.4% MgO are shown in Fig. 3(A) (200:1, etched, monoclinic and cubic phase, pores) and Fig. 3(B) (1000:1, fine-grained monoclinic precipitations within the cubic phase).

Reference Systems

In oxygen sensors for metallurgical applications, solid metal-metal oxide or gaseous references are used. Problems may arise in the high-temperature range from 1000 to 1700 °C through polarization effects at the electrode/electrolyte interfaces caused by oxygen ion transport across the electrolyte wall. Oxygen ion transport is favored by a high partial electronic conductivity of the electrolyte, a high gradient of the oxygen potential in the cell, and by a high temperature.

Solid Metal-Metal Oxide References

Powder mixtures of metal and metal oxide are preferred reference systems for rapid immersion sensors to determine oxygen in molten steel. It is important to know accurate data on the free energies of formation and their temperature dependence according to the basic chemical reaction

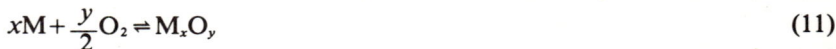

$$x\text{M} + \frac{y}{2}\text{O}_2 \rightleftharpoons \text{M}_x\text{O}_y \qquad (11)$$

422

Selected $\Delta G°$ values of MoO_2, Cr_2O_3, NbO, and Ta_2O_5,[9] applied for emf measurements in molten steel, and of Cu_2O, NiO, CoO, and $Fe_{1-x}O$,[16] applied for emf measurements in other liquid metals, are listed in Table III.

Gaseous References

Gaseous references are commonly used in solid electrolyte gas analyzers. They also seem to be more effective for continuous emf measurements in liquid metals.

Inert gas-oxygen mixtures, e.g. air, or reactive gases may serve as a gaseous reference in oxygen concentration cells. In N_2-O_2 or Ar-O_2 gases the oxygen partial pressure is simply determined by the mixing ratio. In CO-CO_2 or H_2-H_2O, the p_{O_2} is controlled by the chemical equilibrium constants at a given temperature

$$p_{O_2} = K_{CO_2}^{-2} \left(\frac{p_{CO_2}}{p_{CO}} \right)^2 \tag{12}$$

$$p_{O_2} = K_{H_2O}^{-2} \left(\frac{p_{H_2O}}{p_{H_2}} \right)^2 \tag{13}$$

Design of Oxygen Sensors

Short-Term Measurements in Liquid Metals

Three types of sensors have been developed for rapid immersion measurements in liquid metals (Figs. 4(A), (B), (C)). Type (A) is a ZrO_2 tube of 1 mm wall thickness, closed at one end. A cylindrical ZrO_2 plug welded to the open end of an insulating ceramic tube forms the electrolyte of type (B). In both cases a molybdenum contact wire is inserted and a solid metal-metal oxide mixture added to the tube. The needle sensor, represented by type (C), has been developed most recently.[20,21] The core part of this oxygen probe is a metallic conductive needle bearing thin coatings of a reference powder mixture and an electrolyte material.

The contact to the molten metal is usually made by a molybdenum rod. The normal measuring period is 10 to 20 s.

Long-Term Measurements in Liquid Metals

Solid electrolyte tubes closed at one end are normally applied for continuous immersion measurements in liquid metals. The oxygen sensor is either preheated and slowly immersed into the melt (to prevent thermal shock damage) or installed in the wall or bottom of the furnace (Figs. 5(A), (B)). Air or oxygen streaming around a platinum wire in the inner compartment is generally preferred as a reference system. A cermet rod immersed from the top or a metal rod inserted through the bottom may serve as a contact lead.

The measuring period should at least cover the duration of the whole refining and casting process.

Electrochemical Gas Analysis

Commercial ZrO_2 tubes closed at one end are also used for the continuous detection of oxygen partial pressures in gas atmospheres. A deposited platinum layer or a coil of PtRh wire is applied for the outer electrode. The most convenient reference for the inner electrode is streaming air.

Two measuring systems have been developed which are schematically sketched in Figs. *6(A),(B).*[18] The external device (type *(A)*) consists of a separate heating system containing the zirconia probe. The gas is cooled and filtered before it enters the electrochemical measuring device. Empirical correlation functions have to be established to evaluate the p_{O_2} in the primary gas.

The internal device (type *(B)*) is introduced into the furnace and makes it possible to measure the p_{O_2} of the furnace atmosphere in situ. It is customary to fit an outer tube of stainless steel or silicon carbide for protection against corroding fumes.

Service lives vary considerably between six months and several years depending on the measuring conditions.

Use of Oxygen Probes in Steelmaking Processes

The hitherto usual oxygen determination in steel is based on sampling and chemical analysis by vacuum fusion extraction. In this analytical procedure, both dissolved oxygen and oxygen present in precipitated suspensions are determined by smelting reduction through carbon at 2000 to 2500 °C. The total oxygen content

$$\%O(\text{total}) = \%O(\text{dissolved}) + O(\text{precipitated}) \tag{14}$$

can only be set equal to the amount of dissolved oxygen if the precipitated oxides are completely separated from the steel bath. Results on the Al deoxidation of an inductively stirred iron melt are shown in Fig. 7.[22] Initially, the total oxygen content of 0.010% and the dissolved oxygen content of 0.009% which was electrochemically determined are nearly identical. After the addition of aluminum, the dissolved oxygen decreased to 0.0005% while the total oxygen content remained at a level of 0.004%, indicating the presence of residual Al_2O_3 particles in the melt.

In steelmaking practice, the conventional oxygen determination causes a delay of 8 to 10 min which is due to sampling, transport to the laboratory, preparation, and chemical analysis. Compared with this, the electrochemical oxygen determination through immersion sensors requires only ≈ 20 s. This measuring technique indicates the actual degree of oxidation of the steel bath and enables steelmakers to take immediate measures for its regulation.

Oxygen Activity and Oxygen Content in Steel Melts

The content of dissolved oxygen can be derived from the electrochemically determined oxygen activity when the activity coefficient f_O in the steel melt is known

$$\log [\%O] = \log a_O - \log f_O \tag{15}$$

In unalloyed steel melts f_O may be taken approximately as unity. But in alloyed steels deviations of f_O have to be considered according to the effect of the various alloying elements on the oxygen activity. In the dilute solution range this effect can be calculated by Wagner's expression[23]

$$\log f_O = e_O^C[\%C] + e_O^{Si}[\%Si] + e_O^{Mn}[\%Mn] + \cdots + \tag{16}$$

$e_O^X n$ represents the interaction parameter for the change of $\log f_O$ with the content (in wt%) of the added element X_n. Parameters $e_O^X n$ for liquid Fe-O-X_n alloys can be taken from literature, e.g. Ref 19. The activity coefficent f_O in liquid iron alloys can also be determined experimentally by electrochemical measurement of a_O and chemical analysis for %O. Results are presented in Fig. 8 for liquid Fe-O-Cr alloys at 1600 °C.[24]

Alloyed steels of technical importance deviate from the assumption of dilute solution behavior. Since the interaction between the added elements is mostly unknown or uncertain, the experimental determination of f_O remains the only way to ascertain the effect of these elements

$$\log f_O = \log a_{O(emf)} - \log [\%O]_{(chem. anal.)} \tag{17}$$

In Fig. 9 experimentally determined and calculated activity coefficients f_O are compared for a carbon-containing chromium steel melt.[25] It is shown in the experiment that, based on the assumption of a strong interaction between C and Cr, the oxygen activity is slightly increased by additions of carbon.

Deoxidation Equilibrium Constants in Liquid Iron

Equilibria of deoxidizing elements in liquid iron and iron alloys were systematically investigated in past years using the solid electrolyte measuring technique. Typical results are shown in Fig. 10 for the equilibrium of the reaction

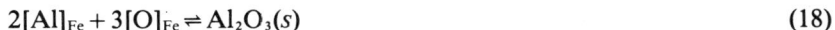

$$2[Al]_{Fe} + 3[O]_{Fe} \rightleftharpoons Al_2O_3(s) \tag{18}$$

in pure liquid iron contained in an alumina crucible representing the equilibrium oxide phase.[26] The equilibrium constant $[\%Al]^2 \times a_O{}^3 = 2.42 \times 10^{-14}$ at 1600 °C derived from the relation

$$\log a_O = -0.633 \log [\%Al] - 4.539 \tag{19}$$

agrees satisfactorily with the earlier value of 2.04×10^{-14} of Gokcen and Chipman[27] who applied the H_2-H_2O gas equilibrium technique.

Oxygen Probes in Steelmaking Operation

As can be seen in Fig. 11 (tubular sensor) and Fig. 12 (needle sensor), the solid electrolyte sensor is incorporated in a measuring device together with a molybdenum rod as a contact lead and a U-shaped quartz glass tube containing a Pt-PtRh thermocouple.[21] The detachable device is mounted to a lance which is electrically connected to an emf recorder (Fig. 13). From this type of probe simultaneous readings of the emf (a_O) and the thermoelectric voltage (temperature) can be obtained. In Fig. 14 two consecutive emf and thermoelectric voltage recordings are shown for the same steel melt, indicating an oxygen activity of 29×10^{-4} and a temperature of 1575 °C.

The application of one-reading probes for the rapid detection of a_O and temperature has become a routine, especially in West European and Japanese steel shops. It is predominantly practiced in refining and casting ladles up to 300 tons of metal weight where the homogenization and final control of the steel composition takes place. One-reading probes are also expected to be regularly used to complete the sublance technique for the dynamic control of the BOF converter process.

Attempts are also made for a continuous oxygen determination in molten steel which is desired for continuous refining and casting processes. But suitable probes still have to be developed and tested under laboratory and operative conditions.

In stainless steel refining units the electrochemical control of the exhaust gas composition has become routine.[28,29] Waste gas analyzers are used in VOD (vacuum oxygen decarburization) refining plants where chromium steel melts are decarburized with an oxygen jet under vacuum conditions. The external analyzer is installed in the waste gas pipe system. The decarburization process is controlled by the emf of the cell in the analyzer indicating the p_{O_2} or CO-CO_2 ratio in the exhaust gas as shown in Fig. 15.[30] At the start of oxygen blow, there is a sharp increase in the emf due to a high CO concentration in the waste gas, indicating a predominant carbon oxidation of the melt. When the critical carbon content of 0.04% is attained, the emf and the CO concentration decrease. In this stage of the process, an undesired chromium oxidation occurs and the oxygen blow is ceased. In such a way, the emf gas analyzer enables a reliable control of chromium oxidation in the process.

References

[1]H. Schmalzried, *Ber. Bunsenges. Phys. Chem.,* **66,** 572 (1962).
[2]H. Schmalzried, *Z. Phys. Chem.,* **38,** 87 (1963).
[3]Emf Measurements in High-Temperature Systems. Edited by C. B. Alcock. Institute of Mining and Metallurgy, London, 1968.
[4]R. A. Rapp and D. A. Shores; in Techniques of Metals Research, Vol. IV, Part 2. Edited by R. F. Bunshah. Interscience Publishers, New York, 1970.
[5]J.W. Patterson, *J. Electrochem. Soc.,* **118,** 1033 (1971).
[6]K. S. Goto and W. Pluschkell, in Physics of Electrolytes, Vol. 2. Edited by J. Hladik. Academic Press, New York, 1972.
[7]T. H. Etsell, *Chem. Rev.,* **70,** 339 (1972).
[8]W. A. Fischer and D. Janke; Metallurgische Elektrochemie. Springer, Berlin/Heidelberg/New York, and Stahleisen, Düsseldorf, 1975.
[9]I. Barin and O. Knacke; Thermochemical Properties of Inorganic Substances. Springer, Berlin/Heidelberg/New York, 1973.
[10]J. Chipman; pp. 621–90 in Basic Open Hearth Steelmaking, 2nd ed. Edited by W. O. Philbrook and M. B. Bever. New York, 1951.
[11]D. Janke and W. A. Fischer, *Arch. Eisenhuettenwes.,* **49,** 425 (1978).
[12]D. Janke, *ibid.,* pp. 217, 413.
[13]D. Janke and W. A. Fischer, *ibid.,* **46,** 477, 683 (1975).
[14]D. Janke and W. A. Fischer, *ibid.,* p. 755.
[15]D. Janke; unpublished work.
[16]J. Moriyama, N. Sato, H. Asao, and Z. Kozuka, *Mem. Fac. Eng. Kyoto Univ.,* **31** [2] 253 (1969).
[17]R. Steffen, *Stahl Eisen,* **94,** 547 (1974).
[18]W. Pluschkell, *ibid.,* **98,** 398 (1978).
[19]D. Janke; in Application of Solid Electrolytes. Edited by T. Takahashi and A. Kozawa. Electrochemical Society of Japan, JEC Press, 1980.
[20]D. Janke and K. Schwerdtfeger, *Stahl Eisen,* **98,** 825 (1978).
[21]D. Janke, K. Schwerdtfeger, J. Mach, and G. Bamberg, *ibid.,* **99,** 1211 (1979).
[22]D. Janke and H. Richter, *Arch. Eisenhuettenwes.,* **50,** 93 (1979).
[23]C. Wagner, Thermodynamics of Alloys. Addison-Wesley, Reading, Mass., 1952.
[24]D. Janke and W. A. Fischer, *Arch. Eisenhuettenwes.,* **47,** 147 (1976).
[25]W. A. Fischer and D. Janke, *ibid.,* p. 589.
[26]D. Janke and W. A. Fischer, *ibid.,* p. 195.
[27]N. A. Gokcen and J. Chipman, *Trans. AIME,* **197,** 173 (1953).
[28]J. Otto, G. Pateisky, and H. J. Fleischer, *Stahl Eisen,* **96,** 939 (1976).
[29]P. Meierling; Dr.-Ing.-Dissertation Technische Hochschule Aachen, 1976.
[30]G. Pateisky, J. Otto, and H. J. Fleischer; in Automatisierung in der Eisen- u. Stahlindustrie. Internationaler Eisenhuettentechnischer Kongress, Brussels, May 17–18, 1976. Centre de Recherches Metallurgiques, Liege, Brussels, 1976.

Table I. Free energies of formation of zirconia, hafnia, and thoria

(a) Referred to pure metal and oxygen

Reaction	$\Delta G°$ (J mol^{-1})	Temp. (°C)	P_{O_2} (Pa) (1600°C)	Reference
$Zr(s) + O_2 \rightleftharpoons ZrO_2(s)$	$-1\,082\,740 + 178.06\,T$	1205–1852	1.31×10^{-16}	9
$Hf(s) + O_2 \rightleftharpoons HfO_2(s)$	$-1\,098\,369 + 169.56\,T$	927–1700	1.74×10^{-17}	9
$Th(s) + O_2 \rightleftharpoons ThO_2(s)$	$-1\,221\,214 + 181.96\,T$	1360–1755	2.88×10^{-20}	9

(b) Referred to metal and oxygen dissolved in liquid iron; standard state hypothetical 1 wt% solution

Reaction	$\Delta G°$ (1600°C) (J mol^{-1})	P_{O_2} (Pa) (1600°C, liquid Fe, 0.0005 Me)	a_O (Fe)	Reference
$Zr_{(Fe)} + 2O_{(Fe)} \rightleftharpoons ZrO_2(s)$	$-85\,705$	1.57×10^{-9}	0.0003	10
$Hf_{(Fe)} + 2O_{(Fe)} \rightleftharpoons HfO_2(s)$	$-94\,713$	2.59×10^{-10}	0.0001	11
$Th_{(Fe)} + 2O_{(Fe)} \rightleftharpoons ThO_2(s)$	$-100\,897$	5.00×10^{-11}	5.8×10^{-5}	12

Table II. Parameters $p_{e'}$ ($t_{ion} = 0.5$) for solid oxide electrolytes at 1200 to 1600°C

Electrolyte	$\log p_{e'}$ ($p_{e'}$, $\times 10^5$ Pa)	$p_{e'}$ (1600°C) (Pa)	Reference
ZrO_2 (14 mol% CaO)	$-(68\,400/T) + 21.59$	1.2×10^{-10}	13
ZrO_2 (2.4 mol% MgO)	$-(74\,370/T) + 24.42$	5.2×10^{-11}	14
HfO_2 (16.5 mol% CaO)	$-(70\,262/T) + 20.35$	6.9×10^{-13}	15
ThO_2 (8 mol% Y$_2$O$_3$)	$-(82\,970/T) + 26.38$	1.2×10^{-13}	13

Table III. Free energies of formation of metal oxides for solid reference systems in oxygen sensors

(a) Emf measurements in liquid steel

Reaction	ΔG° (J mol^{-1})	Temp. (°C)	P_{O_2} (Pa) (1600 °C)	Reference
$Mo(s) + O_2 \rightleftharpoons MoO_2(s)$	$-558\ 791 - 155.37\ T$	1450–1650	3.41×10^{-3}	9
$4/3\ Cr(s) + O_2 \rightleftharpoons {}^2/_3 Cr_2O_3(s)$	$-744\ 484 + 168.06\ T$	1050–1550	1.05×10^{-7}	9
$2\ Nb(s) + O_2 \rightleftharpoons 2NbO(s)$	$-801\ 035 + 158.85\ T$	1030–1730	9.25×10^{-10}	9
$4/5\ Ta(s) + O_2 \rightleftharpoons 2/5 Ta_2O_5(s)$	$-799\ 122 + 158.84\ T$	1030–1730	1.02×10^{-9}	9

(b) Emf measurements in liquid metals of lower melting point

Reaction	ΔG° (J mol^{-1})	Temp. (°C)	p_{O_2} (1000 °C) (Pa)	Reference
$4Cu(s) + O_2 \rightleftharpoons 2Cu_2O(s)$	$-342\ 061 + 148.09\ T$	700–1100	5.02×10^{-2}	16
$2Ni(s) + O_2 \rightleftharpoons 2NiO(s)$	$-477\ 295 + 175.89\ T$	700–1100	4.03×10^{-6}	16
$2Co(s) + O_2 \rightleftharpoons 2CoO(s)$	$-468\ 922 + 141.64\ T$	700–1100	1.45×10^{-7}	16
$2(1-x)Fe(s) + O_2 \rightleftharpoons 2Fe_{1-x}O(s)$	$-532\ 142 + 132.64\ T$	700–1100	1.25×10^{-10}	16

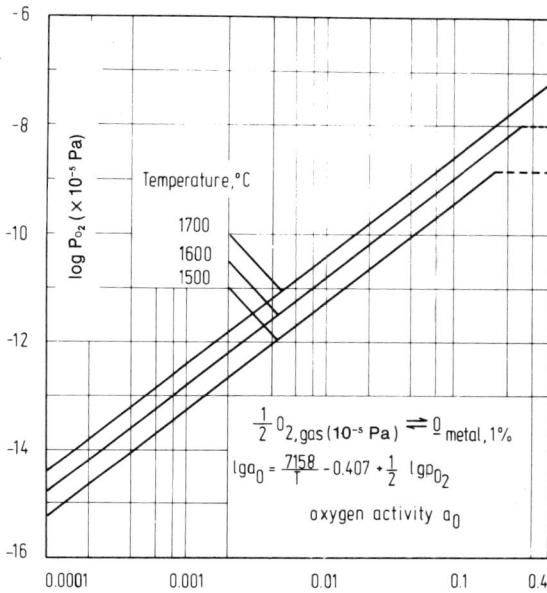

Fig. 1. Equilibrium pressures and activities of dissolved oxygen in liquid iron.

In the figure:

Temperature, °C

1700
1600
1500

$$\frac{1}{2} O_{2,\,gas}\,(10^{-5}\,Pa) \rightleftharpoons \underline{O}_{metal,\,1\%}$$

$$lga_0 = \frac{7158}{T} - 0.407 + \frac{1}{2}\,lgp_{O_2}$$

oxygen activity a_0

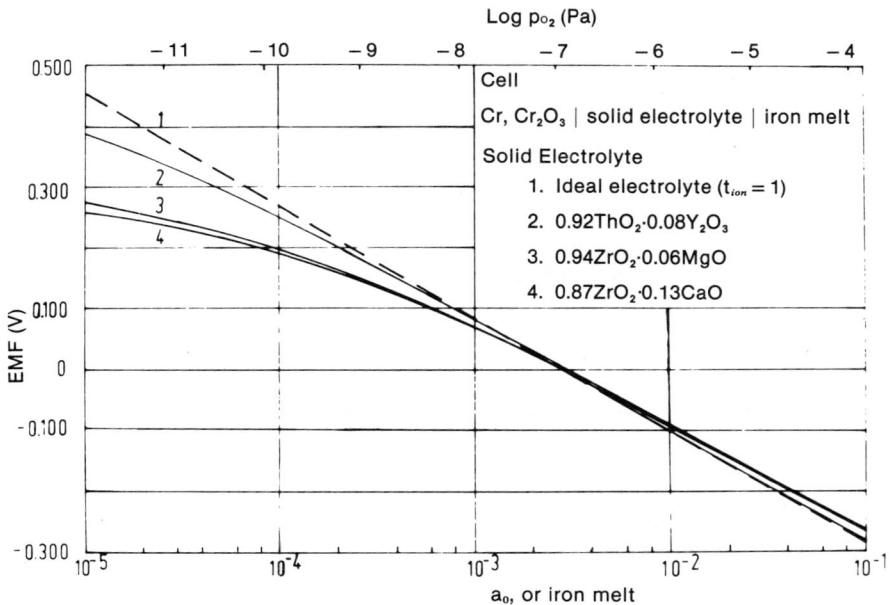

Log p_{O_2} (Pa)

Cell

Cr, Cr_2O_3 | solid electrolyte | iron melt

Solid Electrolyte

1. Ideal electrolyte ($t_{ion} = 1$)
2. $0.92ThO_2 \cdot 0.08Y_2O_3$
3. $0.94ZrO_2 \cdot 0.06MgO$
4. $0.87ZrO_2 \cdot 0.13CaO$

EMF (V)

a_0, or iron melt

Fig. 2. Calibration functions of oxygen probes for iron melts at 1600 °C.

429

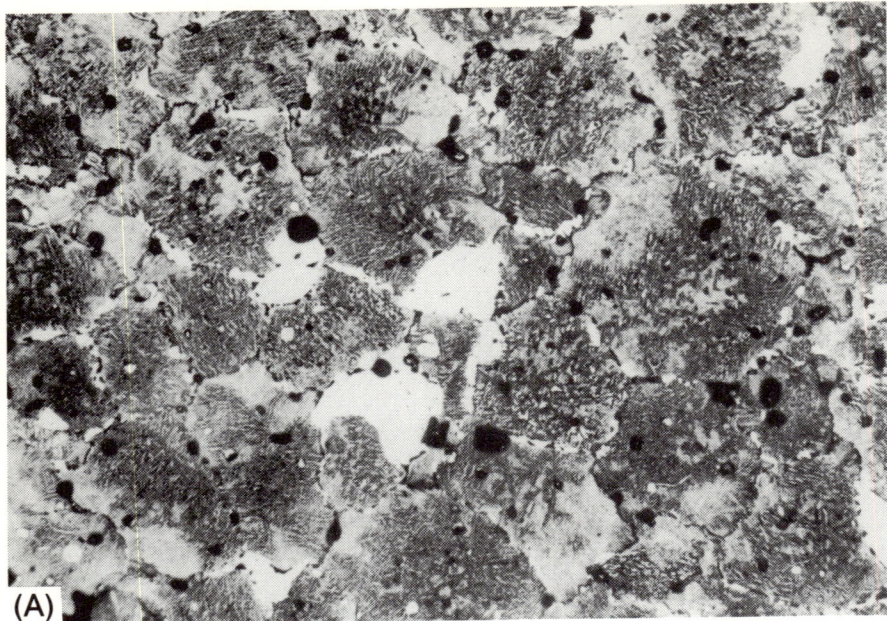

Fig. 3. Structures of a partially stabilized ZrO_2 electrolyte with 2.4 wt% MgO. (A) Optical micrograph. Etching reagent: $H_2O + HF + HNO_3 + HCl$, 20°C/15 min. Light grains, monoclinic phase; dark grains, cubic phase. (B) Scanning electron micrograph. Surface of ZrO_2 tube.

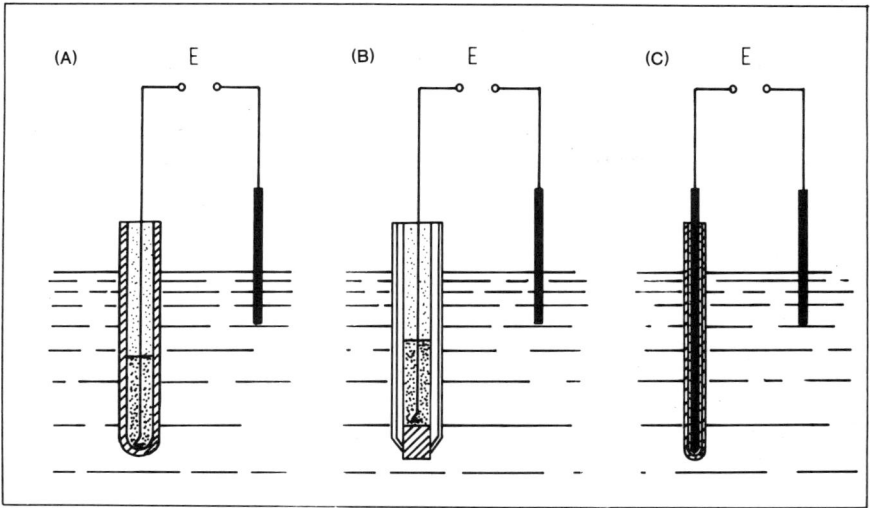

Fig. 4. Design of short-term oxygen sensors. (A) Tubular sensor, (B) plug sensor, (C) needle sensor.

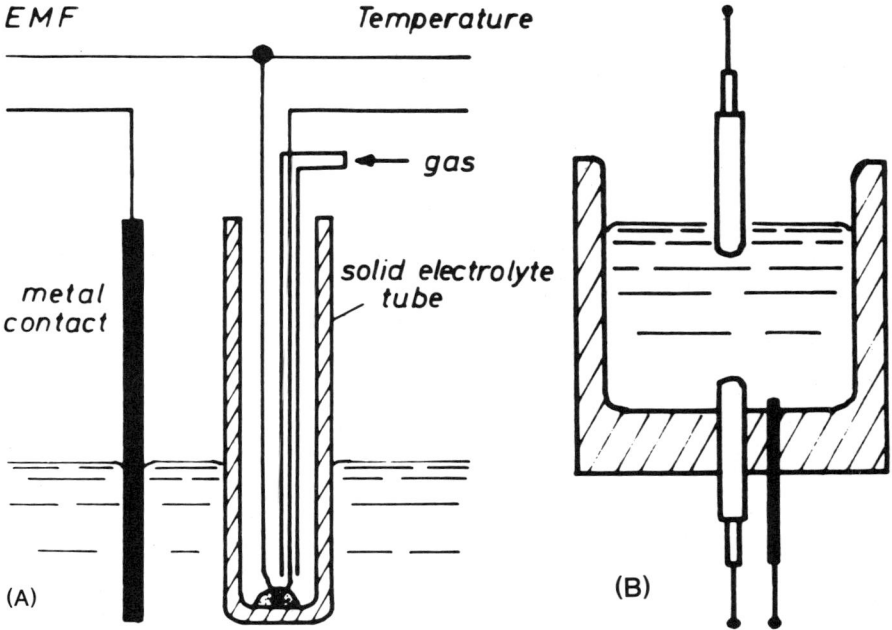

Fig. 5. Long-term oxygen measuring devices.

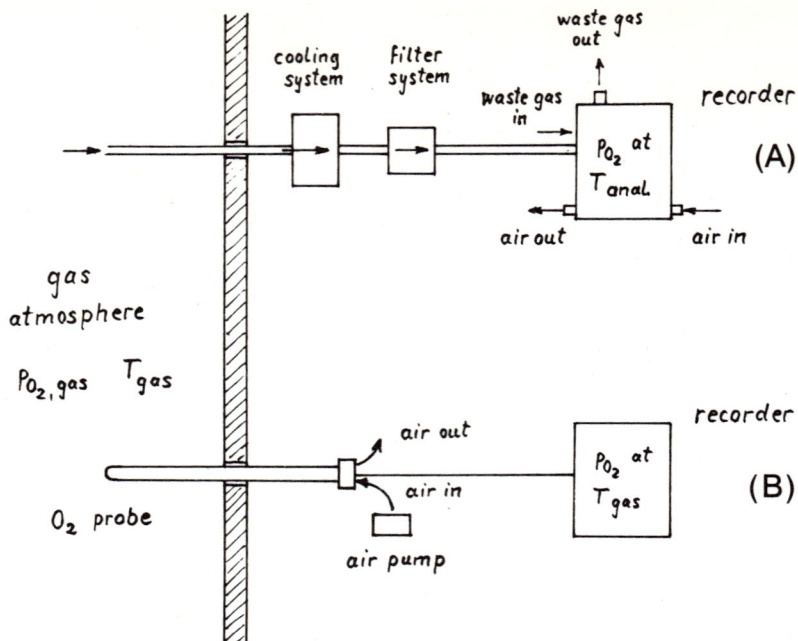

Fig. 6. Electrochemical gas analyzers. (A) External device, (B) internal device.

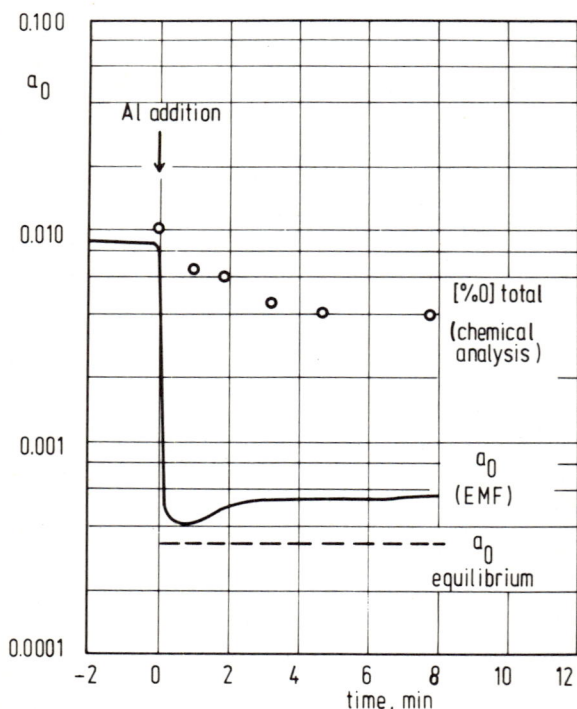

Fig. 7. Dissolved and total oxygen in an aluminum-deoxidized iron melt at 1600 °C (Ref. 22). 1 kg induction furnace, Al_2O_3 crucible, Ar atmosphere, ThO_2 oxygen sensor.

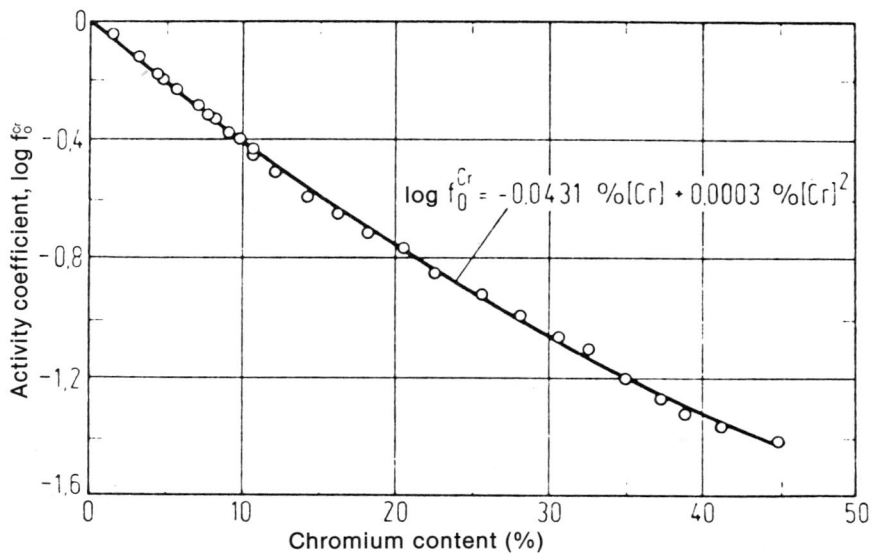

Fig. 8. Activity coefficient of oxygen in liquid iron as a function of chromium content at 1600 °C.

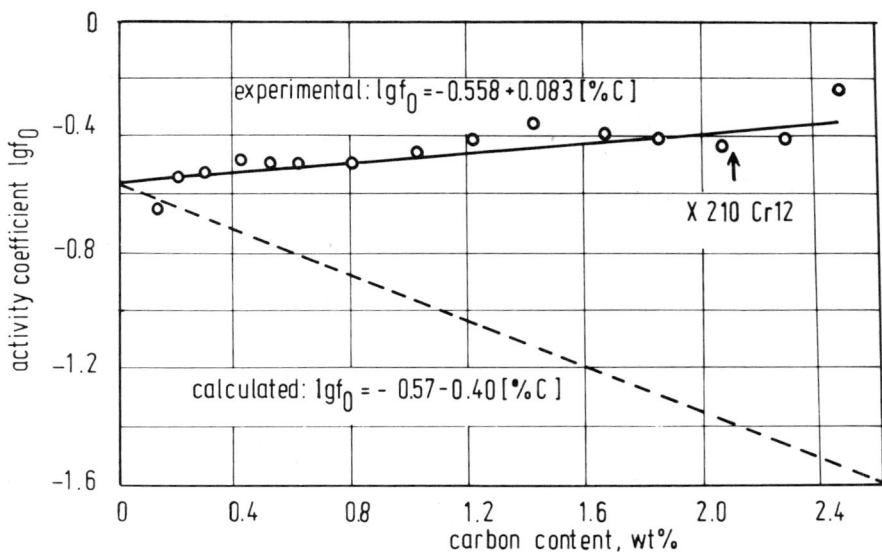

Fig. 9. Log f_O-%C relationship at 1600 °C for alloyed steels with 0.40% Si, 0.35% Mn, 10% Cr, 0.45% W, 0.55% V, and 0.45% Mo.

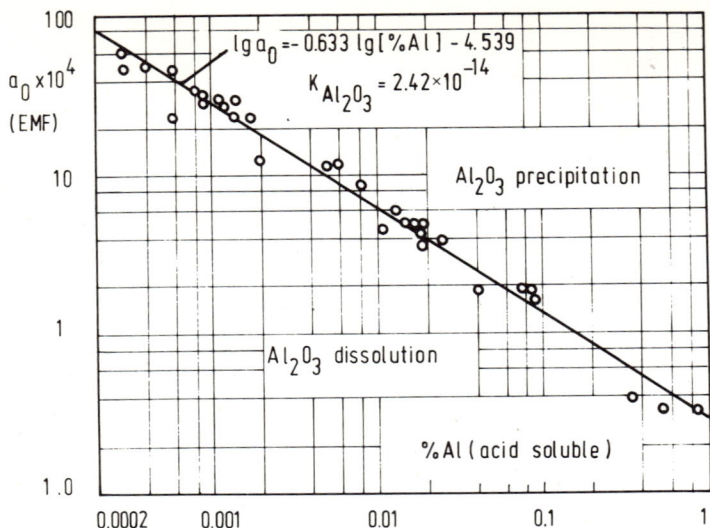

Fig. 10. Aluminum-oxygen equilibrium in pure liquid iron at 1600 °C. (Laboratory experiment, mechanically stirred melt, Al_2O_3 crucible, Ar atmosphere, $ThO_2(Y_2O_3)$ solid electrolyte cell).

$lg\, a_0 = -0.633\, lg\,[\%Al] - 4.539$

$K_{Al_2O_3} = 2.42 \times 10^{-14}$

Al_2O_3 precipitation

Al_2O_3 dissolution

% Al (acid soluble)

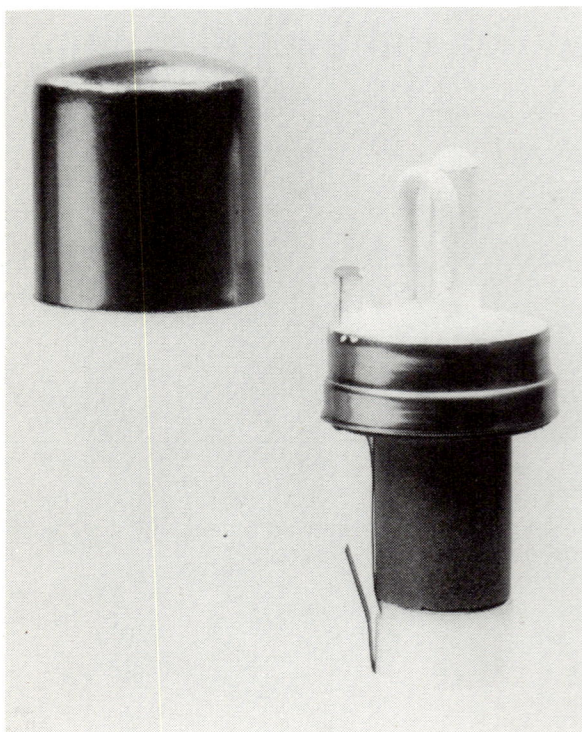

Fig. 11. One-reading probe for oxygen in molten steel (tubular sensor).

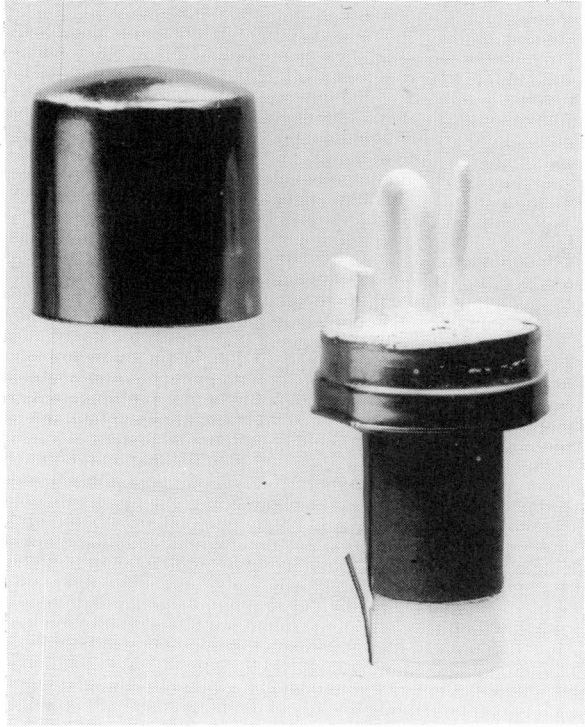

Fig. 12. One-reading probe for oxygen in molten steel (needle sensor).

Fig. 13. Oxygen lance for steel shop applications.

435

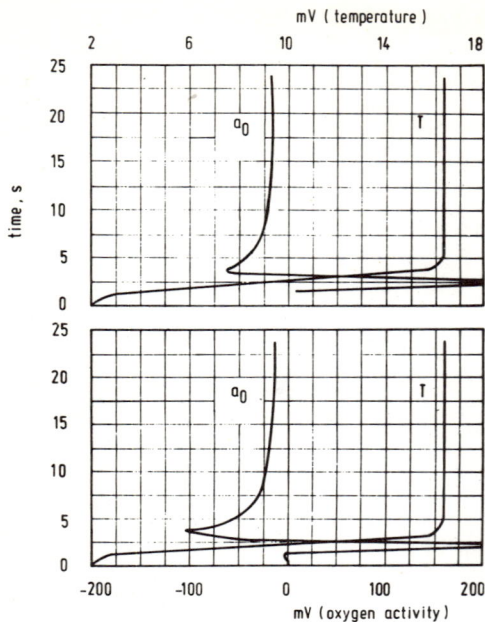

Fig. 14. Two consecutive oxygen and temperature readings (molten steel, $a_O = 0.0029$, 1575 °C).

Type × 2 CoNi 189 steel					
Composition	% C	% Si	% Cr	% Ni	T (°C)
Before vacuum treatment	0,86	0,12	19,20	9,46	1550
At the critical point	0,04	0,10	18,40	9,42	
After vacuum treatment	0,01	0,10	18,70	9,60	1710
After deoxidation	0,01	0,30	18,95	9,58	1720

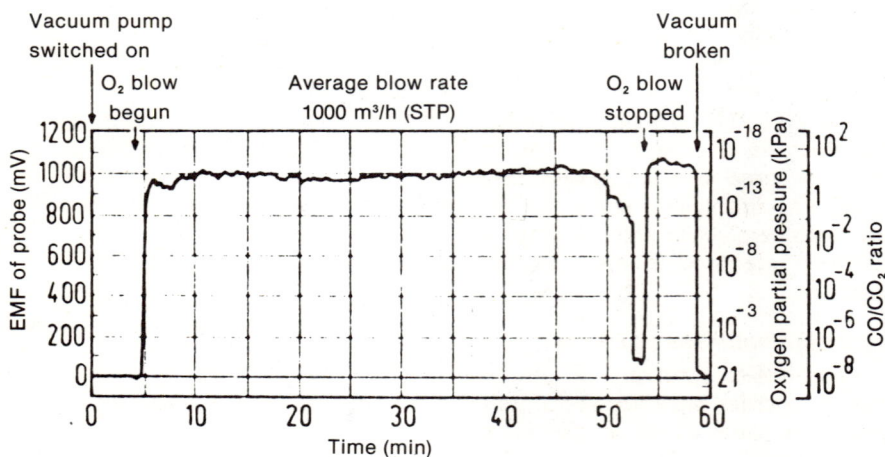

Fig. 15. Measurement of p_{O_2} in exhaust gases to control chromium oxidation in stainless steel refining.

436

High temperature refractory applications of zirconia

A. M. ANTHONY

Centre de Recherches sur la Physique des Hautes Températures
45045 Orléans-Cédex
France

High temperature refractory applications of zirconia are mainly related to its electrical conductivity and thermoelectronic emission. These applications involve the conversion, storage, and conservation of energy. For example, the good electrical conductivity at high temperature and the chemical inertness of zirconia compounds in an oxidizing atmosphere can be utilized in heating elements for furnaces working up to 2500 °C in air. The arc plasma torch working in air is also of considerable use, and preliminary trials with zirconia cathodes are promising. Zirconia electrodes are also convenient for magnetohydrodynamic processes. Zirconia solid electrolytes are used as oxygen sensors or fuel cells in many fields, and they are used in steelmaking to control the oxygen content in the molten metal. For energy storage, hydrogen represents a potentially important medium. While electrolysis of water can be used to produce hydrogen, thermolysis of water is considerably simpler: no electrode, no overvoltage, and no corrosion. Thermolysis requires high temperatures and an oxygen membrane for collecting hydrogen. One of the best materials for this application is zirconia doped so that its electronic conductivity equals its ionic conductivity.

High temperature applications of zirconia are related to its electrical properties, namely its ionic and electronic conductivities and thermoelectronic emission. Such applications often occur in energy technology.

In energy production, zirconia is useful in three applications: (1) in heating elements for furnaces working in oxidizing atmospheres in the range 1500–2500 °C, (2) as a cathode material for an air plasma source, and (3) as an electrode material for the conversion of energy by open-cycle magnetohydrodynamic (MHD) devices. For conservation of energy, zirconia is useful as an oxygen sensor for monitoring the oxygen content in combustion chambers or in steelmaking. In energy storage applications, zirconia electrodes are used in high temperature thermolysis of water to produce hydrogen.

Electrical Properties of Zirconia

The defect structure of zirconia determines its transport properties. These properties are discussed at length elsewhere,[1-5] and our intent is only to summarize some which are pertinent.

Electrical Conductivity

In pure zirconia, the main charge carriers are provided by the thermal generation of defects, which can be described by the pseudoreaction in Kroeger-Vink notation[6]

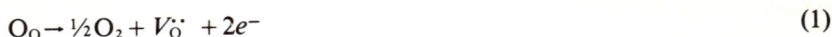

$$O_O \to \tfrac{1}{2}O_2 + V_O^{\cdot\cdot} + 2e^- \tag{1}$$

where O_O is an oxygen on an oxygen site, $V_O^{\cdot\cdot}$ is a doubly ionized oxygen vacancy and e^- is an electron.

The oxygen vacancy in Eq. (1) provides the ionic conductivity σ_i, the electrons the electronic contribution σ_e. The total conductivity σ_T is

$$\sigma_T = \sigma_i + \sigma_e \tag{2}$$

In pure zirconia, the electrical conductivity therefore depends on the oxygen pressure:

$$\sigma_j \propto \begin{bmatrix} \text{predominant} \\ \text{defects} \end{bmatrix} \propto P_{O_2}{}^{1/z} \tag{3}$$

for one type of conductivity σ_j, where z is characteristic of the predominant defect.

Impurities may be easily incorporated into zirconia and strongly influence the defect concentrations. Two types of impurities are important in zirconia. One type, such as calcia or yttria, stabilizes zirconia in the cubic phase, with the generation of charge-compensating oxygen vacancies represented by the reaction

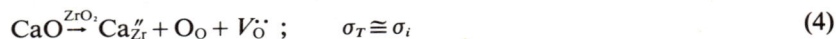

$$CaO \xrightarrow{ZrO_2} Ca_{Zr}'' + O_O + V_O^{\cdot\cdot} ; \qquad \sigma_T \cong \sigma_i \tag{4}$$

The oxygen vacancy concentration increases markedly, and for 15 mol% CaO or 8 mol% Y_2O_3 the total conductivity is fully ionic. A second type of impurity such as CeO_2, which presents a variable valency, increases the electron concentration through reactions of the form

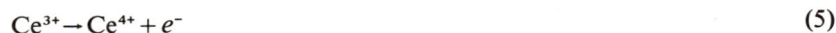

$$Ce^{3+} \to Ce^{4+} + e^- \tag{5}$$

The mobility is characterized by an electronic "hopping" mechanism, and the total conductivity can be mainly electronic.

Figure 1 shows the values of total, ionic, and electronic conductivity for different purities of zirconia at 1600 °C. It is important to note that the total conductivities of pure zirconia and zirconia with 1500 ppm impurity are comparable, but the electronic and ionic contributions are reversed. The first is mainly electronic, the second totally ionic.

Thermoelectronic Emission

The density of electron current emitted from a solid heated to temperature T is given by the Richardson law

$$J_0 = AT^2 \exp(-\phi/kT) \tag{6}$$

where the emission current density J_0 is extrapolated to zero potential between anode and cathode, according to the Schottky law; ϕ is the work function, i.e. the difference between the electron energy in vacuum and the Fermi energy, and A is a constant. For a nondegenerate semiconductor, this law becomes

$$J_0 = BT^{1/2}[e] \exp(-\chi/kT) \tag{7}$$

where $[e]$ is the concentration of electrons in the conduction band, χ is the external work function, i.e. the difference between the electron energy in vacuum and the energy at the bottom of the conduction band, and B is another constant.

Thermal emission, like electronic conductivity $\sigma_e = (n\mu_n + p\mu_p)e$, is simply related to the carrier concentration, and one may easily compare the two kinds of measurements. This technique is very useful for the study of nonstoichiometry in refractory oxides.[7,8] For calcia-stabilized zirconia, for example, the dependence of J_0 on oxygen pressure P_{O_2} can be described by Eq. (3) with

$$[V_O^{\cdot\cdot}] = [Ca_{Zr}''] = [Ca]_{tot} \cong 15 \text{ mol}\% \tag{8}$$

If the total conductivity is determined by the oxygen vacancy concentration and is mainly ionic, there exist some electrons provided by thermal defect reaction Eq. (1), namely

$$[e] = \frac{K^{1/2}}{a_{Ca}} P_{O_2}^{-1/4} \tag{9}$$

where K is a mass law constant including the energy of the vacancy formation, and a is the activity of calcia.

A plot of the current emission against the temperature, given in Fig. 2, demonstrates that calcia-stabilized zirconia is the most efficient emitter material among the zirconia ceramics.

Table I summarizes the many new applications of the last ten years deriving from these electrical properties.

Furnaces

The maximum temperature reached by an electric furnace is determined by the maximum temperature of its heating elements, shown in Fig. 3. A clear difference appears between the temperature ranges accessible in an oxidizing atmosphere and those in neutral, reducing, or vacuum atmospheres.

In an oxidizing atmosphere, zirconia, totally or partially stabilized in its cubic-high temperature form by additions of lime or yttria, appears to be an ideal element material with a melting point of $\approx 2600\,°C$ and reasonable cost. It is an insulator at room temperature, but, by preheating the oxide up to $1000\,°C$, the total conductivity increases dramatically.

Compared to refractory metals, a refractory oxide may be distinguished by good chemical inertness with respect to oxygen but by poor mechanical behavior. Its tensile strength, modulus of elasticity, and thermal conductivity are far below those of metals. It is highly sensitive to any impact or shock whether mechanical, thermal, or electrical. Good design thus involves giving the heating element a shape such that the mechanical stresses are either minimized or accommodated. Two types of heating elements have proved useful. The first is a one-piece ceramic for the laboratory furnace; the second is a heating module which can be stacked to obtain the desired cavity in industrial or induction furnaces. Electrical power is commonly supplied from high frequency (≈ 10 MHz) or low frequency (50–60 Hz) alternating current.

Figure 4(A) shows an induction laboratory furnace and Fig. 4(B) a low-frequency alternating-current, Joule-heating furnace. In the former, the susceptor is provided by small zirconia ceramic rings with a thickness of 20

439

mm appropriate to a skin depth p

$$p(\text{mm}) = 5.10^2 \ \frac{\varrho(\Omega\cdot\text{m})}{f(\text{Hz})} \tag{10}$$

with ϱ the electrical resistivity of zirconia and with a frequency ≈ 10 MHz, $p = 20$ mm. For preheating, a silicon carbide susceptor is convenient. Small pieces of zirconium can be used to increase the temperature of a zirconia susceptor. The zirconia induction furnace does not need electrical contact and the geometry is quite simple. However, the electrical power efficiency is poor and such heating is expensive in energy, although this type of heating is well developed for production of optical fibers at 2100 °C.

For the Joule-effect heating elements, a series of formers in brick, felt, or blanket holds the element and provides thermal insulation. The preheating element is Kanthal Al or gas, and upper contacts are in platinum. The zirconia element is designed so that two conical sections have an electrical resistance gradient such that, at the extremities, the dissipation of electrical energy does not permit temperatures above 1600 °C. The electrical resistance of such a unit can be estimated.[9]

The characteristics of various types of furnaces are compared in Table II for typical laboratory, industrial, or space use. Examples of each type have been built and are in use in many laboratories; Fig. 5 shows examples. The pilot furnace (Fig. 5(A)) has a high-temperature-cavity volume $\approx 10^{-2}$ m^3, and is designed to work at 1800 °C for $> 10\,000$ h in an oxidizing atmosphere.

Cathodes for Plasma Torches

The production of oxidizing plasmas offers potential applications in the field of thermal treatment of minerals and ores, chemical synthesis at high temperature, and cutting of metals. The utilization of oxidizing gases in arc-plasma generators, however, requires the design and study of special electrodes, purposely adapted to operation under oxidizing conditions. Some results have been obtained already for operation of a pure ZrO_2 cathode,[11,12] the design of which is sketched in Fig. 6. A rotating oxide cathode (2), containing zirconia powder, has a small hole in the bottom pierced by a molybdenum needle. The electric arc (5) appears between the cathode and the tubular anode (1). Gas is extracted from the duct (4), and through this hole pyrometric measurements can be made. The apparent temperature of the cathode (measured in the presence of the arc) under argon is between 2400° and 2800 °C. At these temperatures, new phases are observed to form, depending on the gas composition, as indicated in Table III.

One plasma gun equipped with a ZrO_2 cathode was constructed to operate in air. The electrical power is in the range 15–30 kW for an air flow rate between 1.1 and 2.1×10^{-3} kg s^{-1}.

Electrodes for MHD Energy Conversion

Electrical power generation by MHD or magnetogasdynamic (MGD) converters involves driving a conducting fluid at high speed through a perpendicular magnetic field (Fig. 7) and thereby generating a direct current. Coupling MHD or MGD conversion to conventional thermal power generation processes has the potential of increasing the efficiency of overall energy conversion from 45% to 60%. The MHD and MGD processes depend on the fluids used; the open gas-cycle operation uses energy from coal, fuel oil, or natural gas. To make the gas conductive, potassium and cesium are added. The temperature is high and the medium is rich in oxygen.

The materials properties required of converter materials are given in Table IV. A material with all these properties does not exist, so the choice is difficult.

In the channel, ionized gases pass between the two electrodes. Positive ions of the ionized gases are deflected by the induction field to the upper electrode, and these are neutralized by electrons. These electrons are emitted by the upper electrode which is between 1800 and 2000 °C. Thus the current passes through the load. Positive ions accumulate in front of the upper electrode if its electronic emission is not large enough ($\approx 10^4$ A/m^2); a very large electric field then appears, and the field emission increases the electronic current while positive ions bombard the electrode. These electric arcs eventually destroy the electrode. Thus electronic conductivity and thermoelectronic emission by the electrodes must be large; unfortunately, this is not generally the case. Indeed, thermoelectronic emission decreases with oxygen pressure which is high in open-cycle MHD generators. Some $(ZrO_2)_x(LaCrO_3)_y$[13] or $(ZrO_2)_x(CeO_2)_y(Ta_2O_3)_z$[14] compounds, however, are characterized by large electronic contributions to the total conductivity.

For the walls, $SrZrO_3$ ($T_M = 2740$ °C) is a good electrical insulator, but compositions having acceptable properties are only found in a narrow range around stoichiometric composition. Other zirconates, such as $CaZrO_3$ ($T_F = 2340$ °C), $BaZrO_3$ ($T_F = 2700$ °C), and $SrZrO_4$ ($T_F = 2420$ °C), are good candidates.

Solid Electrolytes

Because of the electrical conductivity provided by mobile O^{2-} ions, stabilized zirconia can be used to measure oxygen partial pressure by acting as a conducting membrane separating the unknown oxygen partial pressure from a reference oxygen partial pressure. In practice, one works in a temperature and partial pressure region where t_{ionic} is higher than 0.98. Under these conditions

$$E = \frac{RT}{4F} \log \frac{P'}{P_0}$$

where P' and P_0 are the oxygen partial pressures (fugacities) on both sides of the membrane. Knowledge of E and P_0 allows determination of an unknown P'.

Figure 8 schematically outlines use of the zirconia solid electrolytes.[16] System A is a concentration cell, system B an oxygen pump, and system C a battery or fuel cell. Operation of the device is limited to temperatures above 600 °C, owing either to the conductivity of stabilized zirconia or to electrode polarization. Extending the useful temperature range of operation to lower temperatures requires voltmeters having higher input impedance than presently available. Nevertheless, many systems have been devised for zirconia oxygen sensor control of automobile exhausts at temperatures below 1000 °C.

Apart from purely thermodynamic investigations, many uses of high-temperature oxygen probes using fully stabilized zirconia have been proposed. For instance, high-temperature oxygen measurement can be used to monitor combustion processes. By knowing the oxygen content of the exhaust gas, adjustments can be made to improve combustion efficiency in central heating installations or in steam boilers. An oxygen probe based on stabilized zirconia has also been used to control sintering atmosphere in the manufacture of ferrites. Measurement of oxygen partial pressure in reactive cathodic sputtering is also an interesting possibility. It has been claimed that, under certain conditions, oxygen measurements allow control of carburizing furnace atmospheres, and knowledge of the amount of dissolved oxygen in molten glass and molten metals is important in industrial production.

For direct oxygen measurement in steel, various probes featuring stabilized zirconia are commercially available. In these probes the liquid metal itself is used as one of the electrodes. To protect the solid electrolyte when it is inserted through the slag layer into the molten steel, a thin copper or steel cap covers it. This cap dissolves rapidly in the steel, and measurement is performed after ≈ 15 s when thermal equilibrium is reached. There is, however, uncertainty in the exact limit of the electrolytic domain, and measurements in steel at 1600 °C appear to be limited to between 10^{-8} and 10^{-13} atm. Figure 9 shows one scheme for such a measurement.

Electronic conductivity is responsible for deviation of the solid electrolyte oxygen probe from Nernst behavior and can produce apparent variation in O_2 content of the sample. Knowing the pressure P_- at which ionic conductivity is equal to the electronic conductivity at a given temperature, one can estimate the error made in neglecting electronic conductivity. If P_0 is the reference oxygen partial pressure and P is the pressure to be measured, the emf in the case when there is only electronic conduction is

$$E = \frac{RT}{F} \ln \frac{P_-^{\frac{1}{4}} + P_0^{\frac{1}{4}}}{P_-^{\frac{1}{4}} + P^{\frac{1}{4}}} \tag{12}$$

By neglecting electronic conductivity, one determines a pressure P' from

$$E = \frac{RT}{F} \ln \frac{P_0}{P'^{\frac{1}{4}}} \tag{13}$$

The relation between the true value and the apparent one is then

$$P^{\frac{1}{4}} = P'^{\frac{1}{4}} (1 + \frac{P_-^{\frac{1}{4}}}{P_0^{\frac{1}{4}}}) - P_-^{\frac{1}{4}} \tag{14}$$

It can be seen that the true value will be closer to the measured value using the Nernst relation (Eq. 11) when P_- is very small and the reference pressure much larger than P_-. Taking P_- as 10^{-15} atm, the real and measured activities for oxygen in liquid steel were calculated at 1600 °C (Table V) with air as reference.

There are many sources of errors for the measurements of emf using zirconia electrolytes. This is the reason that the Commission on High Temperatures and Refractory Materials of the IUPAC has promoted a program of international cooperation in which more than ten laboratories have studied zirconia tubes provided from four commercial sources. Establishment of a standard zirconia cell is in progress.

Hydrogen Production by Water Thermolysis

Hydrogen production is important in the future development of nuclear power plants and of the chemical industry. Piped hydrogen is a potential replacement for electrical energy transport which involves energy losses of 15% and pollutes the landscape. Hydrogen gas can be produced in commercial quantities by decomposition of water and piped underground, to be combined with oxygen near a town or factory to produce electrical energy thermally with good efficiency. Table VI summarizes the various mehods using solar energy to decompose water.

It is still difficult to discuss the comparative efficiency of such methods. Photoelectrolysis, however, needs more fundamental research to understand some parameters such as the role of the electron affinity, lifetime, mobility, the flat band potential, and above all the surface states. The efficiency is still very low. For the remaining methods, the main difficulties are technological and economic. The thermochemical cycles present difficult problems, e.g. the kinetics at relatively low temperature and corrosion. Electrolysis of water appears promising for the medium term. The advantages are thermodynamic and kinetic. The decomposition of water requires 0.95 V at 1000° vs 1.23 V at 25 °C, and the overvoltages decrease greatly. Thermolysis of water is quite elegant and clean, but it needs very high temperatures, which can however be realized with solar energy concentration.

The direct thermal dissociation of water needs temperatures in the range 2000–3000 °C. A complex equilibrium between H_2, O_2, H, OH, and O must be considered. To separate species, one needs a membrane. Much work has been done to find a material to produce pure hydrogen by selectively removing oxygen. The combination of high oxygen-ion conductivity, long-term thermal and chemical stability, and relatively low cost has made stabilized zirconia attractive. This material exhibits primarily pure ionic conduction, but at high temperature some electronic contribution is available.

Figure 10 shows a reactor which is completely made of zirconia. Inside a zirconia furnace, which is insulated by zirconia felt, is a mixed-conductor zirconia. Water diluted by argon flows in and oxygen diffuses out through the high temperature part of the tube. The electronic component of the membrane produces an internal short circuit which ensures oxygen exchange between the electrolyte and the gaseous phase and avoids the use of any external electrode.

The oxygen flow through the zirconia wall is given by

$$J_{O2} = (\text{mol m}^{-2}\text{s}^{-1}) = BT \exp\left(-\frac{\Delta E_e}{kT}\right) \Delta P_{O_2}^{-1/4} \qquad (14)$$

where ΔE_e is the electronic activation energy and ΔP_{O_2} the oxygen pressure gradient. At constant temperature and for a given gradient ΔP_{O_2}, this value is only a function of the physical properties of the membrane and mainly of its electronic conductivity. Two carrier gases are used. One contains CO/CO_2 in mixtures which correspond to the water-gas equilibrium

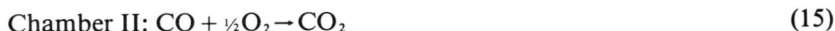

Chamber I: $H_2O \rightarrow H_2 + \frac{1}{2}O_2$

Chamber II: $CO + \frac{1}{2}O_2 \rightarrow CO_2$ \qquad (15)

Thus at a given temperature and steam partial pressure $((P_{O_2})_I$ constant), the evolved hydrogen flow varies as $(P_{O_2}^{-1/4})_{II}$ (Fig. 11). The activation energy of

443

3.9 eV corresponds exactly to the electronic activation energy of the material.

If argon flows in chamber II, the result is the direct thermal decomposition of water

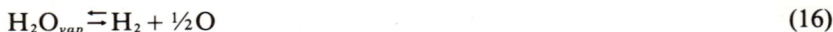

$$H_2O_{vap} \rightleftharpoons H_2 + \frac{1}{2}O \tag{16}$$

The relation $(P_{O_2})_I = \frac{1}{2}P_{H_2}$ gives

$$P_{H_2} = (\sqrt{2} \, K \, P_{H_2O})^{2/3} \tag{17}$$

This dependence is observed (Fig. 12). The hydrogen flow appears to be limited by the characteristics of the membrane. If one dopes stabilized zirconia with a variable valency cation, electronic conduction appears, and the best membrane exhibits equal electronic and ionic conduction. Although many materials appear to possess the required structural stability and conductivity properties, the systems CeO_2-Y_2O_3-ZrO_2 and Cr_2O_3-Y_2O_3-ZrO_2 are most promising.[17]

References

[1]R. W. Vest, N. M. Tallan, and W. C. Tripp, *J. Am. Ceram. Soc.*, **47** [12] 635–40 (1964).
[2]R. W. Vest and N. M. Tallan, *ibid.*, **48** [9] 472–75 (1965).
[3]J. P. Loup and A. M. Anthony, *Phys. Status Solidi*, **38** [1] 499–512 (1970).
[4]A. Guillot and A. M. Anthony, *J. Solid State Chem.*, **15** [1] 89–96 (1975).
[5]J. C. Rifflet, P. Odier, A. M. Anthony, and J. P. Loup, *J. Am. Ceram. Soc.*, **58** [11-12] 493–97 (1975).
[6]F. A. Kroeger and H. J. Vink; pp. 307–435 in Solid State Physics, Vol. III. Edited by F. Seitz and D. Turnbull. Academic Press, New York, 1956.
[7]P. Odier, J. F. Baumard, D. Panis, and A. M. Anthony, *J. Solid State Chem.*, **12**, 324–28 (1975).
[8]P. Odier and J. P. Loup, "Study of Electronic Minority Defects in Stabilized Zirconia by Thermal Emission of Electrons"; this volume, pp. 380–87.
[9]A. M. Anthony, K. Dembinski, K. Dunand, J. L. Dupont, and R. Mottu, *Ceramurgia Int.*, **3**, 29–33 (1977).
[10]A. M. Anthony; pp. 519–26 in Solid Electrolytes. Edited by Paul Hagenmuller and W. Van Gool. Academic Press, New York, 1978.
[11]G. Vallbona, C. Bonet, and M. Foex, *J. Phys. D.*, **8**, 1185–97 (1975).
[12]C. Bonet, *Pure Appl. Chem.*, **52**, 1707–20 (1980).
[13]A. M. Anthony, *Rev. Int. Hautes Temp. Refract.*, **13** [4] 230–36 (1976).
[14]M. Gouet, Thesis, University of Paris–Val de Marne, 1979.
[15]M. Voinov; "Various Utilisations of Solid Electrolytes" in Electrode Processes in Solid State Ionics. Edited by M. Kleitz and J. Dupuy. Kluwer Boston, Hingham, Mass., 1976.
[16]Solid State Chemistry of Energy Conversion and Storage. Edited by J. B. Goodenough and M. S. Wittingham, Advances in Chemistry Series 163. American Chemical Society, Washington, D.C., 1977.
[17]B. Coles, A. Ounalli, J. F. Baumard, and A. M. Anthony; to be published in the *International Journal of Hydrogen Energy.*

Table I. High temperature applications

Type of materials	Electrical properties	Device	Uses
ZrO_2 - 4 wt% CaO + impurities ($\approx 2\%$)	$\sigma_T = \sigma_e + \sigma_i$ σ_T high	Heating element or susceptor	Furnaces 1200–2500 °C in oxidizing atm
ZrO_2 - 4 wt% CaO + impurities ($\approx 2\%$)	σ_T high χ low; J_0 high	Cathode	Plasma torch in air
ZrO_2 - x% CeO_2-y% Ta_2O_5	$\sigma_T \approx \sigma_e$ χ low; J_0 high	Electrode	MHD channel (open cycle)
ZrO_2 - 15 mol% CaO or 8 mol% Y_2O_3	$\sigma_T = \sigma_i$	Solid electrolyte	Monitoring gas mixture in combustion chamber Control of O_2 in steelmaking. H_2 from electrolysis of water (1000 °C)
ZrO_2 - x% CaO_2-y% Y_2O_3	$\sigma_T = \sigma_e + \sigma_i$ $\sigma_e = \sigma_i$	Oxygen semipermeable membrane	H_2 from thermolysis of water (2000 °C)

445

Table II. Performance of furnaces

Type	Laboratory	Industrial	Space
Max temp. (°C)	2450	1900	2000
Lifetime (1800 °C)	10 000 h	5000 h	Few days
Energy consumption (W) (50 cm³ at 2000 °C)	1200	1200	300
Insulation	Felt or blanket	Bricks	Felt and multifoil

Table III. Composition of the ZrO_2 electrode under different arc atmospheres*

	Ar	$Ar + 5\%\ H_2$	N_2	Air
Upper part of cathode	ZrO_2 (M) Zr (c.h.)	ZrO_2 (M) Zr (c.h.)	ZrO_xN_y (C) ZrO_2 (M) ZrN (C)	ZrO_2 (M) ZrN (c) ZrN (c)
Bottom of cathode	ZrO_2 (M) Zr (C.H.)	ZrO_2 (M) Zr (c.h.)	ZrN (C) ZrO_xN_y (C) ZrO_2 (M)	ZrO_xN_y (C) ZrN (C) ZrO_2 (M)
Duration of run (h)	2	1	1	1

*$I = 150$ A; gas flow rate $Q = 10$(STP) L min^{-1}; rotation speed $w = 900$ rpm; M = monoclinic; C = cubic; C.H. = compact hexagonal; m,c,c.h. are the same phases in trace concentrations.

Table IV. Materials properties required for MHD and MGD converters

General
High melting point and low vapor pressure
High resistance to abrasion from plasma jet
High resistance to corrosion by insemination agents
High resistance to oxidation
High resistance to thermal shock

Electrodes
High electrical conductivity with high electronic transfer number even at low temperature
High thermoelectronic emission at high temperature

Insulating walls
Low electrical conductivity

Table V. Actual and measured oxygen activities in molten steel

Real activity (%)	0.005	0.010	0.020	0.040	0.060	0.08	0.1
Measured (Nernst) (%)	0.006	0.012	0.022	0.044	0.064	0.085	0.106

Table VI. Production of hydrogen

Methods	Schematic reaction	Properties of refractory oxides	Examples
Thermochemical cycles	$CaBr_2 + H_2O$ $\downarrow$ $CaO + 2HBr$ $\Big\}$ 800 °C	Chemical inertness (vessels)	Alumina Mullite Sillimanite Zirconia
	$2HBr + Hg$ $\downarrow$ $HgBr_2 + H_2$ $\Big\}$ 200 °C		
	$CaO + HgBr_2$ $\downarrow$ $CaBr_2 + Hg + \frac{1}{2}O_2$ $\Big\}$ 500 °C)		
Photoelectro-chemical	$H_2O \xrightarrow{h\nu} H_2 + \frac{1}{2}O_2$ (25 °C) $(0 < pH < 14)$	Semicon-ductor p or n $E_g > 1.23$ eV $\sigma_T = \sigma_e$	TiO_2, $SrTiO_3$ Ta_2O_2, Fe_2TiO_5
Electrolysis	H_2 (cathode) $\nearrow$ $\uparrow$ H_2O V < 1.23 V $\searrow$ $\downarrow$ (1000 °C) $\frac{1}{2}O_2$ (anode)	Solid electrolyte anionic conductor $\sigma_T = \sigma_i$	Stabilized CaO or Y_2O_3 doped ZrO_2
Thermolysis	$H_2O \rightarrow H_2 + \frac{1}{2}O_2$ (2000 °C)	Mixed conductor $\sigma_T = \sigma_i + \sigma_e$	ZrO_2-Y_2O_3-CeO_2

$1670°K$ $\sigma_T^1 < \sigma_T^2 < \sigma_T^3$ 1 — ZrO_2 purified under vacuum
$\left(\dfrac{\sigma_i}{\sigma_T}\right)_1 < \left(\dfrac{\sigma_i}{\sigma_T}\right)_2 < \left(\dfrac{\sigma_i}{\sigma_T}\right)_3$ 2 — ZrO_2 1500 ppm impurities
3 — ZrO_2 - 15 % mole CaO

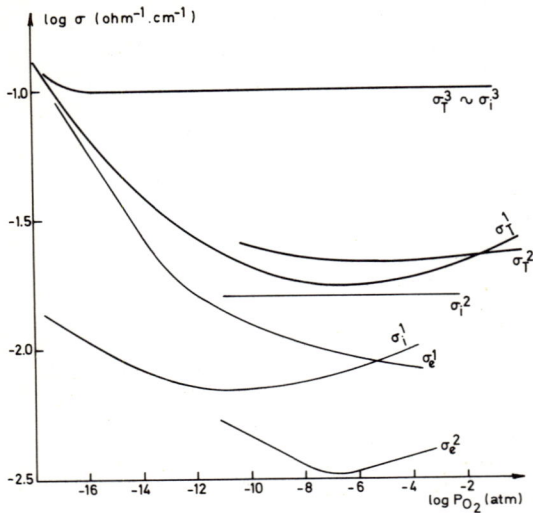

Fig. 1. Conductivity of various zirconia compounds.

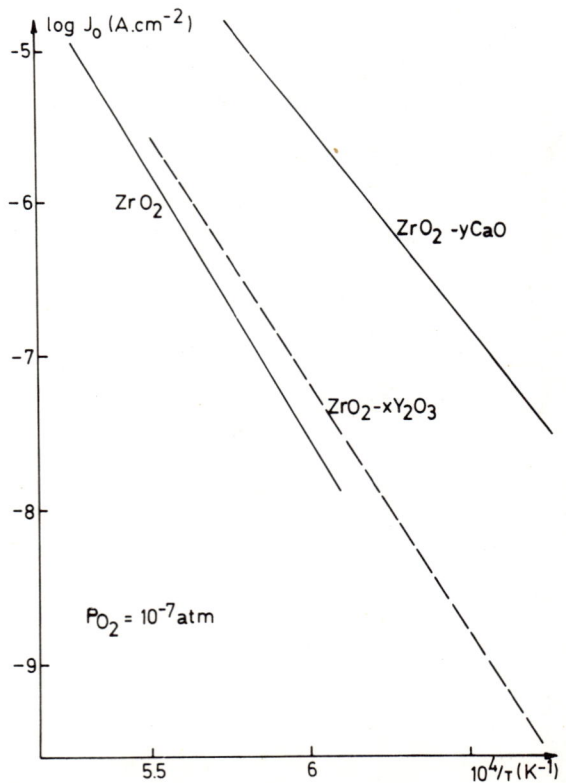

Fig. 2. Thermal emission of various zirconia compounds as a function of temperature.

448

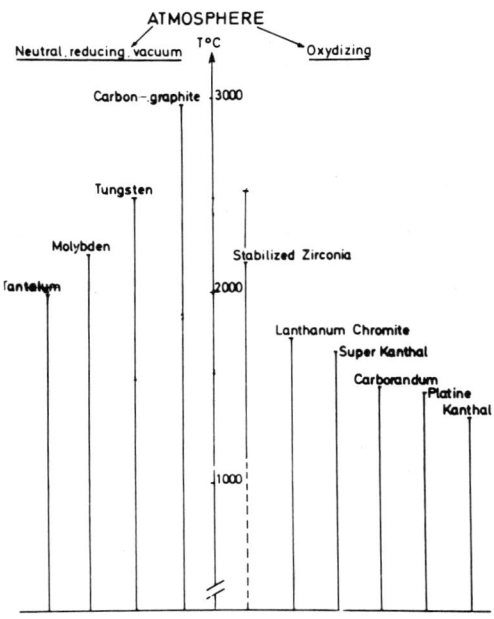

ATMOSPHERE

Neutral, reducing, vacuum   T°C   Oxydizing

Carbon – graphite   3000

Tungsten

Molybden   Stabilized Zirconia

Tantalum   2000

                    Lanthanum Chromite
                    Super Kanthal
                    Carborundum
                                Platine
                                Kanthal

                    1000

Fig. 3. Atmosphere and temperature ranges for the use of heating elements.

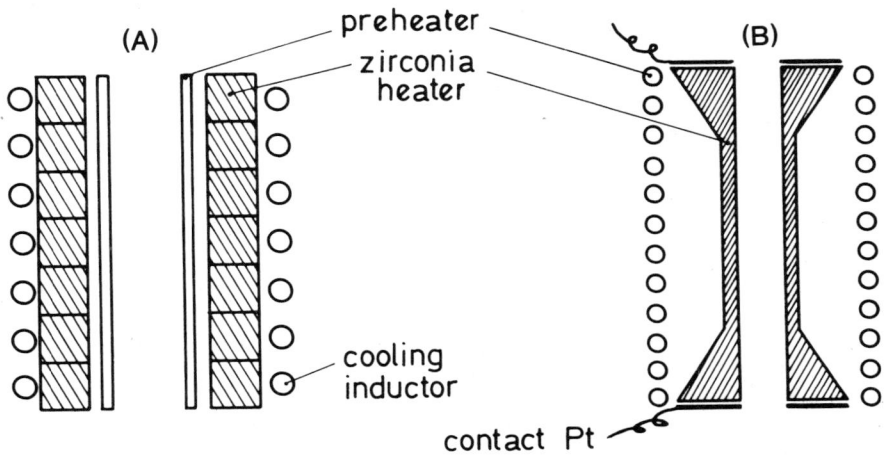

(A)   preheater   (B)
           zirconia
           heater

           cooling
           inductor

contact Pt

Fig. 4. Schematic representation of zirconia heating elements: (A) induction, (B) Joule effect.

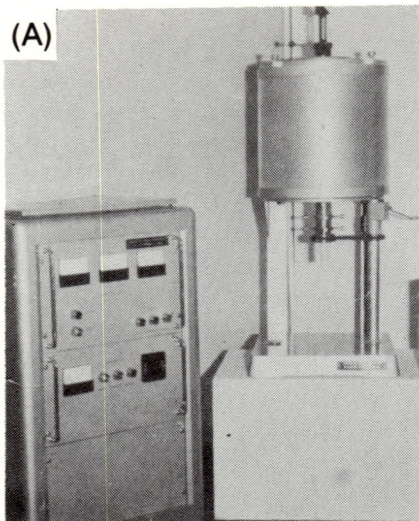

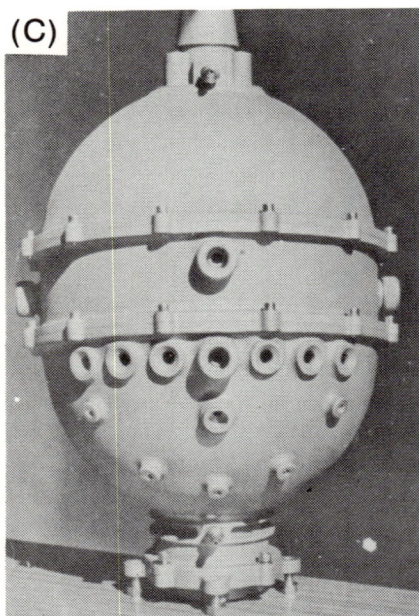

Fig. 5. Types of zirconia furnaces: (*A*) for the laboratory, (*B*) on an industrial scale, and (*C*) for materials science experiments in space.

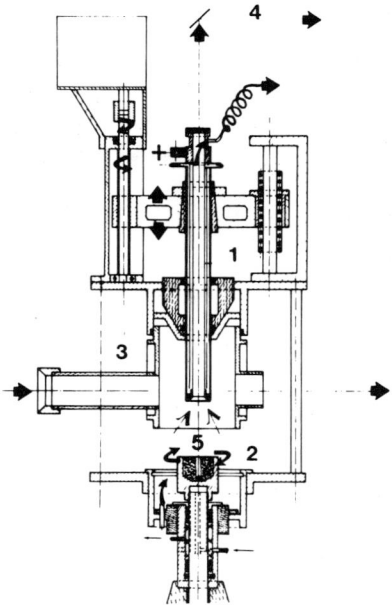

Fig. 6. Experimental arrangement for a plasma torch operating in air.

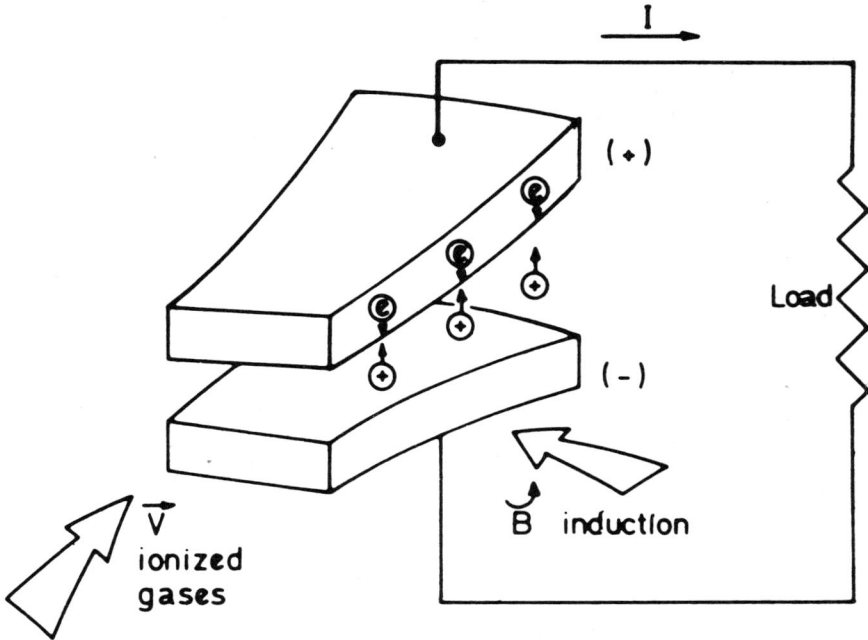

Basic MHD generation configuration

Fig. 7. Schematic representation of an MHD channel.

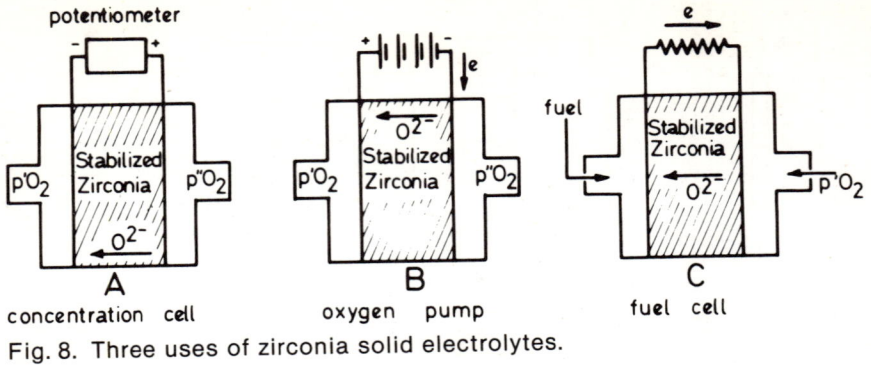

Fig. 8. Three uses of zirconia solid electrolytes.

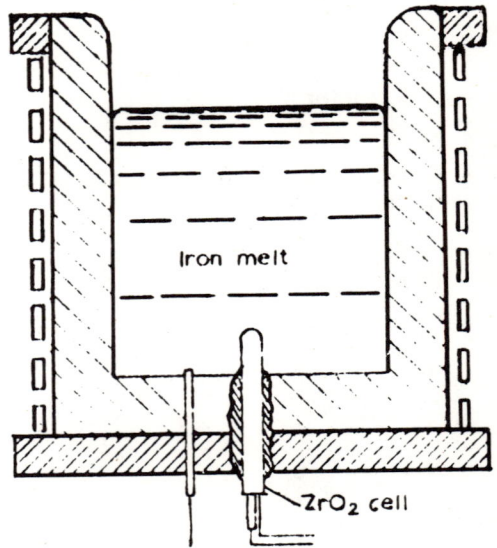

Fig. 9. Direct measurement of oxygen activity in steelmaking.

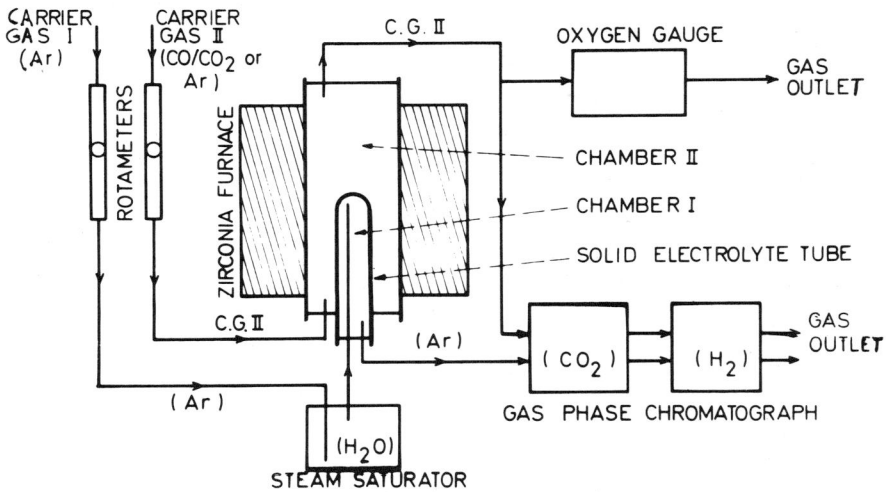

Fig. 10. A reactor for thermolysis of water.

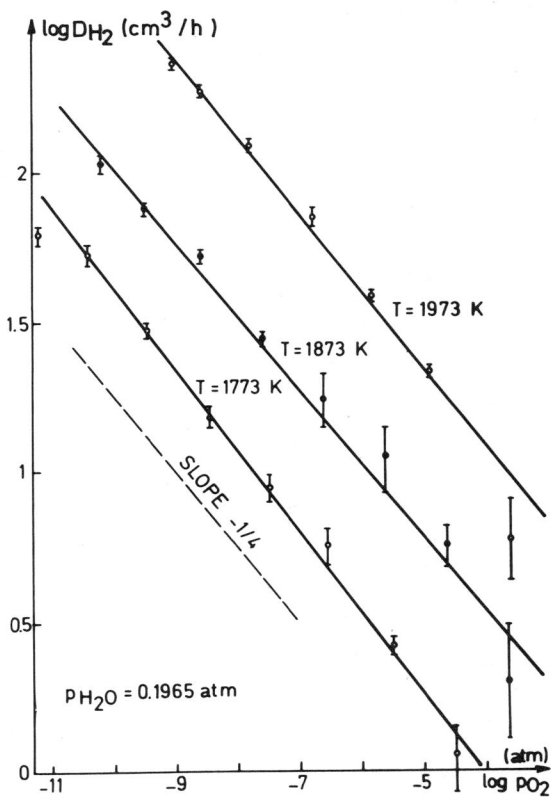

Fig. 11. Hydrogen flow with CO/CO₂ mixtures in thermolysis of water.

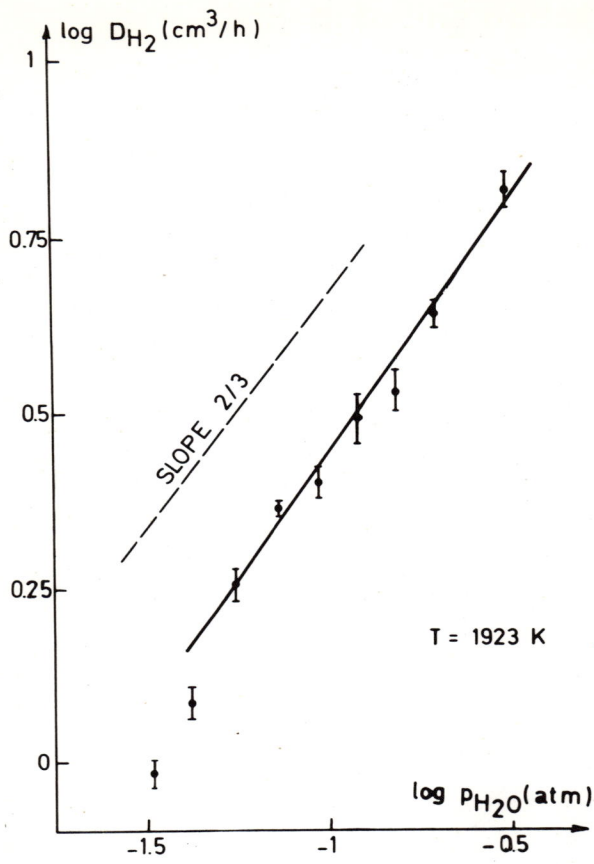

Fig. 12. Hydrogen flow with argon gas in thermolysis of water.

Hydrothermal reaction sintering of monoclinic zirconia

MASAHIRO YOSHIMURA AND SHIGEYUKI SŌMIYA

Laboratory for Hydrothermal Syntheses
Research Laboratory of Engineering Materials
   and Dept. of Materials Science and Engineering
Tokyo Institute of Technology
Nagatsuta, Yokohama 227, Japan

A dense monoclinic zirconia ceramic having a maximum density of 5.85 g/cm³ and grains of 1-5 μm has been prepared from Zr metal powder and water in a Pt capsule by hydrothermal reaction sintering at 1000 °C for 3 h under 98 MPa. The effects of temperatures, pressures, and initial compositions on the densification and sintering were studied. The reaction between Zr metal powder and H_2O started above ≈ 300 °C under 98 MPa, then produced fine grains of monoclinic ZrO_2 and H_2. Sintering of these grains started above 700 °C, where H_2 had permeated the capsule wall. Well-sintered bodies could be prepared at 1000 °C under 98 MPa. Higher pressures or higher temperatures did not produce further densification. The highest densifications were observed in the samples with the ratio of $H_2O/Zr = 2$ in the initial compositions. The samples with the ratio $H_2O/Zr > 2.1$ did not give well-sintered bodies.

Dense monoclinic nonstabilized zirconia ceramics are difficult to fabricate by normal sintering techniques because of cracking due to the phase transformation from tetragonal to monoclinic form during cooling.[1-4] They have been prepared previously by reactive hot-working,[1,2] reactive hot-pressing,[3] or reoxidizing of oxygen-deficient ceramics sintered in vacuo[4] as summarized in Table I. Comparison among them appeared to be difficult since the compositions of the starting materials were different in these studies, and some authors reported relative densities while the others reported bulk densities. Dense monoclinic zirconia ceramics are generally prepared at high temperatures (1600 ° to 2300 °C), then heat-treated at lower temperatures near the transition temperature.

Recently, zirconia ceramics with among the highest densities have been fabricated by the hydrothermal reaction sintering method at 1000 °C.[5] They were prepared at temperatures below the transition temperature of ZrO_2, so that essentially no cracks were expected to have occurred in the specimens. In this hydrothermal reaction-sintering method, Zr metal powder and water were used as starting materials encapsulated in a Pt capsule, as illustrated in Fig. 1. They reacted, then yielded fine grains of monoclinic ZrO_2 and hydrogen gas at high temperatures under pressure. As the hydrogen gas permeated the capsule wall during heating, these monoclinic ZrO_2 grains were compressed more effectively and then allowed to sinter.

In the present study, the effects of the starting compositions, water contents, temperatures, and pressures on the sintering and densification were investigated. Similar hydrothermal reaction sintering of Cr_2O_3 from Cr metal powder and water has already been reported.[6-8]

Experimental Procedure

Zirconium metal powder,* 99.9% pure, containing $\approx 2\%$ Hf, was the starting material. It consisted of particles 1 to 10 μm in diameter (Fig. 2). This zirconium and redistilled water were hermetically sealed with an electric arc in a 2.6–2.7 mm inside diameter platinum capsule 0.15–0.2 mm thick and 35 mm long. The sealed capsule was heated in a test-tube-type pressure vessel or a high-gas-pressure apparatus under the desired pressure of water or argon gas. Water was used for runs up to 700 °C, where temperature was measured by two platinel thermocouples calibrated against the melting point of Zn (419.5 °C) and controlled within ± 3 °C during a run. The pressure was measured by a calibrated Heise gage and controlled within ± 0.2 MPa.

In the experiments using the high-gas-pressure apparatus,[9] the temperature and pressure were held within ± 20 °C and 1.96 MPa, respectively, during a run, which was typically 3 h long. The capsules were weighed to check the leakage and/or the decrease resulting from the loss by permeation of hydrogen gas. Heated specimens were removed from the capsules, then examined by X-ray powder diffraction using a diffractometer† with a rotating Cu anode and by scanning electron microscopy.‡ The density of sintered specimens was measured by a liquid replacement method, where toluene and tungsten wires 0.05 mm in diameter were used to reduce the influence of surface tension.

Results and Discussion

The reaction between Zr metal powder and supercritical water under 98 MPa was observed above 300 °C to yield micrograins of monoclinic ZrO_2. They were much smaller than initial Zr metal grains and formed a crust-like structure, as seen in Fig. 3. The average crystallite size was calculated to be ≈ 24 nm from the measurement of the half-width of the X-ray diffraction peak, using the Warren equation[10] for these monoclinic ZrO_2 grains prepared at 400 °C. This formation of active fine grains is one of the important features in the hydrothermal reaction sintering. At 400° to 700 °C, monoclinic ZrO_2 grains of ≈ 24 nm and hydrogen gas existed in the capsules. Since hydrogen gas started to permeate the capsule wall above 700 °C, the monoclinic ZrO_2 grains thus formed were compressed and sintered.

Well-sintered bodies having bulk densities of 5.17–5.85 g/cm³ were obtained at 1000 °C after 3 h under 98 MPa (Table II). Figure 4 shows a typical scanning electron photomicrograph of the fractured surface of the specimen with the maximum density of 5.85 g/cm³, or relative density of 99.0% compared to the X-ray density of 5.91 g/cm³ for a 98% ZrO_2-2%HfO_2 solid solution. Polyhedral zirconia grains, with a grain size of 1–5 μm, are evident in this figure.

The presence of more water than is required for stoichiometry in the reaction $Zr + 2H_2O = ZrO_2 + 2H_2$ seemed to prevent effective sintering and densification. Figure 5 shows a scanning electron photomicrograph of the fractured surface of the specimen with the initial mole ratio of

$H_2O/Zr = 2.095$ sintered at 1000°C for 3 h under 98 MPa pressure. The microstructure of this specimen, with a density of 5.38 g/cm³ and composed of 0.5-1.0 μm fine grains, aggregated inhomogeneously. With more excess water, no well-sintered bodies could be prepared, even at temperatures as high as 1200°C or under 686 MPa pressure (Table II). Pressures above 98 MPa did not always promote densification, nor did temperatures of 1100° or 1200°C.

The ratio of water to Zr apparently has the greatest effect on the densification. Figure 6 indicates that the highest densities might be attained in the samples with the stoichiometric amount (or slightly less) of water vs Zr and that excess water, even if the excess is small, prevents densification. These results can be explained by the presence of the residual gas(es) in the capsule during sintering; the compression of the monoclinic ZrO_2 particles by outer gas pressure would stop when the inner pressure became equal to the outer ambient gas pressure. The most abundant residual gas appears to be water, which cannot permeate platinum. For example, the initial mixture of 1 mol Zr and 2.10 mol H_2O ($H_2O/Zr = 2.10$, a 5% excess of water over stoichiometry), would yield 1 mol ZrO_2 and 0.10 mol H_2O as the final products in the capsule. This water would occupy a volume of 10.62 cm³ at 1000°C under 100 MPa, which was calculated from the specific volume of water 5.900 cm³/g observed under the same conditions by Kennedy.[11] That volume corresponds to 50.2% of the molar volume of m-ZrO_2, 21.14 cm³.[12] That is, the monoclinic ZrO_2 compacts with a porosity of 33.4%, 50.2/(100 + 50.2) and could no longer be compressed by the ambient gas pressure of 100 MPa at 1000°C. This is qualitatively in agreement with the experimental results, in which no well-sintered bodies could be prepared when the initial compositions had H_2O/Zr ratios larger than 2.10.

Quantitative assessment of the density seems to be difficult, particularly near theoretical density, because other gases such as N_2 in the ambient atmosphere might affect the final sintering stage, and the mechanism of this sintering is not completely understood. It seems likely, however, that sintering and grain growth occurred simultaneously, because the samples with higher densities are generally composed of grains with larger sizes, as seen in Table II with the exception of sample No. 28.

Note that this hydrothermal reaction method enables us to prepare dense fine-grained ceramics at lower temperatures than the other methods, even though the sintering is accompanied by grain growth.

Acknowledgments

The authors are grateful to S. Itoh and H. Kanai for assistance during the study.

457

References

[1]G. K. Bansal and A. H. Heuer, "Transformational Hot-Working of ZrO_2 Polycrystals," *J. Am. Ceram. Soc.*, **58** [1-2] 76–77 (1975).

[2]C. F. Smith and W. B. Crandall, "Calculated High-Temperature Elastic Constants of Zero Porosity Monoclinic Zirconia," *ibid.*, **47** [12] 624–27 (1964).

[3]A. C. D. Chaklader and V. T. Baker, "Reactive Hot-Pressing: Fabrication and Densification of Non-Stabilized ZrO_2," *Am. Ceram. Soc. Bull.*, **44** [3] 258–59 (1965).

[4]H. J. Garrett and R. Ruh, "Fabrication of Specimens from Pure Dense Oxidized Zirconia," *ibid.*, **47** [6] 578–79 (1968).

[5]M. Yoshimura and S. Sōmiya, "Fabrication of Dense, Nonstabilized ZrO_2 Ceramics by Hydrothermal Reaction Sintering," *ibid.*, **59** [2] 246 (1980).

[6]S. Hirano and S. Sōmiya, "Hydrothermal Reaction Sintering of Pure Cr_2O_3," *J. Am. Ceram. Soc.*, **59** [11-12] 534 (1976).

[7]S. Sōmiya, S. Hirano, M. Yoshimura, S. Itoh, and H. Kanai; pp. 267–77 in Proceedings of the International Symposium on Factors in Densification and Sintering of Oxide and Nonoxide Ceramics. Edited by S. Sōmiya and S. Saito. Gakujutsu Bunken Fukyukai. Tokyo Institute of Technology, Tokyo, Japan, 1979.

[8]S. Sōmiya, M. Yoshimura, and H. Kanai, "Effects of Temperature, Pressure, Reaction Period, and Solutions on the Hydrothermal Reaction Sintering of Chromic Oxide"; to be published in *Zairyo Kagaku (J. Mater. Sci. Soc. Japan)*.

[9]S. Sōmiya, S. Hirano, T. Fukuda, and M. Sawada, "Development and Application of a High Gas Pressure and High Temperature Apparatus up to 1500°C and 10 Kb," *Koatsu Gasu*, **10** [6] 368–79 (1973).

[10]L. V. Azároff; pp. 562–71 in Elements of X-ray Crystallography. McGraw-Hill, New York, 1968. Japanese edition by Maruzen Co. Ltd., 1973.

[11]G. C. Kennedy, "Pressure-Volume-Temperature Relations in Water at Elevated Temperatures and Pressures," *Econ. Geol.*, **45**, 629 (1950).

[12]Handbook of Chemistry and Physics. Edited by R. C. Weast. Chemical Rubber Co. Press, Cleveland, Ohio, 1975–76; p. B217.

---

*Kōjundo Chemical Lab Inc., Wako, Saitama, Japan.
†RU-200; Rigaku Electric Co. Ltd., Tokyo, Japan.
‡JSM T-200; JEOL Co. Ltd., Tokyo, Japan.

Table I. Fabrication of dense nonstabilized monoclinic ZrO_2 ceramics

Fabrication method	Bulk density (g/cm³)	Relative density (%)	Composition (%)	Ref.
Reactive hot-pressing 55.8 MPa, 800–1200 °C	5.55 5.45	99.8 98.0	98 ZrO_2 + HfO_2 99.7 ZrO_2	3 3
Reactive hot-pressing 27.5 MPa, 800–1200 °C		99	99.99 (ZrO_2 + HfO_2)	1
Reactive hot-pressing, 165 MPa, 800–1200 °C		98	,,	1
Reactive hot-working, 97.6 MPa, 1000–700 °C, after hot-pressing		95	93.77 (ZrO_2 + HfO_2) 3.90 SiO_2	2
Vacuum sintering at 2300 °C, then oxidized at 1000 °C	5.43–5.52	94–96	High purity ZrO_2	4
Hydrothermal reaction sintering, 1000 °C, 98 MPa or above	5.79–5.85	98–99	99.9 (ZrO_2 + HfO_2) 2 Hf	5

Table II. Selected results of hydrothermal reaction sintering of monoclinic ZrO_2 for 3 h

Sample No.	H_2O/Zr	Deg. of fill (%)	Temp. (°C)	Pressure (MPa)	Bulk density (g/cm³)	Relative density (%)	Avg grain size (μm)
3	1.773	19.3	1000	98	5.85	99.0*	3
5	2.020	15.5	1000	98	5.51	93.2	1
29	2.063	14.8	1000	98	5.17	87.5	0.5
30	2.095	15.0	1000	98	5.38	89.2	0.5
10	2.415	18.4	1000	98	Not sintered		
25	2.038	15.1	1000	196	5.51	93.2	0.3
28	1.985	15.1	1000	294	5.96	100.9†	0.1
8	2.012	22.7	1000	490	5.83	98.7	3
7	2.010	20.4	1000	686	5.77	97.7	3
23	2.038	41.2	1000	686	5.70	96.4	
32	2.016	14.8	1000	98	5.68	96.1	2
40	2.142	14.3	1100	98	Not sintered		
41	2.089	14.2	1200	98	Not sintered		
39	2.091	14.1	1000	490‡	Not sintered		0.5

*With gray skin. †Gray fine body, rare case. ‡Capsule leaked slightly.

$$Zr + 2H_2O \longrightarrow ZrO_2 + 2H_2$$

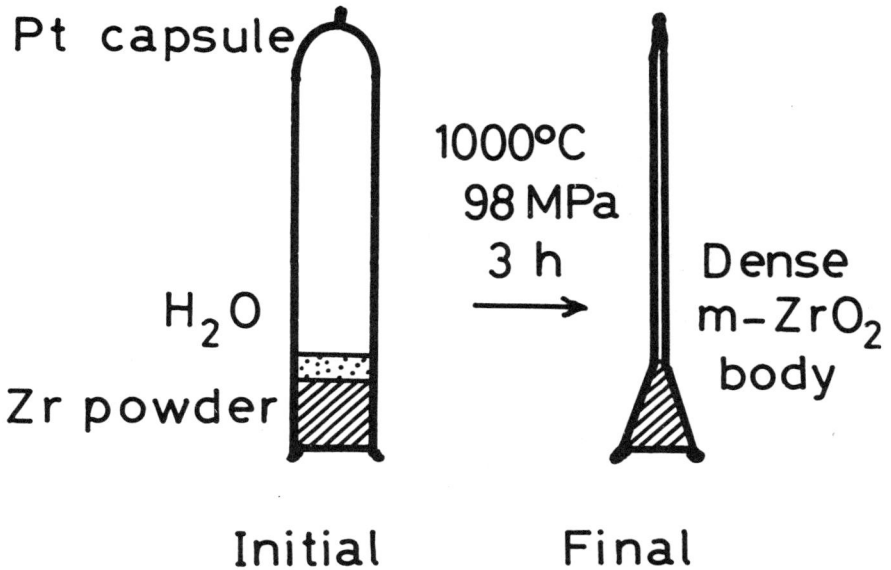

Fig. 1. Schematic illustration of the hydrothermal reaction sintering of monoclinic ZrO_2.

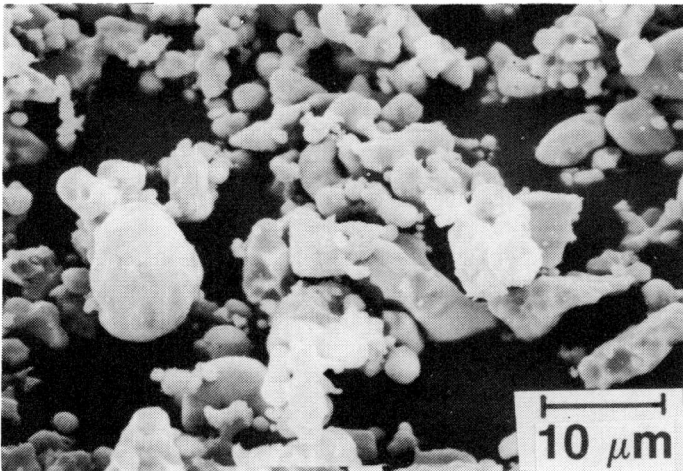

Fig. 2. Starting Zr metal powder.

461

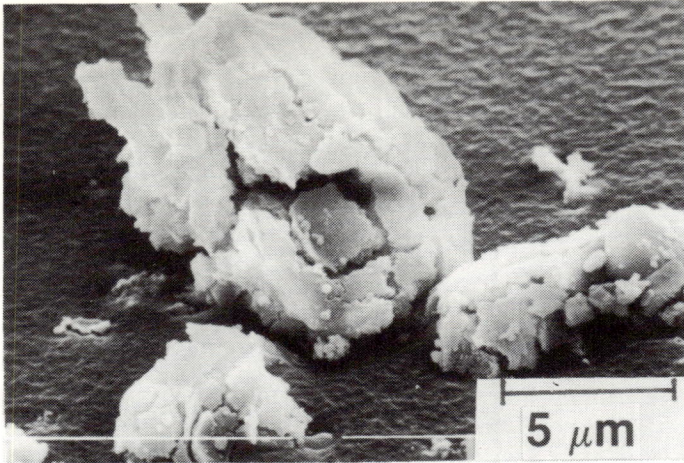

Fig. 3. Zr metal powder reacted with H_2O at 300°C for 3 h under 98 MPa.

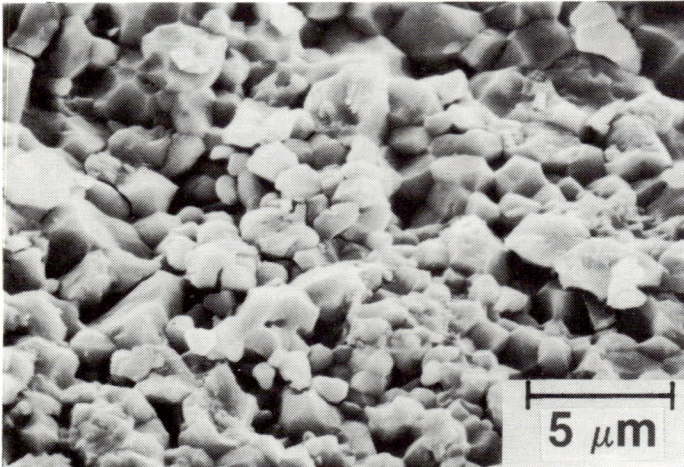

Fig. 4. Fracture surface of m-ZrO_2 specimen sintered hydrothermally; 1000°C, 3 h, 98 MPa, $H_2O/Zr = 1.773$, fill 19.3%.

Fig. 5. Fracture surface of m-ZrO$_2$ specimen sintered hydrothermally; 1000 °C, 3 h, 98 MPa, H$_2$O/Zr = 2.095, fill 15.0 %.

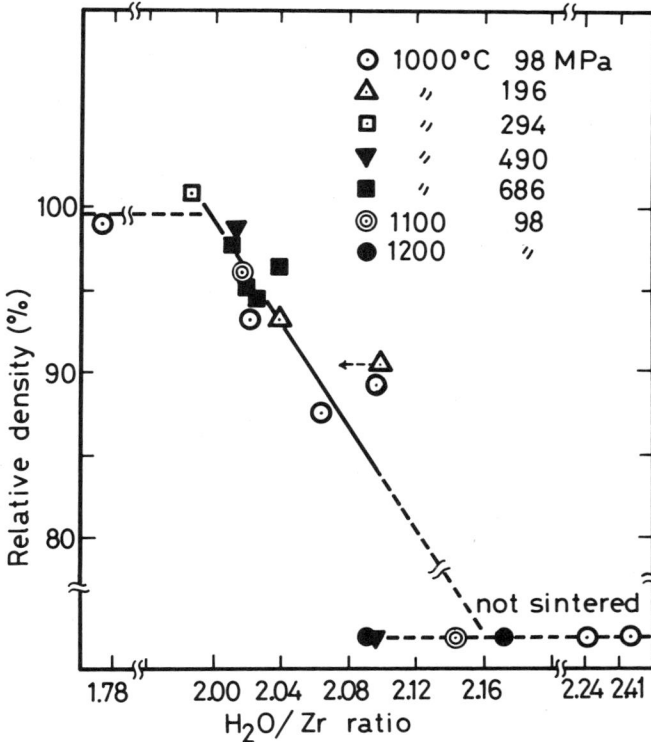

Fig. 6. Relative density of m-ZrO$_2$ ceramics sintered hydrothermally as a function of the H$_2$O/Zr ratio.

Production of high-purity zirconia by hydrothermal decomposition of zircon (ZrSiO$_4$)

P. REYNEN, H. BASTIUS, B. PAVLOVSKI, AND D. VON MALLINCKRODT

Insitut für Gestein Hüttenkunde
Rheinisch-Westfalen Technischen Hochschule Aachen
Aachen, Federal Republic of Germany

The optimization of the hydrothermal decomposition of ZrSiO$_4$ is described, and the impurity level of the final product is considered. Preliminary experiments have shown that good results are obtained with Ca(OH)$_2$ and NaOH. The effects of temperature, pressure, composition, particle size, and solid content on the autoclave reaction have been studied. Optimization of these factors leads to the parameters which have the greatest effect on the reaction.

Among the oxide ceramics, stabilized zirconia is gaining more and more interest due to its outstanding chemical and physical properties: high melting point, low coefficient of thermal expansion, corrosion resistance, and high ionic conductivity.[1] The high price of the pure oxide prevents more general application. The aim of our investigations was to develop a method of producing zirconia and stabilized zirconia from ZrSiO$_4$, which is the principal natural source of ZrO$_2$.[5] Some industrial processes were developed to produce ZrO$_2$ of different grades of impurity. These processes are protected by many patents.[11]

The thermal methods take advantage of the incongruent melting of ZrSiO$_4$ and subsequent removal of free SiO$_2$. This can be achieved in a reducing atmosphere by evaporation of SiO. Another method uses carbon as a reducing agent. Zirconium carbides are formed which are subsequently transformed into volatile zirconium chloride.[12]

Thermal treatment, in the presence of coke, lime, and iron, leads to the formation of ferrosilicon.[13] Direct formation of zirconium chloride is possible at high temperatures in the presence of carbon and chlorine. One process is based on melting in the presence of alkali or alkaline earth hydroxides. More favorable from the energetic point of view are the hydrothermal treatment methods. Russian publications describe the use of LiOH, NaOH, KOH, RbOH, and CsOH, as well as Mg(OH)$_2$, Ca(OH)$_2$, Sr(OH)$_2$, and Ba(OH)$_2$.[5] A full description of the known methods can be found in the literature.[1-3,5] A recent paper describes the reaction of zircon with gypsum (CaSO$_4$) to prepare ZrO$_2$.[4]

No method is completely satisfactory because either the amount of energy required is too high or the final product remains impure. The scope of our investigation was to optimize the hydrothermal decomposition, taking into consideration the impurities of the final product.[9]

Hydrothermal Decomposition of $ZrSiO_4$

Preliminary experiments showed that the hydrothermal decomposition of $ZrSiO_4$ (zircon) in the presence of both $Ca(OH)_2$ and NaOH is advantageous.[5-7] The reaction product of NaOH with $ZrSiO_4$ reacts with $Ca(OH)_2$ to give calcium silicate hydrates and SiO_2, leaving the NaOH unreacted (catalytic effect).

Raw Materials

A commercial finely ground zircon (Table I), reagent grade $Ca(OH)_2$, and NaOH were used.[8] The reactants and water were brought to reaction in an autoclave.

Parameters Influencing the Hydrothermal Reaction

The degree of reaction is defined as the ratio of ZrO_2 formed to the total amount of ZrO_2 available.

The influence of temperature, and consequently of autoclave pressure, is shown in Fig. 1. A drastic increase of α is observed with increasing temperature. All further experiments were carried out at 310 °C and 100×10^5 Pa which were the maxima allowed for the autoclave used. Later a 350 °C autoclave was used. These results are also shown in Fig. 1.

In Fig. 2 the ratio $Ca(OH)_2/ZrSiO_4$ vs α is plotted for constant NaOH concentration and temperature. There is a steady increase of α with increasing $Ca(OH)_2$ up to $Ca(OH)_2/ZrSiO_4 = 2$. Further increase of $Ca(OH)_2$ has only a slight effect. The ratio 2 was chosen in further experiments.

The degree of reaction, α, is strongly influenced by the size of the zircon particles, as shown in Fig. 3. Further experiments were carried out using zircon sand with a particle size of $90\% < 5$ μm.

Figure 4 shows the influence of NaOH. The reaction rate increases with increasing pH.

Another important parameter is the liquid/solid ratio. The highest degree of reaction is obtained for a ratio of 7.5 (curve A, Fig. 5) but the effective yield of ZrO_2 which is the product of α and the amount of ZrO_2 available (curve C) is shown by curve B in Fig. 5. The maximum is obtained at a liquid/solid ratio of 2. It is not possible to apply a factor of < 2 as it is no longer possible to put the solids into suspension.

The influence of stirring is shown in Fig. 6. Without stirring the sedimentation of the solids decreases the reaction rate but also with stirring the initial high reaction rate decreases as the process is controlled by diffusion through the product layer, as will be shown later.

On the basis of these experiments the following conditions were determined for an optimal reaction:

Compound	(wt%)
$ZrSiO_4$	18.43
$Ca(OH)_2$	14.9
NaOH	4.67
H_2O	62

Liquid/solid ratio:	2
NaOH concentration in the solution:	7 wt%
Mol ratio $Ca(OH)_2/ZrSiO_4$:	2

465

Temperature:	350 °C
Vapor pressure:	170×10^5 Pa
Reaction time:	8 h

Reaction Mechanism

The solid $ZrSiO_4$ reacts under hydrothermal conditions with $Ca(OH)_2$ to give ZrO_2 and calcium silicate hydrates. NaOH has a catalytic influence as it increases the pH.

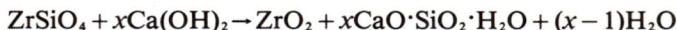

$$ZrSiO_4 + xCa(OH)_2 \rightarrow ZrO_2 + xCaO \cdot SiO_2 \cdot H_2O + (x-1)H_2O$$

Transmission electron microscopy (TEM) shows the geometry and morphology of the reaction products. Figures 7 and 8 show a partly reacted $ZrSiO_4$ particle. The reaction products were disagglomerated in a methanol suspension by ultrasonic treatment. Figure 8 shows agglomerations of needlelike CSH crystals containing spherical ZrO_2 particles. By ultrasonic treatment some ZrO_2 particles can be separated from the needles (Figs. 9, 10). The ZrO_2 modification was identified on isolated particles by small area diffraction (SAD) and appeared to be the tetragonal modification (JCPDS 17–923). This modification was also found by X-ray diffraction (XRD) (Fig. 11). The identification of the CSH needles was more problematic, as the characteristic d values of the CSH phases are very similar and are slightly shifted by composition variations.[14,15]

The hydrothermal reaction between $ZrSiO_4$ and NaOH gives the following reaction product

$$ZrSiO_4 + 2NaOH \rightarrow Na_2ZrSiO_5 + H_2O$$

In the presence of $Ca(OH)_2$, however, Na_2ZrSiO_5 reacts to the more insoluble compounds ZrO_2 and CSH. The reaction products ZrO_2 and CSH form a porous product layer around the $ZrSiO_4$ grains. The diffusion of Na^+, OH^-, and Ca^{2+} through the open channels is rate controlling for the reaction as could be proved by performing the reaction with grains of uniform size.

In case of an isotropic reaction of spherical particles wtih a surrounding medium, one finds

$$(1-\alpha) = \left(1 - \frac{kt}{a_0}\right)^3$$

if the surface reaction is rate controlling.[10] In the case of a diffusion-controlled process,

$$(1-\alpha) = \left(1 - \frac{k\sqrt{t}}{a_0}\right)^3$$

$(1 - \sqrt[3]{1-\alpha})$ were plotted vs $\sqrt{t}$ for $ZrSiO_4$ particles with particle size between 25 and 32 μm; the result is a straight line which slightly deviates for high $\sqrt{t}$ values because the assumption of uniform particle size is not exactly fulfilled (Fig. 3). This shows that the hydrothermal reaction is diffusion controlled.

Separation and Purification of ZrO_2

Figures 7–10 show that the ZrO_2 particles are embedded between the CSH needles. A physical separation is not possible and chemical methods must be applied. The reaction product was treated with sulfuric acid in which the small tetragonal ZrO_2 particles dissolve. The CSH phases react to insoluble $CaSO_4$ and SiO_2. The various methods applied are shown in Fig. 12. The analyses of the various products are given in Table II.

Treatment with Sulfuric Acid

After hydrothermal treatment with $Ca(OH)_2$ and NaOH the solid phase is separated from the NaOH solution by filtration. The sodium hydroxide solutions can be used again for the next run in the autoclave. The solid is treated with sulfuric acid and evaporated to dryness by which the silicic acid becomes insoluble. The product is leached with water. From the filtrate, a very pure zirconium sulfate tetrahydrate (1,2 in Fig. 12) is obtained by crystallization.[16-18] The crystallization velocity depends on the concentration of SO_4^{2-} in the solution and takes place in a time of some days. This process is, however, too tedious to be practical.

Adding sulfuric acid to the autoclave product without evaporating to dryness gives a solution containing considerable amounts of silicic acid, Na^+ ions, and Ca^{2+} ions (3 in Fig. 12). The SiO_2 can be brought to precipitation by adding polyethylene glycol (4,5 in Fig. 12). The filtrate is treated with NH_4OH, giving $Zr(OH)_4 \cdot xH_2O$ which is easily freed from Na^+ ions by washing (9 in Fig. 12). The hydroxide is dissolved in sulfuric or hydrochloric acid from which zirconium salts can be obtained by crystallization (6,7 in Fig. 12).

Doped ZrO_2 is obtained from solutions of zirconium salts (sulfates, acetates, and formates and the corresponding impurities, Mg^{2+}, Al^{3+}, Ca^{2+}). By spray drying or hot kerosene drying[19] and subsequent thermal treatment, homogeneous, doped zirconia is obtained.

Summary

The hydrothermal reaction of $ZrSiO_4$ with $Ca(OH)_2$ in the presence of NaOH is described. The reaction is much faster than with $Ca(OH)_2$ alone because the NaOH addition increases the pH. On the other hand the NaOH remains in solution and can be separated from the reaction product (tetragonal ZrO_2 particles embedded in CSH needles). The optimal conditions for the hydrothermal reaction were determined experimentally. The reaction is controlled by diffusion of Na^+, OH^-, and Ca^{2+} ions through the open channels of the product layer around the $ZrSiO_4$ particles. Several methods are described for producing pure and doped zirconia from the autoclave product. The removal of silicic acid is the main difficulty.

References

[1]Ullman's Encyclopedia of Technical Chemistry (Ullman's Enzyklopaedie der Technischen Chemie) 3rd ed., Vol. 19; p. 179 ff., p. 195 ff. Urban and Schwarzenberg, Munich, Federal Republic of Germany, 1969.
[2]Gmelin's Handbook of Inorganic Chemistry (Gmelin's Handbuch der Anorganischen Chemie), Zirconium, System No. 42, p. 59 ff. Verlag Chemie, Weinhein, Federal Republic of Germany, 1962.
[3]R. H. Neilsen; pp. 190–93 in Fundamentals of Refractory Compounds. Edited by H. H. Hausner and M. G. Bowman. Plenum Press, New York, 1968.
[4]S. B. Hanna, "Studies on the Production of Zirconia from Egyptian Zircon," Tonind. Ztg., **103**, 459–62 (1979).
[5]B. Pavlovski, Optimum Conditions for Hydrothermal Investigation of $ZrSiO_2$ and the Production of ZrO_2; Dissertation, Technishce Hochschule Aachen, Federal Republic of Germany, 1976.

[6]V. G. Chukhlantsev and Y. M. Galkin, "The Hydrothermal Reaction of Zircon with Ca(OH)$_2$ and Mg(OH)$_2$," *Zh. Neorg. Khim.*, **12**, 1730–32 (1967).

[7]V. G. Chukhlantsev and A. K. Shtol'ts, "Sodium Zirconosilicates," *ibid.*, **6**, 684–87 (1961).

[8]Calcium Hydroxide, No. 2047 and Sodium Hydroxide, No. 6462; E. Merck AG, Darmstadt, Federal Republic of Germany.

[9]P. Reynen and D. Steffen, Effects of Stabilizing Additives and Impurities on the Properties of ZrO$_2$ Solid Electrolytes for Measurement of Oxgyen Activity in Steel Melts: Final Rept. (1979). VDEh-Forschungsfoerderung FAA 634 EGKS 7210-GA/1/110.

[10]B. Delmon, Kinetics of Heterogeneous Reactions. Moskwa, 1972; p. 227.

[11]Degussa, Ger. Pat. 1 108 193 (1959); Columbia Nat. Corp., Br. Pat. 876 513 (1960); C. Lorenz, Ger. Pat. 543 675 (1932); C. J. Kinzie, U.S. Pat. 2 072 889 (1937); and O. Ruff, Ger. Pat. 522 702 (1931).

[12]W. J. Kroll, W. W. Stephens, and H. P. Holmes, *J. Met.*, **2**, 1445–53 (1950).

[13](a) A. W. Ballard and D. W. Marshall (Norton Co.), U.S. Pat. 2 535 526 (1950).
(b) O. J. Whittemore and D. W. Marshall, "Fused Stabilized Zirconia and Refractories," *J. Am. Ceram. Soc.*, **35** [4] 85–89 (1952).

[14]B. Franke, *Z. Anorg. Chem.*, **147**, 180–84 (1941).

[15]H. F. W. Taylor, The Chemistry of Cements. Academic Press, London, 1964; pp. 167–232.

[16]M. Falinski, *Ann. Chim.*, **16** [11] 237–325 (1941).

[17]H. Bastius, "Production of Very Pure Zirconium Salts"; to be published in *Berichte der Deutschen Keramischen Gesellschaft.*

[18]G. Brauer; pp. 1232–33 in Handbook of Preparative Inorganic Chemistry, Vol. 2, 2nd ed. Academic, New York, 1965.

[19]P. Reynen, H. Bastius, M. Faizullah, and H. v. Kamptz, *Ber. Dtsch. Keram. Ges.*, **54**, 63–68 (1977).

Table I. Chemical analysis and grain size distribution of ZrSiO$_4$

Composition		Grain size distribution	
Component	Amount (wt%)	Grain size (μm)	Amount (wt%)
SiO$_2$	34.25	62	0.45
ZrO$_2$(+ HfO$_2$)	65.00	30-62	0.66
Al$_2$O$_3$	0.10	10-30	3.96
TiO$_2$	0.09	5-10	3.29
Fe$_2$O$_3$	0.07	1-5	46.15
CaO	0.08	1	45.49
MgO	0.03		
Na$_2$O	0.09		
K$_2$O	0.03		
Ignition loss at 1000 °C	0.26		

Table II. Chemical analysis of purified ZrO_2

Oxid M.-%	1	2	3	4	5	6	7	8	9
Na_2O	0,10	0,11	6,40	6,30	6,30	–	–	6,30	0,003
MgO	0,14	0,10	3,67	0,35	0,30	<0,01	0,01	0,30	0,30
Al_2O_3	0,17	0,39	3,09	0,21	0,21	0,073	0,071	3,00	3,00
SiO_2	0,11	0,43	0,47	0,24	0,040	0,040	0,041	0,043	0,043
K_2O	0,29	0,07	0,07	0,07	0,07	–	–	0,07	–
CaO	2,62*	7,06*	1,63*	0,44	0,43	0,033	0,031	1,56*	0,277
TiO_2	0,02	0,23	0,37	0,16	0,07	0,040	0,043	0,328	0,32
Cr_2O_3	n.b.	n.b.	0,018	0,008	0,008	–	–	0,016	0,01
MnO	n.b.	n.b.	0,003	0,003	0,003	–	–	0,002	0,002
Fe_2O_3	0,03	0,12	0,35	0,12	0,07	0,004	0,003	0,33	0,33
NiO	n.b.	n.b.	0,035	0,01	0,01	<0,01	–	0,033	0,022
CuO	n.b.	n.b.	0,005	0,002	0,002	–	–	0,005	0,004
ZnO	n.b.	n.b.	0,013	0,002	0,002	–	–	0,012	0,012
ZrO_2	96,52	91,49	83,87	92,08	92,48	99,80	99,80	88,03	95,30

*Depending on the volume of the solution (solubility of $CaSO_4$). n.b. = not estimated.
– = <0.001 mol%.

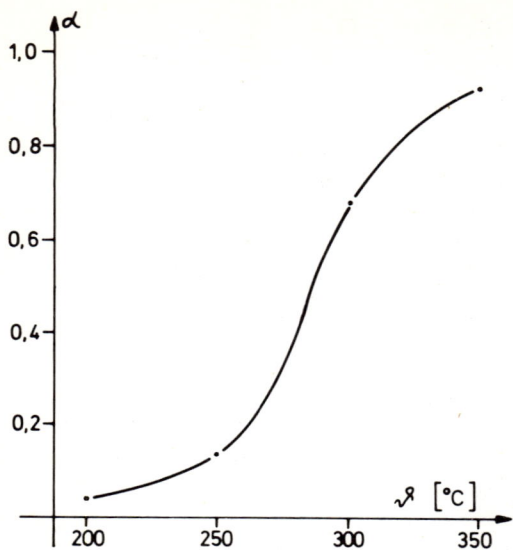

Fig. 1. Influence of temperature. $Ca(OH)_2/ZrSiO_4 = 2$, NaOH 3.6 wt%, liquid/solid 3.1, and time 8 h.

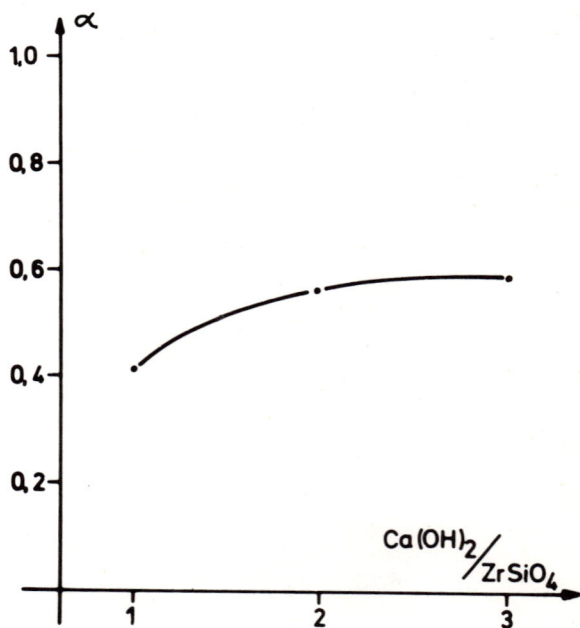

Fig. 2. Influence of the molar ratio. $Ca(OH)_2/ZrSiO_4$. $\theta = 310\,°C$, NaOH 3.6 wt%, liquid/solid 3.1, and time 4 h.

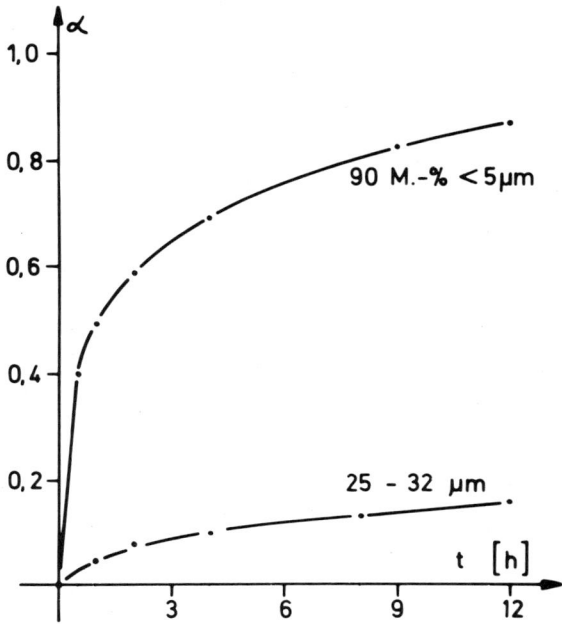

Fig. 3. Influence of grain size. $\theta = 310\,°C$, $Ca(OH)_2/ZrSiO_4 = 2$, NaOH 6.5 wt%, liquid/solid 3.2.

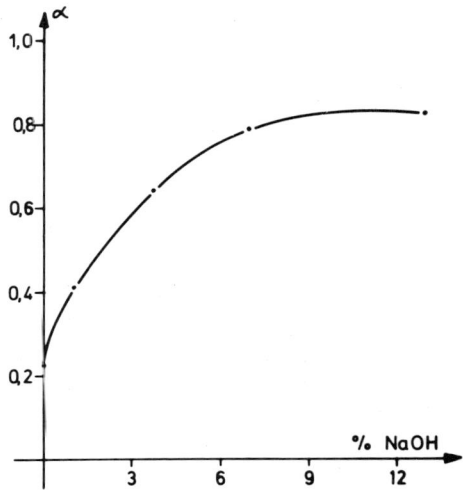

Fig. 4. Influence of the NaOH concentration. $\theta = 310\,°C$, $Ca(OH)_2/ZrSiO_4 = 2$, liquid/solid 3.6, time 4 h.

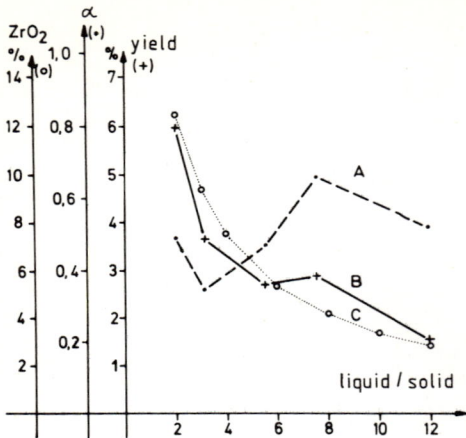

Fig. 5. Influence of the ratio liquid/ solid. A = degree of reaction, B = yield, (A·C), C = ZrO available, $\theta = 310\,°C$, $Ca(OH)_2/ZrSiO_4 = 2$, NaOH 7 wt%, time 4 h.

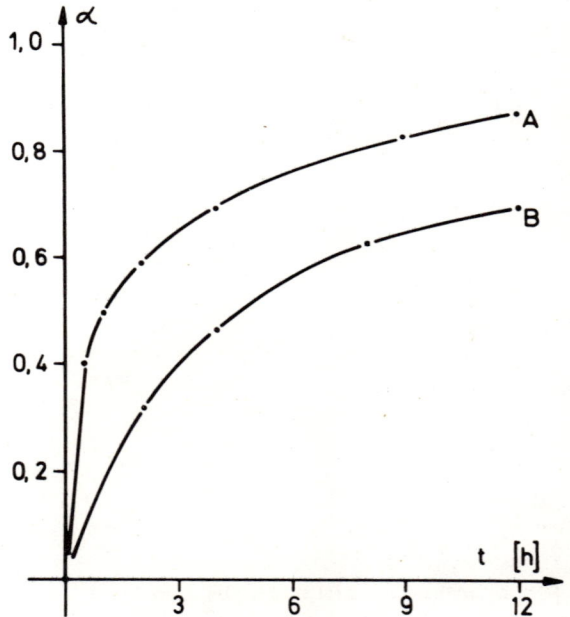

Fig 6. Influence of stirring. A with stirrer, 310°C, $Ca(OH)_2/ZrSiO_4 = 2$, NaOH 6.5 wt%, liquid/solid 3.2.

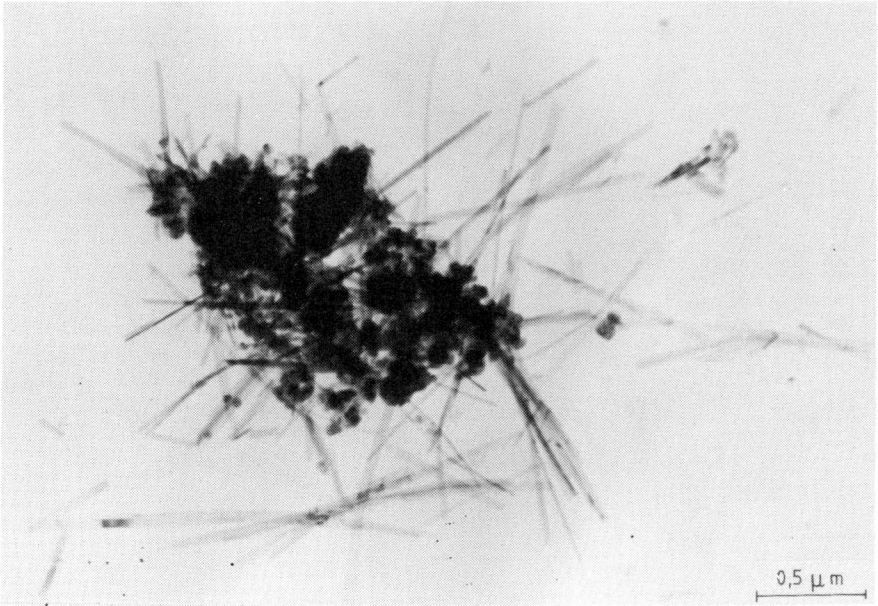

Fig. 7. Partly reacted zircon particle.

Fig. 8. Partly reacted zircon particle.

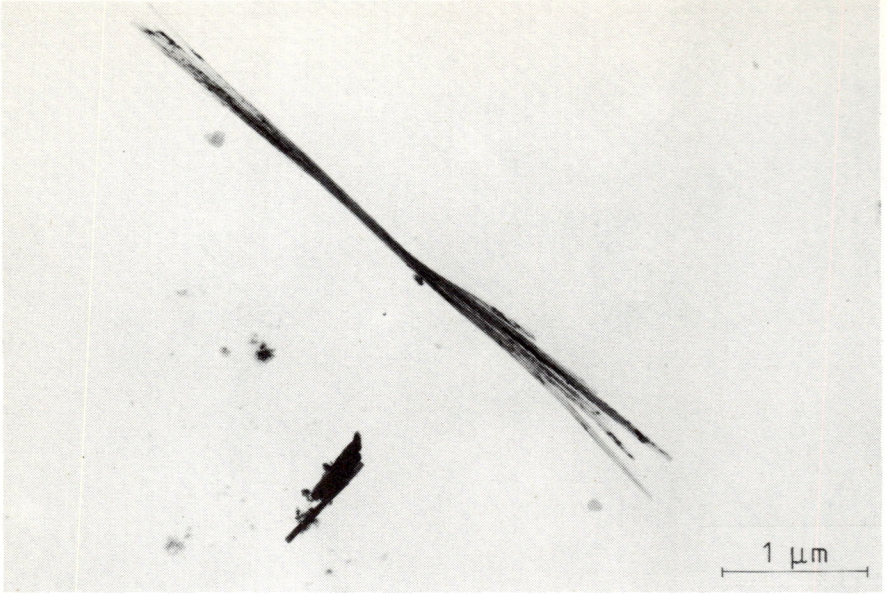

Fig. 9. CSH needles.

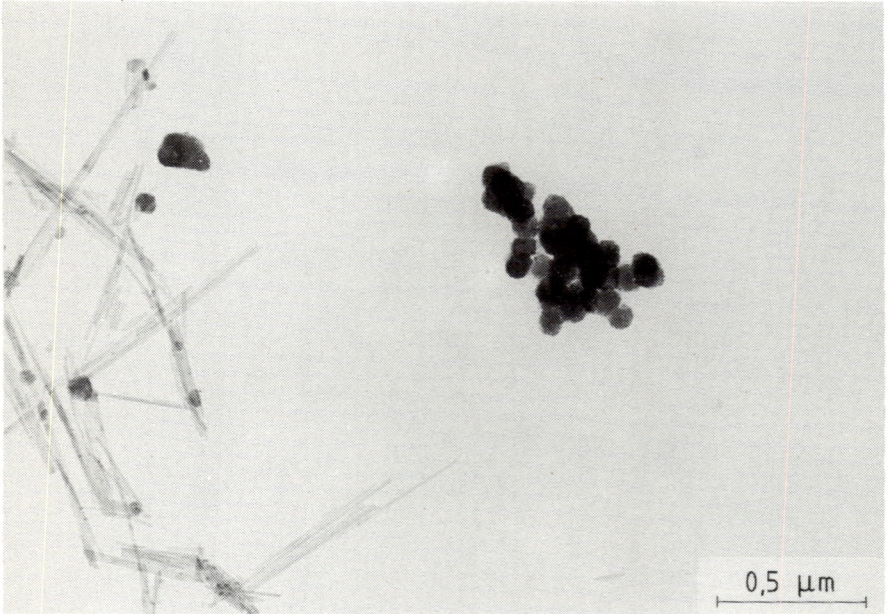

Fig. 10. Spherical ZrO₂ particles.

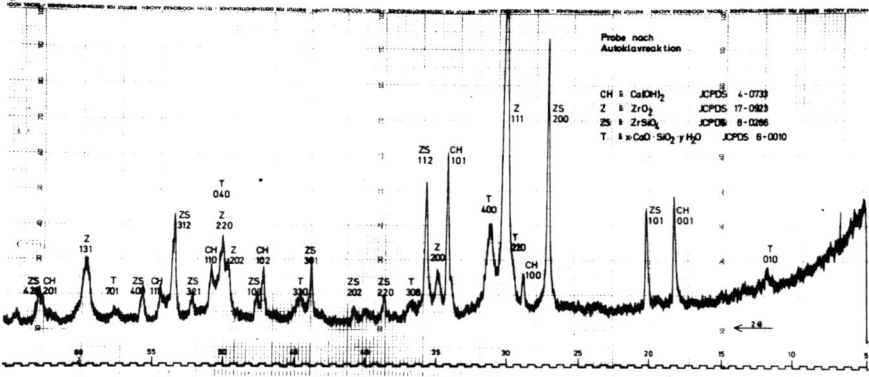

Fig. 11. XRD diagram after hydrothermal reaction.

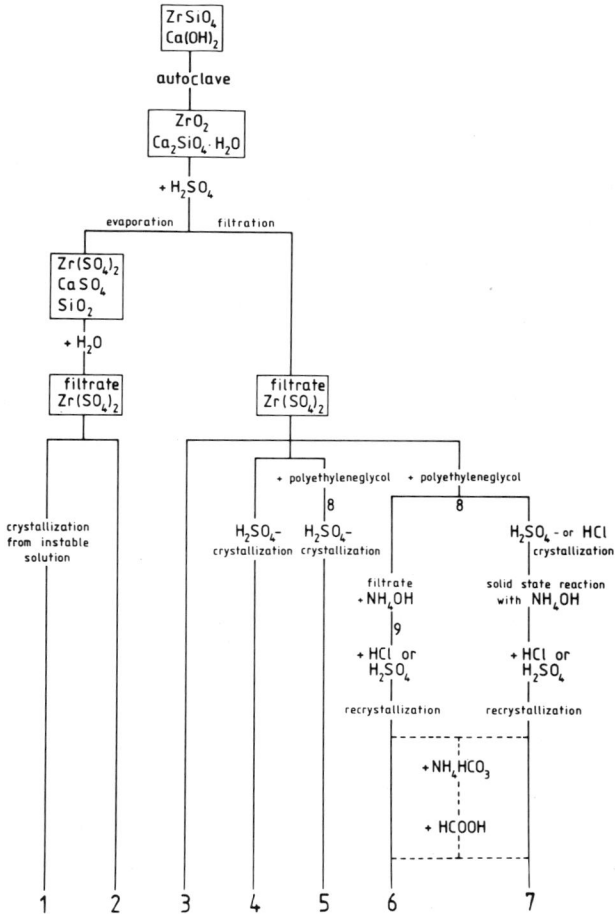

Fig. 12. Preparation methods.

475

Author Index

Subject Index

479